Ancient Egypt, New Technology

Harvard Egyptological Studies

Harvard Egyptological Studies

This monograph series ("HES") was established in 2015 to present scholarly publications in the field of Egyptology. It highlights, but is by no means limited to, sites and selected aspects of the Harvard University-Boston Museum of Fine Arts Expedition (1905–1947). Invited topics include recent PhD dissertations; reports from excavations; specialized studies in ancient Egyptian language, history, and culture; conference proceedings; publications of scholarly archives; and historiographical works covering the field of Egyptology. *Harvard Egyptological Studies* is published by the Department of Near Eastern Languages and Civilizations, and the Department of Anthropology, both of which are in the Faculty of Arts and Sciences at Harvard University.

VOLUME 17

The titles published in this series are listed at *brill.com/hes*

Ancient Egypt, New Technology

The Present and Future of Computer Visualization, Virtual Reality and Other Digital Humanities in Egyptology

Edited by

Rita Lucarelli
Joshua Aaron Roberson
Steve Vinson

TUTA SUB AEGIDE PALLAS 1683

BRILL

LEIDEN | BOSTON

Cover illustration: Lid of the coffin of Irethorerou. Egyptian, Akhmim, ca. 500 BC, Wood with polychromy. Gift of First Federal Trust Company (from the Estate of Jeremiah Lynch), Fine Arts Museums of San Francisco, 42895. Digital render by Kea Johnston. Book of the Dead in 3D project https://3dcoffins.berkeley.edu/content/welcome-book-dead-3d

Library of Congress Cataloging-in-Publication Data

Names: Ancient Egypt - New Technology (Conference) (2019 : Bloomington, Ind.), creator. | Lucarelli, Rita, 1972- editor. | Roberson, JoshuaAaron, editor. | Vinson, Steve, editor.
Title: Ancient Egypt, new technology : the present and future of computer visualization, virtual reality and other digital humanities in Egyptology / edited by Rita Lucarelli, Joshua Roberson, Steve Vinson.
Other titles: Harvard Egyptological studies ; v. 17.
Description: Leiden : Boston ; Brill, [2023] | Series: Harvard Egyptological studies, 2352-7501 ; volume 17 | Includes bibliographical references and index.
Identifiers: LCCN 2022050724 (print) | LCCN 2022050725 (ebook) | ISBN 9789004501287 (hardback) | ISBN 9789004501294 (ebook)
Subjects: LCSH: Egyptology–Data processing–Congresses. | Egyptology–Methodology–Congresses. | Digital humanities–Congresses. | Three-dimensional imaging in archaeology–Congresses. | Information visualization–Congresses. | LCGFT: Conference papers and proceedings.
Classification: LCC DT60 .A755 2019 (print) | LCC DT60 (ebook) | DDC 932–dc23/eng/20221020
LC record available at https://lccn.loc.gov/2022050724
LC ebook record available at https://lccn.loc.gov/2022050725

Typeface for the Latin, Greek, and Cyrillic scripts: "Brill". See and download: brill.com/brill-typeface.

ISSN 2352-7501
ISBN 978-90-04-50128-7 (hardback)
ISBN 978-90-04-50129-4 (e-book)

This book is printed on acid-free paper and produced in a sustainable manner.

Contents

PREFACE

Looking Back, Looking Forward: Ancient Egypt—New Technology

Rita Lucarelli, Joshua Aaron Roberson and Steve Vinson

During the summer of 2015 and summer 2016, Rita Lucarelli was awarded two Collaborative Research Grants from the Mellon Foundation and the Digital Humanities Department of the University of California, Berkeley, to study the "materiality of the ancient Egyptian Book of the Dead," by realizing 3D models of decorated coffins in the Phoebe A. Hearst Museum of Anthropology of the University of California in Berkeley. During this same period, Lucarelli began co-operative efforts with the Computer Cluster of UC Berkeley, concerning issues of sustainability and data sharing, and the Institute of Digital Humanities at UCLA, concerning the annotation of 3D models with Hieroglyphic metadata, et al.[1] During the summer of 2016, Joshua Roberson was awarded a Faculty Research Grant from the University of Memphis, for the purpose of creating 3D models of Egyptian artifacts in the Art Museum of the University of Memphis. From late Summer 2016, Roberson and Lucarelli began to discuss the possibility of collaboration between Berkeley and Memphis, concerning digitization and annotation of Egyptian mortuary objects. These conversations underscored the lack of interconnection among disparate projects in digital Egyptology and the benefits that might result from increased collaboration. In response to this perceived deficit, Roberson and Lucarelli developed a proposal for a large conference and workshop featuring a roster of leading scholars working actively in the digital realm, for the purpose of assessing the current "state of the art" in Egyptology,[2] as well as the directions that the field might be headed. In April of 2017, as the proposal was being finalized for submission, Lucarelli and Roberson met with Steve Vinson at the sixty-eighth annual meeting of ARCE in Kansas City, Missouri. Vinson expressed his interest in the proposed conference and workshop, insofar as he had also recognized the need for assessing the state of digital research in our field. Unfortunately, however, Roberson and Lucarelli's

1 http://digitalhumanities.berkeley.edu/blog/15/10/22/dh-fellow-prof-rita-lucarelli-developing-book-dead-3d, accessed 04-04-2022.

2 For which, see now Zamacona and Ortiz-García, eds. 2021.

proposal went unfunded at that time and the conference and workshop project was stalled.

In the meantime, Vinson had, in the spring of 2016 and 2017, obtained internal funding from Indiana University (New Frontiers program of the Indiana University Office of the Vice President for Research; the Hamilton Lugar School of Global and International Affairs; IU Institute for Advanced Studies; Office of the Vice Provost for Research) to launch a pilot project to produce three-dimensional digital models of the small Egyptian collection in the Indiana University Eskenazi Museum of Art (known then as the Indiana University Art Museum). However, later in 2017, the IU Art Museum received a major gift from the Eskenazi family, a happy development that nevertheless resulted in a two-year closure of the museum for renovation and reinstallation of its collection. When it became clear that the Eskenazi project was going to be delayed until at least 2020, Vinson inquired with contacts at the Brooklyn Museum of Art as to whether the Brooklyn Museum would be interested in a photogrammetry and modeling project in Brooklyn's Egyptian collection. With their agreement, Vinson undertook short, exploratory photogrammetry campaigns in Brooklyn in 2017 and 2018.

Late in 2018, this preliminary work culminated in major additional funding from Indiana University to move the Brooklyn project forward, courtesy of the Indiana University Vice President for Research, and the Ostrom Grant program of the Indiana University College of Arts and Sciences. In 2019, with these resources, Vinson was able to undertake a three-week photogrammetry campaign in Brooklyn, hire a research associate (Mohamed Abdelaziz) to process the resulting models, and to organize and host what was hoped to be the first of a continuing series of international conferences on "Ancient Egypt—New Technology." In the process of organizing this conference, Vinson re-connected with Lucarelli and Roberson, who were invited to serve as advisors and panel moderators. The conference, held March 29–31 2019 in Bloomington, Indiana, was attended by 80 registered participants, including presenters from 14 countries including the U.S., who contributed 34 public presentations and 10 posters, as well as live displays of Virtual Reality technology.

The present volume, co-edited by Lucarelli, Roberson, and Vinson, presents the results of this conference. It offers a snapshot of the sorts of digital projects operating within the field of Egyptology at that time. Chapter 1, by Wendrich, considers both philosophical and practical questions surrounding the ethics of Egyptological research and publication in the digital realm. As such, it was selected to serve as both an introductory essay for the volume and as a backdrop for the studies that follow. Many of those chapters focus on digital techniques, such as photogrammetry, as applied to answer questions that have arisen in specific archaeological contexts (Abdelaziz and Elsayed; Janzen and Nichols; Lang et al.), or to work with specific sorts of artifacts in museum collections (Manieri; Lucarelli and Nederhof). Other contributors consider innovations in online interfaces and digitization of collections (Tomorad and Zlodi; Wilbrink and Roberson). Novel, computer assisted approaches to traditional philological investigation are well represented (Amin et al.; Harel et al.; Martin; Stauder-Porchet; Puglisi and Dakota). The possibilities and pitfalls of the virtual reconstruction of ancient landscapes—a perennial favorite for its obvious pedagogical value, among other reasons—are explored from various perspectives, in chapters by Danelon and Zielinski; Noc; Picardo; and Sykora et al. Along similar lines, Troche and Weston discuss the creation of an Egyptian-themed computer game as a tool for the instruction of schoolchildren. Navratilova examines the case for a robust research infrastructure in the study and analysis of the under-appreciated text genre of so-called "secondary epigraphy," known more traditionally as graffiti, while Moroney discusses the use of computer-assisted topographical analysis, to calculate the most efficient travel routes ("least cost paths") to ancient building sites. On top of all that, we even have a robot (Li et al.). As a snapshot of current work in digital and computer-enhanced Egyptological projects from the US and abroad, the present volume helps to fill a significant lacuna. While these techniques have been employed in our field for the better part of a decade (longer, in some cases; for a helpful summary, see discussion by Navratilova, in Chapter 14), the field itself has yet to pause and reflect on where the technology has taken us and where we are headed for the future. It is the editors' hope that this work, along with similar projects being developed in Europe and elsewhere (see most recently Zamacon and Ortiz-Garciá eds. 2021), will help to lead the field into the next phase of its evolution and to provide something of a signpost and guide for those who are preparing to begin their work on ancient Egypt, as viewed through the lens of new technology.

Works Cited

Zamacona, C.G. and J. Ortiz-García, eds. 2021. *Handbook of Digital Egyptology: Texts*. Monografías de Oriente Antiguo 1. Alcalá de Henares: University of Alcalá.

Acknowledgments

The editors offer their most sincere thanks to Peter Der Manuelian, for his kind invitation to include the present volume in the Harvard Egyptological Studies series. In addition, we are very grateful to the publisher and staff at Brill, in the Netherlands and the United States, for their continuous support and guidance throughout the process. We would also like to thank the anonymous peer reviewers who kindly took time to offer their insightful comments and critiques on a quite diverse range of topics that, despite their common thread of incorporating new and emerging technologies in Egyptological research, nevertheless cover quite a broad range of subject matter. We are also very grateful to the many authors and contributors for their patience, collegiality, and diligence. In addition, we offer our thanks to Paige Nehls (M.A. candidate in Egyptian Art & Archaeology, University of Memphis) and Amalee Bowen (M.A. candidate in Egyptology, Indiana University Bloomington) for their assistance with copy editing select chapters during the busy Spring semester of 2022. Finally, it is important to note that one of the editors' chief goals, since the project's inception, was to ensure that this volume would be made available in both traditional print format and as a freely available, Open Access digital edition—a goal which has only become more relevant in the wake of the global COVID-19 pandemic and the resulting interruptions of services and access to in-person resources. Funding for Open Access to the present volume was made possible through generous grants from the Michela Schiff Giorgini Foundation, the University of Memphis Institute of Egyptian Art & Archaeology, and the University of Indiana, Bloomington, for which the editors wish to express their profound gratitude and appreciation.

CHAPTER 1

Ethics of Digital Representation in Egyptology

Willeke Wendrich

Abstract

Interest in digital approaches to Egyptology started in the second half of the 20th century, for instance with the establishment of the *Manuel de codage* at the 1984 international round table "Informatique appliquée à l'Égyptologie." Since that time, many large textual, visual, and spatial digital projects have been developed by teams from several countries for a variety of audiences. The developments of practical and theoretical approaches of digital Egyptology have subsequently grown within the broader context of the Digital Humanities. Although digital scholarly projects are in the first place content driven, ethical questions on, among other aspects, representation, access and sustainability should always be included and can only be an effective part of the project design when considered and integrated right from the start.

Keywords

Egyptology – Digital Humanities – ethics – access – sustainability – environment – collaboration

Egyptology, similar to many other scholarly endeavors, increasingly makes use of digital tools for recording, studying and presenting original materials and research results. This volume gives an overview of several instances, but the list of digital projects is long and the variety large. In our efforts to explore the capabilities of digital Egyptology and mediate the drawbacks, it is important to explicitly address ethical concerns. With the recent emphasis on decolonizing museums and scholarship, with the attention for diversity, equity and inclusion in the field, ethics are an inherent part of Egyptological scholarship.[1]

1 Daly 2007; Abdel Maguid 2014; Tully 2015; Stevenson 2016; Carruthers and Van Damme 2017.

Digital Egyptology, however, deals with additional sets of ethical questions related specifically to the particularities of digital authorship, access, representation and sustainability (Figure 1.1). The interplay of ethical questions around digital applications and methods, as well as Egyptological questions, makes for a complicated and many-sided set of concerns that I will lay out in this chapter. In many ways this is a thought experiment and an effort to be aware of ethical consequences of our work, every step of the way.

1 Introduction

Many digital Egyptological projects collect and build upon prior scholarship: publications such as dictionaries and excavation reports, often dating to the 19th and early 20th century. Much of this information is valuable, but we should recognize that the research context has changed. Data are no longer considered value free, while research questions and concerns have found different foci. Moreover, digital publication differs from traditional print publication not only in form, but especially in character. Not just the research interests, methods, and theory—in short, the contents—have changed, but publication has become a continuous process reflecting knowledge in flux.

Digital Humanities has developed from "computer-based humanities" to a critical approach of research and representation. As Egyptologists we should use the insights of what I call *data critique* in the work done in the fields of digital humanities, digital archaeology, and digital cultural heritage.[2] I will consider important aspects of "digital Egyptology" and then outline how these principles have been incorporated in existing projects. There are many forms of digital publication of Egyptological content, and a critical approach to these helps to formulate the principles of ethical digital representation. In the course of this discussion, I consider successes, problems and best practices, outlining several principles that are rarely considered within Egyptology, but are part of an ongoing debate in Digital Humanities.

A thoughtful, yet admirably concisely formulated outline of best practices is the *London Charter for the Computer-Based Visualization of Cultural Heritage*,[3] that defines six principles, outlined in Table 1.1

The Charter explicitly does not limit itself to academia but has a broad focus and is concerned with the research and dissemination of cultural heritage

---

2 Bentkowska and Denard 2012; Burdick et al. 2012; Watrall 2016; Bonacchi 2017; I.W.N. Jones and Levy 2018; Richardson 2018; Lercari et al. 2021.

3 EPOCH 2009.

← **Authorship** →	
Collaboration	Single author

← **Access** →	
Open Access	Commercial limitations
Inclusive	Exclusive

← **Transparency** →	
Explicating bias in:	Implicit bias in:
Data, maps, plans, worldviews	Data, maps, plans, worldviews
Space in 3D	Space in 3D
Searches and algorithms	Searches and algorithms
Representation	Representation
Argumentation	Argumentation

← **Sustainability** →	
Data preservation	Data loss
Financial stability	Financial uncertainty
Digital security	Digital vulnerability
Environmental awareness	Large carbon footprint
Equity in labor	Rights abuses

FIGURE 1.1 Scales on which to balance an ethical approach of digital research, publication and representation. The chapter structure (sections 2. through 5.) follows the aspects listed under the four principles of authorship, access, transparency, and sustainability.

across academic, educational, curatorial, and commercial domains. Although the London Charter focuses on computer-based visualization, most of these principles are valid for digital research projects in a broad sense and even to a certain extent for traditionally published research in the humanities and interpretative social sciences. This is illustrated by Figure 1.1 which outlines several spectra that may serve to illustrate and concretize the charter's principles. They are not meant to necessarily present positive or negative features of digital Egyptology or of one particular type of project or product. There are indeed very few inherent pros or cons of a digital approach. The quality of information, and ease of access depend on the choices made in the process of building

TABLE 1.1 The London Charter for the Computer-based Visualization of Cultural Heritage

Principle 1 Implementation	The principles of the London Charter are valid wherever computer-based visualization is applied to the research or dissemination of cultural heritage.	p. 3
Principle 2 Aims and Methods	A computer-based visualization method should normally be used only when it is the most appropriate available method for that purpose.	p. 6
Principle 3 Research Sources	In order to ensure the intellectual integrity of computer-based visualization methods and outcomes, relevant research sources should be identified and evaluated in a structured and documented way.	p. 7
Principle 4 Documentation	Sufficient information should be documented and disseminated to allow computer-based visualization methods and outcomes to be understood and evaluated in relation to the contexts and purposes for which they are deployed.	p. 8
Principle 5 Sustainability	Strategies should be planned and implemented to ensure the long-term sustainability of cultural heritage-related computer-based visualization outcomes and documentation, in order to avoid loss of this growing part of human intellectual, social, economic and cultural heritage.	p. 10
Principle 6 Access	The creation and dissemination of computer-based visualization should be planned in such a way as to ensure that maximum possible benefits are achieved for the study, understanding, interpretation, preservation and management of cultural heritage.	p. 11

EPOCH 2009

a digital research project. What Figure 1.1 represents are the scales on which we need to balance an ethical approach to digital publication and representation in Egyptology.

The spectra of Figure 1.1 can be mapped onto several of the principles as defined in the London Charter, albeit not at a one-to-one coverage. Some of these aspects of digital publication and representation are organizational, others are intellectual, while most have aspects of both. This is quite common for digital projects, which usually combine practice, method, and theory. Best practices in digital publication and representation should, therefore, not just focus on the practicalities of digital projects (ease of use, cost, sustainability), but on the scholarly and societal values of transparency, explicating bias, and environmental sustainability. Design and user experience are important for usability, but also for access to the underlying information. A user should be able to understand the information that visualizations are based on and how much is reconstruction or conjecture. These issues will be explored in the next pages, based on a discussion of digital Egyptological projects, including lessons learned from the work that my collaborators and I have done in the past 15 years.

2 Authorship

2.1 *Collaboration—Single Author*

Digital Humanities has defined itself as different from "traditional" humanities, because of its collaborative nature, its networked environment, which changes both the culture of knowledge creation and the types of questions that can be asked of cultural heritage.[4] Egyptology as a humanistic endeavor has long been and still is characterized by single authorship, even in publications of teamwork such as archaeology. The excavation director used to be responsible for the publication of data and interpretation, while specialists authored their own reports and team members were perhaps mentioned. Sometimes a surveyor or architect, who created maps, plans and elevations of the excavation, was credited, but often not. Local excavators usually remained anonymous.

In digital Egyptology the technical contributions are even greater than on excavations. Digital projects are by definition teamwork and this goes much further than a "tech team" lending "support." Every step in the development, from the construction of a database to user experience, requires experience, knowledge, insights and creativity of content creators, designers, and programmers responsible for search, maintenance, and security structure. It is surprising, therefore, that the London Charter does not address credit for contributions to digital visualization. Even if part of the team is delivering "work for hire," credit is due. Working in teams in which each person contributes specific knowledge, skill and ideas means that the traditional authoritative relations are less impor-

4 Burdick et al. 2012, 3.

tant and preferably are done away with. The roles of faculty, staff and students in a digital project depends on every individual's particular input in the process and ranges from disciplinary to technical contributions. The strength of interdisciplinary projects, of which digital projects are an excellent example, is the fact that everybody learns. With the new ways of communicating, design and user experience testing have become more important than ever to enhance writing.[5] Tasks of digitizing, curating digital data, classifying, and describing digital assets, adding metadata and providing documentation for each step of the process are fundamental tasks in digital scholarship, often done by different team members. Students involved in the complexity of the project gain experience in producing scholarly products but are also challenged to consider how digital representation differs from traditional publication.

There are excellent examples of giving credit to various participants in a project. The most obvious is the "about" page of a web-based project, where all contributors can be recognized for their contributions past and present. Excavation data are increasingly presented with full authorship recognition for original notes, drawings, photographs and recording forms. This not only gives credit where credit is due, but also stimulates a sense of responsibility and a possibility to check the quality and consistency of work by several people. It, furthermore, demonstrates scholarly contributions that are traditionally expressed as class credit, performance evaluations, promotion, and tenure.

3 Access

Access to Egyptological knowledge traditionally depended on well-equipped libraries. "The "holdings" of the library were just that: holdings held for the initiated who had the privilege of access and use."[6] Access to digital Egyptological content seemingly has enlarged access, by making materials available online. Still, that access is not equitable: differences in financial capability (subscriptions or payments for access), ownership of a powerful computer, availability of stable and fast internet, suitability for people with disabilities to negotiate the functionality of a website are all aspects that can help or hinder inclusion in the broadest sense. It may be clear, therefore that statements on access (London Charter principle 6) are closely related to the aim of a digital project but go far beyond.

5 Burdick et al. 2012, 12.
6 Burdick et al. 2012, 45.

3.1 *Open Access—Commercial Limitations*

The term Open Access (OA) originally was defined as access to scholarly or scientific literature that is "digital, online, free of charge, and free of most copyright and licensing restrictions."[7] Over the past decade the term has gradually expanded to comprise not only literature, but archival information, original research results, photographs, video, audio, in short data in the broadest sense. Large scale digitization has provided almost instant access to what used to be the purview of researchers who were able to spend considerable time in archives. The "discovery" of an archived gem that changes our insights has been enhanced by the capability of combining and contrasting large amounts of information from multiple sources. Originally defined and defended by academic librarians, OA is one of the principles of FAIR, which stands for findability, accessibility, interoperability, and reusability.[8] For digital projects beyond articles or books, this means access to the full web functionality, without the hurdle of a pay wall. In most cases it also means that the data underlying the online presentation should be accessible and downloadable. An excellent example is Open Context, which makes granular archaeological information available for online searches, as well as download for re-use. Egyptological examples of online open archives are the Giza archives of Reisner,[9] the archive of the Griffith Institute,[10] and the publicly available photographs and documentation of the Metropolitan Museum[11] and the British Museum.[12] Although many good arguments have been made in favor of open access, there are financial aspects that need to be taken into account: creating and maintaining a digital resource requires considerable funds (see below).

3.2 *Inclusive—Exclusive*

A different side of access to digital resources is that of accessibility: is a web resource inclusive or exclusive? Accessibility includes accommodations for physical and cognitive disabilities, such as low vision, blindness; hearing loss and deafness; limited movement, speech disabilities, photosensitivity, and learning disabilities. Website functionality that has become standard, such as a mouse-over to provide additional information, is inaccessible to someone who is not able to use a mouse. This aspect of access goes further than ensuring that

7 Suber 2004; Holley 2018.
8 FAIR 2018.
9 Digital Giza 2017.
10 Griffith Institute 2021.
11 The Met 2021.
12 BM 2021.

every functionality is made available through different methods, so that there is an alternative way of accessing information. Several initiatives inform website creators on good practices and allows developers to test the accessibility and usability of their site. The Web Accessibility Initiative, which studies how people with a disability use the web, is linked to the World Wide Web Consortium, an international community that develops open standards to ensure the long-term growth of the Web.[13] The standards developed under the title Web Content Accessibility Guidelines[14] have become leading principles for various national governmental regulations such as the U.S. General Services Administration Government-wide IT Accessibility Program[15] and legal requirements in Italy and Germany.[16]

If we take equity, diversity, and inclusion seriously, however, then we should assess all elements of our research, from the fundamental basis, the type of questions we ask, to the reception of our research output. Many aspects are obvious: the language in which a site is written determines who can access it. The projects of the *Center for Documentation of Cultural and Natural Heritage*, in collaboration with the *Bibliotheca Alexandria* are presented on a website that is accessible in both Arabic and English[17] Some German Egyptological websites are bilingual German/English, e.g., TLA 2004. The UCLA Encyclopedia of Egyptology has abstracts in Arabic and keywords in four languages: English, Arabic, French and German,[18] while the digitally born *Rivista del Museo Egizio*[19] is published in multiple languages and has abstracts in English and Arabic. Ancient Egyptian Architecture Online, a temporarily defunct online resource, set out to develop a tri-lingual illustrated resource for architectural terminology, in English, German and Arabic. While the *Getty Online Thesaurus of Art & Architecture* provides terminology in most European languages.[20] A herculean effort to provide a multilingual resource specifically for Egyptology, the *Multilingual Egyptian Thesaurus*, was published online originally with European languages only, to which more recently Arabic was added.[21]

Even though language access has somewhat improved with browser translation capabilities, machine translations of discipline-specific texts can be con-

13 W3C 2020.
14 WCAG 1.0 1999; WCAG 2.1 2018.
15 Section 508 2018.
16 BITV 2019; AgID 2020.
17 CultNat Arabic 2021; CultNat English 2021.
18 UEE, PDF Only 2008; UEE 2010.
19 RIME 2017.
20 AAT 2017.
21 MET 2007.

fusing or misleading. With the importance of community archaeology gaining ground, archaeological projects increasingly involve local excavators in the development of research questions, interpretation, and output. Making reports available in Arabic is just one step towards increasing access to Egyptian cultural heritage in Egypt. Publication online in formats that can be understood by a broad part of the population, in Egypt or elsewhere provides immediate access to research activities that otherwise remain invisible. Incorporating videos and photographs, including those produced not by professional archaeologists, but by local participants, further increases inclusion.

Fundamental in the ethics of access is the implicit (dis)incentive of using a website based on the perception whether a visitor is actually welcome, whether she is part of the target group or not. In Egyptology this is not just linked to modern languages, but also to whether and to what extent one reads the ancient Egyptian language, knows the technical terminology, or understands how an archaeological excavation is documented.

Sometimes access is explicitly restricted. There are excellent reasons to make certain data available to researchers only. One example is the restriction that is put on geographic coordinates of antiquities sites, which cannot be made public because advertising locations enables the exact position of vulnerable sites to become known. This can be highly problematic because of potential damage caused by larger number of visitors. In the worst, but not uncommon, case publishing site locations results in targeted looting. Although most archaeologists are aware that it is often necessary to restrict access to geographic data, very little has been published about it. It is, for instance, not emphasized in the "guides to good practice," developed through a collaboration of several national repositories of archaeological digital data.[22] In Egypt, many sites are officially under protection of the Ministry of Tourism and Antiquities, but only a limited number are under permanent watch.

Research output has various objectives and audiences. Different criteria are required for information that is geared towards specialists, versus the presentation of results in an interpretative environment for a general audience. The aim of a digital project is thus not just defined by the content, but also by the targeted audience. The *UCLA Encyclopedia of Egyptology*,[23] for instance, is geared to colleagues, but also to an advanced undergraduate level general audience and the editors discuss the content and terminology that authors use in order to ensure that specialized jargon is avoided as much as possible. In addition, in

22 *Archaeological Data Service* 2009.

23 UEE, PDF Only 2008; UEE 2010.

the full version of the work the terms that cannot be found in a regular dictionary are clarified in the side bar, activated by clickable links (Figure 1.2). This version of the UEE differs in lay-out from the (also freely available) PDF version that is published through eScholarship, the online publication platform of the University of California. The printable form of the article is designed in two columns, a format that is unsuitable for online reading, especially on tablets and telephones.

A project that is accessible to an English-speaking general audience is the *Theban Mapping Project*, which focuses on the tombs in the Valley of the Kings.[24] The project of mapping, measuring, and photographing all tombs in this famous area started in 1979, under direction of Kent Weeks, American University in Cairo. In 1989 the results were published online as an interactive site which allowed the user to access the location, configuration, and decoration of the tombs. The site also provided references. This compilation of information was made available in a very intuitive way and widely used by a general audience. The website remained online for 21 years, a remarkable feat considering the many changes in browser configurations which results in most online resources needing an overhaul approximately every five years. From 2010 to 2020 the website remained inaccessible, but at present it is live again thanks to a grant from the American Research Center in Egypt. A map-based interface allows a very intuitive way of accessing materials and the clustering of information at different scales, from overview to detail.

A similarly long running project is *Digital Giza*, directed by Peter Der Manuelian.[25] This project combines three-dimensional reconstructions of the architecture of the Giza Plateau with a vast archive of archaeological research materials, based on a wide variety of American, European, and Egyptian expeditions, starting with the Harvard University—Boston Museum of Fine Arts Expedition from about 1903 to 1947. The online presentation of interlinked archival materials draws upon the Giza Consolidated Archaeological Reference Database. The open access of literature, made available for download on the website, is an example of the collaborative effort of this project.

An even longer history underlies the *Thesaurus Linguae Aegyptiae*, developed in Germany and incorporating a large number of German and Belgian initiatives,[26] but was based on and inspired by a standard work in Egyptology: the *Wörterbuch der ägyptischen Sprache*, published between 1926 and 1950,

24 TMP 2020.
25 Digital Giza 2017.
26 TLA 2004.

but for which the work started in 1897.[27] The TLA is enhanced with the results of several in-depth research projects. The English introduction of the website states that "All texts come with running translations to assist particularly non-specialists and scholars of neighboring disciplines in their work" (TLA Introduction). Nevertheless, a person with an interest in ancient Egypt, and even Egyptologists who are not very familiar with the site will have difficulty negotiating this important source of information.

Non-Egyptologists will have an easier time with another standard work in Egyptology, that has transferred from a book series to an online version: The *Online Egyptological Bibliography*[28] provides titles and in many cases abstracts of books, articles, and reviews of Egyptological publications from 1822 to the present. It combines the Annual Egyptological Bibliography,[29] with the *Bibliographie Altägypten*,[30] and the *Aigyptos* database, with keywords, and more than 40,000 further items. This important resource is subscription-based and that is a hindrance for independent scholars, or students and faculty from smaller universities that do not have institutional subscriptions. It also excludes members of the audience with a casual interest in ancient Egypt.

Most of the entries in the *Ancient World Online* (AWOL), which started as a blog by librarian Charles E. Jones to highlight open access digital content, point to traditional journals and articles that are made available as PDF.[31] Although extremely useful, such forms of online journal publication do not benefit from the possibilities that the medium potentially offers, which includes enhanced searching, analysis, imbedding of different data formats and full data availability. The format of the AWOL site itself is very basic, and in terms of access, is relatively easy to maintain, but runs the risk of linking to online projects that are no longer available. That is the final aspect of access that needs to be mentioned and has been referred to above: online presentation of digital information that are well-designed with complicated functionality are very difficult and costly to maintain. The website of the UCLA projects of *Digital Karnak* and *Ancient Egyptian Architecture Online*[32] are good examples. Digital Karnak is now only available through the internet archive,[33] while AEGARON is seeking funding for restoration of functionality.

27 Erman and Grapow 1926.
28 OEB n.d.
29 AEB 1947–2001.
30 BA 1822–1946.
31 Jones and Elliott 2015.
32 AEGARON 2010.
33 Digital Karnak 2008.

UCLA ENCYCLOPEDIA OF EGYPTOLOGY

About Subjects Articles Time Map Glossary

TABLE OF CONTENTS

ABSTRACT
ABSTRACT IN ARABIC
INTRODUCTION
EVOLUTION AND DOCUMENTATION
INTERPRETING ATENIST ART
THE AFTERMATH OF AMARNA ART
MINI MAP
BIBLIOGRAPHIC NOTES
REFERENCES
IMAGE CREDITS
RELATED IMAGES

REFERENCE LINKS
PLACE LINKS
PERIOD LINKS
TERM LINKS
ARTICLE LINKS

Amarna Art

DOWNLOAD PDF

Dimitri Laboury

فن العمارنة
Amarna Kunst
Art amarnien

Keywords: evolution, pharaoh-centrism, talatat, Atenism

Abstract

The art that developed in the reign of Amenhotep IV/Akhenaten, known as "Amarna art," has largely been considered revolutionary in the history of ancient Egyptian art. As such, it has been the subject of much debate and has generated numerous theories, often contradictory or controversial, and, in fact, deeply influenced by the history of its modern reception. Nevertheless, the remaining evidence still permits us to investigate Amenhotep IV/Akhenaten's conception of images and art, as well as the artistic evolution under his reign. From a hermeneutic point of view, Amarna art can be interpreted as a multi-causal phenomenon, involving an internal evolution of 18th Dynasty society and art, pharaoh-centrism, and purely aesthetic factors.

Abstract in Arabic

« فن العمارنة » يعتبر الفن الذى تطور خلال عهد امنحتب الرابع/أخناتون والمعروف باسم يعتبر الى حد كبير تطور ثورى فى تاريخ الفن المصرى القديم فيمثل هذا التطور محور الكثير من النقاشات وبنى عليه العديد من النظريات التى غالبا تكون متناقضة أو مثيرة للجدل وعلى أى حال فإن هذه النقاشات والنظريات متأثرة بشدة بتاريخ استقباله الحديث، مع ذلك فإن الادلة المتبقية لا تزال تسمح بفحص مفهوم امنحتب الرابع/اخناتون للفن والتصوير، فضلا عن التطور الفنى فى ظل حكمه. يمكن تفسير فن العمارنة باعتباره ظاهرة متعددة الاسباب والتى تشمل تطور الذاتية للمجتمع والفن خلال الاسرة الثامنة عشر، دور الفرعون والعوامل الجمالية البحتة.

Aten
Deity representing the actual sun disc; raised to the position of sole state god under the rule of Amenhotep IV/Akhenaten (1349 - 1332 BCE). Cult center: Akhetaten.

ı traditionally used to refer to the unconventional art developed in the reign of Amenhotep
:E). It was coined after the modern Arabic name for the site of el-Amarna, where the king
en, in the first third of his year five. So-called Amarna art had nevertheless been initiated
ar four, when the cult of the solar god Aten was officially launched at Karnak, a period that
nating the "proto-Amarna stage" (Vergnieux 1996; 1999: 201 - 202; Vergnieux and Gondran
able link between the art instituted by Amenhotep IV/Akhenaten and the new ideology of the
cult of Aten, the designation "Atenist art" thus appears more appropriate.

FIGURE 1.2 A page from the UCLA Encyclopedia of Egyptology with multilingual main words, a dual abstract in English and Arabic and clarifications of discipline-specific terms in the margin. The text is rendered in one column and the table of contents links to various sections of the article.

4 Transparency

Transparency in digital Egyptology starts with the basic questions of who did the research, who built it, who paid for it, what is the purpose, who is the audience. These are usually addressed in the footer or the "about" section of the webpage. Beyond that there are a many other aspects that determine whether a digital Egyptological project is constructed in a transparent way or not.

Perhaps the main contribution to transparency is whether the purpose and research questions have been made explicit. This is not just a matter of presentation but lies at the heart of the development of the project. The London Charter principles that are central here are aims and methods (principle 2), research sources (principle 3) and documentation (principle 4). Sub principle 3, the charter states: "Particular attention should be given to the way in which visual sources may be affected by ideological, historical, social, religious and aesthetic and other such factors."[34]

In Figure 1.1, I proposed that the two extremes of the transparency scale are implicit bias and explicating bias. Humans are inherently biased,[35] and even though bias in itself is not unethical, it definitely is highly problematic to consider biased perceptions as "objective," or to use inherently biased considerations as the basis for social, economic or other forms of inequality. Small steps to understanding our own biases are based on listening and self-reflexivity. We should at least aim to be as explicit as possible in defining our goals, building our methods, and formulating our arguments. Some of the tools we have for this are data and critical thinking. In technical terms transparency is effectuated through the documentation of data, meta-data, and para-data.[36]

The line between these three types of data is unclear at best and depends on the research question. Data makes up the information that we seek or develop to address our research. Meta-data comprise information about the data, for instance the name of the photographer who took an image we are using as a source, and the date on which the photo was taken. Para-data describe the process of information gathering. While co-teaching a workshop on meta-data at UCLA, I found it quite liberating to realize that this distinction is not fixed, perhaps not that relevant, and can change at a moment's notice. When a researcher is focused on the contribution of photographer Harry Burton to the discovery and registration of the tomb of Tutankhamun, the photographer's name is part of the data, rather than the meta-data.

34 EPOCH 2009.

35 Eberhardt 2020.

36 Bentkowska and Denard 2012.

The relationship between different data formats (texts, numbers, maps, plans, 3D models, still images, moving images, sound) and different types of knowing (visual, aural, tactile) is complex. Still, the digital format enables connections between most of these, perhaps with exception, for now, of the tactile aspects of perception. Recording through these data formats results in, for instance, measurements, locations, descriptions, illustrations, photographs, videos, references, or oral histories. What is important at every step, is to define what the status of knowledge is: what is the degree of certainty in an observation. I would argue that digital representation requires an even more rigorous explication to ensure intellectual integrity, with clear indications whether the information is uncertain, ambiguous, or hypothetical. If a complex combination of data and datatypes is presented as evidence, it should be explicated how these contribute to addressing the underlying questions.

Each type of data presentation has its own forms of inherent bias and manners to explicate these. In the following sections we will consider those of spatial data, in two and three dimensions, quantitative data, search structures, representation and argumentation.

4.1 *Data, Maps, Plans and Worldviews*

Ancient Egyptian Architecture online was started with the express purpose to provide transparency in the sources used to produce architectural plans, following defined drawing standards. In Egyptological publications plans of temples, tombs or settlements are often copied without specifying whether the plan represents a reconstruction or the actual state of a monument. In addition, the "actual state" in 1909 is not the same as that in 2021.[37] A further consideration of mapping should address the fundamental basis of how space is understood. Spatial orientation is not universal and cultural as well as intracultural differences in perception of maps and mapping have been recognized in geographic and anthropological communities, but not so much within Egyptology.[38]

Archaeologists' use of maps to indicate the geographical context of the site where they work and detailed plans of excavations. Maps are, by definition an abstraction or manipulation from the situation on the ground. The projection of the globe on a two-dimensional surface (paper or screen) results in a distortion of spatial relations. The equal earth map projection,[39] is a recent attempt to solve a number of projection problems, for instance that of the Mercator

37 Fauerbach et al. 2010.

38 Chang and Antes 1987; Perkins 2008; Miller 2012.

39 Šavrič, Patterson, and Jenny 2019.

projection, which is maligned, because of its distortion of the landmasses. In this map projection Europe is looming large over Africa, while this is a gross misrepresentation of the actual size of the two continents. The purpose of the Mercator projection was, however, not to represent landmasses correctly, but to provide the most economic ("straight") sea routes for navigation before the time that ships had the availability of using the Global Positioning System (GPS). It may seem that the globe (as represented, for instance by Google Earth) is, therefore, the most "neutral" spatial representation, but looking from space towards earth is a very particular way of thinking about human lived space.[40] This approach to spatial context is fundamentally different than the experience of being in the world. Especially if our goal is to understand or represent an ancient world view, the birds-eye view as exemplified by mapping is incongruous.

The liability of accidentally (or purposefully) using statistics, charts and graphs to misrepresent quantitative data has been described in a large body of literature.[41] It outlines how both the design and the underlying data can be biased. Maps and charts always contain an argument. Whether they set out to be argumentative or are meant to provide a "value-free" image, their design and the conventions used are particular to a certain way of understanding and representing the information. Much has also been written about maps and mapping, mostly regarding their design, what information they contain, how they are used to make arguments and how they can mislead by accident or on purpose.[42] Egyptian archaeology and history make extensive use of maps, often without the critical approach or explicit statements that clarify the purpose, or underlying data.[43] For digital mapping and geographic information systems, the selection of layers of information and the manner in which these are combined and displayed are even more influential.

The digital turn allows for interactive maps, where users can explore information, a development that has led to greater attention to representation, instead of communication.[44] It should be noted, however, that users' agency in exploration, is bounded by the pre-defined content, layering and structure, as well as the underlying research questions and suppositions. In addition, there are cultural differences in how populations understand and interact with spatial information.[45]

40 Oliver 2015.
41 e.g., Huff 1954; Tufte 1997; 2001.
42 Monmonier 1991; 2018; Tufte 1997; 2001; Wood 2010.
43 Gillings, Hacıgüzeller, and Lock 2020.
44 Perkins 2008, 151.
45 Alberti and Marshall 2009.

4.2 *Space in Three Dimensions*

The context for human domestic, social, and divine interaction is created, built, and shared three-dimensional space. The London Charter is specifically meant to define good practices for cultural heritage computer visualizations. It does not, however, explicitly separate the visualization of what exists from three-dimensional modeling of what does not or is only partly present. The first we will call three-dimensional recording, the second three-dimensional reconstruction. Critique of three-dimensional representation is not new, as expertly and concisely outlined by Paul Miller and Julien Richards,[46] but there are many good reasons that both three-dimensional recording and three-dimensional reconstructions are used in archaeology.

Archaeological hand-drawn plans and elevations have been the standard of recording length, width, and height/depth of excavated remains, or complete buildings or rock-cut tombs. These drawings are rapidly replaced by three-dimensional recording through 3D-scanning or photogrammetry.[47] The main reasons are speed and precision of recording, as well as the ease of three-dimensional representation, which provides an intuitive understanding of spatial relationships.[48] In Egyptology digital techniques are increasingly used for epigraphy, to locate inscribed spaces into their three-dimensional context and to record the three-dimensional nature of texts and imagery in relief. In addition, digital applications are used to enhance photographic recording.[49] The second edition of Krisztián Vértes' 2014 book *Digital Epigraphy*, is available in open access on a website dedicated to training digital epigraphers.[50] The site is a perfect example of collegial collaboration and the establishment of standards by the large epigraphic projects working in Egypt: the Oriental Institute of the University of Chicago, the Epigraphic Survey (Chicago House), the Giza Project at Harvard and the KU Leuven.

An argument against replacing hand drawing with digital registration is that epigraphy and field drawing are forms of visual interpretation and as such an integral part of analysis.[51] The lengthy process of spending time and focus can be replaced by an intense process aided by digital enhancement. Here the processes of epigraphy and archaeological field drawing differ. The archaeologist cannot return to what has been excavated away.

---

46 Miller and Richards 1995.

47 Kamermans et al. 2016.

48 Schnabel and Kvan 2003.

49 Lima et al. 2018; Evans and Mourad 2018; Vértes 2020.

50 Digital Epigraphy 2021.

51 Morgan and Wright 2018.

The form of three-dimensional visualization that is usually comprised by that term is reconstruction, rather than recording. If recording includes interpretation, reconstruction does so exponentially. The role of para-data is to describe the process of reconstruction, including the source material, the level of interpretation, speculation, and uncertainty.[52] The London Charter specifies that the choice of the type of visualization, whether a reconstruction is presented photorealistic or schematic, for instance, should be made deliberately. The influence of visualizations, especially in the Virtual Reality realm is persuasive, even seductive. It is very difficult to maintain a critical attitude, unless fuzziness and uncertainty are pointed out in no uncertain terms, within the reconstruction.[53] Although considered especially useful to inform a general audience, the process of creating a three-dimensional reconstruction is extremely helpful to generate additional research questions and formulate argumentations (see below).

4.3 *Searches and Algorithms*

Databases or spreadsheets can be used to organize information in a fully transparent manner. By making the data sources, objectives, and search criteria explicit, the user can trace search results to the source, including the underlying presuppositions. Unfortunately, as with maps and charts, databases are not designed to allow for ambiguity or uncertainty of information. This is problematic in all historical disciplines: a drop-down menu requires a clear, positive choice, which ancient sources hardly allow. Historical or archaeological data are generally uneven, with missing information. When relationships and search terms are organized in implicit hierarchies, the database becomes even more opaque.

Transparency requires that the sources are identified and evaluated, tested for level and kind of bias and search protocols. When these aspects are not in place, the search algorithm becomes a black box. It is what I call "the computer as connoisseur": by an undefined, yet somehow measured combination of qualities, a result is produced. Connoisseurship is the result of years of experience-based building of knowledge, but simply "knowing" that an object is e.g., genuine, is based on a tacit, complex, weighing of attributes. It should, then, be possible to explicate what these are. Apart from requiring real effort, that would diminish the mysticism of connoisseurship. The digital equivalent of the connoisseur is the algorithm. Algorithms are part of proprietary business

52 Bentkowska and Denard 2012.

53 Miller and Richards 1995.

secrets, and their lack of transparency is a given. Yet they are a source of bias that penetrates society, as well as scholarship.[54]

4.4 *Representation*

The maps and charts discussed above are forms of visualization and representation. In this section we consider ethical considerations of representation *sensu stricto*: how do we represent persons, ages, genders, occupations, and cultures in a digital (re)construction. Before the digital age Egyptological studies have produced representations of knowledge mostly in the form of descriptions based on the study of texts, imagery, material culture and the archaeological context. Apart from textual representations can have many forms, including visualizations—image based interpretations. What these have in common is that they are based on a mental image, that is the result of tacit understandings of various aspects of ancient Egypt. Since the data are ambiguous, uncertain, and incomplete, historical information is open to interpretation. Digital representations often require unequivocal choices: drop down menus, points on a map, color spectra in Virtual Reality reconstructions. An ethical digital Egyptology will be aware of the inherent biases that are especially sensitive when concrete choices need to be made about ancient society, spaces, and people. Where an image says more than a thousand words, an interactive three-dimensional Virtual Reality model subconsciously determines our understanding in an underhanded manner. The responsibility of those who create Virtual Reality models, games, or Augmented Reality applications, especially when it includes people, is enormous.

Ancient Egypt has both been claimed as European and African. Petrie's theories that a dynastic race built the pyramids after colonizing primitive early Egypt was part of the racist ideas that African people would not have been capable to build the pyramids.[55] Pushing back against the whitening of ancient Egypt as part of Eurocentric claims are scholars that maintain that the ancient Egyptians were black.[56] Depicting the skin color of people in VR reconstructions directly brings up these ongoing debates. Similarly, the orientalist and male gaze that is at play in depictions of Cleopatra and other female historical figures is of influence on both the creation and reception of digital representations.[57] Independent of how people are represented in reconstructions,

54 Pariser 2012; Noble 2018; Christian 2020.

55 Petrie 1879; Challis 2016.

56 Diop 1989.

57 Godon 2018.

photographs or 3D scans of human remains (skeletons, skulls and mummies) can be a sensitive issue among particular audiences, although not among others.[58]

Apart from the depiction of humans, whether a temple or village is represented in simple outlines, or "enhanced" with atmospheric details makes a difference in how the model as representation of the past will be experienced. Here the London Charter outlines most detail: in the Aims and Methods section it specifies that scholars should be explicit about the purpose of a model, as well as its execution. Should the model be schematic or photo realistic? Does the model represent a state of knowledge or a hypothesis? Lighting, colors, movement, and sound all contribute to the impression that a VR model conveys. What a digital model evokes, but does not actually contain are emotions, expectations, and relationships. What underlies a model is the ontology of the creators, their often-implicit world view. Ancient built environments are a complex material expression that is multi-layered: the same space might reflect shelter, safety, power, piety, posturing, negotiations, justice, or threat. Sharing space is a sign of social or ritual belonging. Location, orientation, context, building materials, decoration, re-use, and cleanliness all potentially inform us on what the space might be about. Is an ancient Egyptian village represented as grimy and the temple as pristine? Such choices should be made consciously and expressed in the documentation through the para-data.[59] For scholarly projects it is important that this documentation is available, for instance in the form of in-project annotations, such as exemplified by the digital publication of Elaine Sullivan and Lisa Snyder.[60]

It may be clear that with the power that digital creators have on the perception of ancient Egypt, the question of who represents Egyptological knowledge is critical. The systematic exclusion of Egyptian scholars in the representation of ancient Egypt, has been outlined by Reid.[61] The website of the Center for Documentation of Cultural and Natural Heritage features the work and archive of only one famous Egyptian Egyptologist,[62] while the work of Egyptian scholars on ancient Egypt may present the opportunity to foreground alternative viewpoints.[63]

58 Colwell 2017.

59 Bentkowska and Denard 2012.

60 Sullivan and Snyder 2017; also Sullivan, Nieves, and Snyder 2017.

61 Reid 2003; 2015.

62 Selim Hassan 2019.

63 Elgewely 2014; Elgewely and Wendrich 2015.

4.5 *Argumentation*

Above it was stated that maps and charts almost by definition contain an argument. Ethical considerations are important, especially when that argumentation is implicit, problematic, or both. An (in)famous example are the nationwide maps created in the USA between 1935 and 1940 by the federal government's Home Owners' Loan Corporation, known as "redlining."[64] In the third edition of Monmonier's *How to Lie with Maps*[65] he added a new chapter on animated, interactive, or mobile maps, in which he states "Possibilities abound when a viewer can explore the fuzziness of the data by changing a map's time frame or definitions, or question the reliability of a choropleth map by experimenting with cut points or number of categories."[66] Digital media allow for the creation of transmedia modes of argumentation:[67] an argument can be formed by or built into a digital representation other than text. An example is the Virtual Reality reconstruction of Karnak Temple, which shows the temple complex in multiple phases of development. The 3DVR model has been used in several publications, which I will briefly outline here. Two online publications use the model to discuss the construction phases of the Karnak complex. The first one is an encyclopedia article, that makes use of screenshots of the model,[68] the second is a video and PDF, part of the Digital Karnak website's educational materials.[69] In an online article, which includes the actual model Elaine Sullivan and Lisa Snyder discuss the process of authoring and peer-reviewing a three-dimensional model.[70] A multi-authored book chapter discusses how the production of 3DVR models is scholarly production, including the statement that "Models can offer new techniques to investigate questions of how gender, ethnicity, and power are conceptualized by a society and inscribed into the very space that structures such relationships."[71] Additionally, the Karnak model was used in an argument on monumental architecture and cultural memory. The digital model can actually not only show the building phases but can clarify both the addition and the removal of architecture. This enables demonstrating the creation, as well as *damnatio memoriae* and usurpation of memory.[72]

---

64 Mapping Inequality 2020.
65 Monmonier 2018.
66 Monmonier 2018, 201.
67 Burdick et al., 10.
68 Sullivan 2010.
69 Sullivan 2008.
70 Sullivan and Snyder 2017.
71 Sullivan, Nieves, and Snyder 2017, 301–302.
72 Wendrich 2014.

5 Sustainability

5.1 *Data Preservation—Data Loss*

Digital Egyptology has shown to enable the aggregation of information, by combining various formats (texts, photograph, videos) around a similar subject, such as the archives of excavations mentioned above. Amassing and analyzing information is known as the field of cultural analytics, which uses computational analysis and data visualization interpret cultural data on a large scale.[73] The counterpoint of amassing data is loss of information: we are all aware of the fickleness of digital preservation in a time when floppy discs are a distant memory. The London Charter's principles 2 (aims and methods) and 5 (sustainability) address concerns of preservation of data. Perhaps the best way to think about this is that the creation of digital Egyptological resources is conceptualizing prospective memory: thinking through how the reception of the present will be in the future.

Using standardized formats for the digital files that can be archived and updated is but one step in the process. Documentation of the relationships, processes and considerations is central to data sustainability. Initiatives such as the Digital Archaeological Record in the US,[74] the Archaeology Data Service in the UK,[75] and Ariadne in the European Union[76] provide the archival power and knowledge base to ensure long-term preservation of data and stimulate the FAIR principles (findable, accessible, interoperable and re-usable). An important ethical question in relation to this is: who has the power to decide what should be preserved, what can be deleted and who has access? The Indigenous Data Governance project[77] has been created to address traditional power inequalities. Apart from the FAIR principles this international alliance promotes the CARE principles (Collective benefit, Authority to control, Responsibility and Ethics).

Digital archiving is an increasingly specialized field, where the academic libraries have taken the forefront. The UCLA Encyclopedia of Egyptology has been designed from the very start for long-term preservation, by storing all assets, article texts, images and the data tables that drive the interactive maps and timelines with copious meta-data in the UCLA Digital Library. Whenever the front end of the website is outdated, on average once a decade, but often

73 Burdick et al. 2012, 40.

74 TDAR 2018.

75 ADS 2012.

76 Meghini et al. 2017; Ariadne 2020.

77 CARE 2021.

sooner, the information is safely stored and the relationships between all the different parts have been defined so that the functionality can be updated or rebuilt.

Data sustainability is closely related to continued access and the problem of dead links. The Multilingual Egyptological Thesaurus was created by the Center for Computer-aided Egyptological Research at Utrecht University, The Netherlands in the 1980s. In 2021 the acronym and the corresponding URL CCER.nl no longer refers to an Egyptological entity, but to computational economics research. The rights for the Thesaurus were transferred to the Center for Documentation of Cultural and Natural Heritage (CultNat), which licensed the web-access to a French initiative, Projet Rosette.[78] The difficulty of sustaining resources that were often built through the research and efforts of a large group of people over many years is illustrated by two projects that both focus on tomb reliefs. *MastaBase* is a database developed in 1998 at Leiden University, which enables searches through the decorative programs of Old Kingdom tombs. Originally appearing as a CD-ROM, the database was transferred to an open access website in 2014.[79] The database of the Oxford Expedition to Egypt likewise organizes and archives tomb reliefs from the Old Kingdom.[80] It is one of eight Egyptian data sets comprised in the Archaeology Data Service. Both initiatives partially cover the visual and intellectual information that has been generated during the lifetime of these research projects. Both have an enormous potential for growth, but they are stagnant, a situation that is related to the costs of both creating and sustaining digital projects.

5.2 *Financial Sustainability—Financial Uncertainty*

Creating a digital resource is expensive, and yet there is an unspoken expectation that everything that is online should be available for free. In the business world the real costs are, therefore, covered with advertisements or user data, by enticing users to continue their foray on the website. Presenting users more of what they are apparently interested in is a means of retention.[81]

Open Access stands in direct conflict with licensing of online publications or images. The copyright laws differ world-wide, but the common tendency is one of expansion of the time that resources stay under copyright.[82] That in itself

78 MET 2007.
79 MastaBase 2014.
80 OEE 2006.
81 Krunic 2020.
82 Lessig 2000.

is an ethical question: what is more important: the rights of the author or the author's heirs or broad access to the public? Related to that is the question how far fair use can be stretched.[83]

Many digital Egyptological resources have been created with grant money, in the US for instance from the Mellon Foundation or the National Endowment for the Humanities, which demands that projects created with public funds are publicly accessible. The conundrum is that there are funds available for new projects, but hardly for maintenance, refurbishment, or recreation of an existing resource. The expectations for digital resources adapt to the functionality that is developed by large companies with thousands of developers ("you may also like ..."). Commercial products that enable such functionality are not only expensive, but they do not guarantee longevity, or the capability to retrieve all data that were entered. A large community of developers that works on alternatives for commercial products, is supported by a non-profit organization, the Open Source Initiative.[84] Open Source does not necessarily mean that the software or product is free, but it indicates that the code is made available for others to build on. All digital resources require regular upkeep, a task which can be onerous, is quite thankless, and requires specialized knowledge.

Most Egyptological resources aim to provide open access and work on the basis of open-source software. As indicated above, that does not mean they are free to create. The term "sweat equity" indicates the many hours that are volunteered for the creation or upkeep of a resource. The UCLA Encyclopedia of Egyptology is fully dependent on the time donated by its editors, authors, reviewers. Technical staff and copyeditors are paid by the university through internal grants. In academia, where peer review is considered part of professional service, this is of course nothing new. As soon as an organization needs to hire staff to manage the project, however, it is no longer feasible to exist without a form of income. The Online Egyptological Bibliography has chosen for a subscription model, others are looking for private donors. One of the ethical considerations related to finances is where the money is originating and whether that origin is made public to the users.

5.3 *Digital Security—Digital Vulnerability*

Data security and data governance are ethical issues: even though humanities' research does usually not generate data as sensitive as medical or social science data, all work that includes humans should be treated with ethical

83 Aufderheide and Jaszi 2018.

84 OSI 2021.

considerations. Ethno-archaeological research directly involves human subjects (as research partners, they share sensitive information) and in the US requires permission from the Institutional Review Board of the researcher's university.

5.4 *Environmental Awareness—Large Carbon Footprint*

The ethics of digital Egyptology includes awareness that computers, datacenters, screens, all our instruments to create, use and archive important Egyptological resources, require large amounts of energy.[85] The carbon footprint of digital scholarship is large. Outsourcing to the cloud does not solve the problem, because large data centers are notorious energy consumers.

5.5 *Equity in Labor—Rights Abuses*

In a similar vein, much of the coding that is required to develop and maintain software is done in low-wage countries. Not only that, the machines on which we work, the laptops, computer chips, external hard drives, etc. are produced in low-wage countries and often involve child labor. These aspects of the digital turn in Egyptology often remains invisible and is rarely discussed.

6 Conclusion

In the above I stated that the continuum of Figure 1.1 is not an assessment, but that is actually not accurate. The ethics of digital Egyptology lean towards the left side of the chart: digital projects are inherently collaborative, rather than single authored works, and should be fully credited as such. Access to content should be as inclusive as possible, especially in countries where resources are lacking. This means that we should perhaps not design the latest and the greatest functionality but create something that provides a simple way of getting at important information, that downloads also at low bandwidth. Digital projects should be explicit about aims and purpose. While recognizing that it is impossible to exclude bias, we should give every effort to create an inclusive resource, based on sensitivity to different viewpoints. This requires a collaborative thoughtful process, in which aspects of inclusion and transparency are taken into serious consideration. The strongest and at the same time most difficult ethical requirements are the ones related to sustainability. We should strive to create sustainable projects that avoid data loss, and we should try to

85 Pendergrass et al. 2019.

make content available for free. Most importantly, an ethical approach to digital Egyptology should be aware of the costs to the environment and harm to human life and dignity. In the industrial systems of the global economy these detrimental and toxic realities are not easily addressed, but that does not mean they should be taken out of consideration. An ethical approach means that we should look at the system as a whole.[86]

An ethical digital Egyptology promotes the democratization of knowledge and is explicit in how argumentation links to data. In the light of the Trump-era disinformation and lack of authority of scholarship, we should not give in to the temptation to simplify "truth" as something that is true and unambiguous. Ultimately, digital Egyptology is all about content, centered around data in the broadest sense of the word. Its form is transient and will change, with the development of new forms of digital communication and information access. Digital formats allow us to combine and search content in exciting new ways, allowing for serendipitous new insights. Then again, the main form of human communication has been focused on giving meaning to bits of information in the form of a narrative, and even if we agree on the particulars, the story might still spin out of ethical control.

7 References

Abdel Maguid, Ossama. 2014. "Ethics of Egyptology and Collecting: Who Needs the Past? National Values and Egyptology." *Shedet* 1: 44–51.

Alberti, Benjamin, and Yvonne Marshall. 2009. "Animating Archaeology: Local Theories and Conceptually Open-Ended Methodologies." *Cambridge Archaeological Journal* 19 (3): 344–356. https://doi.org/10.1017/S0959774309000535, accessed 03-29-2022.

Archaeological Data Service. 2009. "Guides to Good Practice." 2009. https://guides.archaeologydataservice.ac.uk/g2gp/Main, accessed 03-29-2022.

Aufderheide, Patricia, and Peter Jaszi. 2018. *Reclaiming Fair Use: How to Put Balance Back in Copyright, Second Edition.* 2nd ed. edition. Chicago; London: University of Chicago Press.

Bentkowska, Anna, and Hugh Denard, eds. 2012. *Paradata and Transparency in Virtual Heritage.* Ashgate: Routledge.

Bonacchi, Chiara. 2017. "Digital Media in Public Archaeology." In *Key Concepts in Public Archaeology*, edited by Gabriel Moshenska, 60–72. UCL Press. https://www.jstor.org/stable/j.ctt1vxm8r7.9, accessed 03-29-2022.

86 Drucker 2021.

Burdick, Anne, Johanna Drucker, Peter Lunenfeld, Todd Presner, and Jeffrey Schnapp. 2012. *Digital Humanities*. MIT Press.

Carruthers, William, and Stéphane Van Damme. 2017. "Disassembling Archeology, Reassembling the Modern World." *History of Science* 55 (3): 255–272. https://doi.org/10.1177/0073275317719849, accessed 03-29-2022.

Challis, Debbie. 2016. "Skull Triangles: Flinders Petrie, Race Theory and Biometrics." *Bulletin of the History of Archaeology* 26 (February). https://doi.org/10.5334/bha-556, accessed 03-29-2022.

Chang, Kang-tsung, and James R. Antes. 1987. "Sex and Cultural Differences in Map Reading." *The American Cartographer* 14 (1): 29–42. https://doi.org/10.1559/152304087783875345, accessed 03-29-2022.

Christian, Brian. 2020. *The Alignment Problem: Machine Learning and Human Values*. 1st edition. New York, NY: W.W. Norton & Company.

Colwell, Chip. 2017. "The Long Ethical Arc of Displaying Human Remains." *Atlas Obscura* (blog). 2017. http://www.atlasobscura.com/articles/displaying-native-american-remains, accessed 03-29-2022.

Daly, Okasha El. 2007. *Egyptology: The Missing Millennium*. Routledge.

Diop, Cheikh Anta. 1989. *The African Origin of Civilization: Myth Or Reality*. Chicago Review Press.

Drucker, Johanna. 2021. "Sustainability and complexity: Knowledge and authority in the digital humanities." *Digital Scholarship in the Humanities* 36 (2): ii86–ii94, https://doi.org/10.1093/llc/fqab025, accessed 03-29-2022.

Eberhardt, Jennifer L. 2020. *Biased: Uncovering the Hidden Prejudice That Shapes What We See, Think, and Do*. Illustrated edition. Penguin Books.

Elgewely, Eiman. 2014. "3D Reconstruction of Furniture Fragments from the Ancient Town of Karanis." *Studies in Digital Heritage* 1 (2): 409–427.

Elgewely, Eiman, and Willeke Wendrich. 2015. "Reviving Karanis in 4D: Reconstruction of Space through Time." *Conference on Cultural Heritage and New Technologies*.

EPOCH. 2009. *The London Charter for the Computer-Based Visualisation of Cultural Heritage*. http://www.londoncharter.org/, accessed 03-29-2022.

Erman, Adolf, and Hermann Grapow, eds. 1926. *Wörterbuch Der Ägyptischen Sprache*. Leipzig: J.C. Hinrichs'sche Buchhandlung. https://www.ancientegyptfoundation.org/worterbuch_der_aegyptischen_sprache.shtml, accessed 03-29-2022.

Evans, Linda, and Anna-Latifa Mourad. 2018. "DStretch® and Egyptian Tomb Paintings: A Case Study from Beni Hassan." *Journal of Archaeological Science: Reports* 18: 78–84. https://doi.org/10.1016/j.jasrep.2018.01.011, accessed 03-29-2022.

Gillings, Mark, Piraye Hacıgüzeller, and Gary Lock, eds. 2020. *Re-Mapping Archaeology: Critical Perspectives, Alternative Mappings*. 1st edition. S.l.: Routledge.

Godon, Stephen. 2018. "Cléopâtre Au Service Des Théories Raciales Aux États-Unis et Symbole de l'émergence d'une Nouvelle Femme Américaine Dans La Seconde

Moitié Du XIXe." In *Cléopâtre En Abyme: Aux Frontières de La Mythistoire et de La Littérature*, edited by Sydney Hervé Aufrère and Anaïs Michel, 365–376. Paris: L'Harmattan.

Holley, Robert P. 2018. "Open Access: Current Overview and Future Prospects." *Library Trends* 67 (2): 214–240. https://doi.org/10.1353/lib.2018.0034, accessed 03-29-2022.

Jones, Charles E., and Tom Elliott. 2015. "AWOL Index, Institute for the Study of the Ancient World (NYU)." https://isaw.nyu.edu/publications/awol-index/, accessed 03-29-2022.

Jones, Ian W.N., and Thomas E. Levy. 2018. "Cyber-Archaeology and Grand Narratives: Where Do We Currently Stand?" In *Cyber-Archaeology and Grand Narratives: Digital Technology and Deep-Time Perspectives on Culture Change in the Middle East*, edited by Thomas E. Levy and Ian W.N. Jones, 1–17. One World Archaeology. Cham: Springer International Publishing. https://doi.org/10.1007/978-3-319-65693-9_1, accessed 03-29-2022.

Kamermans, Hans, Wieke de Neef, Chiara Piccoli, Alex G. Posluschny, and Roberto Scopigno, eds. 2016. *The Three Dimensions of Archaeology: Proceedings of the XVII UISPP World Congress (1–7 September 2014, Burgos, Spain). Volume 7.* http://www.archaeopress.com/ArchaeopressShop/Public/download.asp?id=%7B4D879534-2D7C-4B48-AA7A-642EFD11753B%7D, accessed 03-29-2022.

Krunic, Veljko. 2020. *Succeeding with AI: How to Make AI Work for Your Business*. 1st edition. Shelter Island, NY: Manning Publications.

Lercari, Nicola, Willeke Wendrich, Benjamin W. Porter, Marie Burton M., and Thomas E. Levy, eds. 2021. *Preserving Cultural Heritage in the Digital Age. Sending Out an S.O.S. Edited by Nicola Lercari* [+] *University of California, Merced Willeke Wendrich* [+] *University of California, Los Angeles Benjamin W. Porter* [+] *University of California, Berkeley Margie M. Burton* [+] *University of California, San Diego Thomas E Levy* [+] *University of California, San Diego*. New Directions in Anthropological Archaeology. Archaeology and History. Sheffield: Equinox Publishing Ltd.

Lessig, Lawrence. 2000. "Copyright's First Amendment Melville B. Nimmer Memorial Lecture." *UCLA Law Review* 48 (5): 1057–1074. https://heinonline.org/HOL/P?h=hein.journals/uclalr48&i=1074, accessed 03-29-2022.

Lima, Roberto de, Toon Sykora, Marleen De Meyer, Harco Willems, and Maarten Vergauwen. 2018. "On Combining Epigraphy, TLS, Photogrammetry, and Interactive Media for Heritage Documentation: The Case Study of Djehutihotep's Tomb in Dayr al-Barsha." In *GCH 2018: Eurographics Workshop on Graphics and Cultural Heritage. Vienna, Austria, November 12–15, 2018*, edited by Robert Sablatnig and Michael Wimmer, 225–229. Goslar: Eurographics Association. https://doi.org/10.2312/gch.20181367.

Meghini, Carlo, Franco Niccolucci, Achille Felicetti, Paola Ronzino, Federico Nurra, Christos Papatheodorou, Dimitris Gavrilis, et al. 2017. "ARIADNE: A Research Infras-

tructure for Archaeology." *Journal on Computing and Cultural Heritage* 10 (August): 1–27. https://doi.org/10.1145/3064527, accessed 03-29-2022.
Miller, J. Hillis. 2012. "How To (Un)Globe the Earth: In Four Easy Lessons." *SubStance* 41 (1): 15–29. https://www.jstor.org/stable/23261100, accessed 03-29-2022.
Miller, P., and J. Richards. 1995. "The Good, the Bad, and the Downright Misleading: Archaeological Adoption of Computer Visualisation." In *CAA Online Proceedings, CAA94. Computer Applications and Quantitative Methods in Archaeology*, edited by J. Huggett and N. Ryan, 19–22. BAR International Series 600. Oxford: Tempus Reparatum. https://proceedings.caaconference.org/paper/03_miller_richards_caa_1994/, accessed 03-29-2022.
Monmonier, Mark S. 1991. *How to Lie with Maps*. Chicago: University of Chicago Press.
Monmonier, Mark S. 2018. *How to Lie with Maps*. Third edition. Chicago: The University of Chicago Press.
Morgan, Colleen, and Holly Wright. 2018. "Pencils and Pixels: Drawing and Digital Media in Archaeological Field Recording." *Journal of Field Archaeology* 43 (2): 136–151. https://doi.org/10.1080/00934690.2018.1428488, accessed 03-29-2022.
Noble, Safiya Umoja. 2018. *Algorithms of Oppression: How Search Engines Reinforce Racism*. Illustrated edition. New York: NYU Press.
Oliver, Kelly. 2015. *Earth and World: Philosophy After the Apollo Missions*. Columbia University Press.
Pariser, Eli. 2012. *The Filter Bubble: How the New Personalized Web Is Changing What We Read and How We Think*. Reprint edition. Penguin Books.
Pendergrass, Keith, Walker Sampson, Tim Walsh, and Laura Alagna. 2019. "Toward Environmentally Sustainable Digital Preservation." *The American Archivist*, June. https://doi.org/10.17723/0360-9081-82.1.165, accessed 03-29-2022.
Perkins, Chris. 2008. "Cultures of Map Use." *The Cartographic Journal* 45 (2): 150–158. https://doi.org/10.1179/174327708X305076, accessed 03-29-2022.
Petrie, W.M.F. 1879. "On Metrology and Geometry in Ancient Remains." *The Journal of the Anthropological Institute of Great Britain and Ireland* 8: 106–116.
Reid, Donald M. 2003. *Whose Pharaohs? Archaeology, Museums, and Egyptian National Identity from Napoleon to World War I*. Berkeley, CA: University of California Press. http://www.ucpress.edu/book.php?isbn=9780520240698, accessed 03-29-2022.
Reid, Donald M. 2015. *Contesting Antiquity in Egypt: Archaeologies, Museums & the Struggle for Identities from World War I to Nasser*.
Richardson, Lorna-Jane. 2018. "Ethical Challenges in Digital Public Archaeology." *Journal of Computer Applications in Archaeology* 1 (1): 64–73. https://doi.org/10.5334/jcaa.13, accessed 03-29-2022.
Šavrič, Bojan, Tom Patterson, and Bernhard Jenny. 2019. "The Equal Earth Map Projection." *International Journal of Geographical Information Science* 33 (3): 454–465. https://doi.org/10.1080/13658816.2018.1504949, accessed 03-29-2022.

Schnabel, Marc A., and Thomas Kvan. 2003. "Spatial Understanding in Immersive Virtual Environments." *International Journal of Architectural Computing* 1 (4): 435–448. https://doi.org/10.1260/147807703773633455, accessed 03-29-2022.

Stevenson, Alice. 2016. "Conflict Antiquities and Conflicted Antiquities: Addressing Commercial Sales of Legally Excavated Artefacts." *Antiquity* 90 (349): 229–236. https://doi.org/10.15184/aqy.2015.188, accessed 03-29-2022.

Suber, Peter. 2004. "Open Access Overview." *Open Access Overview* (blog). 2004. http://legacy.earlham.edu/~peters/fos/overview.htm, accessed 03-29-2022.

Sullivan, Elaine. 2010. "Karnak: Development of the Temple of Amun-Ra." *UCLA Encyclopedia of Egyptology* 1 (1). https://escholarship.org/uc/item/1f28q08h, accessed 03-29-2022.

Sullivan, Elaine A. 2008. "Digital Karnak: Experience Karnak." 2008. http://wayback.archive-it.org/7877/20160919161550/http://dlib.etc.ucla.edu/projects/Karnak/experience/DevelopmentAndChronologicalChanges, accessed 03-29-2022.

Sullivan, Elaine A., Angel David Nieves, and Lisa M. Snyder. 2017. "Making the Model: Scholarship and Rhetoric in 3-D Historical Reconstructions." In *Making Things and Drawing Boundaries*, edited by Jentery Sayers, 301–316. Experiments in the Digital Humanities. University of Minnesota Press. https://doi.org/10.5749/j.ctt1pwt6wq.38, accessed 03-29-2022.

Sullivan, Elaine A., and Lisa M. Snyder. 2017. "Digital Karnak: An Experiment in Publication and Peer Review of Interactive, Three-Dimensional Content." *Journal of the Society of Architectural Historians* 76 (4): 464–482. https://doi.org/10.1525/jsah.2017.76.4.464, accessed 03-29-2022.

Tufte, Edward R. 1997. *Visual Explanations: Images and Quantities, Evidence and Narrative*. Cheshire, Conn: Graphics Press.

Tufte, Edward R. 2001. *The Visual Display of Quantitative Information*. 2nd edition. Cheshire, Conn: Graphics Press.

Tully, Gemma. 2015. "Contemporary Egyptian Dialogues: Collaborative Approaches to Museum Displays of Ancient Egypt." In *The Management of Egypt's Cultural Heritage*, edited by Okasha El Daly, F.A. Hassan, G.J. Tassie, J. van Wetering, L.S. Owens, and A. De Trafford, 57–66. London: ECHO; Golden House.

Vértes, Krisztián. 2020. "Tradition and Innovation in Digital Epigraphy." In *The Oxford Handbook of Egyptian Epigraphy and Palaeography*, edited by Dimitri Laboury and Vanessa Davies, 370–386. New York: Oxford University Press. https://doi.org/10.1093/oxfordhb/9780190604653.013.25, accessed 03-29-2022.

Watrall, Ethan. 2016. "Archaeology, the Digital Humanities, and the 'Big Tent.'" *Debates in the Digital Humanities 2016*. https://dhdebates.gc.cuny.edu/read/untitled/section/abddodb0-2eaa-4af7-aac3-7f24ca70ed3b, accessed 03-29-2022.

Wendrich, Willeke. 2014. "Visualizing the Dynamics of Monumentality." In *Approaching Monumentality in Archaeology. Proceedings of the IEMA Postdoctoral Visiting*

Scholar Conference on Theories and Methods in Archaeology, edited by James Osborne, 409–430. Buffalo, NY: IEMA.

Wood, Denis. 2010. *Rethinking the Power of Maps*. Illustrated edition. New York: The Guilford Press.

8 Cited Web Resources

AAT	Art & Architecture Thesaurus (Getty Research Institute). 2017. https://www.getty.edu/research/tools/vocabularies/aat/, accessed 03-29-2022.
ADS	The Archaeology Data Service. re3data.org—Registry of Research Data Repositories. 2012. http://service.re3data.org/repository/r3d100000006, accessed 03-29-2022.
AEGARON	Ancient Egyptian Architecture Online—Development Site. 2010. http://drupaldev.aegaron.ucla.edu/, accessed 03-29-2022.
AgID	Accessibilità |Agenzia per l'Italia digitale. 2020. https://www.agid.gov.it/it/design-servizi/accessibilita, accessed 03-29-2022.
Ariadne	Ariadne plus Portal. 2020. https://ariadne-infrastructure.eu/portal/, accessed 03-29-2022.
BITV	Barrierefreie-Informationstechnik-Verordnung. 2019. https://www.bitvtest.eu/bitv_test.html, accessed 03-29-2022.
BM	The British Museum Collection. 2021. https://www.britishmuseum.org/collection, accessed 03-29-2022.
CARE	Global Indigenous Data Alliance. 2021. https://www.gida-global.org/care, accessed 03-29-2022.
CultNat Arabic	مركز توثيق التراث الحضاري والطبيع https://www.cultnat.org/ProgramDetails?ID=63&%d9%85%d8%b5%d8%b1_%d8%a7%d9%84%d9%82%d8%af%d9%8a%d9%85%d8%a9, accessed 03-29-2022.
CultNat English	Ancient Egyptian Heritage 2021 https://www.cultnat.org/ProgramDetails?ID=63&%d9%85%d8%b5%d8%b1_%d8%a7%d9%84%d9%82%d8%af%d9%8a%d9%85%d8%a9, accessed 03-29-2022.
Digital Epigraphy	Digital Epigraphy, Groteszk Kreatív Társulat. 2021 https://www.digital-epigraphy.com/, accessed 03-29-2022.
Digital Giza	Digital Giza. 2017. http://giza.fas.harvard.edu/, accessed 03-29-2022.
Digital Karnak	Digital Karnak. 2008 http://wayback.archive-it.org/7877/20160919152116/http://dlib.etc.ucla.edu/projects/Karnak/, accessed 03-29-2022.

FAIR	GO FAIR Initiative. 2018. https://www.go-fair.org/, accessed 03-29-2022.
Griffith Institute	Griffith Institute | Online Resources. 2021. http://www.griffith.ox.ac.uk/archive/onlineresources/index.html, accessed 03-29-2022.
Mapping Inequality	Mapping Inequality. Redlining in New Deal America. 2020 https://dsl.richmond.edu/panorama/redlining/, accessed 03-29-2022.
MastaBase	MastaBase—Database of Old Kingdom Tomb Scenes. 2014. https://digitalegyptology.org/mastabase/, accessed 03-29-2022.
MET	Multilingual Egyptian Thesaurus. 2007. http://projetrosette.info/page.php?Id=290, accessed 03-29-2022.
OEB	Online Egyptological Bibliography. 2021. http://oeb.griffith.ox.ac.uk/, accessed 03-29-2022.
OEE	Oxford Expedition to Egypt: Scene-Details Database. 2006. https://archaeologydataservice.ac.uk/archives/view/oee_ahrc_2006/, accessed 03-29-2022.
OSI	Open Source Initiative. 2021. https://opensource.org/, accessed 03-29-2022.
RIME	Rivista del Museo Egizio. 2017. https://rivista.museoegizio.it/, accessed 03-29-2022.
Section 508	U.S. General Services Administration Government-Wide IT Accessibility Program. 2018. https://www.section508.gov/, accessed 03-29-2022.
Selim Hassan	The Archive of the Egyptologist Selim Hassan. 2019. https://www.cultnat.org/ProjectDetails?ID=1121&The_Archive_of_the_Egyptologist_Selim_Hassan, accessed 03-29-2022.
TDAR	The Digital Archaeological Record. 2018. https://core.tdar.org/, accessed 03-29-2022.
The Met	The Metropolitan Museum of Art. 2021. https://www.metmuseum.org/art/collection, accessed 03-29-2022.
TLA	Digital Project. Thesaurus Linguae Aegyptiae. 2004. http://aaew.bbaw.de/tla/, accessed 03-29-2022. TLA Introduction https://aaew.bbaw.de/tla/servlet/S05?d=d001&h=h001, accessed 03-29-2022.
TMP	The Theban Mapping Project. 2020. https://thebanmappingproject.com/, accessed 03-29-2022.
UEE	UCLA Encyclopedia of Egyptology https://uee.ucla.edu
UEE, PDF Only.	https://escholarship.org/uc/nelc_uee, accessed 03-29-2022.
W3C	World Wide Web Consortium. 2020. https://www.w3.org/, accessed 03-29-2022.

WCAG 1.0. Web Content Accessibility Guidelines 1.0. 1999. https://www.w3.org/TR/WCAG10/, accessed 03-29-2022.

WCAG 2.1. Web Content Accessibility Guidelines (WCAG) 2.1 2008.https://www.w3.org/TR/WCAG21/, accessed 03-29-2022.

CHAPTER 2

The Contribution of Photogrammetry and Computer Graphics to the Study and Preservation of Monuments in Alexandria, Egypt

Mohamed Abdelaziz and Mohamed Elsayed

Abstract

This chapter presents the results of archaeological project documentation through digital photogrammetry at different sites and for a variety of objects in Egypt. We demonstrate its promise for both documentation and restoration. We describe a study in which photogrammetry and digital modeling techniques were used to obtain a three-dimensional surface digital model of an important underwater site, and we shall see also how photogrammetry with the aid of computer graphics is an important tool for virtual restoration of fragmentary statues lifted from the sea. Finally, we demonstrate the use of 3D scanning data sets to reveal an inscription, invisible to the naked eye in natural light, on a part of a deteriorated granite column.

Keywords

photogrammetry – 3D modeling – underwater archaeology – virtual anastylosis – reconstruction – inscriptions – digital surface model – Alexandria – Pharos – lighthouse

1 Introduction

Archaeological excavations are often irreversibly destructive, so it is important to accompany them with detailed documentation reflecting the accumulated knowledge of the excavation site.[1] Graphical representations of archaeological sites such as drawings, sketches, photographs, topography, and photogrammetry are indispensable for such documentation and are an essential part of

1 Drap 2012.

an archaeological survey. Photogrammetry has been extensively used in heritage documentation in many countries; it offers a rapid, accurate method of acquiring three-dimensional information, especially for large complex sites or objects, with relatively little time required in the field for data acquisition. With photogrammetry, we can obtain accurate 3D metric and descriptive object information from multiple digital images.[2] These photos contain information about surface details of the site and can provide information on the condition of a monument, before, during, and after excavation and restoration, which is difficult to achieve by graphic documentation alone.

In campaigns in Alexandria, Egypt, undertaken from 2014–2016 we used several techniques for the first time to document archaeological sites or objects using photogrammetry either on land or underwater in collaboration with Centre d'Études Alexandrines (CEALex) and the Egyptian Ministry of Antiquities. In this paper we present our results in underwater and terrestrial archaeological photogrammetry, based on practical experience and using low-cost materials. We shall explain some examples of this documentation, beginning with a study in which underwater photogrammetry methods and digital modeling processing techniques were used to obtain a three-dimensional georeferenced digital surface model (DSM) for the submerged site of the lighthouse of Alexandria. Secondly, we shall see how photogrammetry with the aid of computer graphics was an important tool to perform the virtual restoration of a broken colossal female statue. Finally, we demonstrate the use of 3D scanning data sets to reveal an invisible inscription on a deteriorated archaeological object.

2 The Submerged Site of the Lighthouse of Alexandria

The underwater site of the lighthouse of Alexandria, lies to the east of Qaitbay Fort in the open sea. It is affected by environmental factors, mainly waves, current, and sewage discharge which changes the water quality. In addition, the prevailing wind is N-NW, but is variable from December to May.[3] The rough sea, current, and visibility rarely exceeding three meters provide substantial challenges to, but also valuable experience in, site documentation by photogrammetry.[4] In 1994, CEAlex launched the first scientific excavation in the field of underwater archaeology in Egypt on the submerged site of the lighthouse of Alexandria, well known as one of the Seven Wonders of the Ancient World.

2 Al-Ruzouq 2012.
3 El-Gindy 2000, 144.
4 Abdelaziz, Elsayed 2019.

FIGURE 2.1
Fragments of a female colossal statue discovered by Kamel Abou Elsaadat and raised from the sea in 1962 at the foot of the Qaitbay Fort
MOHAMED ELSAYED

This is situated on the eastern extremity of the ancient island of Pharos at the foot of Qaitbay Fort, which was constructed at the end of 15th century by the Mamluke Sultan Qaitbay. The submerged site was discovered in 1960 thanks to the pioneering work of Kamel Abou Elsaadat.[5]

In 1968, Honor Frost undertook a UNESCO mission (Figure 2.2) on the site which led to the publication of a preliminary report revealing the importance of the site.[6]

The underwater site of Qaitbay Fort holds the ruins of the lighthouse of Alexandria, which stood for almost seventeen centuries. It was built towards the beginning of the 3rd century BCE and was accessible until the end of the 14th century. The last mention of the visible presence of its ruins dates to 1420, almost 60 years before the construction of Qaitbay Fort.[7]

5 These discoveries were due to the curiosity of Kamel Abou Elsaadat, who enjoyed fishing in Silsileh. He had located some of the sunken monuments in the eastern harbor of Alexandria in 1962, with the help of the Egyptian Navy under the supervision of Dr. Henry Riad, at that time director of the Greco-Roman Museum (Riad 1964). Abou Elsaadat recovered an anthropoid sarcophagus lid near Silsileh and a female colossal statue of Isis near Qaitbay Fort in Alexandria (El sayed 2012).

6 Frost 1975; Empereur 2000.

7 Hairy 2007.

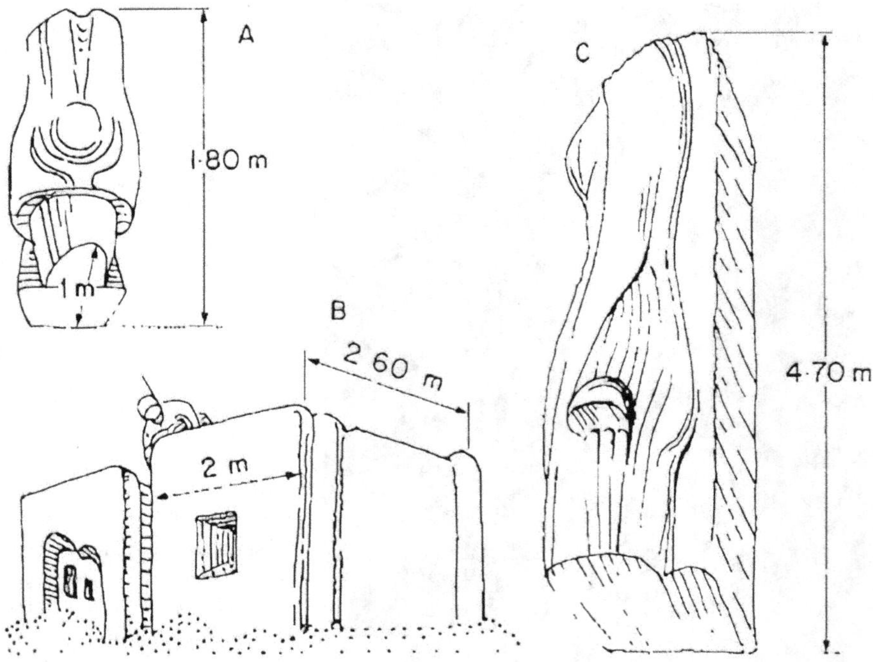

FIGURE 2.2 Sketches produced by Honor Frost, during the UNESCO mission in 1968 (after Frost 1975, 129)

3 Excavations and Documentation

Underwater excavations began in 1994 under the direction of Jean Yves Empereur,[8] director of CEAlex, in the hope of shedding new light on the question of the appearance of Alexandria's lighthouse. These excavations led to the reconstruction of certain parts of the lighthouse and to an understanding of its design process. Nevertheless, the study of the site ran up against the limitations of traditional data recording methods. The extent and unique nature

8 Thanks to the alert given by the Egyptian filmmaker Asmaa Elbakry in 1993, the Egyptian authorities stopped the construction of a breakwater of concrete blocks to protect the Qaitbay Fort built in 1477 from the storms and waves of the Mediterranean Sea. A few hundred concrete blocks had already been thrown at the foot of the building before the alert (Corteggiani 1998, 25). In the fall of 1994, the Supreme Council of Antiquities of Egypt asked Jean-Yves Empereur, director of CEAlex, to undertake a salvage underwater excavation on the submerged site near Qaitbay Fort in collaboration with the SCA, the Arab Maritime Academy and the Egyptian Navy, but the project was interrupted by bad weather. One year later, in 1995, the Centre d'Études Alexandrines undertook, with the support of the Institut français d'archéologie orientale, the first scientific mission on the submerged site.

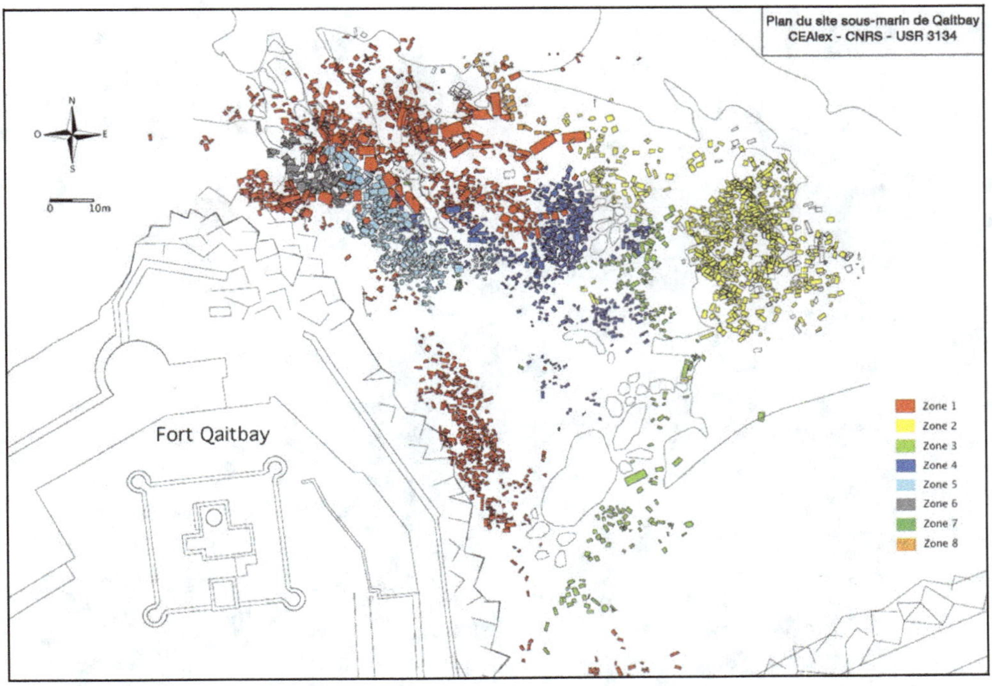

FIGURE 2.3 Plan of the underwater archaeological site of Qaitbay Fort (CEAlex)

of this sunken site encouraged innovation in data gathering, both as regards the ancient material of more than 3,525 blocks (Figure 2.3) and the overall site itself, whose size and uneven contours make any analysis complicated to say the least.[9]

In 2009/2010, a 3D photogrammetry data-gathering program was launched, particularly focusing on the broken statues that surrounded the lighthouse.[10] The research and development department of EDF (Electricité De France) had begun this work in 1998, and it was continued between 2009 and 2012 as part of the AR-Search program (Agence National de la Recherche). In 2012–2013 the aim of the campaign was to continue the virtual restoration work on the pair of colossal royal statues that once stood next to the lighthouse. This work demonstrated that photogrammetry was the preferred solution for rapidly providing quality "digital duplicates" of either sunken or lifted objects.[11] With around 30

9 Hairy 2009.
10 Hairy 2011.
11 Reuter et al. 2011.

FIGURE 2.4 Equipment used for photogrammetry on the Pharos site
MOHAMED ELSAYED AND YASSER GALAL

blocks removed from the sea and more than 3,000 pieces still underwater—some weighing almost 40 tons—this was an important step forward.

From 2013, with the support of the Honor Frost Foundation (HFF), the gathering and processing of photographic data in this research were improved, which led to substantial development in data gathering techniques and the consequent improvement in the quality of results.[12] Fully manual photographic data acquisition methods were applied, using a digital single-lens reflex camera and low-cost materials (Figure 2.4) to develop a digital surface model (DSM) of the entirety of Qaitbay underwater site, which exceeds 13,000 square meters. One of the objectives of this project was to use photogrammetry to create a digital duplicate of the submerged site.

12 Hairy, Abdelaziz, Elsayed and Soubias 2016.

4 Methods and Techniques of Data Acquisition for the Digital Surface Model

Any underwater archaeological surveying technique must satisfy two competing requirements: speed and accuracy. While many different photogrammetric techniques have been used on underwater archaeological sites around the Mediterranean, it is often difficult to select one that will be suitable for a particular site. In our case the underwater archaeological site of Qaitbay was particularly difficult to record, not only because of the conditions of the sea and the weather, but also because of its uneven, rocky bottom and the variation of the depth in the site, from two to nine meters. Therefore, using heavy materials such as a metal photo tower or frames installed upon the site was impossible. Additionally, use of more modern equipment such as side-scan sonar, autonomous underwater vehicles (AUVs) or remotely-operated vehicles (ROVs) was not promising, due to both cost and the rocky bottom and shallow waters in some areas of the site. Therefore, methods and techniques of data acquisition had to be adapted to the weather condition and the topography of the site. Our technique for preparing the site for photography and topography was completely manual, using simple and cheap materials including a large-scale bar-pole, measuring tapes, buoys, ropes, tags and a compass (Figure 2.4).

The first step towards creating an accurate three-dimensional model of the site was to take a series of good photos using a high-quality camera. However, shadows can cause errors, with picture information lost, especially underwater. Therefore, the most important thing for the precision and the accuracy of the work when using manual methods was the diver's competence in orientating himself underwater and in respecting the level of the flight plan, which could not exceed three meters above the site. The data capture method was generally inspired by aerial photogrammetry. The photographic work on the site was carried out by Mohamed Elsayed (underwater archaeologist—Ministry of Antiquities) using a Nikon D700 DSLR camera (full frame, 12.9 megapixel resolution) with a fixed 24 mm lens, in a hemispheric dome housing (Figure 2.5). This allowed for a wide coverage area and close proximity to the subject, especially in low visibility.

All photos were taken in manual mode without flash whatever the sea conditions. F-stop and shutter speed varied according to underwater visibility and illumination, either f/7.6 with a shutter-speed of 1/60, or f/5.6 and a shutter-speed 1/40, depending on the condition of the water itself. From 2014 until 2016, about 26 weeks were dedicated to the photographic survey, and eventually 50,152 photos (Figure 2.11) were used to create an orthophoto representing a

FIGURE 2.5 Underwater photogrammetry data acquisition by Mohamed Elsayed
PHOTO BY PHILIPPE SOUBIAS, PROCESSING BY MOHAMED ABDELAZIZ

part of a DSM covering an area of 7,200 square meters of the total 13,000 square meters of the site (Figure 2.10).

The flight plan method developed for the Qaitbay site required a period of experimentation to manage the constraints of the underwater context, including the troublesome problems of visibility on the site. Further difficulty was occasioned by the large surface area (1.3 hectares) and the size of certain blocks, requiring the combining of several methods of shooting. For these reasons, it was impossible to gather data in order to process the entire site at one time. The data capture method was adapted to respond to these constraints, and so several methods were applied during the missions of 2014, 2015 and 2016. In order to increase the area covered by photography in previous missions we reduced the materials used in the operation to save preparation time for each zone. The flight plans of the initial experiences in 2014 served as the basis of future work, with the diver moving along lines attached to a rod, doubling back and forth after 20 meters of continuous shooting to capture the images that were to be processed into the surface model. At the end of the lane, the photographer would turn around and cover another swath while maintaining the overlap.

The diver maintained his path by visual orientation, checking his trajectory against the features of the seabed, the many ancient blocks, the uneven surface (Figures 2.6 and 2.7) and scales placed on the bottom. Measuring tapes and buoys marked the boundaries of the area to be covered during the flight plan.

The major difficulty encountered in the topography of the underwater site lay in the significant variation in the elevation of the seabed, and in the varying weather conditions: small or large swell, sunny weather, and sometimes

FIGURE 2.6
Orthophoto and the flight plan
PHOTOS BY ASHRAF HUSEIN AND MOHAMED ELSAYED, PROCESSING BY MOHAMED ABDE-LAZIZ

overcast, but always with sufficient brightness. In some areas, there is a change in depth from eight meters to four meters moving northwards, and from nine meters to five meters while following a south-east direction. The photos taken during these flights were always close to an angle of 45 degrees (Figure 2.5), thus capturing the rugged topography of the site surface. The diver had to move in as straight a line as possible, and at a constant speed in order to ensure sharp images; he was instructed not to exceed a height of three meters from the bottom, and was authorized to approach as close as 1.65 meters above the seabed in conditions of bad visibility. The flight plans required a forward overlap of 70–80 percent and lateral or side overlap of the same order (Figure 2.6), which is superior to conventional aerial photography. This overlap rate reduced mistaken pairing by increasing the number of matching images within the software. The shots were taken in different weather conditions. In 2015 and 2016, the area covered was approximately 7,200 square meters.

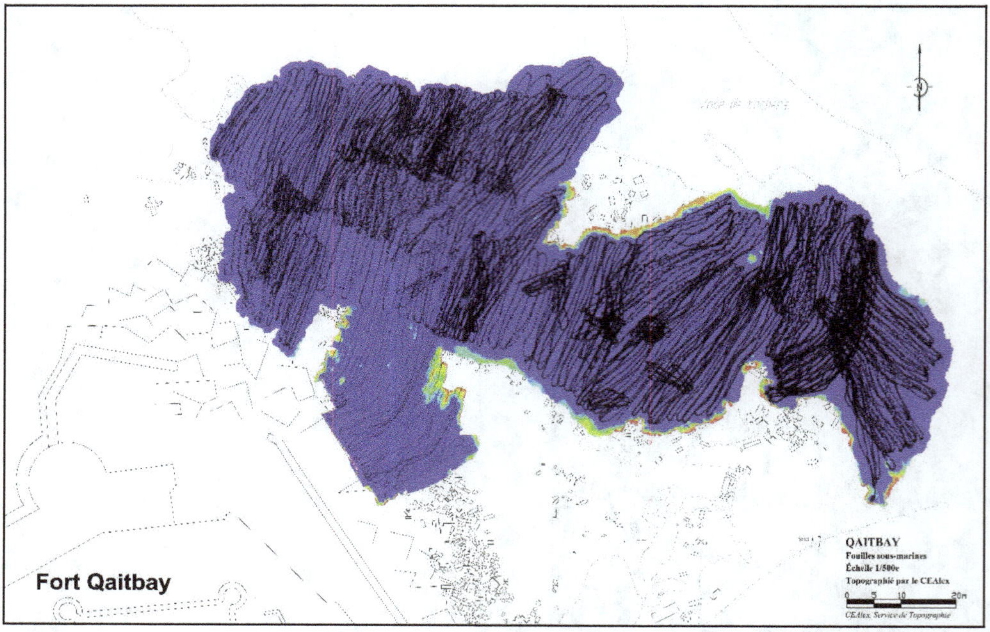

FIGURE 2.7 Diver's flight-paths above the seabed, Alexandria Harbor
PROCESSING BY MOHAMED ABDELAZIZ

5 Data Processing

At the beginning of the project in 2014 the available computer for data processing had limited capability: RAM—16G/GPU—GT630, 2G/CPU—Intel core i3. The final aim of photo processing with photogrammetry software was to build an orthophoto and a textured 3D model. The first step was the preprocessing of the images, which were dominated by blue/green and yellow color because of the turbidity of the water, low visibility and lack of light. These photos could not be used without prior enhancement; image processing was necessary for color correction, contrast, removing shadows and reducing highlights in order to get more matching points between pairs of images (Figure 2.8).

For this we used Adobe Lightroom and Photoshop for editing and for batch processing. All photos were captured in a RAW format and then converted into JPG files to be accepted by the photogrammetry software Agisoft Metashape 1.5 (formerly Photoscan). Metashape generates an accurate 3D model from images which are converted into a textured 3D model in four straightforward processing steps, namely: 1) Aligning Photos, 2) Building Dense Cloud, 3) Build-

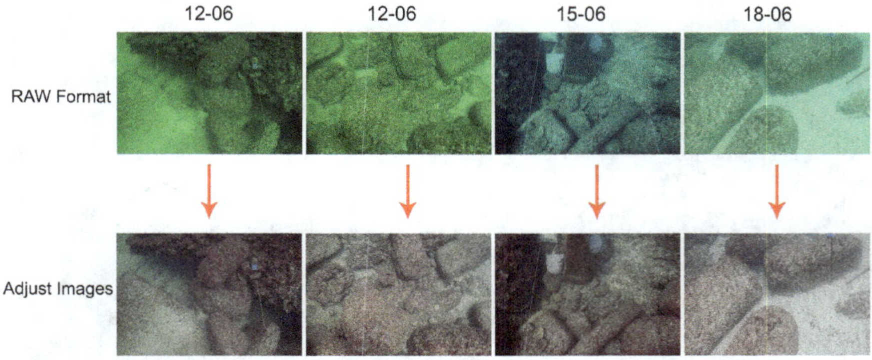

FIGURE 2.8 Images processing
PROCESSING BY MOHAMED ABDELAZIZ

ing Mesh, 4) Building Texture.[13] In order to improve the results, some manual interventions were necessary, including image masking and deletion of erroneous points. Beginning by loading uncalibrated overlapping images into the software, the camera auto-calibration runs automatically through mathematical algorithms. The software detects and then tracks how many points move throughout the series of images. Even the images that are not calibrated will be automatically computed inside the photogrammetry software.[14]

6 Orthophoto and Digital Elevation Model (DEM)

After finishing all the processing, we produced the orthophoto (Figure 2.9) and digital elevation model (DEM) (Figure 2.12), which can be obtained with Metashape in any projection required.[15] From this, we can extract a plan, sections, volumes and elevation.[16] The orthophoto, which was assembled from raw images, had an average pixel size varying between 0.64 mm and 0.50 mm/pixel because of the height variation of the camera and the surface of the seabed (Figure 2.8).

Between the years 2014 and 2016 a massive number of photos, about 50,152, were used to generate the final 3D model/orthophoto covering 7,200 square meters of the submerged site of Qaitbay (Figures 2.9 and 2.10).

13 Van Damme 2015; Yamafune, Torres and Castro 2016.
14 Balletti et al. 2015.
15 Yamafune, Torres, Castro, 2016.
16 Abdelaziz and Elsayed 2019.

FIGURE 2.9 Orthophoto created from 50,152 images
PROCESSING BY MOHAMED ABDELAZIZ

FIGURE 2.10 3D model/orthophoto covering 7,200 square meters of the submerged site of Qaitbay, with additional areas sketched in
PROCESSING BY MOHAMED ABDELAZIZ

Because the characteristics of the PC used in processing, as mentioned above, Metashape could not handle all of the photos at once. The best solution was to process a "chunk" each day (Figure 2.11), so we had in total 39 areas of the site representing the different daily captures on the underwater site during the different campaigns from 2014 until 2016.

Between each chunk there was an overlap of 25 percent which helped to merge the processed areas (Figure 2.7) through the use of manually-positioned digital marker features on each of the paired chunks.

In conclusion, creating a 3D model using photogrammetry was a good tool for documenting such a huge underwater archaeological site. We were also able to study the natural changes of the site. One remarkable observation after three years of work, between the years 2014 and 2016, was the constant movement of sand on the sea bottom. This resulted in the covering and uncovering of dozens of archaeological blocks (Figure 2.11), which appeared on earlier orthophotos and models only to disappear later on.

The DSM of the submerged site of the lighthouse near Qaitbay is not yet completed, but our initial experience in underwater photogrammetry to create such a DSM has provided good and efficient results, using the technical application of low cost materials for data acquisition and data processing.[17] More than 7,200 square meters have been documented, representing about 60 percent of the total surface area of the submerged site. Nonetheless, we still need to make improvements in our work in the coming seasons in areas such as:

- Improved photographic equipment. Current DSLR or digital mirrorless cameras achieve far higher resolution than the Nikon D700 (e.g., 46.9 megapixels for the Nikon D850; or 62.5 megapixels for the Sony Alpha 7r).
- Improved georeferencing of the model by increasing the number of GCPs (Ground Control Points) in the future zones before photographing, in order to increase the accuracy of the model. The DSM, once completed, will make it possible to generate lengthwise and transverse profiles, and to create a digital terrain model (DTM) of the site. It already allows for the production of orthophoto plans with a pixel size adapted to the power of our computers, which can, in absolute terms, reach a pixel resolution equaling between 0.64 mm to 0.50 mm on the seabed (Figure 2.8).
- We can improve and extend our digital elevation model (Figure 2.12) from which we can obtain 'z' points or the elevation for any block in the site without using a GPS, because the site is already georeferenced. In addition, we can also extract a section to see the uneven surface levels and compute the volume of any block in the model.

17 The first results were presented at the International Society for Photogrammetry and Remote Sensing annual meeting in 2019; see Abdelaziz and Elsayed 2019.

FIGURE 2.11 Orthophotos cover the area of work (7,200 square meters) during the seasons of 2014, 2015, 2016
PROCESSING BY MOHAMED ABDELAZIZ

- Producing a 2D drawing from the orthophoto using Auto-CAD, which can be a faster means of drafting a plan (within a few days), rather than spending extensive time on the submerged site.
- With the open-source 3D processing software MeshLab, we can apply an ambient occlusion rendering to the 3D model, thus improving visualization, which can help in counting the archaeological objects (Figure 2.13). MeshLab provides tools for editing, cleaning, healing, inspecting, rendering, texturing and converting meshes. It offers features for processing raw data produced by 3D digitization tools/devices and for preparing models for 3D printing.

- Since the Qaitbay Fort (Figure 2.14) site is not accessible to people who do not dive, and underwater conditions and visibility are not stable, we have created a virtual diving tour from the model of the site which could be developed in the future, after finishing the entire model. Preliminarily, it was possible to create a complete an animated video of the 3D model corresponding to an area of 7,200 square meters.

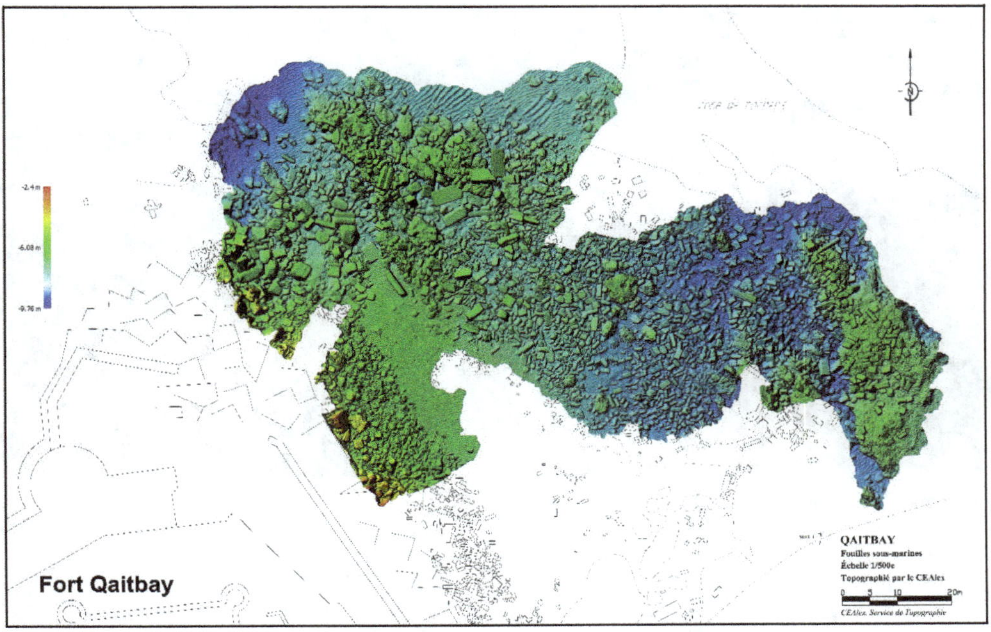

FIGURE 2.12 Digital elevation model
PROCESSING BY MOHAMED ABDELAZIZ

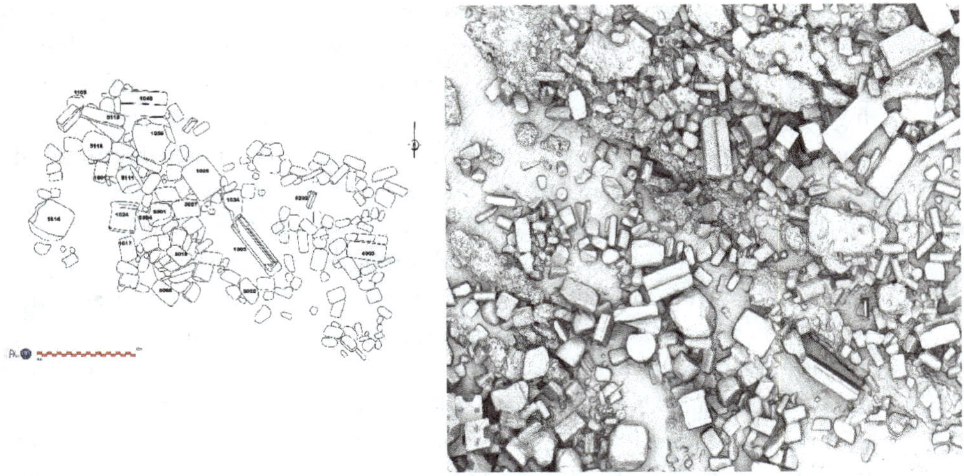

FIGURE 2.13 2D sketch (left) and an ambient occlusion rendering (right) of the 3D model from 2014, showing parts of the monumental door of the ancient lighthouse of Alexandria
PROCESSING BY MOHAMED ABDELAZIZ

FIGURE 2.14 Qaitbay Fort
MOHAMED ELSAYED

7 Virtual Anastylosis

Restoration of physical monuments requires extreme caution and careful study. Archaeologists and conservation experts are very reluctant to proceed to restoration and or reconstruction projects without detailed consultation and planning. Currently, anastylosis (re-erection) executed on a real object is highly challenging. Contemporary technologies have provided archaeologists and other conservation experts with the tools to embark on virtual restorations or anastylosis, thus testing various alternatives without physical intervention on the monument itself, respecting international norms and conventions.[18]

Throughout our archaeological work we often find sculptures that require reassembly of their fragments and the restoration of their lost parts. Traditional restoration work has certain disadvantages: it is expensive because it involves the use of real materials, and it is often difficult or impossible to accurately predict and visualize the final result of the restoration before it is completed. Although virtual restoration cannot replace real restoration for a piece that is in danger of deterioration, it can be a very useful tool to plan interventions and to illustrate the hypothetical original state of an object without having to intervene directly in it, thus respecting the principle of minimum intervention.

18 Stampouloglou et al. 2020.

FIGURE 2.15 The head and torso of the colossal statue; legs in the background. Inset: Ashraf Hussein (CEAlex photographer) takes photos of the statue at the Maritime Museum of Alexandria
MOHAMED ELSAYED

Through virtual restoration we are able to think about the reassembly of the fragments of our sculpture in an unhurried and more systematic way, observing more easily the correspondences between the different pieces that compose an object. With a 3D model of a proposed restoration, we can also contemplate the visual impact of different materials that we may choose when carrying out a subsequent real-world restoration.

We have two important examples from the seafloor near Alexandria. First we discuss the first virtual anastylosis of a colossal statue of the goddess Isis of red granite (Figure 2.17). The sculpture is now composed of three separated parts. Two major fragments, the legs, and the torso and head (Figure 2.15) were raised from the sea in 1962 in front of Qaitbay Fort. The third fragment, the crown of the statue, was recovered in 1995. Then, we discuss the virtual reassembly of a group of three fragments of a sphinx of granodiorite stone, representing the sphinx's base (N. 3002), neck and upper body (N. 1325) and the head (N. 1324) (Figure 2.18).

8 A Virtual Anastylosis of a Colossal Statue of Isis

In 2011, CEAlex launched a project using photogrammetry to carry out virtual anastylosis of the colossal statue of Isis recovered in large fragments from the sea floor of Alexandria Harbor. As the statue fragments weighs more than 20 tons in total (calculated from our 3D analysis of the models), we used a crane to photograph the inaccessible surfaces.

For the virtual anastylosis, a photogrammetric survey of our piece was performed. We created the 3D model of the fragments of the statue using Agisoft Metashape. This operation consisted of the placement of the different pieces in a three-dimensional space as they should have been before its fragmentation (Figure 2.16). For this last operation we used the Autodesk 3ds Max and MeshLab. We used 445 photos to create the 3D models, with the results as follows (Figure 2.16):

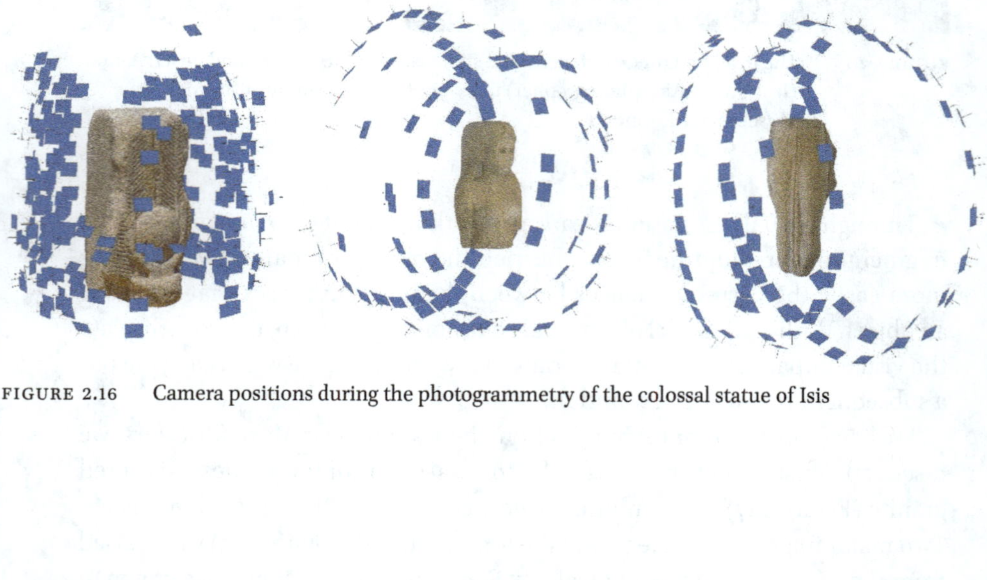

FIGURE 2.16 Camera positions during the photogrammetry of the colossal statue of Isis

FIGURE 2.17 Virtual Anastylosis of the three main fragment of the colossal statue of Isis
PROCESSING BY MOHAMED ABDELAZIZ

9 Virtual Anastylosis of the Sphinx

In 2014, we undertook the restoration of three parts of a sphinx (total weight: 1.5 tons) which had been raised from the sea in 1995. However, as it was difficult to use lifting equipment to be sure that the head and the upper body fit together with the base of the sphinx, we proposed a virtual restoration of the three parts before the intervention of the conservator, to be sure of proper fit of the fragments (Figure 2.18).

A photogrammetric plan was realized for the following parts: (1) the base of the sphinx with hind legs and human arms, of which the hands were holding a removable vase which has disappeared. This fragment fits with the (2) upper

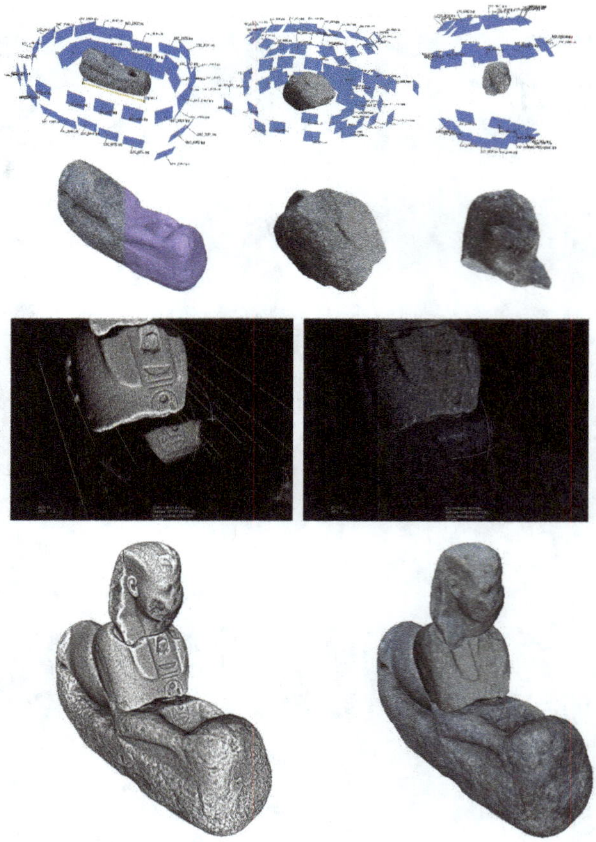

FIGURE 2.18
Virtual restoration of the three parts before restoration, created using 315 images
PHOTOS AND PROCESSING: MOHAMED ABDELAZIZ

body and neck (3) the head (Figure 2.19). A royal cartouche (evidently Ramesses II) is partly inscribed on the base of the sphinx, and partly on the upper body.

10 Unrolling Cylinder Shapes to Facilitate Reading Near-Invisible Inscriptions

Our third example is very important because we were able to recover an all-but-invisible inscription on a piece of a granite column originally from the Roman theatre in Alexandria. With this example, we present the unrolling the cylinder shape, which allows us to show it in an undistorted 2D plane.

A useful way to document archaeological finds is the representation of so-called rollouts, the analysis of rotation-symmetric objects with paintings or inscriptions. The advantage of a rollout is that it can give an overall view of the

FIGURE 2.19 Reassembling and restoration of the three parts of the sphinx by CEALex—Roman theatre, Alexandria
MOHAMED ELSAYED

object's inscribed or decorated content. Typically, rollouts are created either by manual drawing or from photographs. Generating 2.5D rollouts from color and geometry acquisition represents a more reliable method, which assists scholars in the task of interpreting iconography. In this example, we addressed this problem by proposing an automatic method to unroll decorations or inscriptions on 3D cylinder shapes into 2D and 2.5D planar space, using MeshLab 2018.04 and Agisoft Metashape software.[19]

Running the radiance scaling shader in MeshLab also returned quite interesting results. This rendering technique permitted the depiction of object shape through shading via the modification of light intensities around specific features like concavities and convexities. The radiance scaling rendering technique works in real-time on modern graphics hardware, making it feasible for an interactive inspection in MeshLab. By using this filter or renderer on a photogrammetric model, it becomes possible to greatly clarify certain inscriptions, particularly after the color and texture information have been stripped from the model (Figure 2.21). This technique is particularly effective on granite, the surface mottling of which can make even well-preserved surface inscriptions difficult or impossible to detect with naked-eye inspection in natural light. Using a model of our column piece processed with Agisoft Metashape 1.5 and MeshLab 2018.04, we were able to reveal seven lines of a Latin inscription which were hardly visible on the object itself or in daylight photographs.

19 Vergne et al. 2018.

FIGURE 2.20 Unrolled cylinder decoration, showing a 3D surface projected onto a 2D plane
PROCESSING BY MOHAMED ABDELAZIZ

FIGURE 2.21 Unrolling cylinder shape, MeshLab software
PROCESSING BY MOHAMED ABDELAZIZ

Acknowledgments

The authors would like to thank the Centre d'Études Alexandrines for supporting us during these projects and welcoming us as part of the team.

References

Abdelaziz, M. and M. Elsayed. 2019. "Underwater Photogrammetry Digital Surface Model (DSM) of the Submerged Site of the Ancient Lighthouse near Qaitbay Fort in Alexandria, Egypt." *International Archives of the Photogrammetry, Remote Sensing and Spatial Information Sciences* 42–42/W10: 1–8. https://www.int-arch-photogramm-remote-sens-spatial-inf-sci.net/XLII-2-W10/1/2019/. Accessed 2-23-2022.

Al-Ruzouq, R. 2012. "Photogrammetry for Archaeological Documentation and Cultural Heritage Conservation." In *Special Applications of Photogrammetry*, edited by Daniel Carneiro Da Silva, 97–110 (London: INTECH Open Access Publisher). https://www.intechopen.com/chapters/36196. Accessed 2-22-2022.

Balletti C., C. Beltrame, E. Costa, F. Guerra and P. Vernier. 2015. "Underwater Photogrammetry and 3D Reconstruction of Marble Cargos Shipwreck." *The International Archives of the Photogrammetry, Remote Sensing and Spatial Information Sciences* 40:W5: 7–13. https://www.int-arch-photogramm-remote-sens-spatial-inf-sci.net/XL-5-W5/7/2015/. Accessed 3-1-2022.

Corteggiani, J.P. 1998. "Les *Aegyptiaca* de la fouille sous-marine de Qaïtbay." *Bulletin de la Société française d'égyptologie* 142: 25–40.

Drap, P. 2012. "Underwater Photogrammetry for Archaeology." *In Special Applications of Photogrammetry*, edited by Daniel Carneiro Da Silva, 111–136 (London: INTECH Open Access Publisher). https://www.intechopen.com/chapters/36197. Accessed 2-22-2022.

El-Gindy, A. 2000. "Currents and their variability in Eastern Harbour and Qait Bey areas." In *Underwater Archaeology and Coastal Management: Focus on Alexandria.* Edited by M. Hassan Mostafa, N. Grimal and Douglas Nakashima, 144–146. Coastal Management Sourcebooks 2. Paris: UNESCO. https://unesdoc.unesco.org/ark:/48223/pf0000119751?1=null&queryId=51c6372a-71ae-4ad3-a83d-6024442f57b8. Accessed 2-22-2022.

Elsayed, M. 2012. *L'archéologie sous-marine en Egypte. Rappel critique de son histoire et propositions pour une politique de gestion des vestiges immergés en Egypte, à la lumière des autres expériences en Méditerranée.* PhD diss. Université Lumière, Lyon.

Elsayed, M. 2013. "Unterwasserarchäologie in Ägypten." In *Archäologie im Mittelmeer:*

auf der Suche nach verlorenen Schiffswracks und vergessenen Häfen, edited by M. Reinfeld, 127–138. Darmstadt: von Zabern.

Empereur, J.Y. 1998. *Le phare d'Alexandrie: la merveille retrouvée*. Paris: Gallimard.

Empereur, J.Y. 2000. "Underwater archaeological investigations of the ancient Pharos." In *Underwater Archaeology and Coastal Management: Focus on Alexandria*. Edited by M. Hassan Mostafa, N. Grimal and Douglas Nakashima, 54–59. Coastal Management Sourcebooks 2. Paris: UNESCO. https://unesdoc.unesco.org/ark:/48223/pf0000119751?1=null&queryId=51c6372a-71ae-4ad3-a83d-6024442f57b8. Accessed 2-22-2022.

Frost, H. 1975. "The Pharos Site, Alexandria." *International Journal of Nautical Archaeology* 4.1: 126–130.

Green, J.N., E. Baker, and B. Richard. 1971. "Simple Underwater Photogrammetric Techniques." *Archaeometry* 13.2: 221–232.

Hairy, I. 2007. "Pharos, l'Égypte et Platon." *Images et modernité hellénistiques: appropriation et représentation du monde d'Alexandre à César*, edited by F.H. Massa-Pairault and G. Sauron, 61–89. Rome: EFR.

Hairy, I. 2009. "Site sous-marin de Qaitbay, méthodes de fouilles et dernières découvertes." *Bulletin de la Société archéologique d'Alexandrie* 48: 113–132.

Hairy, I., M. Abdelaziz, M. Elsayed and P. Soubias. 2016. "Le Phare, un site immergé." *Dossiers d'Archéologie* 374: 24–25.

Reuter, P. et al. 2011. "Semi-automatic 3D Acquisition and Reassembly of Cultural Heritage: The SeARCH Project." *ERCIM News* 86: 12–13.

Riad, H. 1964. *Récentes découvertes archéologiques en Alexandrie*. Alexandria: Institut français d'archéologie orientale.

Skarlatos, D., S. Demestiha and S. Kiparessi. 2012. "An 'Open' Method for 3D Modeling and Mapping in Underwater Archaeology Sites." *International Journal of Heritage in the Digital Era* 1: 1–24. https://journals.sagepub.com/doi/10.1260/2047-4970.1.1.1. Accessed 2-23-2022.

Stampouloglou, M., Olympia Toska, Sevi Tapinaki, Georgia Kontogianni, Margarita Skamantzari, and Andreas Georgopoulos. "Archaeological Anastylosis of Two Macedonian Tombs in a 3D Virtual Environment." *Virtual Archaeology Review* 11 (22): 26–40. https://doi.org/10.4995/var.2020.11877. Accessed 2-23-2022.

Van Damme, T. 2015. "Computer Vision Photogrammetry for Underwater Archaeological Site Recording in a Low-Visibility Environment." *ISPRS International Archives of the Photogrammetry, Remote Sensing and Spatial Information Sciences* XL-5/W5: 231–238. https://www.int-arch-photogramm-remote-sens-spatial-inf-sci.net/XL-5-W5/231/2015/. Accessed 2-23-2022.

Vergne, R. et al. 2018. "Enhancing surface features with the Radiance Scaling MeshLab Plugin." *Computer Applications and Quantitative Methods in Archaeology (CAA) 2012, March 2012*: 417–422. https://hal.inria.fr/hal-00877158. Accessed 2-23-2022.

Yamafune, K., R. Torres, R. and F.J. Castro. 2016. "Multi-Image Photogrammetry to Record and Reconstruct Underwater Shipwreck Sites." *Journal of Archaeological Method and Theory* 24: 703–725. https://link.springer.com/article/10.1007/s10816-016-9283-1. Accessed 2-23-2022.

CHAPTER 3

The Digital Rosetta Stone Project

Miriam Amin, Angelos Barmpoutis, Monica Berti, Eleni Bozia, Josephine Hensel and Franziska Naether

Abstract

The goal of this contribution is to present The Digital Rosetta Stone, which is a project developed at Leipzig University by the Chair of Digital Humanities and the Egyptological Institute/Egyptian Museum Georg Steindorff in collaboration with the British Museum and the Digital Epigraphy and Archaeology Project at the University of Florida. The aims of the project are to produce a collaborative digital edition of the Rosetta Stone, address standardization and customization issues for the scholarly community, create data that can be used by students to understand the language and content of the document, and produce a high-resolution 3D model of the stone. First, the three versions of the text were transcribed and encoded in XML according to the EpiDoc guidelines. Next, the versions were aligned with the Ugarit iAligner tool that supports the alignment of ancient texts with modern languages, such as English and German. All three texts were then parsed syntactically and morphologically through Treebank annotation. Finally, the project explored new 3D-digitization techniques of the Rosetta Stone in the British Museum in order to enhance traditional archaeological methods and facilitate the study of the artifact. The results of this work were used in different courses in Digital Humanities, Digital Philology, and Egyptology.

Keywords

Rosetta Stone – Textual Alignment – Treebanking – 3D Modelling – Shape-from-Shading

1 Introduction, Aims, and Origin of the Project

This contribution aims to describe the origin, execution, and results of the pilot project *The Digital Rosetta Stone* and shows its usefulness in editing and teach-

DOI:10.1163/9789004501294_004

ing ancient texts. The main techniques involved are alignment, treebanking, 3D imaging, and online visualization.[1]

The Greek text of the Decree of Memphis, better known as the Rosetta Stone, was published as the *Marmor Rosettanum* by Karl Müller in an appendix of the first volume of the *Fragmenta Historicorum Graecorum* (FHG). It includes the addition of the French literal translation and the commentary that Jean-Antoine Letronne produced to help Jean-François Champollion with the analysis of the differences between the Egyptian and the Greek versions of the decree carved on the stone. The reason for the publication in the FHG is due to the importance of the Rosetta Stone as a source of ancient political history and the desire of the French editor Ambroise Firmin-Didot to include it in the FHG together with the text of the *Marmor Parium*.[2]

In 2014, while working on the *Digital Fragmenta Historicorum Graecorum* (DFHG) project, the PI Monica Berti decided to promote two separate initiatives for experimenting with the analysis and representation of the Parian Marble and the Rosetta Stone in a digital environment.[3] Since the FHG edition includes only the Greek text of the Rosetta Stone and the document bears three scripts (beyond the Greek, also a hieroglyphic one in Middle Egyptian language and another one in Demotic Egyptian script and language), Berti approached the Egyptologist Franziska Naether with the idea to create a new digital edition of the Rosetta Stone to experiment with digital technologies applied to epigraphy and develop educational resources for students in Classics and Egyptology.

Initially, the team focused on the alignment of the texts of the stone, which was accomplished by Julia Jushaninowa using the alignment editor of the Alpheios project.[4] The morpho-syntactic annotation of the Greek version was produced by Giuseppe Celano and Polina Yordanova with the tool Arethusa.[5] In order to produce more data and expand the project, Monica Berti and Franziska Naether successfully applied for a grant from the German Federal Ministry of Education and Research (BMBF).[6]

---

1 https://rosetta-stone.dh.uni-leipzig.de.

2 Müller 1841–1873, vol. I, v–viii.

3 On the DFHG project see http://www.dfhg-project.org and Berti 2019. On the Digital Marmor Parium project, see http://www.digitalmarmorparium.org. The DFHG includes the *Marmor Rosettanum*, which is identified as urn:lofts:fhg.1.marmor_rosettanum, and also the index of its Greek words at http://www.dfhg-project.org/DFHG/index_marmoris_rosettani_volumen_primum.php.

4 https://alpheios.net/

5 Preliminary ideas and results were published by Berti, Celano et. al. 2016.

6 See the implementation of the funding line "StiL—Studieren in Leipzig" and "LaborUniversität": www.stil.uni-leipzig.de/7-projektkohorte-der-laboruniversitaet/.

Thanks to this grant, the Egyptologist Josephine Hensel and the Digital Humanist Miriam Amin were hired to work on the alignment and the visualization from October 2017 to October 2018. The treebanking of the Egyptian languages, however, proved more difficult. To address this issue by devoting more resources for developing guidelines for the linguistic annotation of the Demotic part of the text, Berti and Naether were able to obtain an additional grant from the BMBF that allowed Josephine Hensel to compile a test case of the Demotic in the academic year 2019/2020.

Moreover, as the project focuses on advanced visualizations of the artifact, the team invited Angelos Barmpoutis and Eleni Bozia of the Digital Epigraphy and Archaeology Project at the University of Florida to take new high-resolution pictures of the Rosetta Stone.[7] In June 2018, the team traveled to the British Museum in London to take new images of the monument that were used to visualize the text alignment and produce a new high-resolution 3D model.[8]

Several courses in Digital Humanities and Egyptology at Leipzig University were devoted to the language, historical context, and digital edition of the Rosetta Stone.[9] The results of the project were also presented at the Digital Classicist Seminar in London,[10] the Sunoikisis Digital Classics online study course[11] and in an online exhibition.[12] These activities have been generating considerable interest from colleagues, museum curators, and the public in this project.[13]

2 Features and Peculiarities of the Monument

This section conveys the basic features of the monument briefly. Then, it relates its place in the history of archaeological research to demonstrate the signif-

7 http://www.digitalepigraphy.org/

8 Two time-lapse videos of the working session in the British Museum are available on YouTube: see https://youtu.be/soOboUFtNTw and https://youtu.be/of7vVcp3tCk.

9 These courses are taught at Leipzig University as part of the Bachelor of Science and the Master of Science in Digital Humanities and as part of the Master of Arts in Egyptology at the Faculty of History, Arts and Regional Studies. See a full list of modules and classes at https://rosetta-stone.dh.uni-leipzig.de/rs/teaching/.

10 Digital Classicist London Seminars 2018, session 5: https://www.youtube.com/watch?v=Mm_705Zwf9Y.

11 SunoikisisDC Summer Semester 2018, common session 12: https://youtu.be/s2zHi1OiN-A.

12 https://ausstellungen.deutsche-digitale-bibliothek.de/rosetta-stone/.

13 The main results of the project were presented in November 2019 through the Mitteldeutsche Rundfunk (Central German Broadcasting, MDR). One of their outputs is available at https://www.mdr.de/wissen/stein-von-rosette-digital-leipzig-100.html.

icance of the current project's in-depth study, as we cannot fully cover the almost 200 years of scholarship on the Rosetta Stone that is as old as the discipline of Egyptology itself.[14]

Whereas the existence of the Rosetta Stone and its importance in deciphering the Egyptian languages (both the hieroglyphic and the demotic script)[15] are widely known, presumably a smaller number of people would know about the contents of the three versions of its text and the purpose of its erection. The Stone bears a decree with the publication of decisions from a priestly synod—the outcome of a meeting of the pharaoh and select priests from all over Egypt. Most prominently, the items pertain to the funding of temples, construction of shrines, statues, benefactions of priests, gods, and ancestral cults of the Ptolemaic family. Sometimes, political events are mentioned. Apart from the Rosetta Stone (or the Decree of Memphis), about 30 other monuments, temple texts, and papyri carry comparable decrees from the Ptolemaic Period and beyond, such as the Gallus Stele. Whereas the format of the sacerdotal decrees as a text type are modeled after Greek honorary inscriptions,[16] the genre of decrees is attested in Egypt since the Old Kingdom, with famous examples dating to the rule of Pepi II or during the time of the New Kingdom.[17]

With its trilingual inscription, the Rosetta Stone was erected in the first year of the reign of pharaoh Ptolemy V in 196 BC. The dark stela made of granodiorite measures 114.4 × 72.3 × 27.93 cm (45 × 28.5 × 11 in). Perhaps a total height of 200 cm (78 in) is realistic. However, the original layout of the stela with a possible decoration of the lunette and an offering scene with the pharaoh before the gods and accompanying Hieroglyphic labels is speculative. Having been found as part of a wall in a fortress close to the city of Rosette in the Eastern Nile Delta in Lower Egypt in August 1799 by a French general named Boussard, the monument was kept for two years in the mansion of the governor in Alexandria. After Napoleon Bonaparte's unsuccessful military campaign in Egypt (during which the first facsimile in ink of the Stone was produced), it fell into the possession of the major general Turner, who then transferred it to London in 1801. Since 1802, the Rosetta Stone has been an iconic object of the British Museum[18] and served

14 A handy introduction is Ray 2007 or the scholarly peer-reviewed Wikipedia article by Dalby 2019.

15 See the monograph by Thomasson 2013 on Åkerblad, especially 215–288.

16 Clarysse 2000.

17 General introduction: Blumenthal 1974; for Pepi II's decree see Goedicke 1989; for Horemheb's decree see Helck 1955; for Sethi I's decree in Nauri see Edgerton 1947 just to name a few examples.

18 Inventory: BM EA 24; Trismegistos identification no.: 8809; Greek inscription publication no.: OGIS 1, 90.

as one of the objects—though not the decisive one, which was the Bankes Obelisk in Kingston Lacy, Dorset[19]—to decipher ancient Egyptian. While the Danish Johan David Åkerblad and the British Thomas Young worked mainly on the Demotic version, Jean-François Champollion successfully worked with all three versions, including other monuments and Coptic sources. This consultation of other manuscripts is an often-overlooked fact but stated clearly in Champollion's decisive document, the *Lettre à M. Dacier* from 1822. The deciphering of the Hieroglyphs also marks the beginning of Demotic Studies and contributes to Coptology, both sub-disciplines of Egyptology.[20]

The decipherment of the ancient Greek version happened much quicker. In 1803, Richard Porson completed the first edition that includes a reconstruction of the broken lower right edge,[21] followed by others, such as the edition by Letronne mentioned above.

Our transliteration and German translation of the Rosetta Stone is based on two current editions: Günter Vittmann's record of it in the *Thesaurus Linguae Aegyptiae*[22] and an unpublished dossier of the late Heinz-Josef Thissen, who thankfully shared his findings with us at an early stage of this project. The English translation is the one of Stephen Quirke and Carol Andrews from 1988.[23] Autopsy and the new images helped to improve both slightly.

There is a tiny bit of history about the Rosetta Stone and Leipzig University.[24] Georg Steindorff, Professor of Egyptology since 1894, tried to obtain a plaster cast of the monument in 1892 while still a museum curator in Berlin. According to the letters of Steindorff and the British Museum's head of Egyptian antiquities, Sir Ernest Alfred Thompson Wallis Budge, preserved in the archive of the Egyptian Museum in Leipzig, Budge agreed to create a copy of the Stone. However, the project was never completed. Other local institutions, such as the German Museum of Books and Writing of the German National Library in Leipzig or the Museum of Archaeology at the Department of Ancient Studies at Martin Luther University Halle-Wittenberg, own plaster casts.[25]

19 Schenkel 2016. The name "Kleopatra" on the obelisk provided the main clue.
20 Champollion 1822.
21 Published in Porson 1812.
22 See http://aaew.bbaw.de/tla/ (Login required; accessed 02-25-2022).
23 Quirke and Andrews 1988.
24 https://rosetta-stone.dh.uni-leipzig.de/rs/the-rosetta-stone/rosetta-stone-and-leipzig/ (accessed 02-25-2022).
25 Ellen Rehm provided useful insights on the distribution of plaster casts of chiefly Near Eastern Monuments in German collections (Rehm 2018).

FIGURE 3.1 Possible reconstruction of the Rosetta Stone of the Egyptian Museum Leipzig in a special exhibition 2022
PHOTO: FRANZISKA NAETHER[26]

26 Another reconstruction is proposed by Patrick Brose in Hoffmann and Pfeiffer 2021. We wish to thank the authors for sharing their results before publication with us.

3 Word-by-Word or Expression-by-Expression? The Alignment

Alignment in a text edition is a tool for the comparison of texts and languages.[27] By putting corresponding words or expressions (and translations thereof) right next to each other and linking them, the system enables the user to see the aligner's connections. Therefore, a digital alignment—e.g., highlighting corresponding words and expressions by color-coding—facilitates comprehension and learning of foreign languages, both ancient and modern. Moreover, for scholars of Greek, Latin, and Egyptian, which are not usually taught in the same departments and joint study programs, textual alignment provides an excellent opportunity to make use of original sources instead of translations only.

As mentioned above, the alignment with Alpheios was unsuccessful because this tool supports only the alignment of two texts.[28] This is one of the reasons why the Open Greek and Latin project of the Chair of Digital Humanities at Leipzig University assigned Tariq Yousef the implementation of Ugarit iAligner, which is a tool that supports the alignment of three texts, therefore offering an ideal feature for an inscription in three versions.[29] We used Unicode fonts for ancient Greek and the transliteration of the two Egyptian languages. However, it should be noted that Unicode versions of Ptolemaic Hieroglyphic and Demotic are in the process of being developed and not fully available yet. We will include these passages in the future.[30]

The decree on the Rosetta Stone is commonly described as *one* text translated into two languages and written in three scripts. We aimed to analyze how the translation from one language to another was done and address the following questions: How close are the translations of the text in terms of vocabulary? Have mistakes been made in the translations? To this end, we worked on two levels: vocabulary and syntax.[31]

---

27 Textual and translation alignment of historical languages are still experimental due to the complexities of these languages and their sources. For example, see Bamman and Crane 2009, Crane 2019, Cronin 2013, Poibeau 2017.

28 On Alpheios and its use in Perseids, see Almas 2017.

29 See http://ugarit.ialigner.com. The tool was originally developed as part of a MSc dissertation written by Tariq Yousef in the Digital Humanities department at Leipzig University under the supervision of Gregory R. Crane and Monica Berti: Yousef 2015. See also Yousef and Palladino 2017 and Yousef 2020. Our alignment in several versions is available at http://ugarit.ialigner.com/userProfile.php?userid=74. More information on using the Ugarit iAligner is available on the site.

30 See https://hyperallergic.com/441168/soon-you-may-be-able-to-text-with-2000-egyptian-hieroglyphs/; Miagawa 2016; http://hieroglyphseverywhere.blogspot.com/ and the Demotic Palaeographical Database Project (DPDP; http://129.206.5.162/beta/index.html) for recent developments.

31 For the analysis of syntax, see the section below about treebanking.

Initially, we had to separate each text section of the Rosetta Stone in its sentences. Then, we created a synoptic text version based on this sentence level. Our synopsis consisted of 47 paragraphs, which were defined by content, i.e., units of meaning. Therefore, a paragraph can include one or more sentences.

Subsequently, we worked on alignment—a first way of comparing the Hieroglyphic, the Demotic, and the Greek text on word level. During this process, we were able to identify different word relations (1:1, 1:2, 1:n, 2:1, 2:2, 2:n, n:1, n:2, n:n)[32] and also observe different kinds of inconsistencies between the texts, such as omissions,[33] additions,[34] twisted words,[35] differing vocabulary,[36] and "real" mistakes.[37]

---

32 In these cases, "n" stands for "many", i.e., three and more words.

33 One example could be found at the beginning of the text. Within the last title of pharaoh Ptolemy v, we have two different translations. In Demotic, we read: "Son of Re Ptolemy living forever, beloved of Ptah, *the God who appears, whose goodness is perfect, (son of) Ptolemy and Arsinoe, the Gods who love their father*" (line 2). And the Greek part has: "Son of the Sun Ptolemy everliving, beloved of Ptah" (line 3–4). The second part (here cursive) is missing in the Greek text—but why? Has it been forgotten? Has it been deliberately omitted to avoid repetition (cf. the mention of the parents at the beginning of line 2)? Or does this information at this point not correspond to the Greek standard form? Sometimes it is not clear if words or phrases are omitted or if one text is more detailed than the other (see the following case "additions"). In the above-mentioned example, from the perspective of the Demotic text, the Greek shows an omission, but we could also say, that on the basis of the Greek text, the Demotic is extended—maybe following the Egyptian tradition of protocol.

34 There are many examples for this case. The selected one will demonstrate it: At the end of the decree there is a description of how to celebrate the yearly festival for the pharaoh in the first month of inundation. The Hieroglyphic Text writes: "**making festive the altars**, presenting libations and everything it is fitting to do" (line x+12). In Demotic, we find: "and making burnt offerings and libations and the other things it is fitting to do" (line 30) and finally the Greek: "performing sacrifices and libations and the other fitting matters" (line 50). The festive decoration of the altars is only mentioned in one text. Both Demotic and Greek are listing burnt offerings and libations. Either one text is extended, or the other ones rephrased the information (festive altars > burnt offerings).

35 They are rare. In most examples, we have nouns that stand in a reversed order, e.g.: "He [= the pharaoh; J.H.] has taken every care to send infantry, cavalry and ships against those who came **by land and sea** to attack Egypt ..." (Demotic, line 12), cf. "He has provided also that forces of cavalry and infantry and ships be sent out against those who attacked Egypt **by sea and by land**" (Greek, line 20–21).

36 If one compares only the Egyptian versions of the decree, there could be some interesting cases in vocabulary usage. This shows the development of the language in its written and spoken use or different levels of speech, i.e., Hieroglyphs as *mdw-nṯr* "words of Gods" (for priests) and Demotic as popular speech. Some examples: *ḏdb* > *twtw* ("to assemble, to gather"); *rꜣ-pr* or *gs.w-pr.w* > *jrpy* ("temple"); *sjꜣ* > *swn* ("to recognize"); *sḫꜣ.w* > *wt* ("order, decree"), etc.

37 E.g., paragraph 29 about the remission of some residues of the temples. In the Demotic

Altogether, these inconsistencies may have resulted in some text passages being more detailed in their descriptions than the corresponding version(s) in the other language(s), which is then abbreviated. To illustrate this in context, we offer here two synoptic examples:

4 Paragraph 25 "Building Canals and Dams"

Demotic (line 14–15)	Greek (line 24–26)
dj=f tn=w nȝ yꜥr.w r-wn-nȝw dj.t šm mw r tȝ rsȝ.t (n-) rn=s r-bn rḫ nȝ Pr-ꜥȝ.w ḥȝ.ṱ.w jr=s m-qd=s jrj={w}⟨f⟩ ḥḏ ꜥšȝy n he wbȝ=w *jp=f mšꜥ rmṯ-rd.wj=f ḥtr r-rȝ (n) nȝ yꜥr.w n-rn=w r ḥrḥ r-r=w r dj.t wḏȝ=w (r-)ḏbȝ nȝ [mḥ(.w)] n pȝ mw r-wn-nȝw ꜥȳ.w n ḥȝ.t-sp 8.t r nȝ yꜥr.w n-rn=w nȝ ntj dj.t šm mw r jtn ꜥšȝy jw=w mty.w m-šs*	τοῦ τε Νείλου τὴν ἀνάβασιν μεγάλην ποιησαμένου ἐν τῶι ὀγδόωι ἔτει καὶ εἰθισμένου κατακλύζειν τὰ πεδία κατέσχεν ἐκ πολλῶν τόπων ὀχυρώσας τὰ στόματα τῶν ποταμῶν, χορηγήσας εἰς αὐτὰ χρημάτων πλῆθος οὐκ ὀλίγον καὶ καταστήσας ἱππεῖς τε καὶ πεζοὺς πρὸς τῆι φυλακῆι αὐτῶν
79 (+ 1) words "He caused to be dammed up the canals which provided water for the stronghold in question,[a)] *the like of which no former pharaohs had been able to do,*[b)] expending much silver upon it.[c)] He appointed a force of infantry and cavalry *at the mouth of the canals in question* to watch over and guard them,[d)] because of the overflow of water which was great in year 8 at the canals in question,[e)] *which provided water for an extensive area and were (therefore) extremely deep.*[b)]"	42 words "And when the Nile had made a great rise in the eighth year, being wont to flood the plains,[e)] he checked it by damming at many points the outlets of the channels,[a)] expending on it no small sum of money[c)] and setting cavalry and infantry to guard them.[d)]"

text, (line 16–17) we read: "He has remitted the dues of Pharaoh which were charged to the temples up to **Year 9** (*ḥȝ.t-sp 9.t*) …" and the corresponding Greek text (line 28–29) writes: "He has also remitted the dues to the Crown from the temples up to **the eighth year** (τοῦ ὀγδόου ἔτους) …." Which text has the mistake—the Demotic or the Greek one?

4.1 *Commentary*

The Hieroglyphic text is lost.[38] This example clearly illustrates how detailed or abbreviated an ancient translation could be. Comparing both passages, we see that the Demotic text is more extended, as it needs 80 words in contrast to the Greek text, which contains all the important content information in only 42 words. Some observations will be discussed in detail:

[a)] The Demotic text begins this paragraph with the following information: what the pharaoh has done, namely building canals and dams to protect the stronghold with an explanation as to the reasons. The stronghold is not mentioned directly in Greek, in which it is related that in many places the outlets of the canals had to be checked.

[b)] These two subordinate clauses either are added in the Demotic text or are omitted in the Greek text. The first one is an important phrase within Egyptian royal inscriptions. On the one hand, a pharaoh has to legitimize himself and his actions. On the other hand, he emphasizes his superiority and performance over the former kings by giving orders that no one before him had given.[39]

The second addition, a relative clause, repeats what was said at the beginning of this long sequence, thus emphasizing the overflow of the water as the result of the high flood.

The Greek text avoids unnecessary repetitions and concentrates on the present time, i.e., what king Ptolemy is doing to solve problems. Orders of former kings in the same context are probably not noteworthy.

[c)] This phrase is nearly identical in its expression—except for the word order (which is not surprising because Demotic and Greek belong in two different language families) and slightly deviating linguistic phrases. Both texts begin with a verb. The Demotic expression *jr n he* "to incur expenses" (see Erichsen 1954, 267) has its equivalent in the Greek verb χορηγέω "to supply, to furnish."[40] The reference is made by using a preposition and a pronoun (Demotic: *wbꜣ=w*; Greek: εἰς αὐτά). The financial cost, described by the phrase "much silver," is

---

38 A reconstruction of the Hieroglyphic part based on the Nobaireh stela (Egyptian Museum Cairo, CG 22188) is given in Urk. II, 181.3–182.2. But, as Sethe already noted, this text version is very faulty because it is: "eine im Altertum hergestellte Reproduktion eines zertrümmerten Exemplares des Dekretes, (…). Dem Urheber der Reproduktion lagen eine Anzahl Bruchstücke des alten Steins vor, die er, ohne die Lücken anzugeben, hintereinander kopierte" (Urk. II, 167). Therefore, we compare the Demotic and the Greek text only.

39 Such statements can also be found in ancient Egyptian biographies—the only difference being that an individual emphasizes his person over his contemporaries to demonstrate his uniqueness.

40 See https://logeion.uchicago.edu/χορηγέω.

written in an abbreviated in Demotic with noun and adjective, while in Greek, it is formulated more circumstantially as "silver, not a small amount."

d) In Demotic, this sentence is more detailed than in Greek, and we find a rearranging of the nouns "infantry" and "cavalry." The Verb *jp* (Erichsen 1954, 28) "to count, to appoint" has its equivalent in Greek κατάστασις.[41] The objects of the verbs are the troops of infantries and cavalries. In Demotic, the nouns follow each other without a conjunction, while in Greek, the conjunction τε καὶ is used. The purpose of the deployment of these troops was to guard the newly built canals. The Demotic text uses two verbs here (*ḥrḥ* and *wḏꜣ*) to express both guarding and security. Meanwhile, the Greek version mentions only the guarding (φυλακή). The canals are not named twice, as in the Demotic text. Greek establishes a backreference through the use of a pronoun.

e) This part of the paragraph explains the reason why the pharaoh had the canals built. The structure of the sentences is remarkable. While the Demotic text relays this information nearly at the end, the Greek text starts with it. Once more, it is interesting to note how these texts describe the situation, and this time it is the Greek text that shows a slightly different wording. The "great rise" of the Nile (noun and adjective in Greek) corresponds to an "overflow of water, which was great" (indirect genitive followed by a relative construction in Demotic). Additionally, in Greek, the verb ποιέω "to make, do" is also used. To express the unusually high flood, the comparative with the preposition *r* is used in Demotic. In Greek, on the other hand, a phrase that contrasts habit with excess (καὶ εἰθισμένου κατακλύζειν τὰ πεδία) is preferred.

5 Paragraph 34 "Rewards from the Gods"[42]

Hieroglyphic (line x+5)	Demotic (line 20–21)	Greek (line 35–36)
[...] *jsw nn rdj n=f nṯr.w*	*dj n=f nꜣ nṯr.w (n) tꜣ šb.t (n) nꜣy*	ἀνθ' ὧν δεδώκασιν αὐτῶι οἱ θεοὶ
nṯr.wt qnw nḫt ꜥnḫ wḏꜣ s(nb)	*pꜣ ḏrꜣ pꜣ qny pꜣ nꜥš pꜣ wḏꜣ pꜣ snby*	ὑγίειαν, νίκην, κράτος καὶ τἄλλ'
ḥnꜥ ḫ.t nb(.t) nfr(.t) r-ꜣw=sn	*jrm nꜣ ky.w md.t-nfr.t.w ḏr=w (r)*	ἀγαθ[ὰ πάντα], τῆς βασιλείας
r jꜣw.t=f wr.t ḏd.tw ẖr=f ḥnꜥ	*tꜣj=f jꜣw.t (n) Pr-ꜥꜣ smn(.w) ẖr-r=f*	διαμενούσης αὐτῶι καὶ τοῖς
ẖrd.w=f ḏ.t	*jrm nꜣj=f ẖrd.w šꜥ ḏ.t*	τέκνοις εἰς τὸν ἅπαντα χρόνον

41 Uses and references can be checked at https://logeion.uchicago.edu/κατάστασις.

42 Urk. II, 187.6–8. This topic has some parallels in other Ptolemaic decrees. Their comparison shows that the rewards of the gods differ from each other. The parallel texts are discussed in Altenmüller, El-Masry et al. 2012, 115–117 and von Recklinghausen 2018, 131.

29 words	37 (+ 4) words	24 words
"[in] return for these things the gods and goddesses[a)] have given him strength[b)], victory[c)], life[d)], prosperity[e)], health[f)] and all good things in their entirety, his great office[g)] being established[h)] with him and[i)] his children forever[j)]."	"As reward for these things the gods[a)] have granted him might[b)], strength[b)], victory[c)], well-being[e)], health[e)] and all other good things, his office of Pharaoh[g)] being established[h)] with him and[i)] his children forever[j)]."	"In return for which the gods[a)] have given him health[f)], victory[c)], power[b)], and [all] other good things, the kingship[g)] being established[h)] for him and[i)] his children for all time[j)]."

5.1 *Commentary*

In this paragraph we find omissions/additions, reversed word order, and differing vocabulary.

[a)] Pharaoh receives presents from *the gods*—here, only the Hieroglyphic text mentions both male and female gods.

[b)] *qnw* > *qny* / *ḏrꜣ* > κράτος "strength, power": In the Greek text, the version κράτος is the last listed reward. The assignment of Greek κράτος to the Egyptian (Hieroglyphic/Demotic) word *qnw* or Demotic *ḏrꜣ* is not conclusive. Both words have the meaning "to be strong, powerful, victorious" and as noun "power, strength, victory."[43] They could be used as synonyms, but an analysis of their usage in context may also show some differentiation.

[c)] *nḫt* > *nꜥš* > νίκη "victory": The Egyptian words are connected based on their meaning.[44] This relation, though, does not necessarily imply further linguistic dependency. The Greek equivalent is the word νίκη, listed here as the second of the three rewards for the pharaoh.

[d)] *ꜥnḫ* "life": It occurs only in the Hieroglyphic version. This could be explained either as an addition of the Hieroglyphic version or an omission of the other two. However, it is impossible to determine, as we do not know which one the source text is.

43 For *qnw* in Demotic, see Erichsen 1954, 539 (earlier records in Wb. v, 41.5–43.17); for *ḏrꜣ*, ibd. 682–683 (also attested earlier, see Wb. v, 599.1–16). For the Greek word, see https://logeion.uchicago.edu/κράτος (accessed 02-25-2022).

44 For using *nḫt* as verb, adjective or noun in Hieroglyphic and Demotic, see Wb. II, 314.6–317.10 and Erichsen 1954, 226 (this word is also attested in Coptic as ⲚϢⲞⲦ, see Westendorf 1977, 132); for the Egyptian *nꜥš* see Erichsen 1954, 208–209. The corresponding Greek can be consulted at https://logeion.uchicago.edu/νίκη (accessed 02-25-2022).

e) *wḏꜣ* "prosperity": It is only listed in the Egyptian texts in the typical combination with "life" and "health." The group *ꜥnḫ wḏꜣ snb* is very common after pharaoh's name and is an expression of blessing ("Pharaoh NN, may he live, be prosper and be healthy").

f) *snb*(*y*) > ὑγίειαν: The Egyptian word for "health" has its equivalent in Greek ὑγίειαν.[45] In the Greek translation, it is the first reward given by the gods, while the Egyptian ones list it at the end.

g) *jꜣw.t wr.t* > *jꜣw.t* (*n*) *Pr-ꜥꜣ* > βασίλεια: The "great office; office of pharaoh" is translated in Greek with the word βασίλεια "kingship, kingdom."[46] The Egyptian texts show variability in word choice. In Hieroglyphic, we read "his great office," while in Demotic, we have "his office of (being) pharaoh." In Demotic, the possessive pronoun is used, while the possession in Hieroglyphic is expressed by the suffix pronoun 3.sg.m. In Greek, the possession is not mentioned.

h) *ḏdj* > *smn* > διαμένω "to establish, to last." The Egyptian versions use two different verbs with similar meaning.[47]

i) *ḥnꜥ* > *jrm* > καὶ "and": Within the Hieroglyphic text, the old word *ḥnꜥ* is used. Later, this conjunction is nearly fully replaced by *irm*. In Greek, it is translated with καὶ.

j) *ḏ.t* > *šꜥ ḏ.t* > εἰς τὸν ἅπαντα χρόνον "forever; for all time." In Egyptian, there are different ways to express "eternity." The *ḏ.t*-eternity means durability, and its Greek equivalent is χρόνος.[48] In this case, the Greek noun alone cannot capture the complexity of the Egyptian concept. To this end, in the Demotic text, a preposition is added before the noun.

To sum up, these two examples illustrate clearly that ancient translations differ from each other. The synoptic comparison of the three texts of the *Rosettana* has shown that strictly speaking, they are not one and the same text. The content of the decree is the main focus. The majority of the ancient translations are not identical or "1:1" (which, of course, is not possible in two languages with different syntax). Several partial omissions, extensions, and even errors occur in both the Egyptian and the Greek text. Further syntactical and morphological analysis will enable us to study such discrepancies more closely.

---

45 See https://logeion.uchicago.edu/ὑγίεια (accessed 02-25-2022).

46 See https://logeion.uchicago.edu/βασίλεια (accessed 02-25-2022).

47 For *smn*, see Wb. IV, 131–134.7, Erichsen 1954, 433–434; for *ḏdj*, see Wb. V, 628.6–629.15. For the corresponding Greek word, see https://logeion.uchicago.edu/διαμένω (accessed 02-25-2022).

48 See https://logeion.uchicago.edu/χρόνος (accessed 02-25-2022), especially the usage in adverbial phrases after prepositions.

In conclusion, we can say that this multilingual decree consists of three text variants. In the future, a comprehensive study of all multilingual decrees from the Ptolemaic period will provide insights into the ancient methods of translating.[49]

6 Scanning the Stone in 2D and 3D

In addition to the text-based representation of the Rosetta Stone that we used in our text alignment, our project provides an image-based representation of its form to present a holistic view of the text in its original context. In June 2018, with the permission of the British Museum, we photographed the Rosetta Stone to generate high-resolution 2D and 3D maps of its inscribed surface. In our setup, we used a single DSLR camera (Nikon D3400), which was fixed on a tripod in front of the stone and calibrated as follows: exposure time = 5 sec., ISO speed = ISO-100, F-stop = f/25, focal length = 135 mm, and max aperture = 4.5. To reconstruct the tridimensional inscribed surface using the shape-from-shading method,[50] we controlled the lighting of the stone using a handheld light wand (Ice Light) that served as a 15-inch-long light source of 1600 lumen at 5600k color temperature.

We divided the artifact into eight regions (four rows and two columns), photographed individually at 6000- × 4000-pixel resolution. Each region was photographed in four different lighting directions (light from the left, top, right, bottom) by placing the light wand in the corresponding side of the region of interest. This quadridirectional lighting configuration allowed us to capture information related to the local orientation at each point of the surface through the differences of the light reflection observed in the corresponding four photographs.[51]

Thirty-two photographs were taken in total (eight regions x four lighting conditions), which were then processed to compose high-resolution 2D and 3D representations of the surface with 0.08141 mm sampling frequency, which is

49 Simpson 1996, 22–24 furnishes some considerations about nature and manner of text composition. The opinions of the Egyptological community differ and mostly refer to one decree—either the original text was written in Greek (because of a Ptolemaic government, Greek as official language, and the possible origin of the genre in Attic honorary decrees), or it was Egyptian (because of many details within religious aspects or special Egyptian knowledge). Up to now, the question of composition cannot be answered satisfactorily.

50 Barmpoutis, Bozia, and Wagman 2010.

51 Gallen, Eastop et al. 2015.

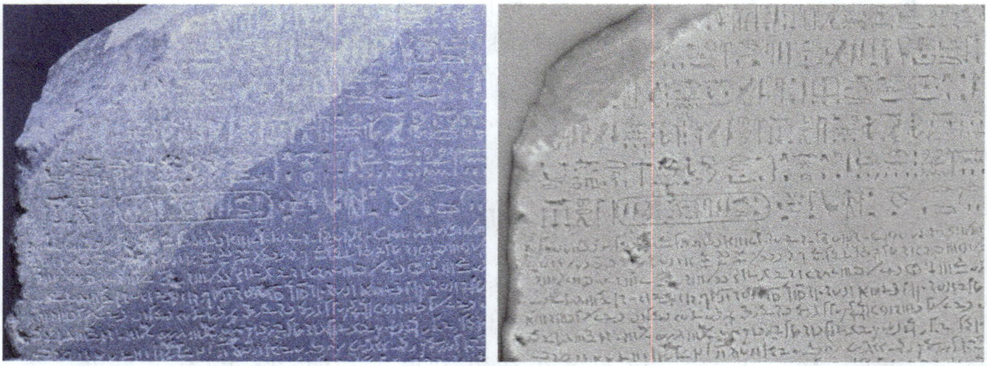

FIGURE 3.2 A+B A sample photograph from our dataset (left). The surface's corresponding reconstructed depth map (right) using the shape-from-shading algorithm
BARMPOUTIS ET AL. 2010

equivalent to 312 DPI resolution. Figure 3.2a shows a sample of our photographs taken with the light source placed on top of the depicted region. The tridimensional details of the inscribed surface were then be captured by the depth map, which was computed by processing the four corresponding images of the same region of interest illuminated with four different lighting orientations. An example of a reconstructed depth map from our photographs is shown on the right panel of Fig. 3.2. This is a small section of the complete reconstructed depth map, which we have published electronically and is available on the project's website.[52] The color intensity of the depth map is proportional to the depth of each point of the surface. As a result, deeper inscribed or weathered locations appear darker, making the inscription more legible than the original photograph.

The depth map contains detailed three-dimensional information of the inscribed surface to be visualized in 3D, as shown in Fig. 3. The 3D reconstructed surface can be rendered as an interactive 3D model that the user can manipulate (move, scale, rotate) and can be inspected under different virtual lighting orientations and shading methods. A 3D visualization is shown on the top left panel that shows the 3D inscription as if it were on white porous material. Non-natural visualization can also be used to enhance the legibility of the 3D reconstructed text. For example, the visualization shown on the top right panel combines 3D rendering with the depth map to generate higher contrast between the virtual stone and the inscribed text.

52 Amin, et al. 2018.

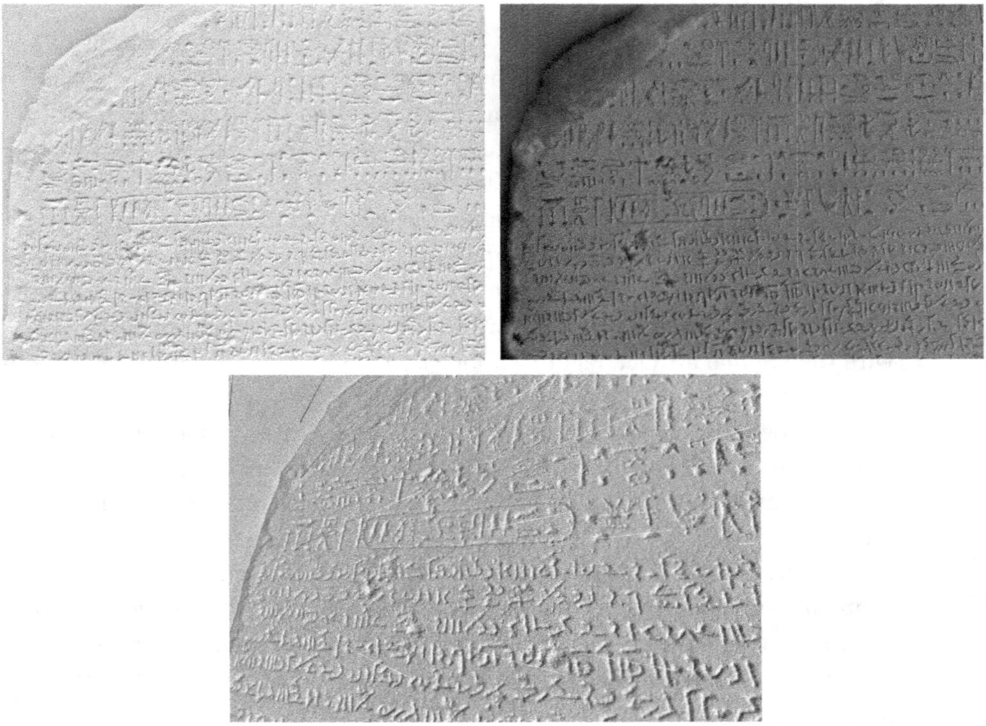

FIGURE 3.3 A–C Different 3D visualizations of the region shown in Fig. 2. The surface can be visualized with natural shading using virtual material properties (top left) or the depth map as the texture of the 3D surface to enhance legibility (top right). The 3D surface can be observed from different points and angles of view and virtual lighting orientations to assist scholars in studying the inscription closely.

Finally, in addition to the 3D reconstruction of the inscribed surface, we used a hand-held laser scanner (Structure Sensor by Occipital) mounted on a tablet computer (iPad Air by Apple) to create a 3D model of the entire stone.[53] Although the 3D model generated by this scanner can depict the overall shape of the whole artifact, it does not have enough resolution to capture the fine details of the inscribed surface.[54] Therefore, the 3D reconstructed surface using shape-from-shading is complementary to the laser-scanned 3D model, as both of these forms can co-exist to serve different needs.

53 Barmpoutis, Bozia and Fortuna 2015.

54 A 3D model of the Rosetta Stone was scanned previously by the British Museum. See https://skfb.ly/6GUQx and https://github.com/BritishMuseumDH/rosettaStone (accessed 02-25-2022).

7 Visualization

To display the parallel versions in the alignment directly on the stone, we had to visualize this within the framework of our website. The user can hover with their mouse over the lines of the stone. While doing this, each passage and its corresponding versions in the other two languages (if preserved) is highlighted on the new 3D image.[55] Transcription of the corresponding text is provided in the tooltip. By clicking on the text passage, the user is directed to the alignment in the Ugarit iAligner that provides the parallelized text versions along with a translation. This way, we connect the abstract textual representation with the original artifact. Figure 3.4 shows a detail of our visualization of the alignment.

To connect the three versions of the same unit of meaning and their position in the photo, we developed an XML scheme that matched our requirements. Therefore, we treated the stone as a hierarchically structured document.

The root element is the Rosetta Stone itself. The following level of structural elements conveys the units of meaning, as defined above. Again, one unit of meaning refers to the abstract idea of the passage rather than the specific realization in one language. Each unit of meaning has two or three child elements that represent the languages as instances of the abstract form. Since some parts of the text are not preserved, several units of meaning convey merely two languages. Fig. 3.5 illustrates the document structure.

The language elements have the three same child elements: transcription, translation, and coordinates. Initially, transcriptions and translations were encoded according to the TEI Epidoc guidelines.[56] This entails e.g., the specification of the lines of the text in the original artifact and the markup of additional epigraphic information of parts of the text, such as preservation state. However, for ease of legibility, the epigraphic information is annotated according to the Leiden conventions for the purpose of visualization. The child element coordinates the link between the text and the actual object, resp. the photograph of it. Each language representation of each unit of meaning comes with one set of coordinates corresponding to the HD photograph's position.

55 https://rosetta-stone.dh.uni-leipzig.de/rs/the-digital-rosetta-stone/visual-alignment/ (accessed 02-25-2022).

56 https://sourceforge.net/p/epidoc/wiki/Home/ (accessed 02-25-2022).

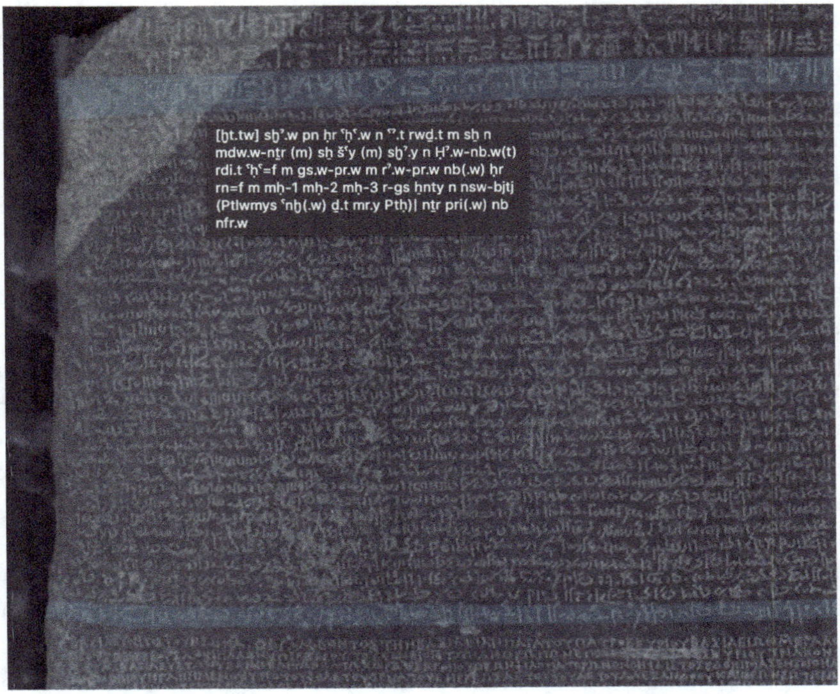

FIGURE 3.4 Detail of the visualization of the text alignment on the high-resolution photograph of the Rosetta Stone

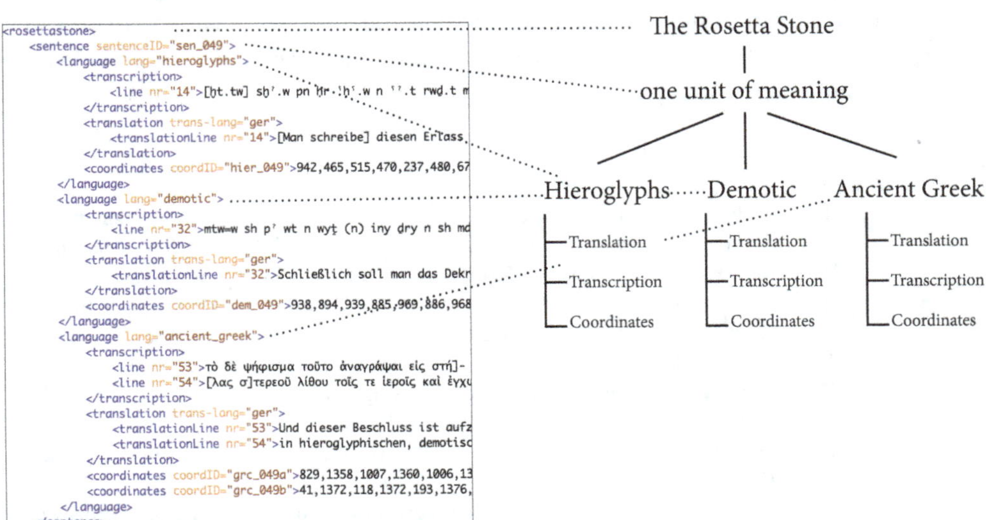

FIGURE 3.5 Document structure by the example of one single unit of meaning

For the frontend of the visualization, we use a JavaScript Query to get the elements from the external XML source and build up an HTML image map. All are implemented within our website in a Typo3 environment.

8 Treebanking

Treebanking is a way of visualizing texts after they have been annotated syntactically in two ways: on the "sentence-level" and the "word-level." While the first means describing the structure of a sentence (main or subordinate clause), type (e.g., interrogative, declarative, circumstantial), and certain elements (e.g., subject, attribute, object), the latter includes information on morphology, syntax, and semantics for each word. The international standard of annotating, i.e., glossing texts, are the "Leipzig Glossing Rules" established by the Max Planck Institute of Evolutionary Anthropology in Leipzig,[57] adapted for Egyptian.[58]

Giuseppe Celano and Polina Yordanova treebanked the ancient Greek version of the Rosettana in 2015 using Arethusa (Fig. 3.6).[59] The statistics show that this part comprises 29 sentences with 1681 tokens of 1514 words—a diameter of 52.28 words per sentence.

Since there has been no treebanked version of an Egyptian text (except in Coptic),[60] we had to add a corpus grammar of our test case, Demotic Egyptian, to the Arethusa treebanking application.[61] This Demotic tagset must include all types of sentences, i.e., main and subordinate clauses, all kinds of word categories ("part of speech"), and their possible functions within a sentence. Each category had to be defined precisely. In the case of Demotic, we provide an example of this analysis:

57 The Leipzig Glossing Rules: Conventions for Interlinear Morpheme-by-Morpheme Glosses, ed. by the Department of Linguistics of the Max Planck Institute for Evolutionary Anthropology (Bernard Comrie, Martin Haspelmath) and by the Department of Linguistics of the University of Leipzig (Balthasar Bickel), https://www.eva.mpg.de/lingua/resources/glossing-rules.php (accessed 02-25-2022).

58 Di Biase-Dyson, Kammerzell, et al. 2009 and Glossing Ancient Languages, Open access Wiki, http://wikis.hu-berlin.de/interlinear_glossing/ (accessed 02-25-2022), edited by Daniel A. Werning, Berlin: Humboldt University Berlin.

59 https://www.perseids.org/tools/arethusa/app/#/perseids?chunk=1&doc=26671 (accessed 02-25-2022).

60 For this, see the project Coptic SCRIPTORIUM: https://copticscriptorium.org/treebank.html (accessed 02-25-2022).

61 The project repository can be accessed at https://github.com/digitalRosettaStoneProject (accessed 02-25-2022).

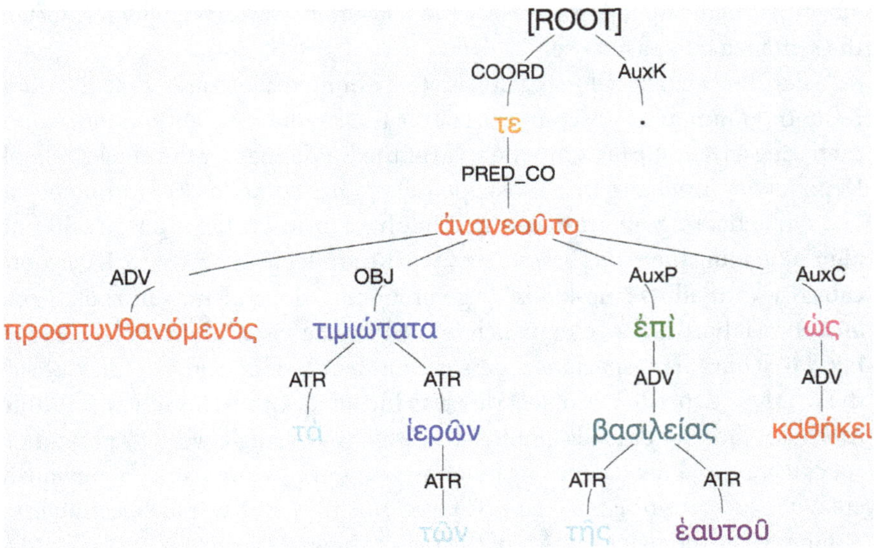

FIGURE 3.6 Visualization of the structure of sentence 16 as a tree (Rosettana, Greek, line 35)

Line 15

jr	*=f*	*jr-sḫy*	(*n*)	*nꜣ*	*sbꜣ.w*
"to do"	"he"	"have power"	"in"	"the"	"enemy"
verb	suffix pronoun	noun	preposition	definite article	noun
PRF	3SG.M	NMLZ.INF	PREP	DEF.ART.PL	M.PL
predicate	subject	direct object	indirect object		

r-wn-nꜣw	(*n*)	*pꜣj*	*=s*	*ẖn*
"who were"	"in"	"this"	"she; her"	"interior"
prefix of participle + imperfect converter	preposition	possessive article	suffix pronoun	noun
REL.PST	PREP	POSS.M.SG	3SG.F	M.SG
relative	predicate			

Literal translation: "He made having-power over the enemies who were in its [= *the stronghold Shekan; J.H.*] interior." > "He prevailed over the enemies who were within it."

This sentence shows a typical Egyptian word order: predicate—subject—object—attribute. The predicate of the main sentence is verbal, followed by a pronominal subject. There are two objects because the verb is a compound one (*jr* + infinitive). The indirect object is the subject of the relative clause. Its adverbial predicate follows directly the converter (*r-wn-nꜣw*), which transfers the sentence into past tense.

Using the Leipzig Glossing Rules, the example illustrates that Egyptian Demotic grammar is very complex. For the text's syntactical annotation, we are preparing a Demotic tagset in the style of the Guidelines for the ancient Greek Dependency Treebank that Giuseppe Celano has compiled.[62] Morphological and syntactical tagsets are necessary. They have to be written in an XML file and after uploading them on GitHub, they will be available in Arethusa. Each word category with all its "sub-kinds" (e.g., pronoun > dependent, suffix, etc.) gets an abbreviation and a color, which is used for the visualized tree (see above, Fig. 3.6). Grammar aspects like person, number, gender, and the functions of the words (e.g., predicate, object) are also included within this schema. Within Arethusa, the user uploads a sentence, assigns each single word its grammatical category, and determines its function within the sentence. The previously created word categories etc. (i.e., the tagset) are now available as selection lists.

Finally, the following extract of the tagset shows its structure:

```
{
  "postagSchema" : [
    "pos",
    ...
  ],
  "styledThrough" : "pos",
  "attributes": {
    "pos" : {
      "long" : "Part of Speech",
      "short" : "pos",
      "values" : {
        "art" : {
          "long" : "article",
          "short" : "art",
          "postag" : "l",
          "style" : {
            "color" : "lightblue"
          },
          "values" : {
```

62 See https://github.com/PerseusDL/treebank_data/blob/master/AGDT2/guidelines/Greek_guidelines.md (accessed 02-25-2022) and Celano 2019.

```
                    "def" : {
                        "long" : "definite article",
                        "short" : "def",
                        "postag" : "dl"
                        }
                    }
                },
    ...
    }
```

9 Teaching Ancient Languages and Cultures

As stated above, new learning applications aligning words and expressions with their respective translations facilitate language learning. Additionally, visualization of grammar and syntax structures by treebanking can serve as teaching tools to foster a deeper understanding.

Classes on political and religious communication in the Ptolemaic Period or lessons in Middle Egyptian/Égyptien de tradition, Demotic, and Greek featuring versions of the inscription are not syllabus mainstream in Egyptology and related disciplines. Some institutions have been teaching in-depth seminars on the decrees, namely the Ancient Studies Center at the University of Trier and its alumni, continuing the tradition in Halle, Tübingen and elsewhere. However, with the new technologies of Digital Humanities and a turn in Egyptology to research its beginnings, we detected a heightened interest of scholars and students in the Rosetta Stone.[63] Additionally, air travel along with museums' social media presence have resulted in increased public awareness of and interest in such artifacts, and the bi-centennial jubilee of the decipherment of Egyptian in 2022 will bring more exhibitions (e.g. in London and Leipzig) and publications

In the past, we have used the alignment of the Rosetta Stone in modules teaching Digital Philology (Berti), Demotic (Naether/Hensel), and in classes on material culture and Egyptian history. Our mission is to teach ancient sources and their functions, historical contexts, provenances, and reception. In some of these seminars, students were required to add new alignments to the Ugarit iAligner as part of their graded assignment. Then, in an oral

63 See the material from the Humboldt Universität Berlin, Creation of a digital transliteration and translation (XML) with interlinear morphemic glossing (Leipzig Glossing Rules): The Rosetta Stone Online project, ed. by Daniel A. Werning, Eliese-Sophia Lincke, http://hdl.handle.net/21.11101/0000-0001-B537-5 (accessed 02-25-2022).

exam, they had to justify the approach of their alignments of the ancient and/or modern languages. During this task, they also provided the developer and us with valuable feedback on the application and the teaching methods.

There are two observations that we would like to share for future endeavors. First, to our experience, students' understanding of ancient languages improved through aligning and treebanking methods. Generally, they worked with more enthusiasm in such project-based classes since their contributions also had an impact beyond the classroom.[64] Students were also more motivated to contribute due to the project's visibility as well as scholarly and historical significance. If successful, they are also more inclined to get involved in such projects, especially when they expect that their input will make a difference and may result in an exhibition, an event, or a publication. One of the challenges in teaching using digital tools is that it is necessary to schedule extra time for IT troubleshooting.

10 Conclusions and Potential for Further Research

With our pilot project on the Rosetta Stone, we tried to combine new technologies for studying and researching ancient texts. The trilingual inscription of this well-known landmark proved ideal to develop a new application for a state-of-the-art digital edition, on account of its brevity, its highly formulaic language, and because it highlights well the challenges in our disciplines. Many corpora could be analyzed similarly, such as the multilingual magical papyri of Greco-Roman Egypt, which can then lead to an in-depth study of multilingualism as a phenomenon in social, cultural, and religious practices.

The Unicode solutions for Demotic and Hieroglyphic and a dedicated tagset for ancient Egyptian languages are still under development. Hopefully, this project and its establishment of the Leipzig Glossing Rules will mark the beginning of more such multidisciplinary digital editions with advanced features that will revolutionize the study and teaching of ancient documents.

64 We have shared our teaching experience at https://www.stil.uni-leipzig.de/teilprojekte/laboruniversitat/ (accessed 02-25-2022).

Acknowledgments

We wish to thank Leipzig University's LaborUniversität framework under which The Digital Rosetta Stone Project was funded through a main grant in 2017–2018 and an extra grant in 2019. The funds were administered by the German Federal Ministry of Education and Research (BMBF). We would like to express our sincere thanks to Prorektor Thomas Hofsäss, Katharina Günther, Beatrice Hartung, and Marit Vissiennon, who have always supported our project, as well as Monika Wohlrab-Sahr, head of the Leibniz program in Leipzig, for providing us with travel funds to London. Giuseppe G.A. Celano and Polina Yordanova thankfully contributed the treebanking of the Greek version, whereas Tariq Yousef's web-based platform Ugarit iAligner was finished just in time for the beginning of our project. Big thanks go to our colleagues Eliese Sophia-Lincke and Daniel Werning from the Humboldt Universität in Berlin to share their results on the glossing of the Rosetta Stone with us and the British Museum (namely Ilona Regulski and her colleagues) for enabling us to take new photographs and 3D images of the monument. Uwe Kretschmer from the Saxon Academy of Sciences and Humanities provided IT support in a tricky tech issue. The Digital Epigraphy Project benefitted through initial funding from the National Endowment for the Humanities, Grant HD-51214-11. Many colleagues helped us with their feedback on our paper and/or poster presentations in Berlin, Bloomington, Cambridge/MA, Heidelberg, Lecce, Leipzig (12), London, Münster, New York, and Stellenbosch (3). In December 2018, Monica Berti and Franziska Naether received the Theodor-Litt-Award for Excellence in Teaching, in part for this project.[65] Last but not least, thanks go out to the editors and to the two anonymous reviewers of this article.

Bibliography

Almas, Bridget. 2017. "Perseids: Experimenting with Infrastructure for Creating and Sharing Research Data in the Digital Humanities". *Data Science Journal* 16(19): 1–17. DOI: 10.5334/dsj-2017–019.

Amin, Miriam, Monica Berti, Josephine Hensel, and Franziska Naether. 2022. "The Digital Rosetta Stone". In *Proceedings of the 29th International Congress of Papyrology, Università del Salento, 28.07.–03.08.2019*, edited by Mario Capasso and Paola Davoli. Lecce: Università del Salento Press.

65 http://gesellschaft.uni-leipzig.de/auszeichnungen/geschichte/ (accessed 01-20-2023).

Amin, Miriam, Angelos Barmpoutis, Monica Berti, Eleni Bozia, Josephine Hensel, and Franziska Naether. "Depth Map of the Rosetta Stone", 2018. https://rosetta-stone.dh.uni-leipzig.de/ (accessed 02-25-2022).

Altenmüller, Hartwig, Yahia El-Masry, and Heinz-Josef Thissen. 2012. *Das Synodaldekret von Alexandria aus dem Jahre 243 v. Chr.* Hamburg: Buske.

Bamman, David, and Gregory R. Crane. 2009. "Discovering Multilingual Text Reuse in Literary Texts". In *Perseus Digital Library*. White Paper. URL: http://www.perseus.tufts.edu/publications/2009-Bamman.pdf.

Barmpoutis, Angelos, Eleni Bozia, and Robert S. Wagman. 2010. "A Novel Framework for 3D Reconstruction and Analysis of Ancient Inscriptions". *Journal of Machine Vision and Applications* 21(6): 989–998. DOI: 10.1007/s00138-009-0198-7.

Barmpoutis, Angelos, Eleni Bozia, and Daniele Fortuna. 2015. "Interactive 3D Digitization, Retrieval, and Analysis of Ancient Sculptures, Using Infrared Depth Sensors for Mobile Devices". *Lecture Notes in Computer Science* 9178 (Part IV): 3–11. DOI: 10.1007/978-3-319-20687-5_1.

Berti, Monica. 2019. "Historical Fragmentary Texts in the Digital Age". In *Digital Classical Philology. Ancient Greek and Latin in the Digital Revolution*, edited by Monica Berti, 257–276. Berlin and Boston: De Gruyter. DOI: 10.1515/9783110599572–015.

Berti, Monica, Giuseppe G.A. Celano, Julia Jushaninowa, Franziska Naether, and Polina Yordanova. 2016. "The Digital Rosetta Stone". In *Altertumswissenschaften in a Digital Age: Egyptology, Papyrology and Beyond. Proceedings of a Conference and Workshop in Leipzig, November 4–6, 2015 (DHEgypt15)*, edited by Monica Berti and Franziska Naether. Leipzig: Universitätsbibliothek. DOI: urn:nbn:de:bsz:15-qucosa-201522.

Blumenthal, Elke. 1974. "'Befehl des Königs' in den königlichen Rechtsurkunden des Alten Reiches". *Zeitschrift für Ägyptische Sprache und Altertumskunde* 100: 72–76. DOI: 10.1524/zaes.1973.99.2.72.

Celano, Giuseppe G.A. 2019. "The Dependency Treebanks for Ancient Greek and Latin". In In *Digital Classical Philology. Ancient Greek and Latin in the Digital Revolution*, edited by Monica Berti, 279–298. Berlin and Boston: De Gruyter. DOI: 10.1515/9783110599572–016.

Champollion, Jean-François. 1822. *Lettre à M. Dacier, secrétaire perpetuel de l'Academie royale des Inscriptions et Belles-Lettres, relative à l'alphabet des hieroglyphes phonétique employés par les Égyptiens pour inscrire sur leur monuments les titres, les noms et les surnoms des souverains grecs et romains*. Paris: Ambroise Firmin-Didot. PURL: https://digi.ub.uni-heidelberg.de/diglit/champollion1822a/0003.

Clarysse, Willy. 2000. "Ptolémées et temples". In *Le décret de Memphis. Colloque de la Fondation Singer-Polignac à l'occasion de la célébration du bicentenaire de la découverte de la Pierre de Rosette*, edited by Dominique Valbelle and Jean Leclant, 41–65. Paris: Diffusion De Boccard.

Crane, Gregory R. 2019. "Beyond Translation: Language Hacking and Philology". *Harvard Data Science Review* 1:2. DOI: 10.1162/99608f92.282ad764.

Cronin, Michael. 2013. *Translation in the Digital Age. New Perspectives in Translation Studies*. London and New York: Routledge.

Dalby, Andrew. 2019. "Rosetta Stone". *WikiJournal of Humanities* 2(1): 1–19. DOI: 10.15347/wjh/2019.001

Di Biase-Dyson, Camilla, Frank Kammerzell, and Daniel A. Werning. 2009. "Glossing Ancient Egyptian. Suggestions for Adapting the Leipzig Glossing Rules". *Lingua Aegyptia. Journal of Egyptian Language Studies* 17: 343–366.

Edgerton, William. 1947. "The Nauri Decree of Seti I: A Translation and Analysis of the Legal Portion". *Journal of Near Eastern Studies* 6(4): 219–230. URL: https://www.jstor.org/stable/542653 (accessed 02-25-2022).

Erichsen, Wolja. 1954. *Demotisches Glossar*, Kopenhagen: Munksgaard.

Gallen, Rachel, Dinah Eastop, Eleni Bozia, and Angelos Barmpoutis. 2015. "Digital Imaging: The Application of Shape-from-Shading to Lace, Seals and Metal Objects". *Journal of the Institute of Conservation* 38(1): 41–53. DOI: 10.1080/19455224.2014.999004.

Goedicke, Hans. 1989. "The Pepi II Decree From Dakhleh". *Bulletin de l'Institut Français d'Archéologie Orientale* 89: 203–212.

Helck, Wolfgang. 1955. "Das Dekret des Königs Haremhab". *Zeitschrift für Ägyptische Sprache und Altertumskunde* 80: 109–136.

Hoffmann, Friedhelm, Stefan Pfeiffer. 2021. *Der Stein von Rosetta. Originalausgabe. Neuübersetzung des Rosetta-Textes*. Stuttgart: Reclam.

Miyagawa, So. 2016. "An Intuitive Unicode Input Method for Ancient Egyptian Hieroglyphic Writing: Applying the Input Technology of the Japanese Writing System". In *Altertumswissenschaften in a Digital Age: Egyptology, Papyrology and Beyond. Proceedings of a Conference and Workshop in Leipzig, November 4–6, 2015 (DHEgypt15)*, edited by Monica Berti and Franziska Naether. Leipzig: Universitätsbibliothek. DOI: urn:nbn:de:bsz:15-qucosa-201695.

Müller, Karl. 1841–1873. *Fragmenta Historicorum Graecorum*. I–V. Paris: Ambroise Firmin-Didot.

Poibeau, Thierry. 2017. *Machine Translation*. Cambridge, MA: The MIT Press.

Porson, Richard. 1812. "Conjectural Restoration of the Lacunae in the Greek Script on the Rosetta Stone". *Archaeologia* 16. Published posthumously. URL: https://digitalcollections.nypl.org/items/510d47e2-6f3a-a3d9-e040-e00a18064a99.

Quirke, Stephen, and Carol A.R. Andrews. 1988. *The Rosetta Stone Facsimilie Drawing*. London: British Museum Press.

Ray, John. 2007. *The Rosetta Stone and the Rebirth of Ancient Egypt*. Cambridge, MA.

Rehm, Ellen. 2018. *Wertvolle Kopien. Gipsabgüsse altorientalischer Denkmäler in Deutschland*. Münster: Zaphon.

Schenkel, Wolfgang. 2016. "Champollions Kleopatra". In *Gedenkschrift für Werner Kai-*

ser, edited by Daniel Polz and Stephan J. Seidlmayer, 393–404. Berlin and Boston: De Gruyter (*Mitteilungen des Deutschen Archäologischen Instituts, Abteilung Kairo*, 70/71: 2014/2015). DOI: 10.11588/propylaeumdok.00004346.

Sethe, Kurth. 1904. *Historisch-biographische Urkunden aus der Zeit der Könige Ptolemäus Soter und Ptolemäus V. Epiphanes*. Leipzig: Hinrichs.

Simpson, Robert S. 1996. *Demotic Grammar in the Ptolemaic Sacerdotal Decrees*. Oxford: Griffith Institute.

Thomasson, Fredrik. 2013. *The Life of J.D. Åkerblad. Egyptian Decipherment and Orientalism in Revolutionary Times*. Leiden and Boston: Brill.

Von Recklinghausen, Daniel. 2018. *Die Philensis-Dekrete. Untersuchungen über zwei Synodaldekrete aus der Zeit Ptolemaios' V. und ihre geschichtliche und religiöse Bedeutung*. Teil 1: Text. Wiesbaden: Harrassowitz Verlag.

Wb.: Erman, Adolf and Hermann Grapow. 1926–1931. *Wörterbuch der ägyptischen Sprache*, vols. I–V. Berlin: Akademie-Verlag.

Westendorf, Wolfhart. 1977. *Koptisches Handwörterbuch*, Heidelberg: Carl Winter.

Yousef, Tariq. 2015. *Word Alignment and Named-Entity Recognition Applied to Greek Text Reuses*. Masterarbeit. Leipzig: Universität Leipzig. Fakultät für Mathematik und Informatik. Alexander von Humboldt Lehrstuhl für Digital Humanities.

Yousef, Tariq and Chiara Palladino. 2017. "iAligner: A Tool for Syntax-Based Intra-Language Alignment". In *Classical Philology Goes Digital. Working on Textual Phenomena of Ancient Texts*, edited by Monica Berti and Karen Blaschka. Leipzig: Universitätsbibliothek. DOI: urn:nbn:de:bsz:15-qucosa2–20952.

Yousef, Tariq. 2020. "Ugarit: Translation Alignment Visualization". In *OSF Preprints. LEVIA'19: Leipzig Symposium on Visualization in Applications 2019*, 1–5. Leipzig. DOI: 10.31219/osf.io/thsp5.

CHAPTER 4

Mythological Landscapes and Real Places: Using Virtual Reality to Investigate the Perception of Sacred Space in the Ancient City of Memphis

Nevio Danelon and David J. Zielinski

Abstract

Our study seeks to address the long-debated problem of the topographical reconstruction of Memphis, the ancient capital of Egypt, through a new approach that combines archaeology, philology and geomorphology in a single framework. The result of this work is a text-augmented map of the city capable to take into account all those philological clues (ancient descriptions of the city and toponyms) that would not fit in an exclusively archaeological map. We present our work-in-progress Virtual Reality experience that combines 3D maps and models, satellite imagery, and excerpts from the original source texts.

Keywords

Memphis – Abaton of Osiris – White Wall – Virtual Reality

1 Introduction

Memphis was one of the most illustrious and cosmopolitan metropolises of the ancient world. Strategically located at the vertex of the Nile delta, it served as the capital of Egypt throughout almost the entire length of the Old Kingdom, from the Third Dynasty (ca. 2686 BCE) to the end of the Eighth (ca. 2160 BCE). Its lifespan ranged from the beginning of the third millennium BCE to the fifth century CE, making it one of the most long-lived cities of the world.[1] Despite this long history of uninterrupted habitation, Memphis has almost completely

1 Its oldest name "White Wall" (*jnb ḥḏ*), is recorded in a rock inscription in Sinai Peninsula, showing evidence that at least a fortified citadel was already in existence at the time of the pre-dynastic king Iry Hor (Tallet and Laisney 2012).

disappeared, leaving the impressive background scenery of the pyramids as the only evidence of its past grandeur.

Most of the information we know about the city topography comes from the classical sources that described Memphis both intentionally and accidentally. These sources include Greek historians and geographers—such as Herodotus, Thucydides, Diodorus and Strabo to name but a few—who captured a vivid picture of the city as it looked in the Late and Greco-Roman Periods. Among the Egyptian sources, the Memphite Theology is a particularly intriguing religious text because it describes places that so far were considered related to the mythological Egyptian imaginary but that, ultimately, may reflect real locations in Memphis. In fact, many temple and place names abound with topographical references. As an example, the name of the temple of "Ptah who is South of His Wall" provides a possible indication about its location in reference to the fort of the White Wall, of which the god Ptah was the lord.[2] Other Memphite temples, such as "Neith North of the Wall" or "Khnum in front of His Wall" also seem to revolve around this important structure which clearly represented a pivot in the sacred topography of this ancient holy city. It is however necessary to notice that many toponyms have lost their original etymological meaning over the centuries and millennia, outliving the original structures from which they took their name.[3]

Collating pieces of spatial information from textual sources, such as those mentioned above, and confronting them with the archaeological layout as well as the geomorphology of the site, a consistent cityscape image resurfaces from the past. Lake Acherusia, the Abaton of Osiris, the fort of the White Wall and the temple of Ptah "South of His Wall" may eventually find their places in a sort of predictive map intended to revive old debates as well as raise new research questions. In short, what we aim at in this project is a way to produce a text-augmented topographical map of the city capable to take into account all those philological clues that would not fit in a conventional archaeological map.

We present our work-in-progress Virtual Reality (VR) experience, first released for the Oculus Go VR headset, and now ported to work with its successor, the Oculus Quest. Our VR experience combines 3D landscape overviews, linked 2D detail maps, hypothetical schematic models of key Memphite sites, and within those sites, hotspots that lead the user to excerpts from the original

2 Erichsen and Schott 1954, 315, 1.28.

3 Plutarch (infra) reported alternate meanings for the name "Memphis" as "Haven of the Good Ones" or "Tomb of Osiris.". Although these translations sound close to the toponym *mn nfr*, it is likely that the Egyptians of the Late and Greco-Roman Period lost the memory of its original etymon relating to the pyramid complex of Pepy I (Spiegelberg 1911).

source texts. We will discuss our motivations, design choices, and suggestions on how these types of experiences can be improved.

2 Historical Sources: Myths vs Facts

When it comes to drawing information from classical sources—and notoriously from Herodotus's histories—we have to face the problem of their reliability.[4] Most of the stories recorded by these ancient Greek authors took place millennia earlier and were reported only by hearsay. This is when the narrative becomes extremely flawed and needs more interpretation. Any attempt to assimilate Egyptian deities, characters, and anecdotes to the Greek mythology—perhaps to make them more familiar to their audience—resulted in a soundalike confusion that prevents scholars from identifying any historical basis. The fact remains that their autoptic descriptions are still highly valuable, at least for the comprehension of the contemporaneous historical and geographical context.

2.1 *The Foundation of Memphis*

The Late Period of Egypt is also when classical historians came in contact with the indigenous Egyptian culture. The account of the foundation of Memphis as narrated by Herodotus[5] (ca. 450 BCE) has long been considered a fabulous legend mainly because the hydraulic engineering works described by the historians were deemed anachronistic for that period:[6]

> *Of Min (Menes), who first became king of Egypt, the priests said that on the one hand he banked off the site of Memphis from the river: for the whole stream of the river used to flow along by the sandy mountain range on the side of Libya, but Min formed by embankments that bend of the river which lies to the south about a hundred furlongs above Memphis, and thus he dried up the old stream and conducted the river so that it flowed in the middle between the mountains (the Libyan and Mokattam heights): and even now this bend of the Nile is by the Persians kept under very careful watch, that it may flow in the channel to which it is confined,*

4 Lloyd 1995.

5 Hdt. II, 99. Cfr. Diod. I, 50,3–5.

6 "The dyking and the diversion of the Nile would have been a colossal project and had, in any case, such limited value that we are justified in doubting that it was ever carried out". Lloyd 1976, 11.

and the bank is repaired every year; for if the river should break through and overflow in this direction, Memphis would be in danger of being overwhelmed by flood.[7] *When this Min, who first became king, had made into dry land the part which was dammed off, on the one hand, I say, he founded in it that city which is now called Memphis; for Memphis too is in the narrow part of Egypt; and outside the city he dug round it on the north and west a lake communicating with the river, for the side towards the east is barred by the Nile itself. Then secondly, he established in the city the temple of Hephaistos (Ptah) a great work and most worthy of mention.*

Diodorus's idyllic description of lake Acherusia[8] near Memphis still matches the picture of Dahshur seasonal lake (fig. 4.7), possibly the southernmost surviving part of this ancient, swampy body of water:

and "Meadows," the mythical dwelling of the dead, is his term for the place near the lake which is called Acherousia, which is near Memphis, and around it are fairest meadows, of a marsh-land and lotus and reeds. The same explanation also serves for the statement that the dwelling of the dead is in these regions, since the most and the largest tombs of the Egyptians are situated there, the dead being ferried across both the river and Lake Acherousia and their bodies laid in the vaults situated there.

These descriptions are consistent with the picture of a countryside shaped by traditional irrigation earthworks that characterized the Nilotic landscape in the past. The fundamental role of the basin irrigation techniques has been documented since the most ancient eras of Egyptian history.[9] The tradition about Menes's foundation of the city by diverting the course of the Nile seems to reflect a local legend originated from the presence of large irrigation earthworks still visible at the time of Herodotus's visit. The existence of seasonally flooded reservoirs and drainage canals, variously called ḥawḍ, birka, buḥayra, baḥr in Arabic toponymy, can be mapped in historical cartography (fig. 4.2). In

7 Several documents from the Late and Greco-Roman Period record excessive Nile floods that destroyed or damaged some cross-dykes (Greek διαπλευρισμοί, Arabic *ṣaliba*) and irrigation basins (Greek περιχώματα, Arabic *aḥwād*). Calderini 1920, 37–62, 189–216, Bonneau 1993 44–45. The Southern and the Northern Dykes of Memphis are mentioned in a stele of Amasis found at Mit Rahina. Daressy 1923.

8 Diod. I, 96.

9 Cf. the depictions on the mace head of king Scorpion (Butzer 1976, 20).

addition, traces of winding earth dykes crossing the Nile valley east to west can still be found in the Egyptian countryside from the observation of aerial and satellite imagery.[10]

Different studies attempted to reconstruct the Memphite land- and waterscapes based on classical accounts. Among the first, Rennell[11] already surmised the presence of a Nile paleo-channel in the lowland at the foothill of Saqqara plateau. In 1982, the Egypt Exploration Society started the Survey of Memphis aiming to contextualize previous explorations into the broader domain of landscape archaeology.[12] The initial results of geological investigations and the scarcity of archaeological evidence prior to the Middle Kingdom in the site of Memphis led to the conclusion that the original settlement was to be sought in the strip of cultivated land close to the elite tombs of Saqqara. Based on these considerations, the investigators hypothesized the transfer of the Memphite toponymy from theoretical structures of the Early Dynastic Period located further west, to the present location following the natural shift of the Nile towards East.[13]

2.2 *The Abaton of Osiris*

Strabo's description of Memphis, visited around 25 BCE, greatly differs from Herodotus' and Diodorus' in being more concise and focused on the topography of the city rather than on history. An interesting excerpt from his description can be confidently related to the citadel of Kom Tuman in the northern part of Memphis:[14]

> *There are lakes situated in front of the city and the palaces, which latter, though now in ruins and deserted, are situated on a height and extend down to the ground of the city below; and adjoining the city are a grove and a lake.*

Here, the imposing substructures of the pharaoh Apries's palace still stand today above the once seasonally flooded lowlands stretching between the city and the western desert plateau. The grove with a lake near the citadel may have been the location of another important religious site, the already mentioned Abaton of Osiris:[15]

10 Willems et al. 2017.
11 Rennell 1800.
12 Jeffreys 1985.
13 Jeffreys and Tavares 1994. See also Bunbury and Jeffreys 2011; Bunbury 2020; Lourenço Gonçalves 2019.
14 Strab. XVII, 1,32.
15 Plut. De Iside, 359,20; Spiegelberg 1911.

> *In Memphis, however, they say, the Apis (bull) is kept, being the image of the soul of Osiris, whose body also lies there. The name of this city some interpret as "the haven of the good" and others as meaning properly the "tomb of Osiris." They also say that the sacred island in front of the gates (sic)*[16] *at all other times is untrodden by man and quite unapproachable, and even birds do not alight on it nor fishes approach it; yet, at one special time, the priests cross over to it, and perform the sacrificial rites for the dead, and lay wreaths upon the tomb, which lies in the encompassing shade of a persea-tree, which surpasses in height any olive.*

Diodorus also refers to an island placed in front of the city:[17]

> *On one of the islands off Memphis there stands even to this day a temple of Daedalus, which is honored by the people of that region.*

The Greeks used to identify foreign deities with theirs on the basis of common divine attributes or rather on the assonance of their names. Daedalus here seems to evoke the archaic Memphite god Tatenen,[18] ruler of the forces of the subsoil. Its name literally means "Risen Land" and is the deification of the primordial mound of earth that, according to the Egyptian creation myths, emerged from the Waters of Chaos as the very first entity of the universe. The Memphite Theology establishes Ptah as the creator of the universe and Memphis as the location where the unification of Egypt took place.[19] The version of this creation myth written on the Shabako stone—a stela originally erected in the great temple of Ptah in Memphis by Shabako (dynasty 25)—narrates about a royal fortress in Memphis the city of Ptah-Tatenen:

Line 13: (...) Tatenen, South-of-His-Wall, Lord of Eternity.
Line 17: This is the land ////// the burial of Osiris in the House of Sokar.
Line 61: (...) The Granary [sic] of Tatenen is the Great Throne that gives joy to the heart of the gods in the House of Ptah.

---

16 This passage from Plutarch is quite controversial. The rough translation of the manuscript is "the islet that stands before the gates". The critics of the text (Griffiths 1970), however, preferred to reconstruct the word Φιλαῖς (Philae) instead of the original πύλαις (doors), since in front of this island there is another abaton of Osiris. However, it is not understandable the reason why, speaking of Memphis and the origin of its name, Plutarch refers rather to the distant island of Philae (Junker 1913, 69–70).

17 Diod. I, 97,6.

18 Wiedemann 1890, 402.

19 Breasted 1901; Erman 1911; Sethe 1928; Junker 1940; Junge 1973; and El Hawary 2010. A new analytical grammar and updated translation of the Shabaqo Stone is currently in preparation by J.A. Roberson.

Line 62: (…) owing to the fact that Osiris was drowned in His water.
Line 64: (…) Thus Horus came into the earth at the Royal Fortress, to the north of this land to which he had come.

In these excerpts we can find recurrent references to important Memphite temples and palaces. Ptah's epithet "South of His Wall" (rsj jnb=f) refers to the White Wall, the royal fortress which is located north of the land where Osiris was buried. According to line 17, this holy ground was probably located in the temple of Sokar. It is meaningful that these three gods gradually merged into a new syncretistic deity called Ptah-Sokar-Osiris whose peculiar statuettes are commonly found in most elite tombs of the Late Period. The epithet "Granary of Tatenen," with reference to the temple of Ptah, is a hapax and has been generically interpreted in reference to the extraordinary fertility of the Memphite region.[20] It is suggestive that the ideogram "granary" of Shabako stone resembles a variant of the dome-shaped hieroglyph jꜣt ()—yet not topped by shrubs—that represents the primordial mound of earth, of which Tatenen was the embodiment. The Egyptian equivalent word for the Greek "abaton" (inaccessible place) is jꜣt wꜥbt (pure mound) or sometimes jw wꜥb (pure island). Osiris was drowned at the feet of the Great Throne of Memphis which was the Granary of Tatenen or, ideally, the Abaton island.[21]

A plausible location for the Abaton of Osiris and the temple of Sokar could therefore be sought between the temple of Ptah and the citadel of Kom Tuman in the north of Memphis.[22] This is the place where a large sunken oval depression—now called Northern Birka—opens in the core of the ruin field. The Arabic word "birka" refers to a seasonal pond that was filled by the Nile waters during the annual flood. There are three birkas within the Memphite ruin field of which the central one is the great temple of Ptah. Indeed, these geomorphological features are not natural but the result of a gradual process of sediment and material accumulation all around an old temple enclosure. In archaeological terms, the birka is the negative of the tell formation process. Many other ancient temples stood on a lower ground than the city such as in Bubastis.[23]

20 Frankfort 1948, 31.
21 Griffiths 1980, 160–161.
22 A possible representation of an Abaton of Osiris could be found in the central part of the famous Nile mosaic of Palestrina where a conical building with a portal lies in the shade of a tree on an island with a grove surrounded by a wall. Meyboom 1995.
23 Hdt II. 59, 60.

2.3 *The White Wall*

As said, the first evidence recording the White Wall place name goes back to the Predynastic Period whereas the latest appears in the accounts of the Persian Wars narrated by Herodotus and Thucydides, spanning over two thousand and five hundred years of uninterrupted history. However, we do not know in which form the fort subsisted throughout this long period of time or whether only the toponym survived either as the original name of the city itself or the name of its northern quarter.[24]

Several archaeological clues lead to the long-debated identification of the historical White Wall with the so-called Northern Enclosure at Kom Tuman, yet still too poorly investigated to take a definitive stance on the issue. The fortress contained a royal palace of pharaoh Apries (589–570 BCE) with barracks and armories. Flinders Petrie[25] called this area "Military Camp" after finding considerable amounts of bronze armor scales of Persian workmanship.[26] A clear insight into this archaeological scenario comes from Thucydides's narration of the siege of the White Wall led by the Athenian troops during the rebellion of Inaros:[27]

> *Abandoning a Cyprian expedition upon which they happened to be engaged with two hundred ships of their own and their allies, they (the Athenians) arrived in Egypt and sailed from the sea into the Nile, and making themselves masters of the river and two-thirds of Memphis, addressed themselves to the attack of the remaining third, which is called White Wall* (Λευκὸν Τεῖχος).

Here, inside this fortress, the Persians established their headquarter in Egypt. The illuminating statement of the White Wall being one third of the entire city also matches the present topography of the site. Three large rectangular enclosures of almost equal size line up north to south all the way across Memphis ruin field. These are the Northern Enclosure at Kom Tuman, the great temple of Ptah in the Central Birka, and a third unexplored enclosure located in the Southern Birka that Herodotus credits to a certain king Proteus.[28] An interesting scholium (comment) tries to elucidate this sentence of Thucydides:[29]

24 Jeffreys 1991; Jurman 2020, vol. 1, 21–96.

25 Petrie 1909b.

26 Petrie 1909a, § II, 5–7 and pls. III–IX; Lopes and Braga 2011.

27 Thuc. 1.104.

28 Hdt. II, 112. It could be the temple of Sekhmet, consort of Ptah. In its vicinity, Herodotus places the Phoenician quarter (Tyrian Camp) whose inhabitants had built a temple of Astarte (Foreign Aphrodite), associated to Sekhmet as a goddess of war and healing.

29 Hude 1973.

> *They say that Memphis had three walls. Since therefore two had already been taken, the battle took place in front of the third one. It was called White because while the others were built of bricks, that one was made of marble.*

More realistically, the White Wall was originally a white-plastered mudbrick fortification (infra), hence its name. It follows that it must have been rebuilt or restored several times over the centuries.[30]

The nature of the Memphite place name "White Wall" caused a sort of long-lasting debate among scholars. The uncertainties related to this toponym are largely due to its use in reference to the original fort, the whole city and the Memphite nomos (district). The symbolic nature of the White Wall as a landscape feature has also been proposed by some scholars.[31] Recently, the hypothesis of the White Wall location at Kom Tuman regained momentum after the excavation led by the Russian Academy of Sciences.[32]

According to the Egyptian textual sources, the fort was a royal palace, a military stronghold, and a sacred ceremonial complex for the celebration of the royal jubilee (Sed-festival) at the same time. Different excerpts from Papyrus Harris concerning the first jubilee of Ramses III (dynasty 20) in Memphis, provide an insight into the role of this royal venue:[33]

> *I celebrated for thee the first jubilee of my reign, as a very great feast of Tatenen.*
>
> *The gods of South and North were gathered in the midst of it (the court). I restored thy temple, the jubilee-houses which were before in ruins, since the (former) kings.*

At the foot of Apries's palace terrace, Petrie found the fragments of a monumental gate decorated with exquisite carved scenes of a king's jubilee whose name in the cartouches was left blank. Relying on stylistic arguments only, he dated the jubilee portal to the Old Kingdom. Today, the beautiful reliefs have been attributed to the Saite Period, perfectly framed within the context of the archaizing style of the Twenty-sixth Dynasty.[34]

---

30 In his Stele of Victory, Piankhi notes that the fort north of Memphis "was strong, and that the enclosure walls were raised by a new rampart". BAR IV § 861.

31 Love 2003.

32 Belova and Ivanov 2016; Krol 2015.

33 BAR IV, 335–336; Leclère 2008, 53.

34 Kaiser 1987; Der Manuelian 1994.

Several settings of the Sed-festival, such as the shrines of the gods or the Run Around the Wall, are found already in the burial complex of Netjerikhet (Djoser), founder of the Third Dynasty (ca. 2686–2613 BCE). The Northern Enclosure in Memphis shares many formal and functional analogies with Djoser's pyramid complex: its size, orientation and asymmetrical distribution of the doors along the perimeter are in fact comparable.[35] Inside, there must have been those originally wooden pavilions (Jubilee Houses) which in Djoser's tomb served only as a scenic backdrop and façade (i.e. Potemkin architecture), whose function was to reproduce the king's jubilee event ad perpetuum. Here we are faced with life size "statues of buildings" whose purpose was to provide an eternal substitute in limestone for the royal palace and pavilions built in perishable materials, just as the statues were of individuals. The repeating motif of the enclosure wall is interrupted by 14 false gates and an actual entrance to the complex located in the southeast corner of the enclosure. The characteristic pattern with recesses and overhangs adorned with flat undecorated niches, the square notches (loopholes?) placed in the upper part of the wall, and the precise asymmetrical arrangement of the doors (five of which along the east side) make this iconography a distinctive mark of a specific enclosure wall. The same identical motif, featured for first time in Djoser and Sekhemkhet funerary enclosures, also recurs as a bottom frieze in some royal and private sarcophagi of the Middle Kingdom[36] as well as in some bases of statues and shrines.[37] The different contexts and periods to which these objects belong indicate that they refer to a common archetype that can be found in the original White Wall complex of Memphis.[38] In this perspective, Third Dynasty royal tombs in Saqqara would parallel the Memphite fort in the same way as the Early Dynastic funerary enclosures (Talbezirke) at Abydos and Hierakonpolis closely reflect the urban palace complex of Hierakonpolis in its layout and wall decoration.[39]

35 Petrie 1909a, 4, §10. For a broader discussion of Egyptian fortifications and their artistic representations see Monnier 2014, Nadali 2006, O'Connor 1989, Van Walsem 2020, De Trafford 2007.

36 Van Walsem 2014.

37 Arnold 1997, 37; Lauer 1927, 128.

38 "think of the White Wall of Memphis, reproduced in Djoser's complex, which no Egyptologist has ever identified as a palace wall, simply because we find the 'palace' dummies inside that wall". Van Walsem 2014, 8. For the architecture of Djoser funerary complex see Schäfer 1986, Smith and Simpson 1998.

39 The northern gates of Djer, Peribsen, and Khasekhemwy funerary enclosures show a striking resemblance to the gate of the Early Dynastic palace complex found at Kom el Gemuwia. Moeller 2016, 97.

2.4 *The Temple of Ptah*

The great temple of Ptah "South of His Wall" in Memphis is the best-known monument of the ancient city. It was one of the greatest national sanctuaries of Egypt together with the temple of Re in Heliopolis and Amun in Thebes. Unfortunately, very few vestiges remain of the temple of Ptah. The ancient Greek sources describe in detail its four main pylons (monumental gates) oriented towards the cardinal points:

> **Eastern Pylon:**[40] After Mycerinus, said the priests, Asychis became king of Egypt. He built the eastern propylaea of Hephaestus' temple; this is by much the fairest and largest of all, for while all have carved figures and innumerable graces of architecture, this court has far more than any.
>
> **Northern Pylon:**[41] This Moeris was remembered as having built the northern propylaea of the temple of Hephaestus, and dug a lake, of as many furlongs in circuit as I shall later show; and built there pyramids also, the size of which I will mention when I speak of the lake.
>
> **Western Pylon:**[42] The next to reign after Proteus (they said) was Rhampsinitus. The memorial of his name left by him was the western propylaea of the temple of Hephaestus; before this he set two statues of twenty-five cubits height; the northernmost of these is called by the Egyptians Summer, and the southernmost Winter; that one which they call Summer they worship and entreat well, but do contrariwise to the statue called Winter.
>
> **Southern Pylon:**[43] Having made himself master of all Egypt, Psammetichus made the southern propylaea of Hephaestus' temple at Memphis, and built over against this a court for Apis, where Apis is kept and fed whenever he appears; this court has an inner colonnade all round it and

---

40 Hdt. II, 136. Asychis (cf. Sasychis in Diod. I,94,3–4 and Sesonchis in Manetho) is Sheshonq I (942–924 BCE) of Dynasty 22, great conqueror and builder of many monuments. However, the Herodotean character of Asychis is also confused with Shepseskaf, builder of a pyramid and successor of Mycerinus (Menkaura).

41 Hdt. II, 101. Moeris is credited to be Amenemhat III who built his pyramid complex—known as the "Labyrinth"—and dug the Great Canal (*mr wr* cf. Gr. Μοῖρις) at Hawara in the Fayyum area.

42 Hdt. II, 121. Rhampsinitus is Ramses II and, in fact, fragments of his colossi have been found in front of the western pylon of Ptah.

43 Hdt. II, 153. The Embalming House of the Apis bulls has been located in the southern end of the temple enclosure.

many carved figures; the roof is held up by great statues twelve cubits high for pillars. Apis is in the Greek language Epaphus.

Sesostris Colossi:[44] Sesostris was the only Egyptian king who also ruled Ethiopia. To commemorate his name, he set before the temple of Hephaestus two stone statues of himself and his wife, each thirty cubits high, and statues of his four sons, each of twenty cubits.

Amasis Colossus:[45] Moreover Amasis dedicated, besides monuments of marvelous size in all the other temples of note, the huge image that lies supine before Hephaestus' temple at Memphis; this image is seventy-five feet in length; there stand on the same base, on either side of the great image, two huge statues hewn from the same block, each of them twenty feet high. There is at Sais another stone figure of like bigness, lying as lies the figure at Memphis.

Little is known about the vast interior of the temple enclosure with the sole exception of the court of Apis (*supra*) and, possibly, the western hypostyle hall described by Strabo.[46]

Many other pieces of textual evidence concur in completing the urban picture of Memphis. The topographical questions that have arisen so far, are ones of great historical significance for the city and Egypt, with relevant implications for the comprehension of the settlement development in relation to the environmental changes that occurred over the millennia.

3 VR Experience: Conceptual Design

Moving beyond an evocative reconstruction based on historians' accounts, an attempt was made to extrapolate and arrange in a spatial scheme all the pieces of topographic information about Memphis that we were able to find in the textual content. These ancient topographies (written descriptions) provide a consistent picture of the late city but, differently than an actual map, are lack

44 Hdt.110, Diod. I,57,5. The Herodotean figure of Sesostris (Diod. Sesoosis) tends to unite the deeds of Ramses II and the name of Senwosret of the Twelfth Dynasty.

45 Hdt. II, 176.

46 Strabo XVII, 1,28.

ing in precision. We know, as an example, that a building was located south of another one or within a given district of the city. Nonetheless, this kind of information is not sufficient to draw a pinpoint map of Memphis but can still be used to establish spatial relations between individual structures in a quite flexible arrangement.

The result of this work is a schematic map (fig. 4.1) made of rectangular shapes and not-to-scale pictures that act as placemark icons for the topographic clues emerging from the interpretation of the above-mentioned literary excerpts. In other words, it is a compromise between an abstract description and a realistic representation. Different kinds of schematic maps have already been used for the topographical reconstruction of sites largely unknown from an archaeological point of view. For instance, K.A. Kitchen proposed a conjectural map of New-Kingdom Memphis mainly based on Egyptian and later classical sources.[47] In our intentions the map would act as a bridge between textual and ground-based evidence.

The next step was to compare the schematic map with the known archaeological evidence. Although largely sketchy, the archaeological layout of Memphis provides some anchors for the geolocation of the schematic map, such as the great enclosure of Ptah, which was brought to light by Petrie in the early twentieth century. Another fundamental source of information for the city topography is an administrative document from Zenon archive, papyrus PSI 5.488,[48] that lists south to north all the sections of Memphis defensive embankments that were built against the floods. Their names and measures provide relevant information about the location of the urban districts within the city. The sequence of the sections can be hooked to the ground at the "Quay of Ptah" located along the east-west axis of the temple.

The scale of the city can be fixed considering the morphology of the site. The edge of the Memphite ruin field, neatly traceable on satellite imagery and elevation maps, consists in a wide formation of elevated ground that rises up to 20 m above the countryside level. This compact ground is the result of an artificial process of accumulation of mud bricks and other anthropic refuse on the same site over thousands of years. The arched and rounded profile of the eastern side reveals the ancient river front of the Nile while, on the western side, it appears more jagged. The three already mentioned oval depressions (birkas) open along the north south axis across the site providing additional spatial constraints for the geolocation of the map.

---

47 Kitchen 1991; Snape 2014, see image at 36 [https://erenow.net/ancient/the-complete-cities-of-ancient-egypt/the-complete-cities-of-ancient-egypt.files/image175.jpg (accessed 30-08-2020)].

48 Crawford 1984, 18, Thompson 1988 10–20 and fig. 3.

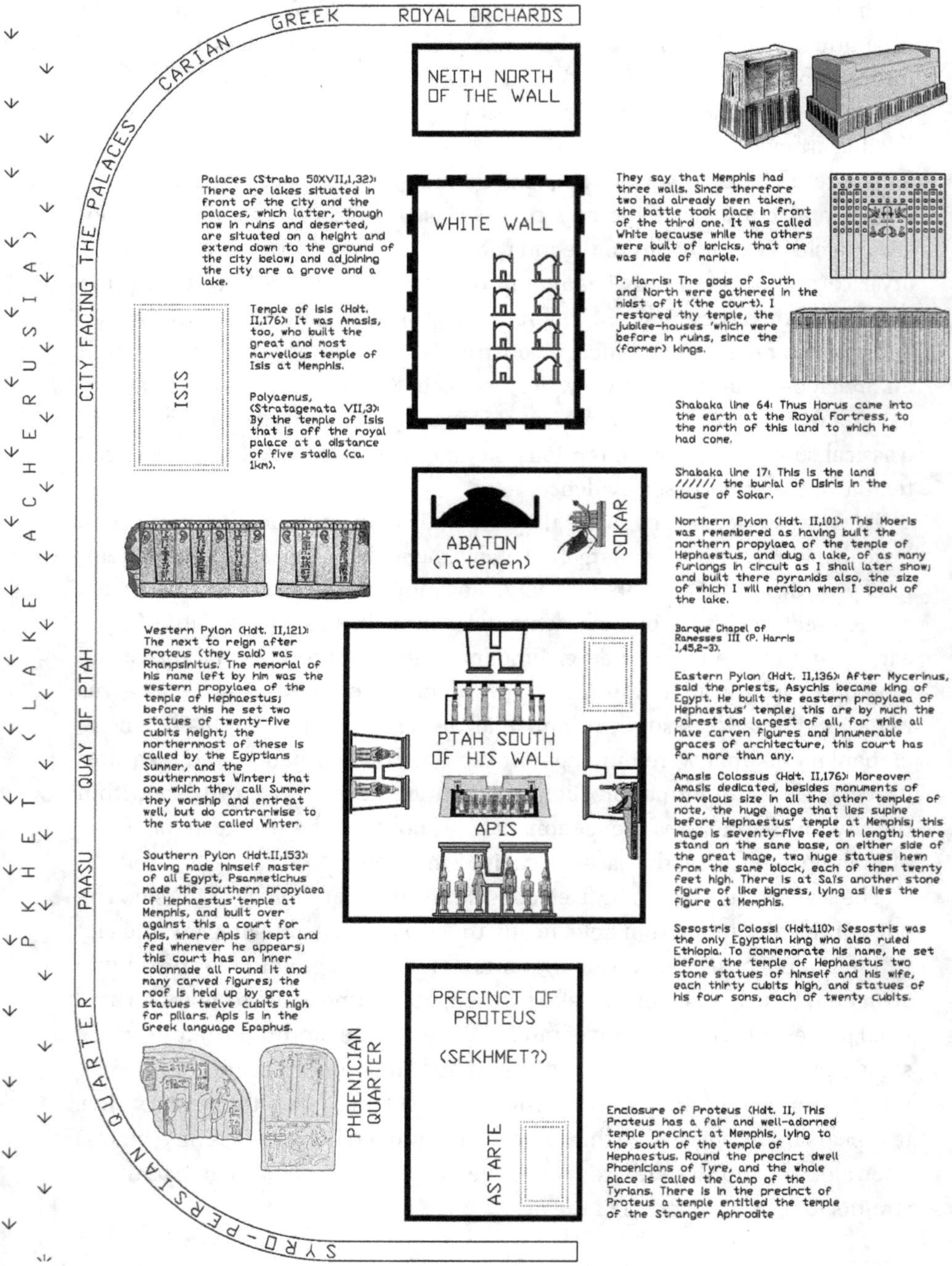

FIGURE 4.1 Schematic map of Memphis based on textual sources

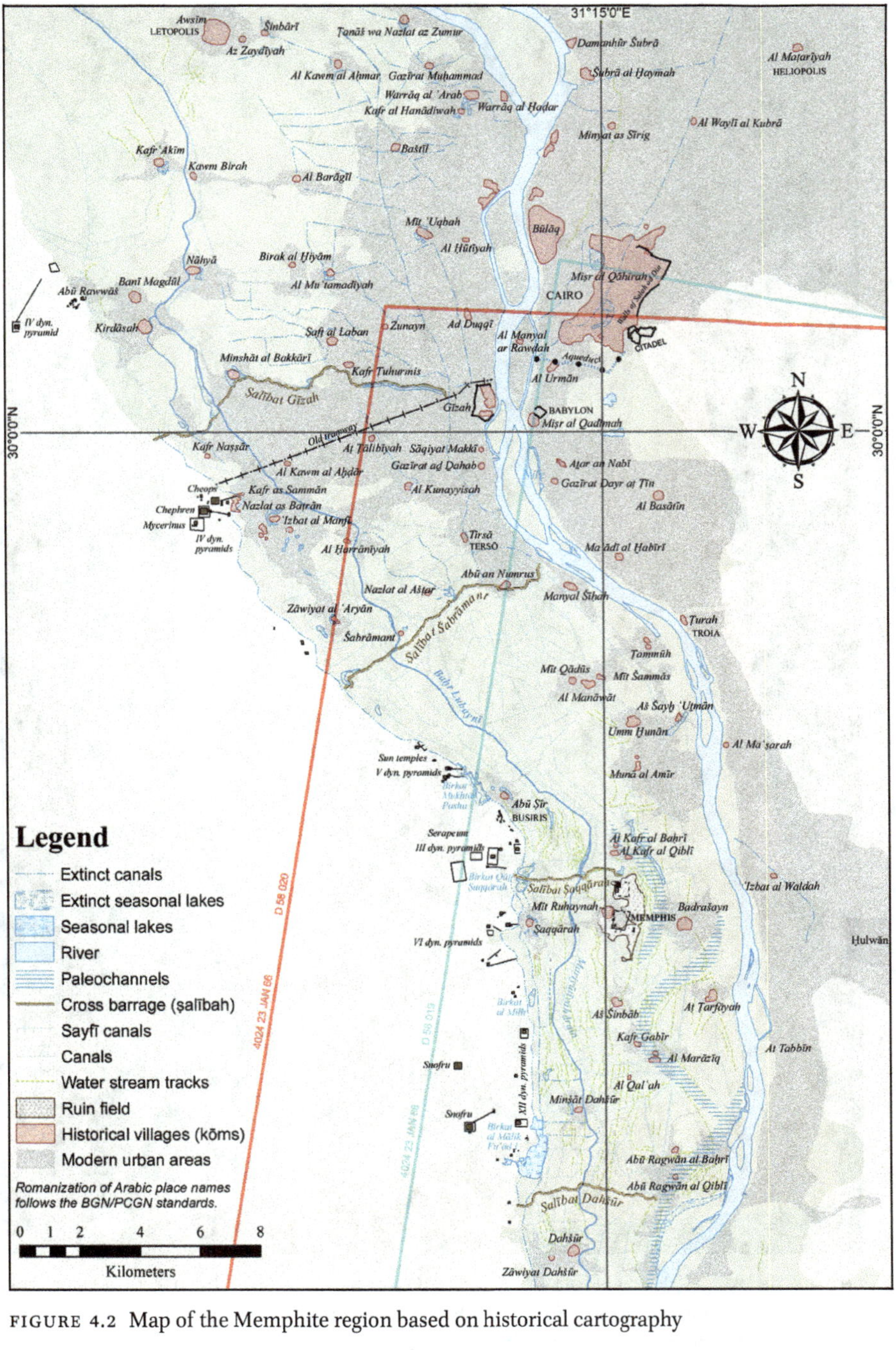

FIGURE 4.2 Map of the Memphite region based on historical cartography

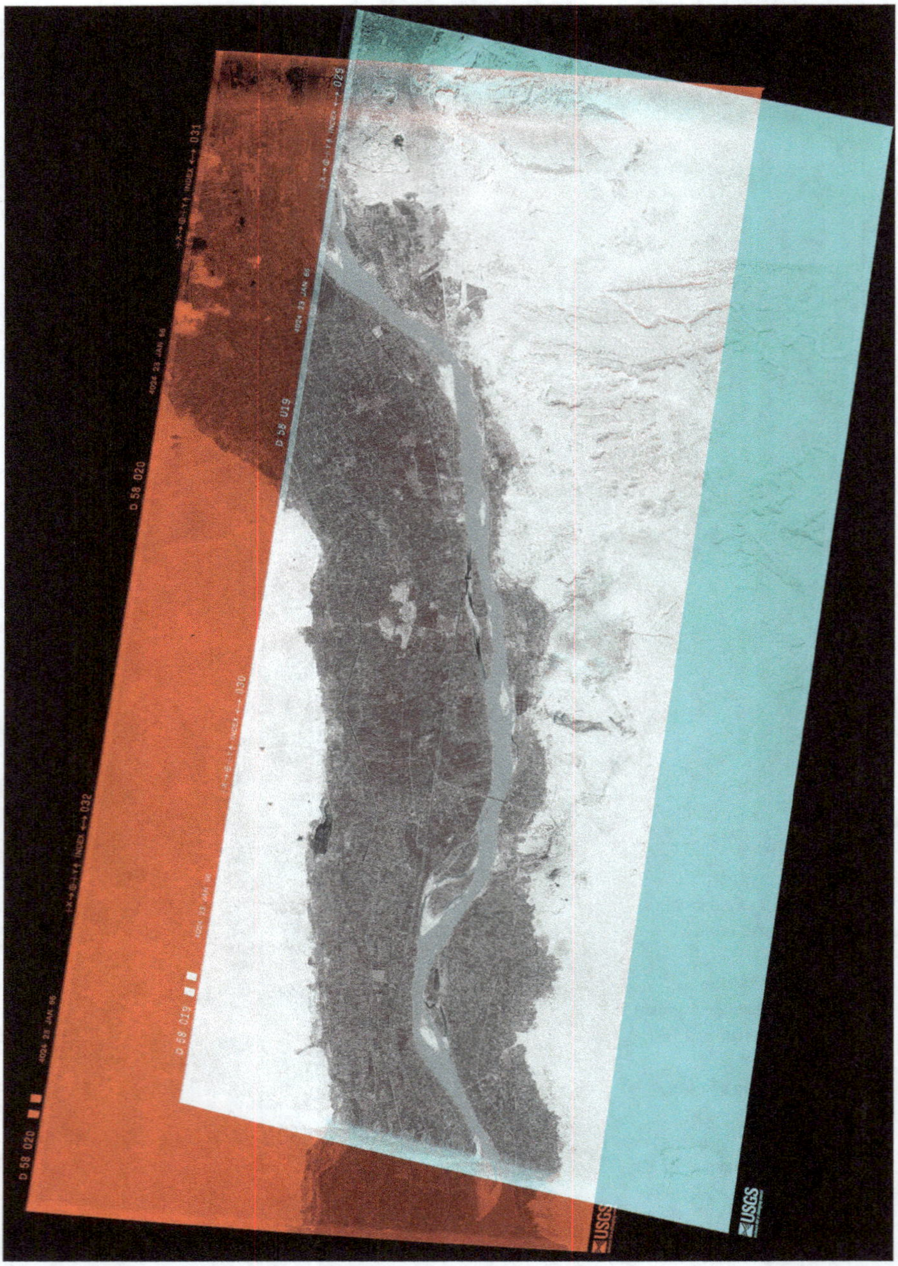

FIGURE 4.3 Anaglyph image combining forward and aft images taken by KH-7 camera of Corona satellite on 23/01/1966. Images have been overlapped and given complementary colors to be viewed with 3D red/cyan glasses (see figs. 4.4–4.7).
COURTESY OF THE U.S. GEOLOGICAL SURVEY

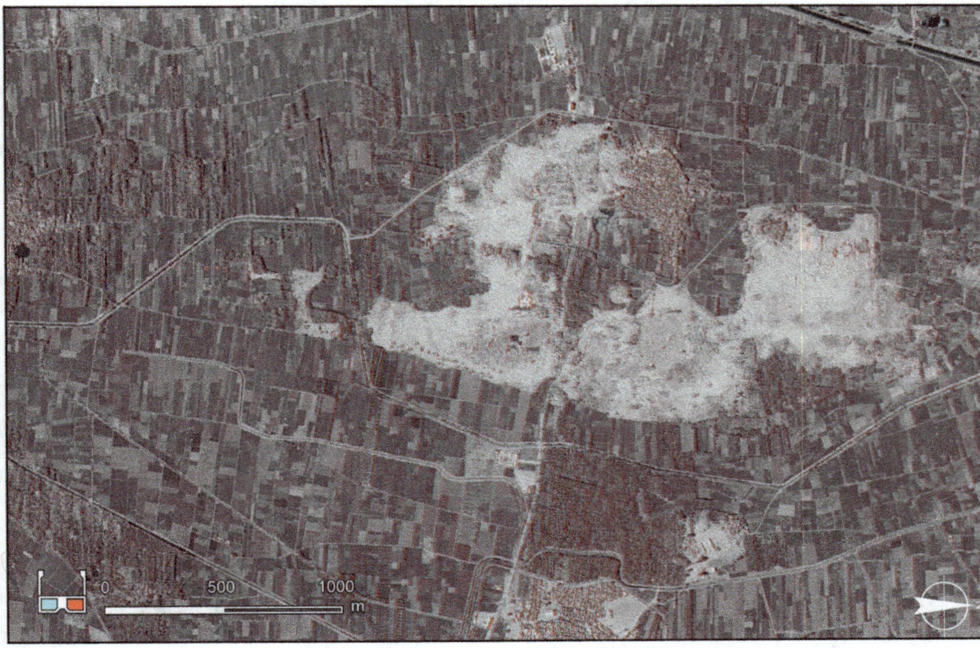

FIGURE 4.4 Clipped anaglyph image showing the Memphite ruin field and the village of Mit Rahina

Declassified Corona satellite imagery (1960–1972) is particularly valuable in archaeology since it preserves a picture of the Near Eastern landscape still largely spared by rural modernization and urbanization.[49] Moreover, most of the photos come in stereopairs suitable for observation through a stereoscope. Thanks to the heightened vertical exaggeration that characterizes Corona stereo camera systems, it is possible to detect slight depressions and bumps in the ground over a regional scale by exploiting depth perception in human vision. Anaglyph images of Memphis (figs. 4.3–4.7) are a precious source of topographic information especially considering that, after the completion of the Aswan dam, the site underwent a rapid urban sprawl. The historical village of Mit Rahina (fig. 4.4) appears still confined within the original boundary of its kōm, one of the many raised villages that were the only dry ground in the valley during the inundation period. Corona satellite imagery is also suitable to be viewed in a stereoscopic display and will be shown in our application (infra).

As a result of this work, we were able to elaborate a new topographic reconstruction of the late city based on archaeological evidence but enriched with

49 Casana et al. 2012; Beck et al. 2017.

FIGURE 4.5 Clipped anaglyph image showing Saqqara North and the Step Pyramid complex

additional information extrapolated from textual sources (fig. 4.8). Many unexplored areas of the site are now shown in a different light that provides valuable information on their archaeological potential. In order to test and communicate this hypothesis, we are creating a tool capable to visualize the spatial content both as a map and an immersive 3D environment. Digital models are made of simple geometric solids to provide schematic representations of the principal city landmarks. By selecting one of these interactive models on the map we can visualize the excerpts of the textual sources that underlie their alleged location in each area of the site shown in a Mapbox viewer. (figs. 4.9–4.10).

FIGURE 4.6 Clipped anaglyph image showing Saqqara South and Shepseskaf's mastaba tomb

FIGURE 4.7 Clipped anaglyph image showing Dahshur with the pyramids of Snefru and Birkat al Malik Fu'ad I

FIGURE 4.8 Hypothetical reconstruction of the Memphite region in the Late and Greco-Roman Period

FIGURE 4.9 Memphis App as viewed in Oculus Go portable VR display system. General overview of the Memphite Region with hotspots on the main sites

FIGURE 4.10 Memphis App as viewed in Oculus Go portable VR display system. Schematic model of the temple of Ptah with satellite view showing the modern area at the southern entrance of the temple. The colossi of Ramses II found here could be those described by Herodotus.

4 VR Experience: Implementation

4.1 *Navigation*

Trying to navigate, or perhaps more precisely engage in locomotion around a 1:1 scale virtual world is an unsolved problem in the field of Virtual Reality. Many techniques have been evaluated, with the most popular being joystick/controller (point to fly), real walking (along with walking in place and redirected walking), and the now popular teleportation. Early research showed that real walking gives the highest sense of presence (of the user being in the virtual space).[50] Recent work[51] has shown that walking and teleportation techniques are better in terms of presence and reduced simulator sickness. Limiting simulator sickness is important if we expect users to spend longer amounts of time immersed in our virtual experience. As an example, one recent and very well-done example is a photogrammetry capture of the Tomb of Nefertari.[52] In this example they use the teleportation navigation method to allow the user to "jump" to new spots around the virtual tomb. While helping with simulator sickness, this technique has drawbacks, such as reduced spatial understanding.[53] Some researchers have tried to modify the flying technique to reduce sickness[54] by dynamically restricting the user's field of view during travel. This type of technique has made its way in modern VR applications such as Google Earth VR, as well as Altspace VR (a social VR platform). We thus decide to adopt a "scholar's studio" paradigm, in which all content is present within a small room. This allows most of the navigation to occur via the user's own walking. This works well especially now that we are using the Oculus Quest which has position tracking (infra). With this decision to use a real-walking navigation technique, we hope to achieve good presence, good spatial understanding, and lower simulator sickness.

4.2 *VR Hardware*

In creating the VR application, we needed to pick which hardware device to target. Our choice was between higher end "tethered" headsets and all-in-one units. A tethered headset is attached to a high-end gaming computer,[55]

50 Usoh et al. 1999.

51 Langbehn et al. 2018.

52 https://store.steampowered.com/app/861400/Nefertari_Journey_to_Eternity/ (accessed 30-08-2020).

53 Langbehn et al. 1999.

54 Fernandes and Feiner 2016.

55 Usually a machine capable to run all the newest games at the highest graphics settings for at least some years.

and allows for more detailed rendering. The all-in-one units have a built in CPU/GPU, which means that no external computer is needed. This gives advantages of being portable, unencumbered (no cables tripping you up), and often lower cost. However, the downside is that for those looking to push the bounds of photorealism, the all-in-one units are going to feel somewhat "cartoony". We decided to go for the all-in-one unit as our goal was to be able to support as many VR users as possible. In 2018, we began work using the Oculus Go. One large limitation of this generation of the headset is that tracking for the head and a single head was rotation only. Position was not tracked, so stepping to the left or right produced an odd sensation, as the scene did not update as the brain expected it to. We have since transitioned to the successor of the Oculus Go, the Oculus Quest.

4.3 *Game Engines*

While there are several game engines worth considering, we felt that our existing knowledge of the Unity game engine would lower our time to implement. Additionally, Unity is known to target low-end mobile graphics, which is useful on the resource constrained all-in-one headset. If we were targeting an aesthetic that demanded more photorealism, it would make sense to evaluate using the Unreal game engine. Other possibilities involve the use of WebVR/WebXR technologies. This involves presenting the VR experience as a webpage with html and java script coding. By adding special WebXR libraries into the webpage, a button can appear that allows us to enter immersive VR mode. This technology is promising in that it makes application distribution easy. We simply send out a URL for users to point their browsers to. This would allow us to bypass "walled garden" app stores, for which acceptance of this kind of academic app is not guaranteed. While in our testing, there was a loss in performance (with peaks below the threshold of 45 frame rate per second) between a compiled Unity application and a browser based WebXR experience, the distribution methods that WebXR are too good to ignore, and we are still evaluating its usage for future applications. Finally, we could consider using some of the VR tools/addons to the popular commercial GIS products. However, we found that while they have some useful pre-made interactions, we desired the interaction design flexibility that comes with using a fully featured, scriptable game engine.

4.4 *Maps*

Our concept (in the above section) involved grounding everything to a real-world base map. We thought about importing terrain, but since we had not conducted on site drone photography in this area, the existing sources of the terrain and aerial photography would have to do. We decided to utilize Mapbox, which is a widely used mapping service. This is often used for 2D mobile apps, such as food delivery and ride sharing. However, they do provide a 3D terrain mode. Mapbox provides a Unity SDK, which is a library that interfaces to their web API. This allows us to specify a lat/lon coordinate and a zoom level, and then Mapbox does the work of pulling down the needed tiles of terrain (geometry) and aerial photographs (the textures). We can also load GIS shapefiles into the Mapbox web portal, which can then be pulled down into our application. Utilizing Mapbox gave us a quick start by easily providing the needed basemap and zoomed in views. We can also easily switch views to anywhere on the planet, which could be useful if the application expands into other areas of Egypt.

4.5 *Content vs Application*

When making a Unity VR application the first tendency seems to be to have all the content inside the application. This means every change—perhaps even just the wording of some text—requires the programmer to intervene and modify scripts. In our application, we made heavy use of the JSON file format to act as a sort of external database of our content. JSON is a now common web standard. Data is presented in name/value pairs and can consist of hierarchical/nested lists. By keeping our content out of the application, this means we can work on the application and content separately. We can also submit the JSON files for independent review, where even non computer expert scholars could read the content and look for any errors. Finally, we could host the JSON files on a server, and pull down the data on startup. This strategy would allow us to update content, or even add new sites of interest, without forcing the user to download a new version of the application.

5 Discussion

The contribution of classical authors in understanding the past is a precious source of information but needs to be interpreted and framed in the proper historical context. Recent discoveries and cross-disciplinary approaches to long debated research questions mostly seem to rehabilitate the ancient historical sources, helping modern scholars challenge some given translations and sometime arguing in favor of a more "verbatim" interpretation of the ancient texts.[56]

---

56 For the interpretation of the historical sources on the Memphite landscape see the funda-

The randomness of many archaeological discoveries in Memphis did not help understand the big topographic picture. The location of the Early Dynastic and Old Kingdom urban settlement in relation to the eastward shifting of the Nile and the development of the necropolis is still a matter of debate among scholars, mostly due to a lack of archaeological evidence directly attributable to these earlier phases. In general, little is known about alluvial settlements in the valley both because of the difficulty for archaeologists to operate in modern urban contexts and the general lack of comprehensive studies on the ancient Egyptian city. More recently, the emergence of settlement archaeology studies in Egyptology is steering the focus of research from the necropolises towards the lesser-known urban areas.[57]

Most of the textual sources refer to the Late Period and Ptolemaic Kingdom, which is also when the city experienced its apogee as a metropolis in terms of population and multi-ethnicity.[58] However, many city landmarks seen and described by the Greek travelers are attributed to the earlier kings and, in particular, the temple of Ptah would date back to Menes, the first king according to the Egyptian tradition (*supra*). This statement is in contrast with the hypothesis of a different location for the Early Dynastic and Old Kingdom city center.[59] The idea of a settlement shift, although supported by some geological clues, relies on indirect archaeological assumptions such as the existence of an older enclosure of Ptah to be sought further west than the Ramesside one.[60] Many of the ancient descriptions of Memphis would find more consistency if the ongoing archaeological investigations at Kom Tuman[61] confirm the existence here of Early Dynastic/Old Kingdom structures which could prove that the later Northern Enclosure (Thucydides's Leukon Teichos) rests on the remains of the original White Wall fort.

The site of Memphis shows a largely undisclosed archaeological potential since it might be still well-preserved under thick layers of alluvium. So far, excavations on the site have been hindered by many factors such as the rising of the water table, the expansion of the modern town of Mit Rahina, the rapid urbanization of the farmland but maybe also an unfounded common belief having it that Memphis, so long plundered of its stones since antiquity, has nothing else

mental work described in Bunbury and Jeffreys 2011. Also noteworthy for the discussion on Herodotus's reliability is the discovery of a *baris* type of ship in Aboukir Bay in 2003 that corroborated his description (Hdt.II,96). Belov 2018, Vinson 1998.

57 Moeller 2016.

58 Thomson 1988; Bresciani 1958.

59 Jeffreys and Tavares 1994.

60 Giddy 1994, 191–192.

61 Belova and Ivanov 2016.

to disclose. The topographic questions so far posed can be addressed only by long-term investigations on the ground, be they archaeological or geophysical. Meanwhile, scholars have now an extraordinary opportunity to test new and old reconstructive hypotheses using the novel technologies available today.

The value of 3D models and Virtual Reality as a tool of knowledge—not just for mere edutainment—has been demonstrated in several cases. The intervisibility between funerary monuments on the western desert plateau and the city in the valley was recently investigated by innovative studies.[62] Moving further, experiencing a fully immersive vision in a "frameless" viewport, such as in a stereoscopic headset, would add more accuracy and realism to the perception of human and natural spaces in a simulated historical environment. In such a virtual laboratory, scholars would be able to test their hypotheses or, rather, find previously unseen perspectives suitable to raise new research questions.

In conclusion, this project intends to provide a virtual environment where scholars and students can visualize, validate, and share both spatial data and textual information in interactive ways and from a human-scale (stereoscopic) viewpoint. We present the complex and long debated case study of Memphis topography with the twofold aim to revive the old discussion and stimulate new ways of looking at the problem of its reconstruction. According to the principles of cyber-archaeology,[63] we do not present here an imaginative and arbitrary reconstruction of Memphis, but rather an interactive non-photorealistic simulation of the Memphite urban environment that allows the observer to contemplate in a single glance the complex and varied sources of data that concur in outlining a potential topography of the city in the Late and Greco-Roman Periods. Thanks to the JSON open format (*supra*), we intend to keep our application open to external contributions and revisions without the need to rebuild and release it with each change.

6 Future Directions

The VR application presented here is still a work in progress, and we therefore have many ideas on how the experience could be expanded. Here we will detail several ideas.

Change over time. At the moment we present static views of both the landscape and the schematic maps. One technique that other research projects

62 Sullivan 2017.

63 Forte and Danelon 2019.

have utilized is to give the user a timeline slider to manipulate.[64] This would add the possibility to broaden the timeframe to previous periods of the city history by showing/hiding topographical features whose coexistence would be anachronistic.

6.1 *Corona Satellite Imagery Integration and Landscape Reconstruction.*

As we mentioned above, Corona satellite stereo imagery focusing on the Memphite region is a precious source of information about the landscape. Moreover, it can be easily viewed in an Oculus Quest display. We intend to broaden the horizon of the research from the city boundaries to the surrounding region to better investigate the visual and spatial relationships between some funerary monuments located on the desert edge and the main city points of interest. For instance, Lepsius's map of Memphis[65] shows an interesting panoramic diagram showing the pyramid silhouettes as seen from Kom Tuman palace mound, a location which seems to act as a privileged point of view overlooking the necropolis.[66]

6.2 *Application Distribution*

When we first started working with the Oculus Go, the Oculus app store seemed more open to accepting different types of content. This is important because getting content onto these all-in-one units can be challenging. However, with the release of the Oculus Quest they detailed in a blog post[67] that they would be restricting access and accepting only well-made polished titles. This basically locks out small proof of concept / work in progress projects such as ours. Being locked out the official app store means we have to explore other ways to distribute the app. Currently Oculus has not completely locked down the headsets, thus it is possible to employ the technique of "sideloading" an application. The user can connect the headset to a desktop computer and using the right tools, can upload a non-official app to the Quest. There is even a software package to facilitate this (SideQuest). Currently sideloading is the technique we are using to load our application onto Quests in our lab. Other options will be discussed below.

---

64 Ozludil 2016.

65 Denkmäler aus Ägypten und Äthiopien, Plates, sec. 1, vol. 1, pl. 9, available online at http://edoc3.bibliothek.uni-halle.de/lepsius/page/abt1/band1/image/01010090.jpg (accessed 30-08-2020).

66 Jeffreys 2009; Sullivan 2017.

67 https://developer.oculus.com/blog/submitting-your-app-to-the-oculus-quest-store/ (accessed 30-08-2020).

6.3 *Social VR*

One big problem with our application is that it is still a single user experience. Much of time is humorously (or perhaps frustratingly) asking "what do you see? Are you seeing the map?" only to find out the user has pressed the wrong buttons and is staring at one of the Oculus menu screens. Now that headsets (especially the all-in-one units) are cheap enough to purchase multiple units, we look towards the paradigm where a group, perhaps the instructor and some students, can visit the experience together. Each user seeing the avatar representation of each other in the space, being able to gesture and point at objects, and even if separated by continents, able to use voice chat to communicate. These visions are not some sci-fi future, but they are the situations we are currently (as of 2020) evaluating in this new paradigm of "Social VR".

The quickly evolving landscape of Social VR would need a full article unto itself, given the quickly evolving number of platforms (at our current count over 20!), which is likely to leave this section looking out of date in the near-future. Still, we will discuss what we have been exploring. One idea we have explored is adding a library to our existing Unity app that will help facilitate multi-user / Social VR. Specifically, we have been looking at the "normcore" networking library. Our initial tests are promising, but one problem is that any app we create will be still left with the distribution issues mentioned above.

We have also been exploring the now Microsoft backed "AltspaceVR" platform. This free service provides rooms that support up to 70 participants along with a robust events system. All major headsets are supported, and there is also a Windows PC desktop client for those lacking a headset. We have so far attended academic talks and musical performances. AltspaceVR is an official app in the Oculus app store, so no sideloading is required. For making our own experiences, there is a process available to import graphical models, and even a type of scripting that allows us to create interactive objects. While we are still in the early stages, we have so far created a proof-of-concept scenario that pulls down Mapbox tiles to create a base map.

Another platform worth considering is the Mozilla backed "Hubs" platform. This is an opensource platform that provides rooms of up to 30 people. This utilizes WebXR technologies, so it runs on almost every desktop and every headset. The experience is "distributed" by sending people a link. Thus, the onboarding process is very easy, no accounts to create or complicated menuing systems to learn. We have been able to use their browser based "Spoke" editor to easily create worlds. One limitation that we are currently investigating is how to add interactable content. On the surface, there is not an easy way to add in user scripts to objects. However, since the whole project is opensource, it should theoretically be possible to add new components into a forked version of the

Hubs platform. Whatever platform we utilize, we look forward to the future of being able to have talks, lectures, and meetups inside our scenarios and no longer requiring VR to be a solitary endeavor.

References

Arnold, D. 1997. "Royal Cult Complexes of the Old and Middle Kingdom" in B.E. Shäfer (Ed.), *Temples of Ancient Egypt*, 31–85. Cornell University Press, Ithaca.

Beck, A., G. Philip, M. Abdulkarim, D. Donoghue 2007. "Evaluation of Corona and Ikonos high resolution satellite imagery: archaeological applications from the Near East" *Antiquity* 82: 732–749.

Belov, A. 2018. *Ship 17: a baris from Thonis-Heracleion*. Oxford Centre for Maritime Archaeology.

Belova, G., S. Ivanov 2016. "Kom Tuman. Season 2016", *Egypt and Neighbouring Countries* 4: 23–30.

Bonneau, D. 1993. *Le règime administratif de l'eau du Nil dans l'Egypte Grècque, Romaine et Byzantine*, E.J. Brill.

Breasted, J.H. 1901. "The Philosophy of a Memphite Priest," *Zeitschrift für ägyptische Sprache und Altertumskunde* 39: 39–54.

Bresciani, E. 1958. "La satrapia d'Egitto" in *Studi Classici e Orientali* 7: 132–188.

Bunbury, J. and D. Jeffreys 2011. "Real and Literary Landscapes in Ancient Egypt" *Cambridge Archaeological Journal* 21(1): 65–75.

Butzer, K.W. 1976. *Early Hydraulic Civilization in Egypt: A Study in Cultural Ecology*. Chicago.

Calderini, A. 1920. "Ricerche sul regime delle acque nell'Egitto greco-romano" *Aegyptus* 1: 37–62.

Casana, J., Jackson Cothren and T. Kalayci 2012. "Swords into Ploughshares: Archaeological Applications of CORONA Satellite Imagery in the Near East" *Internet Archaeology* 32.

Crawford, D.J. 1984. "Hellenistic Memphis: City and Necropolis" in *Alessandria e il mondo ellenistico-romano: Studi in onore di Achille Adriani* 4: 16–24.

Daressy, M.G. 1923. "La crue du Nil de l'an XXIX d'Amasis" *Annales du Service des Antiquités de l'Égypte* 23: 47–48.

De Trafford, A. 2007. "The Palace Façade Motif and the Pyramid Texts as Cosmic Boundaries in Unis's Pyramid Chambers" *Cambridge Archaeological Journal* 17: 271–283.

Der Manuelian, P. 1994. *Living in the Past. Studies in Archaism of the Egyptian Twenty-sixth Dynasty*. London: Kegan Paul International (Studies in Egyptology).

El Hawary, A. 2010. *Word creation: The Memphite theology and the victory stele of Pije—*

two witnesses of cultural representation in the 25th dynasty. Friborg, Switzerland / Göttingen, Germany: Academic Press / Vandenhoeck & Ruprecht.

Erichsen, W. and S. Schott 1954. *Fragmente memphitischer Theologie in demotischer Schrift* (*Pap. demot. Berlin 13603*) Verlag der Akademie der Wissenschaften und der Literatur, Mainz.

Erman, A. 1911. "Ein Denkmal memphitischer Theologie" in *Session reports of the Prussian Academy of Sciences* (*Berlin*) 43: 916–950.

Fernandes, A.S. and S.K. Feiner 2016. "Combating VR sickness through subtle dynamic field-of-view modification," in *2016 IEEE Symposium on 3D User Interfaces* (*3DUI*), *Greenville, SC, 2016*, 201–210.

Forte, M. and N. Danelon 2019. "Cyber-Archaeology". In Oxford Bibliographies in Anthropology, Ed. M. Forte. New York: Oxford University Press. https://www.oxfordbibliographies.com/view/document/obo-9780199766567/obo-9780199766567-0207.xml (accessed 30 Aug. 2020).

Frankfort, H. 1948. *Kingship and the gods: A Study of Ancient Near Eastern Religion as the Integration of Society and Nature*, University of Chicago Press.

Giddy, L. 1994. "Memphis and Saqqara during the Late Old Kingdom: Some topographical considerations" in C. Berger, G. Clerc and N. Grimal (Eds.), *Hommages à Jean Leclant* 1: 189–200. IFAO, Cairo.

Griffith, J.G. 1970. *Plutarch's De Iside et Osiride*, University of Wales Press, Cardiff.

Griffith, J.G. 1980. *The Origins of Osiris and His Cult*. Studies in the history of religions 40. Brill.

Heagy, T.C. 2014. "Who was Menes?" *Archéo-Nil* 24: 59–92.

Hude, K. 1973. *Scholia in Thucydidem ad optimos codices collata*, Leipzig, New York.

Jeffreys, D.G. 1985. *The Survey of Memphis I: The archaeological report*. London: Egypt Exploration Society.

Jeffreys, D. 2009. "Size wasn't everything: the Memphite pyramids as scale models?" in J. Málek, D. Magee, J. Bourriau, & S. Quirke (Eds.), *Sitting beside Lepsius: studies in honour of Jaromir Malek at the Griffith institute*, Orientalia Lovaniensia analecta 185: 257–265. Leuven.

Jeffreys, D. and A. Tavares 1994. "The Historical Landscape of Early Dynastic Memphis" *Mittelungen des Deutschen Archäologischen Instituts, Abteilung Kairo* 50: 143–173.

Jeffreys, D.G. 1991. "Some problems of topography at Achaemenid Memphis" in H. Sancisi-Weerdenburg and A. Kurt. (Eds.), *Papers presented to the Eight Achaemenid History Workshop: Asia Minor and Egypt. Groningen. 27 and 28 May 1988*. Achaemenid History 6: 48–53. Leiden.

Junge, F. 1973. "Zur Fehldatierung des sog. Denkmals memphitischer Theologie oder der Beitrag der ägyptischen Theologie zur Geistesgeschichte der Spätzeit" *Mittelungen des Deutschen Archäologischen Instituts, Abteilung Kairo* 29: 195–204.

Junker, H. 1913. *Das Götterdekret über das Abaton*. A. Hölder, Wien.

Junker, H. 1940. Götterlehre von Memphis, *Treatises of the Prussian Academy of Sciences (Berlin)* 23.

Jurman, C. 2020. *Memphis in der Dritten Zwischenzeit: Eine Studie zur (Selbst-)Repräsentation von Eliten in der 21. und 22. Dynastie.* Widmaier Verlag, Hamburg.

Kaiser, W. 1987. "Die dekorierte Torfassade des spätzeitlichen Palastbezirkes von Memphis" *Mittelungen des Deutschen Archäologischen Instituts, Abteilung Kairo* 43: 123–144.

Kitchen, K.A. 1991. "Towards a Reconstruction of Ramesside Memphis" in E. Bleiberg, R. Freed, (Eds.), *Fragments of a Shattered Visage: The Proceedings of the International Symposium of Ramesses the Great.* Monographs of The Institute of Egyptian Art and Achaeology, 1, Series Ed., W.J. Murnane, Memphis State University, 87–104.

Krol, A. 2015. "White Walls; of Memphis at Kom Tuman" in Kousoulis, P. and Lazaridis, N. (Eds.) *Proceedings of the Tenth International Congress of Egyptologists, University of the Aegean, Rhodes, 22–29 May 2008.* Orientalia Lovaniensia Analecta 241/1: 295–304. Leuven-Paris-Bristol.

Langbehn, E., P. Lubos, and F. Steinicke 2018. "Evaluation of locomotion techniques for room-scale VR: Joystick, teleportation, and redirected walking" *Proceedings of the Virtual Reality International Conference-Laval Virtual.*

Lauer, J.-P. 1927. "Études sur quelques monuments de la IIIe dynastie" in *Annales du Service des Antiquités de l'Égypte* 27: 112–133.

Leclère, F. 2008. *Les villes de Basse Égypte au Ier millénaire av. J.-C.: analyse archéologique et historique de la topographie urbaine*, 2 vols. Bibliothèque d'étude 144.

Lloyd, A.B. 1995. "Herodotus on Egyptian Buildings: A Test Case" In *The Greek World*, Anton Powell (Ed.), 273–300. London and New York.

Lloyd, A.B. 1976. *Herodotus, Book II: Commentary 99–182*, EPRO, 43/3. E.J. Brill, Leiden.

Lourenço Gonçalves, P.M. 2019. *Landscape and environmental changes at Memphis during the dynastic period in Egypt* (Doctoral thesis). https://doi.org/10.17863/CAM.35048

Love, S. 2003. "Questioning the Location of the Old Kingdom Capital of Memphis, Egypt" *Papers from the Institute of Archaeology* 14: 70–84.

Lopes, M.H.T., and Braga, S.F. 2011. "The Apries Palace, Memphis/Kôm Tumân: The First Portuguese Mission in Egypt" *Journal of the American Research Center in Egypt*, 47: 247–258.

Meyboom, P.G.P. 1995. *The Nile Mosaic of Palestrina*, Leiden-New York-Köln.

Moeller, N. 2016. *The Archaeology of Urbanism in Ancient Egypt. From the Predynastic Period to the End of the Middle Kingdom.* Cambridge University Press.

Monnier, Fr. 2014. "Une iconographie égyptienne de l'architecture défensive", *Égypte nilotique et méditerranéenne* 7: 173–219.

Nadali, D. 2006. "Esarhaddon's Glazed Bricks from Nimrud: The Egyptian Campaign Depicted" in *Iraq* 68: 109–119.

O'Connor, D. 1989. "New funerary enclosures (Talbezirke) of the Early Dynastic period at Abydos" *Journal of the American Research Center in Egypt* 26: 51–86.

Ozludil, B. 2016. "Prototyping a Temporospatial Simulation Framework: Case of an Ottoman Insane Asylum" in *Complexity & Simplicity: Proceedings of ECAADe.*

Petrie, W.M.F. 1909a. *Memphis I*, British School of Archaeology in Egypt, London.

Petrie, W.M.F. 1909b. *The Palace of Apries (Memphis II)*, British School of Archaeology in Egypt, London.

Rennell, J. 1830. *The Geographical System of Herodotus examined and explained by a comparison with those of other ancient authors, and with modern geography* 2: 115–146. London, G. & W. Nicol.

Sethe, K. 1928. "Dramatische Texte zu altägyptischen Mysterienspielen. Das Denkmal memphitischer Theologie, der Schabkostein des Britischen Museums" in *Studies on the history and archeology of Egypt* Bd. X, I. Leipzig.

Schäfer, H. 1986. *Principles of Egyptian Art*, Griffith Institute.

Smith, W.S. and W.K. Simpson 1998. *The Art and Architecture of Ancient Egypt*, Yale University Press.

Snape, S.R. 2014. *The Complete Cities of Ancient Egypt*, American University in Cairo Press.

Spiegelberg, W. 1911. "Zu der Etymologie von Memphis bei Plutarch, Isis und Osiris" *Zeitschrift für ägyptische Sprache und Altertumskunde* 49: 129–130.

Sullivan, E.A. 2017. "Seeking a Better View: Using 3D to Investigate Visibility in Historic Landscapes" *Journal of Archaeolgical Method and Theory* 24: 1227–1255.

Tallet, P. and D. Laisney 2012. "Iry-Hor et Narmer au Sud-Sinaï (Ouadi 'Ameyra): Un complément à la chronologie des expéditions minières égyptiennes" *Bulletin de l'Institute Francais d'Archéologie Orientale* 112: 381–398.

Thomson, D.J. 1988. *Memphis Under the Ptolemies.* Princeton University Press.

Usoh, M., K. Arthur, M.C. Whitton, R. Bastos, A. Steed, M. Slater and F.P. Brooks 1999. "Walking > Walking-in-Place > Flying, in Virtual Environments" *Proceedings of the 26th Annual Conference on Computer Graphics and Interactive Techniques*, 359–364.

Van Walsem, R. 2014. "From Skin Wrappings to Architecture: The Evolution of Prehistoric, Anthropoid Wrappings to Historic Architectonic Coffins/sarcophagi; Separate Contrasts Optimally Fused in Single Theban 'stola' Coffins (±975–920 BC)." In Sousa Rogério (ed.), *Body, Cosmos and Eternity: New Trends of Research on Iconography and Symbolism of Ancient Egyptian Coffins*, 1–28. Oxford: Archaeopress.

Vinson, S. 1998. "Remarks on Herodotus' Description of Egyptian Boat Construction (II, 96)." *Studien Zur AltÄgyptischen Kultur* 26: 251–260.

Wiedemann, A. 1890. *Herodots zweites Buch mit sachlichen Erläuterungen*. Leipzig.

Willems, H., H. Creylman, V. De Laet and G. Verstraeten 2017. "The Analysis of Historical Maps as an Avenue to the Interpretation of Pre-Industrial Irrigation Practices in

Egypt" in H. Willems, and J.M. Dahms. (eds.), *The Nile: natural and cultural landscape in Egypt. Proceedings of the International Symposium held at the Johannes Gutenberg-Universitat Mainz, 22 and 23 February 2013*. Transcript Verlag.

Wilson, J.A. 1969. "Egyptian myths, tales and mortuary texts" in James B. Pritchard (Ed.), *Ancient Near Eastern Texts Relating to the Old Testament*, Princeton.

Abbreviations

BAR J.H. Breasted (Ed.), *Ancient Records of Egypt*. Chicago 1906–1907.

CHAPTER 5

"MYTHOPHOR": A Digital Tool for the Collection and Analysis of Mythical Metaphor in Ancient Egypt

Katja Goebs

Abstract

The paper introduces MYTHOPHOR, a digital tool for collecting and analyzing materials relating to the use of mythical metaphor in ancient Egypt. Basing itself on recent findings in both, the study of myth and in Cognitive Science, MYTHOPHOR endeavors to highlight the significance of mythical imagery for human imagination, representation, and thought by—among other things—contextualizing and interpreting evidence for divine metaphor in ancient Egypt in line with theories in metaphor and semantic field studies. The MYTHOPHOR database is structured with a view to answering questions such as what kind of metaphors, similes, or analogies make use of divine imagery, which deities and myths are most often invoked in them, and which specific aspects of these deities predestine them to function as the basis of figurative speech. Besides explaining the structure and use and of the database, the paper also presents some preliminary results.

Keywords

Myth – Metaphor – Figurate Speech – Semantic Fields – Classification in Cognition

Myths are a central feature of cultural expression and tap into the human mind's inherent tendencies—and in fact need—to classify, model, and narrate.[1] They moreover play an important role in representing and explaining the world—in Egypt as well as in essentially all other cultures around the globe.[2] Both these functions, I would argue, find expression in the use of mythical actors or other mythical materials in visual language and metaphor. As

1 Goebs 2013, esp. 132–133.
2 Goebs and Baines 2018, esp. 676–677.

exemplified by the use of divine figures as classifiers in the determinative system,[3] the characteristics associated with particular deities on the basis of their appearance and actions in the mythical stories told about them predestined such divine actors, as well as the objects and situations associated with them in the narratives, to function—in line with a system originally outlined by Charles Sanders Peirce—as mental icons, indices, and symbols.[4] As such, they were able to carry and convey a wide range of culturally important information, which could be evoked simply by referring to such a deity—their physique, mythical activities, functions, and derived epithets—in text, image, and the spoken word.[5]

While conceptual metaphor has been a major subject of research in a number of disciplines especially since the 1980s, when George Lakoff and Mark Johnson first published their foundational work *Metaphors we Live by*,[6] there is a relative dearth of metaphor studies in Egyptology.[7] What is more, divine metaphor has not to date been the focus of an in-depth investigation. While Hermann Grapow, in his necessarily rudimentary 1924 volume on the *Bildlichen Ausdrücke des Ägyptischen*, collected a number of divine metaphors, his analysis did not advance beyond some rather general observations. Thus, he concluded that, even though an analogy with a specific Egyptian deity as expressed in texts could at times be based in aspects that were perceived as characteristic for the divine persona, it was in most cases evocative of very general traits that seem to fit essentially every god (Grapow 1924, 179).[8] Since Grapow's time, many more textual and iconographic sources have become available, and these form the basis for the MYTHOPHOR database, developed as part of a project entitled "Myth, Metaphor, Mythical Thinking—Functions and Uses of Myth in Ancient Egypt" currently underway at the University of Toronto. Drawing on recent

3 See e.g. Goldwasser 1995; 1999; McDonald 2002; 2007.

4 E.g. Peirce 1885: W5, 243; Peirce 1903. A contribution outlining how mythical actors may function in this system is in preparation.

5 Earlier arguments in this regard: e.g. Englund 1989; Frandsen 1997; also Kurth 1977 for the use of secondary deities as "classifiers" in determining specific functions of a superordinate deity in so called "hyphenated", syncretistically fused gods and goddesses such as Hathor-Tefnet.

6 Lakoff and Johnson 1980; see also Lakoff and Turner 1989; Glucksberg et al. 1997.

7 Exceptions are Hsu 2017, who focuses on metaphor in royal inscriptions, and Di Biase-Dyson 2016, 2018 who has explored a number of metaphors as found in particular in Late Egyptian literary texts; some singular concepts are covered e.g. in Haykal 1994, Herrmann 1954, Hsu 2014, Nyord 2017. A survey is presented in Di Biase-Dyson 2017.

8 Thus, the concept of youth is stated to be most commonly associated with the gods Khepri and Horus, intelligence as predominantly attributed to Ptah, Shu, or Thoth, while strength is seen as embodied in particular by Min, Montu, and Horus (see Grapow 1924, 179 for further examples).

findings in both the study of myth and Cognitive Science, this project endeavors to highlight the significance of mythical imagery for human imagination, representation, and thought by collecting, contextualizing, and interpreting evidence for divine metaphor and other figurative language in ancient Egypt. The sources mined for pertinent data range from the earliest texts displaying these traits, the Pyramid Texts, to materials dating to the end of the Late Period, i.e. covering a time span from c. 2350 BCE to c. 332 BCE. In the current phase of the project, Demotic texts are not included, however. Data are sourced by consulting publications of primary texts in conjunction with, where available, translations, and augmented by tools such as the *Thesaurus Linguae Aegyptiae* of the Berlin-Brandenburgische Akademie der Wissenschaften.[9]

MYTHOPHOR and its parent-project are funded by an Insight Grant of the Canadian Social Sciences and Humanities Research Council that is envisaged to run until March 2023. The database structure was developed by Katja Goebs (PI) and Andrew Murphy[10] and programmed by Andrew Murphy, who is also in charge of maintenance and updates. The data is stored in a PostgreSQL database, while the website itself is written in HTML, PHP and Javascript. In March of 2020, the system was transferred to the University of Toronto server, which is going to be its final host and from where the index will eventually be made accessible to the public.[11] Data entry is executed by Katja Goebs and graduate students at the University of Toronto's Department of Near and Middle Eastern Civilizations.[12] Implications of the project's results from the perspective of Cognitive Science are reviewed by psychologist Susanne Ferber, who is a collaborator on the grant.[13]

The MYTHOPHOR database is the project's primary tool for collecting and analyzing pertinent materials relating to divine and mythical metaphor and other figurative language. It is structured with a view to answering questions such as:

– What kind of conceptual metaphors make use of divine imagery and in which linguistic metaphors are they expressed?
– Which deities and myths are most often invoked in metaphor, or in related figures of speech/visual language?[14]

9 http://aaew.bbaw.de/tla/—a part of the project "Strukturen und Transformationen des Wortschatzes der ägyptischen Sprache".

10 katja.goebs@utoronto.ca; andrew.murphy@utoronto.ca.

11 Final URL to be confirmed.

12 At the time of submission, these include Carla Mesa-Guzzo, Jordan Furutani, and Jun-Yi Wong; supporting work was moreover completed by Undergraduate Student Jordan Nemec.

13 ferber@psych.utoronto.ca.

14 At this stage of the project, analogies, metaphors, metonymies, synecdoches, and similes

- Which specific aspects of these deities predestine them to function as the basis of metaphor?
- Which genres of texts are particularly rich in mythical metaphors, and in what types of contexts do they most commonly appear within these texts?
- Can an evolution in the use and forms of divine metaphor be observed over time?
- What can the uses of mythical metaphor / figurative language tell us about the Egyptian view, and in particular classification, of aspects of the world, and how may we relate the Egyptian approach to other, potentially even modern ones?

MYTHOPHOR is organized around **Metaphorical Phrases**,[15] which are partially (pre-)analyzed at data-entry level by identifying the underlying **Myth**, **Mytheme(s)**,[16] and **Deities**[17] upon which the figurative expressions occurring in the phrase are based. Each Metaphorical Phrase is moreover attributed to one or more **Semantic Fields**, and one or more conceptual **Metaphors** (**Superordinate** and **Subordinate**) are identified as being expressed in each phrase. Each of the cited analytical items have their own table, in which further information regarding the individual entries is stored and which is augmented any time a new Metaphorical Phrase is entered. Moreover, for each Metaphorical Phrase, all **Lemmata** (with associated data) occurring in it are recorded, as well as—for the **Source** in which it occurs—the title, text type (e.g. "medical text"), format/script (e.g. "Papyrus, Hieratic (middle)"), date (e.g. "Amenemhat III" or, more broadly, "12th Dyn" or "MK"), and provenance (e.g. "Thebes, West Bank—Tomb Context"). This structure permits, besides simple searches for individual **Metaphorical Phrases**,[18] **Lemmata**, or **Deities** (through the "**Search**" button on

referring to deities are collected and differentiated only as part of the respective item's entry in the **Metaphor** table (for which see below).

15 As stated (fn. 14), expressions including other items of figurative speech referring to a deity, such as simile or synecdoche, are included under this header.

16 The term "mytheme" is here used to denote mythical episodes—segments or components of a full-blown mythical story (= "myth"). See e.g. Goebs 2002, esp. 32ff (not in bibliography); Goebs and Baines 2018, 646–647, where alternative terminology for this concept is discussed.

17 In line with the hypothesis underlying the project, that mythical metaphor can reveal the classifying and categorizing function and use of mythical actors, the classifying determinative used in the writing of the deity is also recorded.

18 Searches for **Metaphorical Phrases** can be executed by typing any lemma or word contained in a given phrase, in either English or Egyptian transliteration (following the rules of the *Manuel de Codage*, as outlined at http://www.catchpenny.org/codage/). For example, typing either "Thoth" or "*Ḏḥwty*" into the search box for **Metaphorical Phrases** results in the same output.

the start page), combined searches (option "**Advanced Search**") that allow to answer a number of potential research questions.

The following three examples with screenshots represent a few sample searches executed in MYTHOPHOR in October 2019.

1 Advanced Search (1)

Search for all **Metaphorical Phrases** in a single source:
Add Query: **Source** = "Seti I"[19]

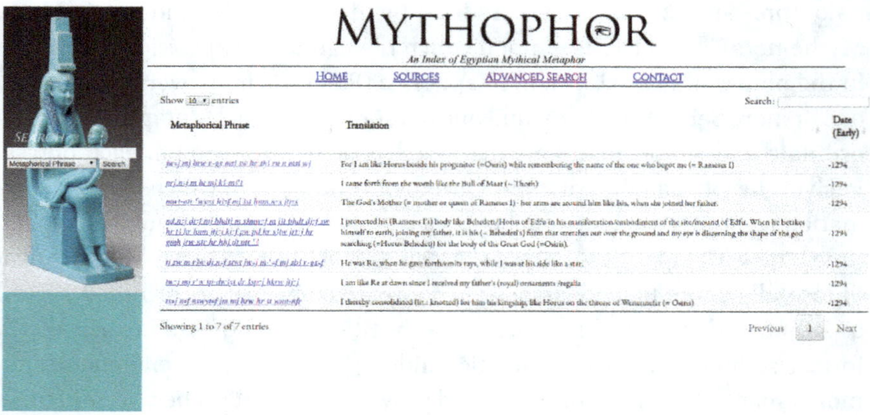

FIGURE 5.1

The results show that this text contains seven **Metaphorical Phrases**—each can be clicked to reveal further associated information (for examples see below).

2 Advanced Search (2)

Search for all **Superordinate Metaphors** featuring:
Add Query: **Primary Deity** = "Eye of Re"
AND
Add Query: **Date** = "MK"

19 In MYTHOPHOR the abbreviated title for the Stela of Seti I for his father Rameses I, from the latter's mortuary temple at Abydos (Schott 1964; *KRI* I, 110–114; *PM* VI, 31, 33).

At present, this search yields four **Superordinate Metaphors** fitting these criteria:

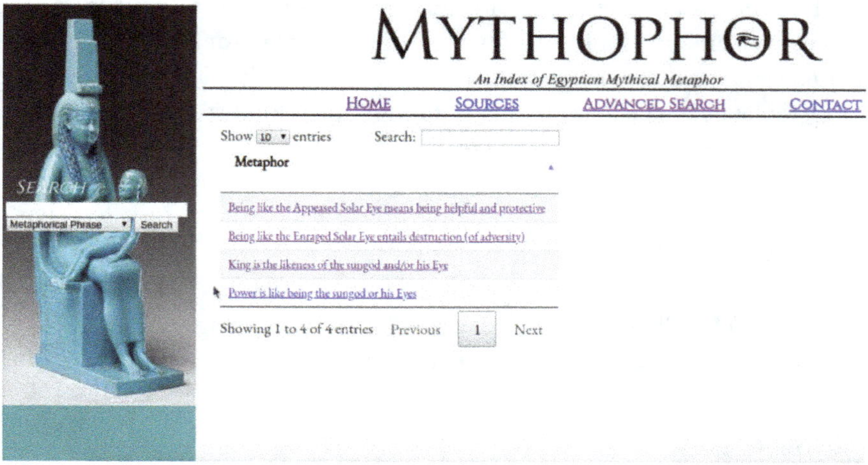

FIGURE 5.2

By clicking on any one of these conceptual metaphors, e.g. [*Being like the Appeased Solar Eye means being helpful and protective*], additional information about the type of metaphor, its associated **Subordinate Metaphors, Semantic Fields**, and **Deities**[20] can be found:

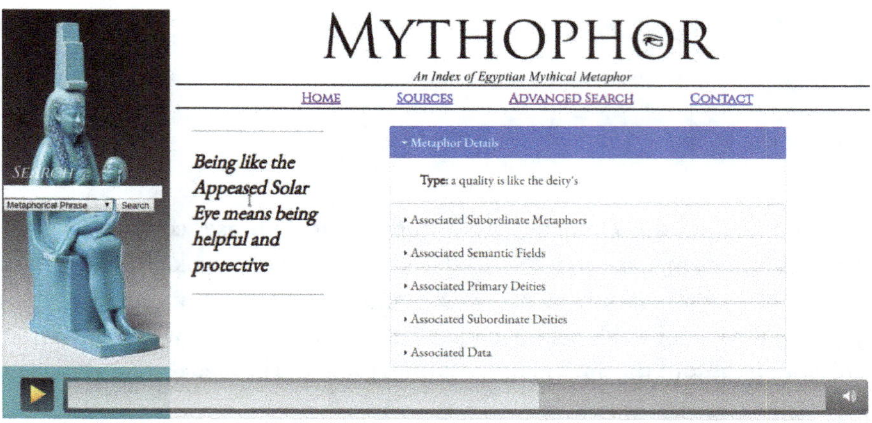

FIGURE 5.3

20 MYTHOPHOR distinguishes "Primary" and "Subordinate Deities", where the former cate-

Each of these categories can be clicked to reveal relevant materials. For example, some of the associated **Semantic Fields** include: *Kingship and Rule*, *Maat—Maintaining Order / Justice / Propriety*, and *Motherhood*; a click on any of these opens another table with information and associated data for the respective **Semantic Field.** The final field on the **Metaphor** results page, **Associated Data**, moreover permits the user to display the data stored in any of the other tables as associated with this **Metaphor**, such as all associated **Metaphorical Phrases, Sources, Lemmata**, etc.

3 Advanced Search (3)

Search for all **Metaphorical Phrases** included in a single source:
Add Query: **Source** = "Eloquent Peasant"

FIGURE 5.4

In this text, there are a total of twelve **Metaphorical Phrases** containing mythical figurative language (ten are displayed here). Several evoke the moon-god Thoth, in charge not only of the waxing and waning moon and time-keeping but also the god of wisdom and justice. This role is especially well known from representations of the judgment of the dead, where Thoth oversees the weigh-

gory would represent a god like Horus, while the latter is used for subforms that are either derived from specific functions or mythical roles of a deity (e.g. Horus Son of Isis / Harsiese) or are the result of syncretistic fusion (e.g. Horus-Sopdu).

ing of the heart process and records the outcome.[21] As such, he represents the perfect divine persona to inform mythical metaphors in a text that has the despondent protagonist fighting for his right against a corrupt official by using ever more colorful, and figurative, language. A good example is the third phrase displayed on the results page—an invocation of the unsympathetic high official Rensi in terms of the tools-of-the trade, and thus symbolic mythical objects, of Thoth:

> *ʿr šfdw gstj Ḏḥwty ḥr.t(j) r jrt jyt*
>
> (You) reed (pen), papyrus roll, and scribal palette of Thoth—far be it from you to do evil![22]

A click on this phrase displays additional information, including the determinative used to write the divine name Thoth, the exact section of the text where the phrase appears, and a brief outline of the context:

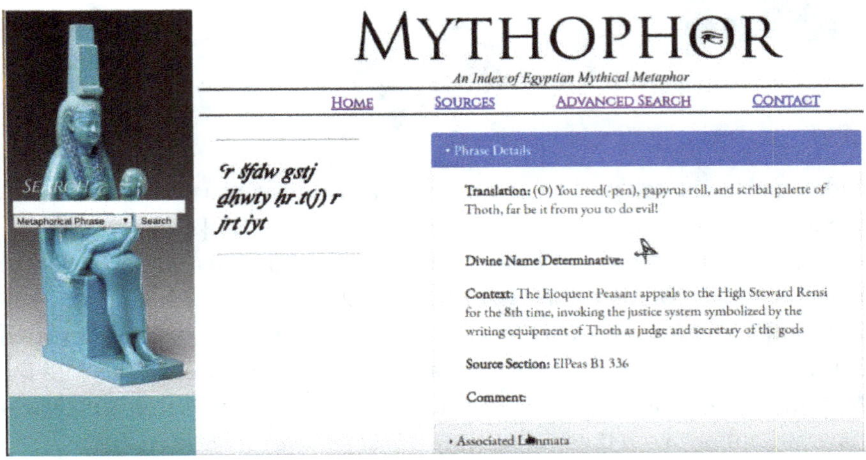

FIGURE 5.5

---

21 For a brief survey of the many roles and functions of Thoth see Stadler 2012; Id. 2009 for a more extensive account.

22 Freely rendered; more grammatical translation: "you should be far/removed from doing evil." See also Herrmann 1954 for this phrase.

A click on categories of further associated materials can reveal, for example, all Egyptian **Lemmata** included in the phrase:

FIGURE 5.6

Individual lemmata can then in turn be clicked to reveal possible translations, dictionary and grammatical information, and further links—to display, for example, all **Metaphorical Phrases** that contain this lemma. Selecting **Source** displays all information relating to the text in question, including text carrier, script, date, provenance, and publications, while a click on **Associated Data** permits to search for any other category of materials associated with the **Metaphorical Phrase**. Below is shown the example of associated **Superordinate Metaphors**, with the result (Being) [*Just and wise as Thoth*]:

FIGURE 5.7

A click on that conceptual **Metaphor**, in turn, renders visible or otherwise accessible (by selecting the appropriate choice from the drop-down menu) all data—of the sort discussed above for the other searches—associated with this metaphor.

In sum, the above description and sample searches in MYTHOPHOR are intended to introduce the reader to this digital tool and its potential for future research on Egyptian myth, religion, and literature, but also on Egyptian systems of classification and, thus, cognition. The system permits, among other things, to evaluate materials according to questions:

(a) Relating to content—in searches for specific deities, myths, or other objects,
(b) Relating to format—in searches for text type, text carrier, or script,
(c) Relating to diachronic development—in searches involving date/reign, or
(d) Relating to geography—in searches involving provenance.

MYTHOPHOR moreover presents a relatively exhaustive collection of an important subsection of Egyptian figurative language and as such should be of interest to scholars of Egyptian and other literature and rhetoric. What is more, the system may be used to illustrate and highlight differences and overlaps between Egyptian classification and that of other cultures—as expressed in conceptual metaphor and semantic fields.

Works Cited

Di Biase-Dyson, Camilla. 2016. "Wege und Abwege: zu den Metaphern in der ramessidischen Weisheitsliteratur." *Zeitschrift für ägyptische Sprache und Altertumskunde* 143 (1): 22–33.

Di Biase-Dyson, Camilla. 2017. Metaphor. In UCLA *Encyclopedia of Egyptology* 1 (1). https://escholarship.org/uc/item/4z62d3nn accessed 12 September 2020.

Di Biase-Dyson, Camilla. 2018. "The figurative network: Tracking the use of metaphorical language for 'hot' and 'cold' in Ramesside literary texts." In *The Ramesside period in Egypt: studies into cultural and historical processes of the 19th and 20th dynasties*, edited by Sabine Kubisch and Ute Rummel, 33–43. Berlin; Boston: Walter de Gruyter.

Englund, Gertie. 1989. "Gods as a Frames of Reference: On Thinking and Concepts of Thought in Ancient Egypt." In *The religion of the ancient Egyptians—cognitive structures and popular expressions. Proceedings of symposia in Uppsala and Bergen, 1987 and 1988* edited by Gertie Englund, 7–28. Uppsala: Almqvist & Wiksell Internat.

Frandsen, Paul John. 1997. "On categorization and metaphorical structuring: Some remarks on Egyptian art and language." *Cambridge Archaeological Journal* 7:71–104.

Glucksberg, Sam, Manfredi, Deanna Ann, McGlone, Matthew S. 1997. "Metaphor comprehension: How metaphors create new categories." In *Creative thought. An investigation of conceptual structures and processes*, edited by Thomas B. Ward, Smith, Steven M., Jyotsna, Vaid, 327 ff. Washington, D.C.: American Psychological Association.

Goebs, Katja. 2013. "Egyptian *mythos* as *logos*. Attempt at a redefinition of 'mythical thinking'." In *Decorum and experience. Essays in ancient culture for John Baines*, edited by Elizabeth Frood and Angela McDonald, 127–134. Oxford: Griffith Institute

Goebs, Katja and John Baines. 2018. "Functions and uses of myth." *Revue de l'Histoire des Religions* 235 (4): 645–681.

Goldwasser, Orly. 1995. *From icon to metaphor: studies in the semiotics of the hieroglyphs*, Orbis biblicus et orientalis 142. Fribourg and Göttingen: Universitätsverlag; Vandenhoeck & Ruprecht.

Goldwasser, Orly. 1999. "The determinative system as a mirror of world organization." *Göttinger Miszellen* 170: 49–68.

Grapow, Hermann. 1924. *Die bildlichen Ausdrücke des Aegyptischen: vom Denken und Dichten einer altorientalischen Sprache*. Leipzig: J.C. Hinrichs.

Haikal, Fayza. 1994. "L'eau dans les métaphores de l'Égypte ancienne". In *Les problèmes institutionnels de l'eau en Égypte ancienne et dans l'antiquité méditerranéenne: Colloque* AIDEA *Vogüé 1992*, edited by Bernadette Menu, 205–211. Cairo: Institut français d'archéologie orientale.

Herrmann, Siegfried. 1954. "Steuerruder, Waage, Herz und Zunge in ägyptischen Bildreden". *Zeitschrift für ägyptische Sprache und Altertumskunde* 79: 106–115.

Hsu, Shih-Wei. 2014. "The images of love: The use of figurative expressions in ancient Egyptian love songs". *Orientalia* 83(4): 407–416.

Hsu, Shih-Wei. 2017. *Bilder für den Pharao: Untersuchungen zu den bildlichen Ausdrücken des Ägyptischen in den Königsinschriften und anderen Textgattungen*. Vol. 36, Probleme der Ägyptologie. Leiden; Boston: Brill.

Kurth, Dieter. 1977. "Götter determinieren Götter." *Studien zur Altägyptischen Kultur* 5: 175–181.

Lakoff, George, and Mark Johnson. 1980. *Metaphors we live by*. Chicago: University of Chicago Press.

Lakoff, George, and Mark Turner. 1989. *More than cool reason: A field guide to poetic metaphor*. Chicago: University of Chicago Press.

McDonald, Angela. 2002. "Animal metaphor in the Egyptian determinative system: three case studies." Doctoral thesis, University of Oxford.

McDonald, Angela. 2007. "A metaphor for troubled times: the evolution of the Seth deity determinative in the First Intermediate Period." *Zeitschrift für ägyptische Sprache und Altertumskunde* 134: 26–39.

Nyord, Rune. 2017. "Analogy and metaphor in ancient medicine and the ancient Egyptian conceptualization of heat in the body." In *The comparable body: Imagination and analogy in ancient anatomy and physiology*, edited by John Wee, 12–41. Vol. 49, Studies in Ancient Medicine. Leiden: Brill.

Peirce, Charles Sanders. 1885. *One, Two, Three: Fundamental categories of thought and of nature*. Robin Catalogue MS [R] 901. http://www.commens.org/bibliography/manuscript/peirce-charles-s-1885-one-two-three-fundamental-categories-thought-and, https://rs.cms.hu-berlin.de/peircearchive/pages/view.php?ref=29632, accessed 15 June 2020.

Peirce, Charles Sanders. 1903. *Logical tracts. No. 1. On existential graphs*. Robin Catalogue MS [R] 491 http://www.commens.org/dictionary/entry/quote-logical-tracts-no-1-existential-graphs-3, accessed 15 June 2020.

Schott, Siegfried. 1964. "Der Denkstein Sethos I. für die Kapelle Ramses I. in Abydos." In *Sitzungsberichte der Akademie der Wissenschaften Göttingen* 1964: 1–84.

Stadler, Martin Andreas. 2009. *Weiser und Wesir. Studien zu Vorkommen, Rolle und Wesen des Gottes Thot im ägyptischen Totenbuch*. Tübingen: Mohr Siebeck.

Stadler, Martin Andreas. 2012. Thoth. *UCLA Encyclopedia of Egyptology*, 1(1). https://escholarship.org/uc/item/2xj8c3qg, accessed 10 September 2019.

Van den Berg, Hans. 1997. " 'Manuel de Codage'—A standard system for the computer-encoding of Egyptian transliteration and hieroglyphic texts", http://www.catchpenny.org/codage/ accessed 17 June 2020.

CHAPTER 6

Mapping the Ancient Mind: *iClassifier*, a New Platform for Systematic Analysis of Classifiers in Egyptian and beyond

Haleli Harel, Orly Goldwasser and Dmitry Nikolaev

Abstract

The research effort behind the creation of the *iClassifier* digital platform is the accurate description of the emic mental landscape of ancient Egypt, as reflected in the *classifier system* of the Egyptian script. Our central hypothesis is that each *semantic classifier* in the script system heads a conceptual category.[1] Following this hypothesis, collecting as many lemmas, tokens, and their classifiers as possible will present us with an estimated, expanding map of the *emic categories in the ancient Egyptian mind*. Classifiers allow us to trace central members and marginal members, interrelations, and diachronic developments in each emerging category, taking into consideration classifier combinations, inverse correlations, and the lack of classifier use. Since classifiers are evident in the Egyptian script from its early stages, through Demotic, and up until the demise of Ptolemaic, the overall corpus should comprise millions of examples. The digital platform *iClassifier* is designed for large-scale data collection, analysis, and study of classifiers in complex scripts.

Keywords

Egyptian Script – classifiers – determinatives – computer analysis

1 State of the Art: Classifier Studies in Egyptology

Understanding the essence of Egyptian classifiers was a necessary step in the decipherment of the Egyptian script. Jean-François Champollion was the first to have acknowledged and defined the essence of the signs playing a role he

1 Goldwasser 2002, and 2003, and recently Winand 2021, 12.

described as *déterminatif*. He then offered a detailed survey of these unpronounced signs that head generic categories in the Egyptian script.[2] Since Champollion's publication in 1836, the term "determinative" has been continuously used by Egyptologists to refer to signs playing this functional role in the Egyptian script. Determinatives were regarded by post-Champollion Egyptological research as silent "reading-aids." Egyptologists considered them as informative, associative, meta-linguistic additions, optionally added to words to help the reader choose the correct meaning from a few possible signified concepts represented by the Egyptian vowelless script. Their role as word dividers (played in other early alphabetic scripts by strokes, dots, or spaces) in a text was also acknowledged. Egyptologists had always recognized the apparent semantic value of the determinatives, but they were never regarded as a rule-governed linguistic system.

In the two recent decades, research exposed how the phenomenon of "determinatives" shows up in various complex writing systems: Cuneiform, Anatolian hieroglyphs, Chinese (ancient and modern), and others.[3] Moreover, it gradually became clear that determinatives play a functional role very similar to the well-known pronounced classifier systems in hundreds *of classifier languages around* the world, e.g., Mesoamerica, East Asia, or Australia.[4] In Egyptology, the study of Egyptian determinatives as graphemic classifiers has developed as a new research field. It took a concerted effort to set the grounds to study this unique script phenomenon as a linguistic phenomenon on its merit and to question the possible "semantic domain" or category each semantic classifier governs.[5] A theoretical framework was gradually established[6] and applied in

---

2 Probably under the influence of Chinese studies, see Lefebvre 1955, 18, n. 6. On Champollion's definitive, long-forgotten contribution to the role of classifiers, see Goldwasser 2006a, 17–20. For a recent contribution on this topic, Polis 2018. For an overview of classifiers in the Egyptian scripts, Goldwasser, in press a.

3 Selz, Grinevald, Goldwasser 2018; Selz and Zhang, in press; Payne 2017, for Chinese, Chen 2016; Handel, in press and Myers 2019. For a general survey of the history of classifier studies, see Kilarski 2014 and Bauer 2017.

4 For general discussion on classifiers, see Senft 2000; Kilarski 2013, 319–173; Grinevald 2015; Bisang 2017. For building bridges between pronounced classifier systems and graphemic classifier systems, see Rude 1986; Kammerzell 1999; Goldwasser 2006a; Goldwasser & Grinevald 2012; Lincke & Kammerzell 2012; Nyord 2015; Bauer 2017.

5 For classifiers as markers of "semantic domains," see Denny 1976; Senft 2000, 2.3.2; Croft 1994. For ingenious early steps in this direction in Egyptology, see Te Velde 1967, in his analysis of the Sethian "determinatives."

6 Goldwasser 2002, Werning 2011, 98–110, Lincke 2011, Goldwasser and Grinevald 2012, Lincke and Kammerzell 2012, Polis and Rosmorduc 2015, Kammerzell 2015, Chantrain 2014; 2021.

several case studies.[7] At the same time, while studying this script phenomenon, its similarity to systems of pronounced classifiers became more and more evident. Hence, the suggestion to identify these signs using the word "classifier," a linguistic term more closely reflecting the semantic essence of this sign type and its link to classification and categorization.[8] Thus, graphemic and pronounced classifiers seem to result from the same human categorization effort, emerging in different media. The last decade gave rise to a series of studies about classifiers in sign languages, including a comparative article on the use of graphemic classifiers in Egyptian and the use of classifiers in German sign language.[9] These findings establish classifiers as a unified cognitive and linguistic phenomenon surfacing in three different media: speech, script, and sign language. Once the Egyptian graphemic classifiers came to be regarded as a system, the next step was to establish and define the functional roles of that system. In the last decade, classifiers were recognized as an essential level of analysis in the "semantic-trail" of every Egyptian lexeme.[10] Additionally, scholars surveyed specific lemmas and corpora with linguistic tools, highlighting the diachronic changes in the use of classifiers along the *long durée* of Egyptian written discourse in hieroglyphs, cursive hieroglyphs, and hieratic.[11] These studies strongly support the analysis of classifiers as reflecting semantic categories.

Up to this day, Gardiner's "Determinative List,"[12] published in 1928/1957, is followed as a reference for the classifier corpus, and an up-to-date comprehensive classifier list is yet to be created. An early attempt to list sign functions per hieroglyphic sign was made in Berlin by J. Spiegel und E. Lüddeckens during the 1930s and published as *Zeichenliste des Wörterbuches der ägyptischen Sprache*.[13] While impressive for their time, these ingenious lists lack consideration of classifiers' frequency, category size, or diachrony.

---

7 E.g., Goldwasser 1999, 2002, 2005, 2006b, 2010, 2017; Shalomi-Hen 2000, 2006, 2008; David 2000; Müller 2002; Allon 2007; 2010; Werning 2011, 98–110 and 323–326; Lincke 2015; Pommerening 2017.

8 A result of a series of collaborative research projects, starting from "*Classifiers and Classification in Ancient Egypt*" (1997–1999) sponsored by the Niedersächsisches Vorab der Volkswagen Stiftung. PIs Orly Goldwasser, Friedrich Junge and Frank Kammerzell. Our most recent research project will be discussed below.

9 Lincke and Kutscher 2012.

10 See Polis and Winand 2015; Grossman and Polis 2012.

11 See Lincke 2011; Werning 2011, 98–110 and 323–326; Peust 2012; Chantrain 2014, 2021; Kammerzell 2015; Winand 2016, 2019, Pommerening 2017 and others.

12 Gardiner 1957, 438–548.

13 Accessible at: http://aaew.bbaw.de/archive/berliner-zeichenliste-1935-39

A new standard for sign-function annotation is offered in the *Thot Sign List* (TSL), a project that aims at creating a digital attestation record of ancient Egyptian hieroglyphs in their various sign functions. "Classifier" as a sign function is marked in the new datasets.[14] Alongside traditional Gardiner sign numbers, TSL entries refer to a Gardiner number and a comprehensive glyph list created by Dimitri Meeks.[15] For later stages of the Egyptian script, the Demotic Paleographical Database Project digitizes signs in the role of classifier in Demotic texts.[16]

Over the last two decades, linguists have examined the impact of pronounced classifier systems on speakers of various classifier languages. One of the research tracks follows how classifiers affect the performance of cognitive tasks. These studies show a constant tension between the language-specific and cross-linguistic effects tied to classifier assignment.[17] Experimental studies are also conducted, e.g., studies of classifier languages have demonstrated that the speaker considers objects belonging to a particular classifier category more similar.[18] As linguists continuously collect data for modern classifier languages, we suggest addressing similar questions by mapping the Egyptian classifier system and looking at how this system reflects culture-specific vs. universal categorization schemes.[19]

2 Primary Goals: Systematic Transliteration of Classifiers and a Modern Classifier List

As of today, there is no standard technique in Egyptology for transcribing classifiers. Unlike other ancient complex scripts (e.g., Cuneiform),[20] most transcriptions of Egyptian texts have neglected the marking or glossing of classifiers.

---

14 Polis and Rosmorduc 2015; Hafemann 2018.

15 Meeks 2013.

16 Attestation-based classifier lists for Demotic are created by the Demotic Palaeographical Database Project (DPDP), available at http://129.206.5.162/beta/index.html and a preliminary classifier list has been compiled by the Demotic Word List database, accessible at: https://www.dwl.aegyptologie.lmu.de/index.php.

17 See Zhang and Schmitt 1998; Saalbach and Imai 2007.

18 Kemmerer 2017; for the links between object concepts and classifier categories, see Zhang and Schmitt 1998; Saalbach and Imai 2010, 2012. For the grammatical essence of classification on a continuum of nominal gender marking, see Corbett and Fedden 2016.

19 For a survey on the typology of classifier systems, see Grinevald 2015. The first steps in the study of culture-specific versus universal categorization were taken already in Goldwasser 2002, 2003, and recently in Goldwasser, in press b.

20 For example, ePSD (University of Pennsylvania) records sign functions in Sumerian in its

Some older valuable lexicographical resources listing lemmas and compounds sometimes even omit the classifiers completely from their hieroglyphic transcription (e.g., Ranke 1935). In some recent publications, when transcribed, they may be marked in an upper case by their "Gardiner no." from Gardiner's general list of hieroglyphic signs, or by transcribing in upper case their reconstructed semantic value as analyzed by the scholar.[21] Just as transliterations of Egyptian are not unified across different academic schools of Egyptology, the transcription of sign functions does not yet adhere to a unified method.[22]

This situation deprives the student of Egyptology of a very substantial layer of information that appears in most Egyptian lemmas—verbs, nouns, and sometimes adverbs and prepositions. Moreover, this partial representation of lemmas makes it impossible to reconstruct the original spelling of the lemma from the transliteration.[23] Should the method of transliteration be changed, it would save students and researchers precious time, freeing them from dependence on libraries or digital resources, which in many cases are incomplete or not available at all. As not all students of Egyptology always have access to original publications in libraries, this classifier marking target is of great importance.

Egyptology is in dire need of a user-friendly, modern list of classifiers, ranked by frequency and co-occurrence. The pioneering list of Gardiner[24] offers no clue to the student of Egyptian if a certain hieroglyph is of very high frequency in the system or low frequency (e.g., A2 𓀁 vs. A9 𓀋) or if a classifier heads a category of many members or only a few members. One can create a current list by marking classifiers for each example (tokens) of a lemma (or "host" word). Then all classified meanings (CMs), calibrated statistically by their frequency of appearance with the classifier, would define a classifier's semantic scope, variability, and frequency through time. A user of such a resource would get a clear picture of the spectrum of words occurring with a classifier. Gradually, infor-

transcription. Signs that are considered by the compilers as classifiers (semantic or phonetic determinatives) are represented in an upper-case script, see http://oracc.museum.upenn.edu/epsd2/sux.

21 E.g., Schneider (1992), marking in uppercase [F] words with the classifier T14 for the [FOREIGN], or transcribing the phonetic value of a classifier in small caps and uppercase, as in Schneider 2008. For a proposal for an advanced transcription system for classifiers in Egyptian, see Kammerzell 2015.

22 Some advancements are published online in the digital project **Thot Sign List**. For an elaborate proposal for glossing ancient Egyptian, see Di Biase-Dyson et al. 2009.

23 Transliteration also ignore phonological interpretants, for a transliteration that includes phonetic classifiers, see Kammerzell 2015 with discussion.

24 Gardiner 1957, 31–33.

mation about the central and peripheral lemmas in the category will emerge. Such a "reverse dictionary" would also be helpful for philological work of word reconstruction in broken or partially lost inscriptions.

3 How to Reach Our Goals: *iClassifier*, a Collaborative Digital Research Environment

iClassifier is a data collection and analysis tool specifically created to investigate the extensive use of classifiers in their contexts (©Goldwasser, Harel, Nikolaev). The platform aims to serve as a digital space to apply uniform marking to classifiers in Egyptian texts, store and display the results, and offer classifier analysis reports for data of any existing projects.[25] By using this tool, a comprehensive picture of classifier categories in ancient Egypt will gradually take shape. After classifiers are marked, an evolving image of their emergence, distribution, and patterns of variation will emerge, making it possible to put the Egyptian data in a general typological perspective.

In *iClassifier*, each classifier analysis has two axes: the **lemma axis** and the **classifier axis**. The lemma-axis aims to study a host lemma and the range of classifiers it may "take" in synchronic and diachronic perspectives and show variation according to scripts or genres. The classifier axis focuses on a particular classifier. It attempts to collect lexical items that occur as its hosts (e.g., collect all lexical items with a 𓄛 [HIDE & TAIL] classifier).

4 Workflow Guidelines for Data Input in *iClassifier*

The following guidelines outline the main input tasks and research emphasis of *iClassifier*.

4.1 *Witness (Text)*

The user provides information about the text (witness) in which a token appears (see the witness form in fig. 6.1 below, colored green). The user may annotate metadata for each text using the witness form. Possible annotations

25 Data will be published for re-use and linked to TLA, TSL, Ramses, and other relevant datasets. Semantic fields are linked to IDs of the *Concepticon* database 'concept' and 'lexical field.' Each language portal in *iClassifier* offers links to datasets in its area of expertise. By default, each project's data is private, and a user can choose when to publicly share their data reports.

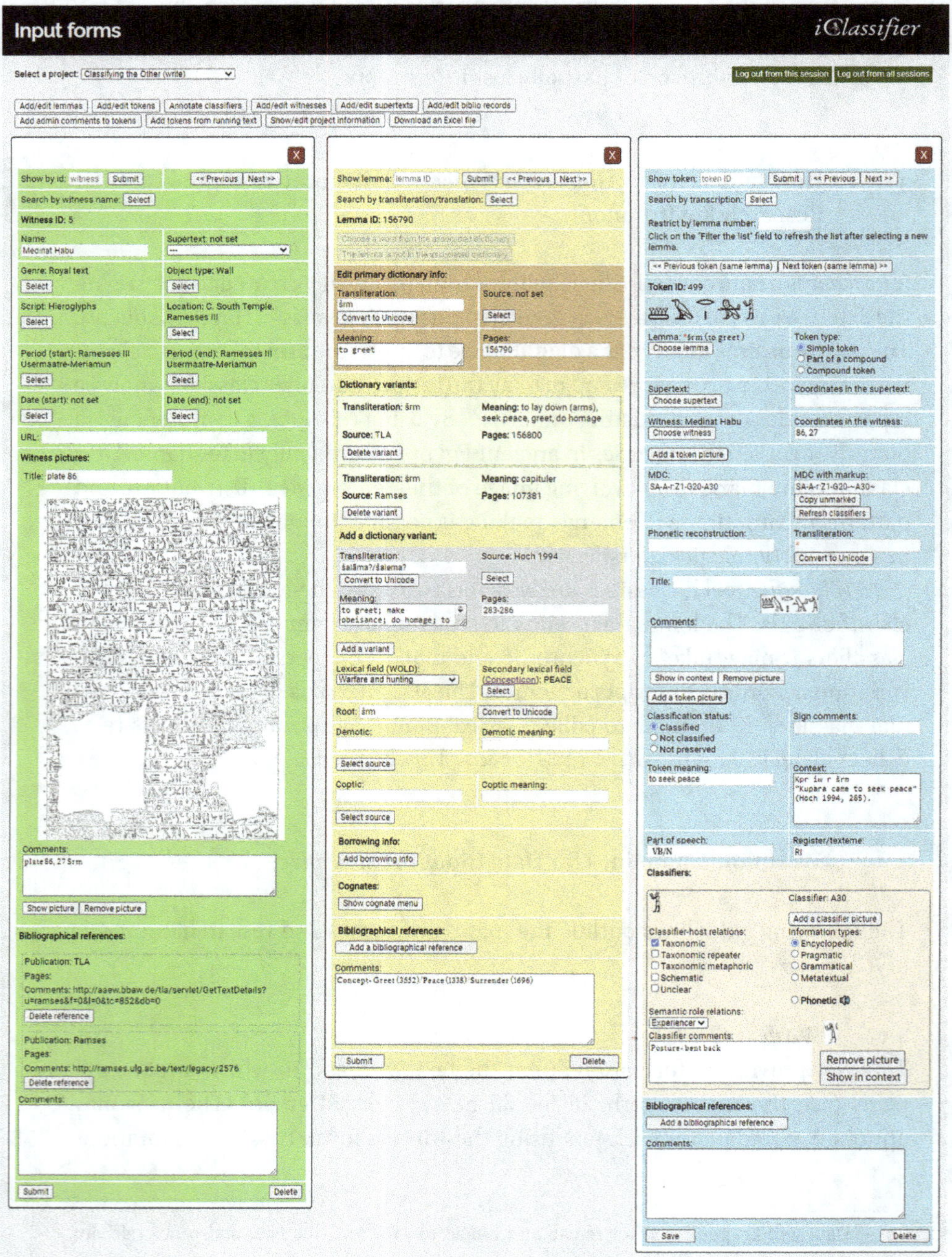

FIGURE 6.1 An example of WITNESS, LEMMA, and TOKEN annotation forms in an instance of the [ADORATION/GREETING] classifier, A30 in Gardiner's list, with the lemma *šrm* (to greet, to seek peace)

are text genre, object type, script,[26] location, and date. The metadata entries feature the values of the THOT thesauri.[27] Later, using queries, users may examine the classification patterns of a specific text (e.g., a particular Coffin Text) or study any variation that was recorded and automatically order their examples by script type, date, geographical location, etc. Additionally, a user can add a complete witness transcribed either in transliteration or Manuel-de-Codage (see below, § 5.4).

4.2 *Lemma* (*Word*)[28]

The *iClassifier* lemma list is based on the TLA lemma list. The user simply starts typing the transliteration or translation and chooses the correct lemma. If a lemma does not appear in the TLA list, the user can add it as a new lemma, referring to another dictionary.

4.3 *Token* (*Example*)

Then, one goes on marking classifiers in a specific token of the lemma. The minimal information for each token is which lemma it attests to, a spelling of the token (its complete transcription in Manuel-de-Codage = MdC), the text and line where it occurs. In each MdC example, the user marks classifier transcriptions with a standardized marking—a tilde symbol (~)—before and after each classifier, e.g., the classifier in the example  (SA-A-r:Z1-G20-A30) is marked as SA-A-r:Z1-G20-~A30~.

Then, an analysis box for each classifier is generated. *iClassifier* offers a uniform platform for the analysis of classifiers.

4.3.1 Classifier Analysis (CA)

This analysis is optional and could be used partially or not at all. Users may leave the analysis to later stages of their work and choose only to mark the classifier/s (or lack thereof).

26 See Van der Moezel 2018 for recent advancements in the codification of hieratic signs.

27 See Polis and Razanajao 2016 for a theoretical framework for recording ancient Egyptian texts' metadata.

28 Or compound. A compound token is marked, and its elements are also analyzed by linking them to a lemma and marking their classifiers, see below in § 5.2.

4.3.2 Classifier-Host Relations

Classifier-host relations define the semantic relationship of a classifier to its host lemma: taxonomic, taxonomic-repeater,[29] taxonomic-metaphoric,[30] or schematic (various, meronymic or metonymic, e.g., "made of," "part of," etc.).[31] If necessary, the user can resort to the "unclear" option. (See in fig. 6.1 above)

4.3.3 Classifier Information Types

iClassifier's "Information types" are meant to distinguish between various classification types. This additional analysis scheme follows distinctions suggested in the "Ebene" analysis by Lincke and Kammerzell 2012 and Lincke 2011 and was adapted to correspond to the terminology of Contini-Morava and Kilarski 2013, used for describing types of classifier functions in classifier languages.

4.3.3.1 *"Encyclopedic/Lexical" vs. "Pragmatic" Classification.*

While the *lexical* meaning suggests that a classifier refers to the encyclopedic meaning of a lemma, the term *pragmatic* refers to cases where a classifier of a specific token of a lemma has a specified meaning in a particular *context*. In such cases, it represents a specific referent in the text. For example, in a historical hieroglyphic text.[32] The lemma *wr*, "ruler" is written with a pragmatic classifier that signifies a particular foreign ruler: . In the original, the hieroglyph presents the Libyan ruler, not as an upright man holding a staff, but a somewhat bent man wearing a long dress. His hands are hanging to his sides. This variation occurs throughout this text as the same hieroglyph is used as a logogram referring to a foreign ruler. In such cases, the user marks the classifier as pragmatic and comment about its particular referent.

4.3.3.2 *Grammatical Classification.*

This grammatical/derivational type, also referred to as "word-form" classification, refers to the additional morpho-syntactic role of some classifiers,[33] e.g.,

29 Cases of unique classifiers should be marked as repeaters. The definition of the unique i.e., a classifier explicitly related to a single referent, entails studying all its occurrences. For unique classifiers, see, Goldwasser and Grinevald 2012, Werning 2011, 100.

30 Goldwasser 2005; Chantrain and Di Biase Dyson 2018.

31 Terms are adapted after Goldwasser 2002; Goldwasser & Grinevald 2012.

32 Medinet Habu, Ramssess III, pl. 27; KRI V, 27, 28. Compare here the discussion in Kammerzell 2015, 1397–1398.

33 Lincke 2011; Lincke and Kammerzell 2012; Werning 2011, 102–104, discussing the term "Grammato-Klassifikator," referring to plural classification and to the classification of pronouns. We mark these sub-types of classification in order to track and study them independently of semantic classification.

in constructions such as names of trades, a classifier marks nominalization, for example, the lemma *wḥꜥ.w*, “fowler,” below.

4.3.3.3 *Metatextual Level*

This level was suggested by Allon 2010.[34] For example, in rare cases, the classifier 𓌙 (T14) or 𓀁 (A2) marks a lemma as “foreign” in Egyptian, i.e., belonging to a foreign language, without any connection whatsoever to its lexical meaning.[35]

4.3.4 Phonetic Classifiers

“Phonetic classifiers” (phonological interpretants) are found in various complex scripts.[36] Marking this information-type will expose the distribution of phonetic classification versus semantic classification in the Egyptian script. It would enable us to survey this script phenomenon with quantitative tools for the first time. Such “phonetic classifiers” have been shown to occur also in classifier languages.[37] Once marked in *iClassifier*, phonetic classifiers would appear as a separate, non-semantic classification list, and their report is excluded from the semantic classification report.

4.3.5 Classifier Valency (Semantic Role Relations)

Users may analyze semantic relations marked by a classifier. Semantic relations apply mainly to verbs but also to deverbals (see the *wḥꜥ.w* example in fig. 6.2 below). Semantic relations may include EXPERIENCER, PATIENT, INSTRUMENT, SOURCE, GOAL, LOCATION, MOVER, ZERO, CAUSEE, and ABSENTEE. The list of semantic relations may expand as data is gathered. Our current list of options (see the analysis of classifier 𓅯 (G41) in fig. 6.2 below) follows the study of semantic relations of verbs and their classifiers in “The Tale of Wenamun,” undertaken in Kammerzell 2015.

4.4 *Image Input*

iClassifier allows users to add images of source inscriptions by uploading an image file from their hard drive and then marking and cropping a specific example. Users must cite the copyright status of an image, and all images offered for reuse will publicly appear in search results.

---

34 See here additional examples for ‘Metasprachlich’ classification in Werning 2011, 103.

35 For another possible type of metatextual classification, see Chantrain and Di Biase Dyson 2018.

36 Werning 2011, 104; Polis and Rosmorduc 2015.

37 see Grinevald in Grinevald & Goldwasser 2012, 46–50.

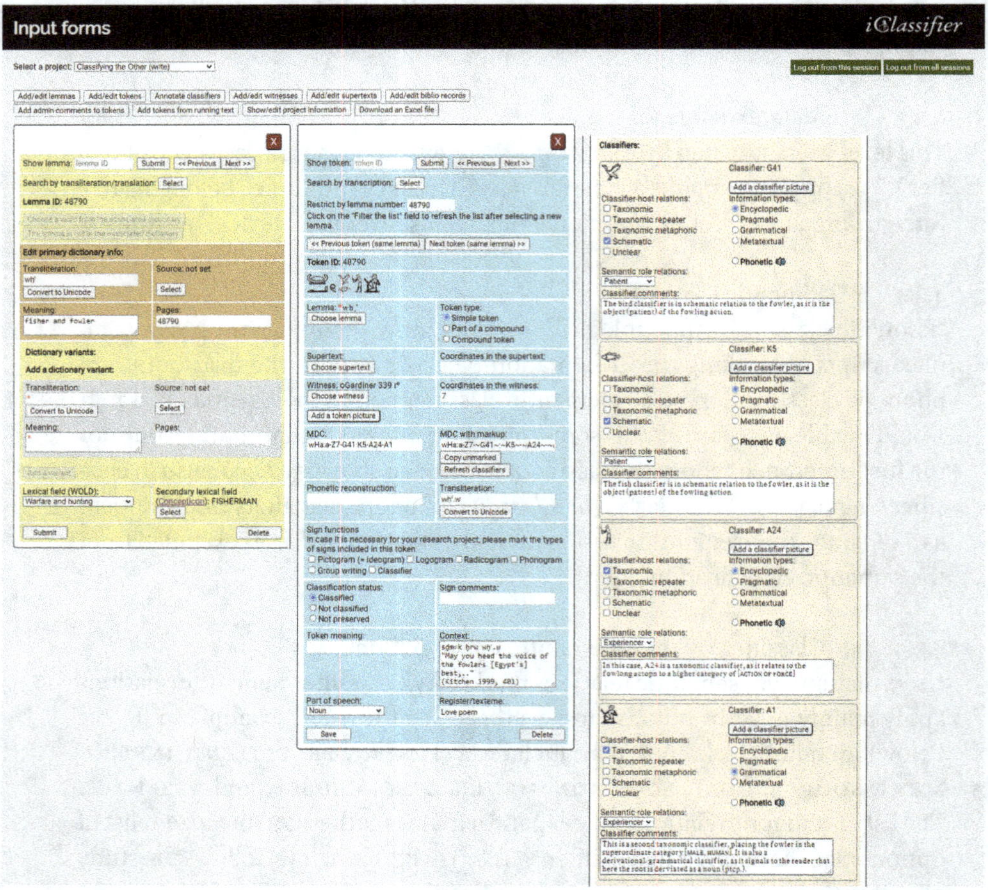

FIGURE 6.2 Marking 'classifier level' and 'classifier type' of a token of *wḥꜥ.w* 'fowler' in *iClassifier* token input form. These analyses are optional in *iClassifier*.

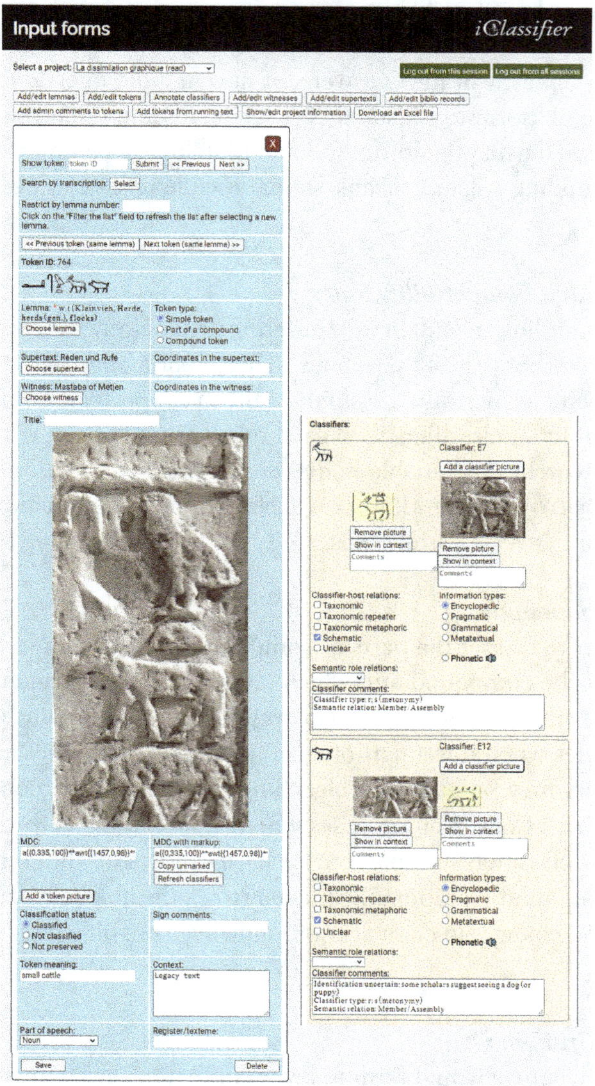

FIGURE 6.3 Sample token with a source image, created by Simon Thuault. An example from the mastaba of Metjen at the Berlin Museum

Note: Photo copyright S. Thuault. See Thuault 2020, 60. For analysis of the classification of "dissemblance graphémique" in Egyptian writing, especially ibid., 261–326. A supplementary study of the data using *iClassifier* allows adding source images to all examples collected for the study and create a digital paleographic inventory.

5 Additional Queries by *iClassifier*

iClassifier offers its users an array of new answers for new and old questions. Researchers may use the platform to pursue their specific research interests. Users conduct their research using the forms and add philological comments for each entry. After compiling a list of tokens, some possible research questions are:

5.1 *Classifier Order and Compatibility Rules*

Cases of multiclassification show a tendency for ordering first metonymic classifiers and then taxonomic ones (e.g., fig. 6.2 above). This phenomenon and its functional rules have been hardly studied systematically in Egyptology. Compatibility vs. incompatibility of classifiers with each other has also been very rarely studied. Complex searches also enable studies of classifier combinations, such as [WATER/BODY-OF-WATER](N35A-N36:N23), and their statistics, compatibility patterns, and inverse correlations.

5.2 *Classifiers in Compounds*

iClassifier allows the user to dissect the parts of compounds. Users can mark which classifiers classify the compound and which classify a lexical element within a compound. This annotation enables us to examine how the classification of a lexical item varies when it is a part of a compound. We hypothesize that the loss of classifiers may point to the bleaching of a compound constituent's original meanings. Conversely, in cases where a compound is "broken" into its components, classifiers may point to lexical awareness of the constituents that make a compound. Additionally, one may track the link between phonological changes and changes in classification, showing a possible loss of word boundaries.

5.3 *Classifiers and Grammar*

Classifiers may point to what we would term today "grammatical awareness." As already noted by Gardiner 1957, in most cases, classifiers appear directly after the root in verbal forms, thus functioning as semantic root markers, e.g. ii.n=(*i*) vb:to come [CL:D54]-pst-1sg.m[CL:A1]-"(that) I came."[38] Note that old (and modern, e.g., Modern Hebrew) Semitic scripts featuring root+suffix systems for verb declination always show the word division (vertical line, dot, or space) after a suffix.[39]

38 Goldwasser and Grinevald 2012, 28.

39 See Werning 2011, 100, 102 fn. 55 and *passim*; Goldwasser and Grinevald 2012, § 2.4.

5.4 *Classifier Distribution across Entire Text and Datasets*

iClassifier features an input scheme in which a user can feed in a complete text, either in Manuel de Codage or transliteration, separating tokens with spaces. The user either adds the tilde signs on both sides of each classifier (e.g., ~A2~) in the transcription or marks classifiers later using *iClassifier*'s token-edit menu. As a result, the user gets a *tokenized version* of their text and can further analyze each token and its classifiers. This feature allows researchers to track classifier tendencies of a specific text or group of texts (e.g., by a certain scribe, from a particular coffin, or geographic area) as well as to compare classifier assignments across texts and datasets. The *iClassifier* platform is currently used to annotate classifiers for fully transcribed and analyzed texts, courtesy of the TLA project, with a pilot analysis of the entire papyrus Ebers.[40]

6 Technical Overview of the System: Database, Website, API

The *iClassifier* annotation is compatible with the TLA data scheme, and users annotate tokens and link them to respective TLA-ids. Classifiers are marked with a tilde sign (i.e., ~CL~) wherever they appear and then further annotated and analyzed within the framework of Classifier Analysis, as described above (§ 4.2). The back end (data storage and web server) of the current version of the database, *iClassifier* BETA (Released May 2020), is implemented as an array of SQLite databases (with a separate database file for each project), with Python (Flask+Gunicorn) and Golang web stack. Its user interface is based on Mithril.js. Maps are drawn using the JavaScript library Vis.js. Hieroglyphs are drawn using Serge Rosmorduc's JSesh library (Java), adapted as an HTTP API (https://github.com/macleginn/jsesh-web), and the database supports Unicode hieroglyphs in most text fields.

7 A Case Study in *iClassifier*: Network Analysis Map of Classifiers of Lexical Borrowings in Egyptian Texts of the New Kingdom

The corpus displayed in fig. 6.4 is a study of approximately 500 lemmas and their tokens, comprising lexical borrowings as well as foreign names and topo-

---

40 The *eBersClassifier* research project is conducted by Prof. Tanja Pommerening and Svenja Stern, thanks go to Lutz Popko and the "Strukturen und Transformationen des Wortschatzes der ägyptischen Sprache" project, Sächsische Akademie der Wissenschaften zu Leipzig.

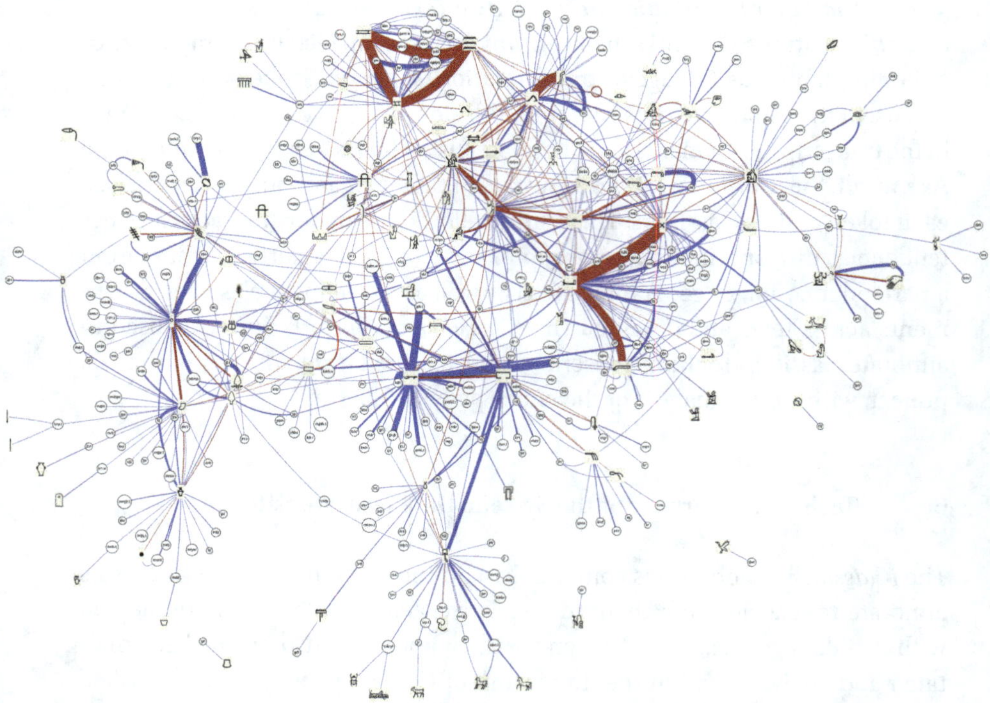

FIGURE 6.4 A Network analysis map of lexical borrowings by their classifiers in *iClassifier*. "Classifying the Other" map. The "big picture" of lexical borrowings and foreign proper names and their classifiers.
Blue lines link a classifier with all lemmas it occurs with.
Red lines link classifiers that co-occur in a specific token
The width of a line represents the number of examples featuring a certain link.
Note: The sample corpus presented here was collected by Haleli Harel as a research corpus for the doctoral dissertation "A network of lexical borrowings in Egyptian texts of the New Kingdom," Hebrew University, Jerusalem, in the framework of an ISF grant 735/17, PI Prof. Orly Goldwasser, titled: "*Classifying the Other: The Classification of Semitic Loanwords in the Egyptian Script.*"

nyms. The lemma list includes lexical items that are possible foreign loans. The corpus follows the list published by Hoch in his *Semitic Words in Egyptian Texts of the New Kingdom and Third Intermediate Period* (1994). As Hoch's list does not include all tokens of his enumerated lemmas, additional tokens are added.[41] Additional examples not included in Hoch are of foreign place names and divine names. Hoch's lemma list was linked to its corresponding entries in the *Thesaurus Lingua Aegyptia* (TLA) and *Ramses* databases via their lemma IDs using *iClassifier*. Transliterations are consonantal after the TLA lemma list.

All classifiers of lexical borrowings and proper names of foreign origin were marked using the *iClassifier* data input scheme (see above, § 4). According to the network map of this specific corpus, one can track immediately classifier communities in the network of categories- e.g., the 𓄛 [HIDE & TAIL] (F27) classifier, the 𓈔 [METAL] (N34) and the 𓏅 [VESSEL] (W7/W6) classifiers.

8 Reading the Network Maps: Center and Periphery

At this stage, the most significant categories in the emerging classifier map of lexical borrowings are containers for commodities, e.g., 𓏅 [VESSEL] (W7), or classification of the commodities themselves by material, e.g., 𓈔 [METAL/MADE OF METAL] (N34), 𓆱 [WOOD/MADE OF WOOD] (M3). The other dominant categories are the classifier 𓌛 [THROW-STICK] (T14) and 𓈉 [FOREIGN LANDS] (N25), which identify toponyms or people as "FOREIGN." These results coincide with historical information and archaeological findings pointing towards strong economic relations between Egypt and Canaan in the New Kingdom. During the Ramesside period, the Southern Levant was rather heavily exploited by the Egyptian administration.[42] Other classifiers and their host words are more isolated "islands" of categorization and occur only with one lexical item. An example is a lexical borrowing meaning "snow," *srḳ*, classified with a 𓇲 [SKY+WATER/RAIN] (N4) classifier.[43] The loanword for "snow" is attested in the records of Ramses II, when the king ponders how his army and of-

41 A comprehensive analysis of the classifiers of lexical borrowings is to appear in Harel, Forthcoming.

42 See most recently Naaman 2020; Goldwasser 2016.

43 TLA lemma no. 139820; Hoch 1994, no. 375, additionally it is attested several times in lexical lists.

ficials will reach the Syrian frontier in days of rain and snow to meet the king's Hittite bride.[44]

An interactive version of the release of the project classifier maps and reports in figs. 6.4–6.9 is accessible at: https://iclassifier.pw/reports#classifyingt heother

8.1 *Glimpses on the Classifier Axis: A Sample Classifier,* 𓍻 (V19) *in the "Classifying the Other" Dataset*

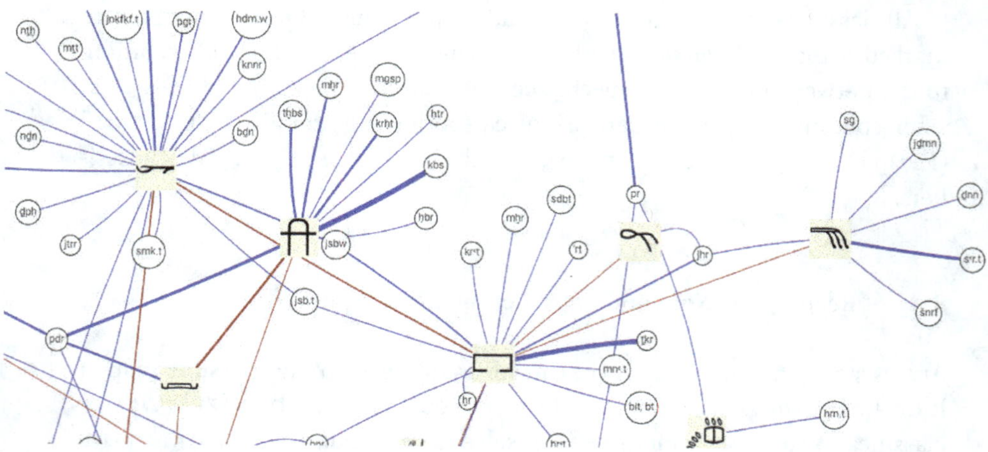

FIGURE 6.5 A detail of network analysis map of lexical borrowings by their classifiers. In the center, the classifier 𓍻 (V19)

Fig. 6.5 shows which other classifier categories are linked to lemmas classified by 𓍻 (V19). The user can query each classifier and produce for it a detailed quantitative distribution report. The graphs in figs. 6.6–6.7 depict the classifier *axis*- lemmas and classifiers that occur with the 𓍻 classifier. Thicker lines show more frequently co-occurring lemmas. All host words seem to belong to the [PACKING][45] category and represent various measurements for commodities.

As seen in the above figure, all examples of the 𓍻 (V19) classifier appear in hieratic sources.[46] These occurrences are mainly in administrative texts but

44 KRI II, 249, 7–15, RITA II, 94.

45 The category's label is a preliminary term, and various semantic aspects of this category are gradually delineated.

46 See discussion in Manassa 2012 regarding the classification of the term *ṯḫbš.t* with V19 throughout the 18th and 19th dynasties.

also in literary compositions such as *The Taking of Joppa*, where there are mentioned "200 *ṯḫbš.t*-baskets full of men."[47]

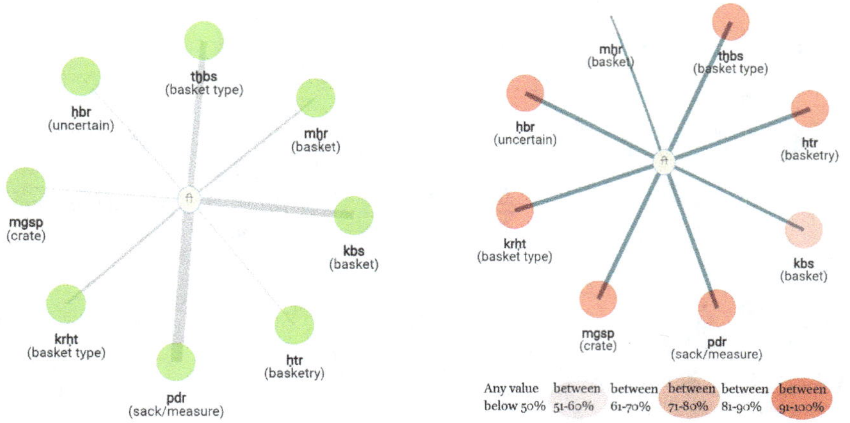

Lemma by no. of examples with V19 classifier

Lemma	Count
pdr 'sack/measure'	12
kbs 'basket'	8
ṯḫbs 'basket type '	7
mḫr 'basket, box '	3
krḥt 'basket type '	3
mgsp 'crate, basket'	1
ḫbr 'uncertain, perhaps 'rope' '	1
ḫtr 'basketry'	1

Tokens for this classifier

- t:x-b-Z7-Aa18-Z1-Ff1-V19 (930, oTutankhamun: no. 59, 2)
- t:x-b-s-ti-i-V19 (931, H.O 47: 2, 5-6)
- t:x-b-s-ti-i-V19 (932, oDeM 102: 4)
- t:x-b-s-ti-i-V19 (933, oDeM 367: 6)
- t:x-b-s-ti-i-V19 (934, oDeM 446: 15)
- p:d-Z4:r-Z1-T9:V19 (935, pHarris I: 16a,10)
- p:d-Z4:r-Z1-T9:V19 (936, pHarris I: 19b,14)
- p:d-Z4:r-Z1-T9:V19 (937, pHarris I: 65b, 8)
- p:d-Z4:r-Z1-T9:V19 (938, pHarris I: 65b, 10)
- p:d-Z4:r-Z1-T9:V19 (939, pHarris I: 71b, 8)
- p:Z5-d-Ff1-V19 (941, oUC 31922: I, 10-11)
- p:d-Z4:r-Z1-T9:D54-V19 (942, pAnastasi IV: 7,5)

Lemma centrality rank statistics with V19 classifier

Lemma	Percentage	Counts
mgsp 'crate, basket'	100	1 / 1
ḫbr 'uncertain, perhaps 'rope'	100	1 / 1
ḫtr 'basketry'	100	1 / 1
krḥt 'basket type '	100	3 / 3
ṯḫbs 'basket type '	100	7 / 7
pdr 'sack/measure'	92	12 / 13
kbs 'basket'	80	8 / 10
mḫr 'basket, box '	60	3 / 5

POS co-occurrence statistics

Part of speech	Count
N	35

Order statistics

Classifier position	Count
1	25
2	7
3	4

Script statistics

Script	Count
hieratic	36

FIGURE 6.6 The category governed by the (V19) classifier in the "Classifying the Other" dataset
Left: Lemmas by frequency rate. The width of lines represents the number of tokens occurring with this classifier. **Right:** Lemmas by frequency rate. Turquoise lines represent the number of examples a lemma has with a classifier. The color of the node represents the percentage of examples a particular lemma has with this classifier. Six different shades of pink are used; each color represents a range of percentages.

47 Taking of Joppa, LES, 83,12.

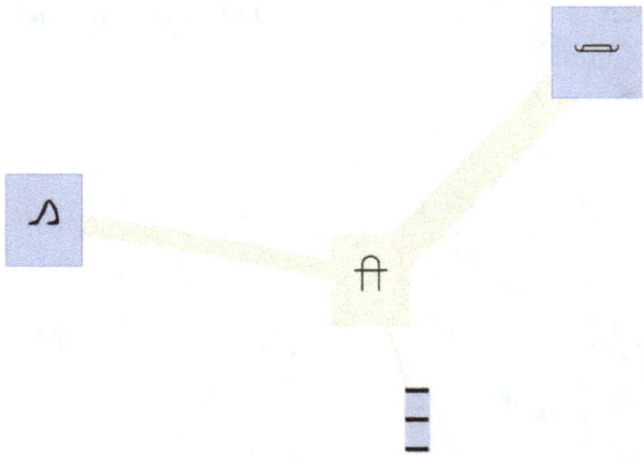

FIGURE 6.7
Classifier co-occurrence graph for (V19)

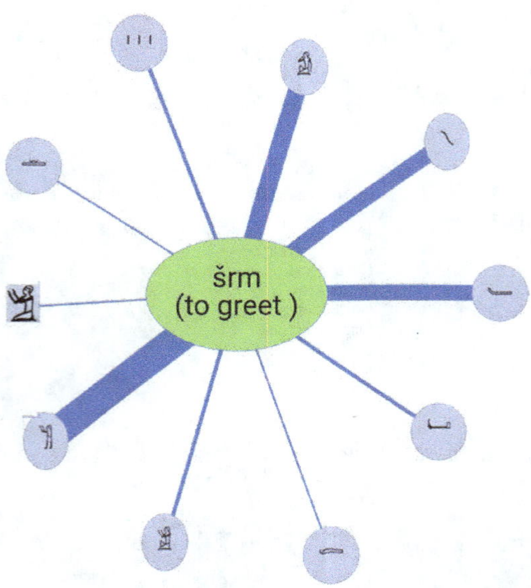

FIGURE 6.8
The lemma axis: Lemmas *šrm* and its classifiers

For each classifier category, *iClassifier* presents classifier co-occurrence (Fig. 6.8). In the case of , its co-occurrence with the 𓂻 (D54) [MOVEMENT] classifier probably suggests the close connection of packing and transport categories. Its co-occurrence with (T9) is when classifying foreign commodities, e.g., pomegranates and grapes. This classifier combination may reflect a particular stretched feature of the classified *pdr*, "sack" (or "measure").[48]

48 TLA Lemma no. 63090, Hoch 1994, no. 159.

8.2 *The Lemma Axis, a Preliminary Map, and Analysis of a Sample Lemma:* šrm, *from "Greeting" to "Surrender"*

In *iClassifier*, each lemma is an abstract entity represented by concrete tokens. An example of the lemma axis is the lemma *šrm*, which has examples as a noun and a verb. All tokens added to *iClassifier* are assigned a lemma and a root when available.[49] The part of speech of each token is marked, and grammatical analysis is added to each token.[50] *iClassifier* allows the user to overview together all tokens of a lemma and look for tokens analyzed as pertaining to a specific part of speech or occurring in a particular text. Examples of *šrm* are classified into a few classifier categories. As can be seen in the report in Figs. 6.8–6.9 below, tokens of this lemma appear with various classifiers, yet most frequently with [SUBMISSION/ADORATION] classifiers, A30 and A4/A4C.

The preliminary results show the first occurrences of *šrm* to be in hieroglyphic sources in monumental royal inscriptions (c. 1200 BCE). There, one sees a consistent use of the 𓀢 (A30) and/or (A4C) or (A4), a kneeling man with a raised arms classifier. It is not entirely clear if the hieroglyph that appears in this word is the same hieroglyph that appears as a classifier in typical Egyptian words of praise/adoration.[51] It may be a different sign that features a kneeling (+SUBMISSION? + ADORATION?) person, as in the Amada [52] or Medinet Habu [53] examples.

Later, examples in the hieratic script show an additional classification of *šrm* into the 𓀁 (A2) category. The magical pBoulaq 6 (21st dynasty) portrays an idiosyncratic repetitive spelling ([54] 𓂝, 𓀁)  (A30-F18-A2).[55] This spelling includes the composite use of the two "speech"

49 Linking to the 'root' list of the TLA.

50 This lemma appears in Hoch 1994 as several separate entries—nos. 406, 407, and 408. Hoch separates the entries into three lexical categories, 'to greet, make obedience, do homage,' a second meaning 'to lay down (arms), seek peace' and a noun 'peace, greetings.' Here we show together all entries of the root *šrm*. Tokens are marked for their specific grammatical role in each example in the database.

51 Gardiner 1957, 445.

52 Merneptah Nubian victory stela at Amada, Youssef 1964; KRI IV, 35,1. Hoch (1994, 285) mistakenly transcribes the following hieroglyph as a second classifier for this token, yet it is probably a logogram for *wr* 'chief.'

53 Medinet Habu, Ramses III, 1st Libyan War, 28, 56.

54 Möller II, no. A2, Möller does not differentiate between a standing A4 and a seated A30 prototype.

55 The examples occurring in pBoulaq 6 include additionally a hieratic stroke. Using *iClassifier* we aim to expose the usage patterns of strokes, and define their possible role as word dividers.

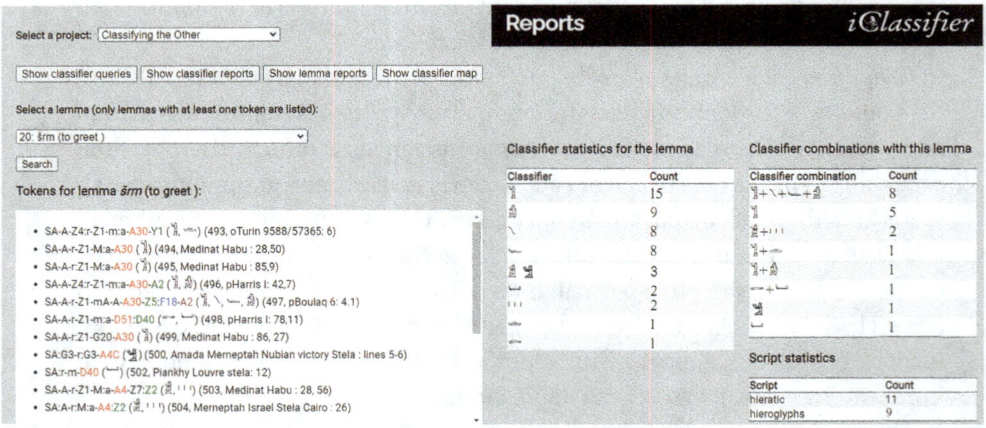

FIGURE 6.9 The classification report of lemma *šrm* in *iClassifier*

classifiers . The emergence of the (A2) classifier for this word in the hieratic texts of this period may be linked to a new metatextual level of as marking in hieratic texts the notion of "foreign word" (see above, § 4.2.2).[56] Another variation of classification which appears in Ramesside royal hieratic texts is with the (D40) classifier (pHarris I) 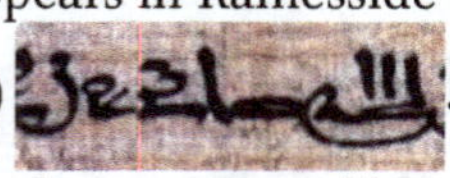.[57] An attestation with the same classifier surfaces years later in a royal hieroglyphic text (dated to the reign of Pianchy).

What emerges is a clear illustration of classifier differences in hieratic and hieroglyphs, as well as possible genre variation in classification. However, further context-sensitive research is needed to analyze this variation and perhaps to tie it to the scribal tendencies of royal vs. non-royal texts.

To sum up, the Egyptian scribes of the official texts of the 19th and 20th dynasties assign *šrm* to their category of submission and adoration—exposing the various semantic aspects it represents—from *greeting* to *surrender*. The one who seeks *šrm* from the Egyptian king is in no way in an equal position, and *šrm* is not peace between equals. The (D40) classifier may further support such analysis. A single appearance of the (Y1) classifier probably reflects a residual classification in hieratic.[58] In modern Hebrew and Arabic, the terms

56 Allon 2010.

57 Hoch, as well as Erichsen, transcribes this classifier as A5. However, it highly resembles A30. See Möller II, no. A2 vs. A39, and compare discussion in Werning 2011, 99.

58 Kammerzell 2015.

'shalom' and 'salaam' no longer refer to 'surrender' or 'submission,' but to a willful cessation of hostilities, preferably between equals.

9 A Final Note for Our Readers

iClassifier allows for and wishes to advance collaborative scholarly research. Anyone interested can join in and use the platform for their research project. Contributors are invited to feed their data into an *iClassifier* private project portal. Our vision is to gradually collect classifier markings over all public datasets to compile a new, extended and corpus-based, growing and changing classifier list for ancient Egyptian. In all cases, contributors receive full credit, and data reused is credited.[59] Gradually, we hope to create together with many other Egyptologists a growing, comprehensive network mapping of Egyptian conceptual organization, as manifested in the script classifier system of the analyzed corpora. Further details are available at https://archaeomind.huji.ac.il. *iClassifier* serves as a common discussion ground and a crowd-sourced portal for studying and comparing classifiers within Egyptology and between other complex scripts (Sumerian, ancient Chinese, et al.) and classifier languages.[60] Our vision and hope are to contribute a new research tool that will allow us to approach the conceptual knowledge and organization of the human mind, as reflected in the elaborate classifier system of ancient Egypt.

Bibliography

Allon, Niv. 2007. "Seth is Baal—Evidence from the Egyptian Script." *Ägypten und Levante* 17: 15–22.

Allon, Niv. 2010. "At the Outskirts of a System: Classifiers and Word Dividers in Foreign Phrases and Texts." *Lingua Aegyptia* 18: 1–17. https://archaeomind.huji.ac.il/publications/outskirts-system-classifiers-and-word-dividers-foreign-phrases-and, accessed December 25, 2021.

Bauer, Brigitte. L.M. 2017. *Nominal Apposition in Indo-European Its Forms and Functions, and Its Evolution in Latin-Romance*. Berlin/Boston: De Gruyter.

59 Each dataset created using *iClassifier* is subject to copyright rules of its creator, and that of the ©*iClassifier* research platform.

60 The platform's development was financed by the Israel Science Foundation Grant 735/17 "*Classifying the Other.*" Preliminary results of network analysis map of Sumerian using *iClassifier* are in Selz & Zhang, in press.

Bisang, Walter. 2017. "Classification between Grammar and Culture—a Cross-Linguistic Perspective." In *Classification from Antiquity to Modern Times*, edited by T. Pommerening & W. Bisang, 199–230. Berlin/Boston: De Gruyter.

Champollion, Jean-François. 1836. *Grammaire égyptienne: ou, principes généraux de l'écriture sacrée égyptienne appliquée à la représentation de la langue parlée*. Paris: Didot.

Chantrain, Gaëlle. 2014. "The Use of Classifiers in the New Kingdom. A Global Reorganization of the Classifiers System?" *Lingua Aegyptia* 22: 39–59.

Chantrain, Gaëlle. 2021. "Classification Strategies from the End of the Ramesside Period until the Late Period: A Living System" *Zeitschrift für Ägyptische Sprache und Altertumskunde* 148 (1): 50–64.

Chantrain, Gaëlle and Camilla Di Biase-Dyson. 2018. "Making a Case for Multidimensionality in Ramesside Figurative Language." *Lingua Aegyptia* 25: 41–66.

Contini-Morava, Ellen, and Marcin Kilarski. 2013. "Functions of Nominal Classification." *Language Sciences* 40: 263–299.

Corbett, Greville G., and Sebastian Fedden. 2016. "Canonical Gender 1." *Journal of Linguistics* 52 (3): 495–531.

Chen, Yongsheng. 2016. "The Prototypical Determinatives in Egyptian and Chinese Writing." *Scripta 8*: 101–126.

Croft, William. 1994. "Semantic Universals in Classifier Systems." *Word* 45 (2): 145–171.

David, Arlette. 2000. *De l'infériorité à la perturbation: l'oiseau du « mal » et la catégorisation en Égypte ancienne*. Wiesbaden: Harrassowitz.

Denny, Peter J. 1976. "What are Noun Classifiers Good For?" In *Papers from the Regional Meeting. Chicago Linguistic Society* 12: 122–132.

Di Biase-Dyson, Camilla, Frank Kammerzell & Daniel A. Werning, 2009. "Glossing Ancient Egyptian Suggestions for adapting the Leipzig Glossing Rules." *Lingua Aegyptia* 17: 343–366.

Gardiner, Alan H. 1957. *Egyptian Grammar: Being an Introduction to the Study of Hieroglyphs*. Oxford: Griffith Institute.

Goldwasser, Orly. 1999. "The Determinative System as a Mirror of World Organization." *Göttinger Miszellen* 170: 73–93.

Goldwasser, Orly. 2002. *Prophets, Lovers and Giraffes. Wor(l)d Classification in Ancient Egypt*. Wiesbaden: Harrassowitz.

Goldwasser, Orly. 2003. "The Building Blocks of a Conceptual Category: The Ancient Egyptian Evidence." In *Philosophers and Hieroglyphs*, edited by L. Morra & C. Bazzanella, 219–236. Torino: Rosenberg & Sellier.

Goldwasser, Orly. 2005. "Where is Metaphor?: Conceptual Metaphor and Alternative Classification in the Hieroglyphic Script." *Metaphor and Symbol* 20 (2): 95–113.

Goldwasser, Orly. 2006a. "A Comparison between Classifier Language and Classifier

Script: The Case of Ancient Egyptian." In *A Festschrift for Hans Jakob Polotsky*, edited by G. Goldenberg, 16–39. Magnes Press: Jerusalem.

Goldwasser, Orly. 2006b. "On the New Definition of Classifier Languages and Scripts." *Lingua Aegyptia* 14: 473–484.

Goldwasser, Orly. 2010. "The Aten is the 'Energy of Light': New Evidence from the Script." *Journal of the American Research Center in Egypt* 46: 159–165.

Goldwasser, Orly. 2016 "From Iconic to Linear—The Egyptian Scribes of Lachish and the Modification of the Early Alphabet in the Late Bronze Age." In *Alphabets, Texts and Artefacts in the Ancient Near East, Studies presented to Benjamin Sass*, edited by I. Finkelstein, C. Robin & T. Römer, 118–161. Paris: Van Dieren.

Goldwasser, Orly. 2017. "What Is a Horse?—Lexical Acculturation and Classification in Egyptian and Nahuatl." In *Classification from Antiquity to Modern Times*, edited by T. Pommerening & W. Bisang, 45–66. Berlin/Boston: De Gruyter.

Goldwasser, Orly. In press a. "Des déterminatifs aux classifcateurs: la catégorisation dans l'écriture des anciens Égyptiens." In *Guide des écritures de l'Égypte antique*, edited by S. Polis, 164–171. Le Caire: coll. Les Guides de l'Ifao, 1.

Goldwasser, Orly. In press b. "Was There an "Animal" in Ancient Egypt? Studies in Lexica and Classifier Systems, with a Glimpse Towards Sumer." In *Seen Not Heard: Composition, Iconicity, and the Classifier Systems of Logosyllabic Scripts*. March 2–3, 2017 Proceedings. Chicago: Oriental Institute.

Goldwasser, Orly & Colette Grinevald. 2012. "What Are Determinatives Good For?" In *Lexical Semantics in Ancient Egyptian*, edited by E. Grossman, S. Polis & J. Winand, 17–53. Hamburg: Widmaier.

Grinevald, Colette. 2015. "Classifiers, Linguistics of." In *International Encyclopedia of the Social & Behavioral Sciences*, edited by J.D. Wright, 1973–1978. Amsterdam: Elsevier.

Grossman, Eitan, and Stéphane Polis. 2012. "Navigating Polyfunctionality in The Lexicon. Semantic Maps and Ancient Egyptian Lexical Semantics." In *Lexical Semantics in Ancient Egyptian*, edited by E. Grossman, S. Polis & J. Winand, 175–225. Hamburg: Widmaier.

Hafemann, Ingelore 2018. "Die beschreibende und kommentierte hieroglyphische Zeichenliste als offenes System." In *Ägyptologische "Binsen"-Weisheiten III: Formen und Funktionen von Zeichenliste und Paläographie. Akten der internationalen und interdisziplinären Tagung in der Akademie der Wissenschaften und der Literatur, Mainz, im April 2016*, edited by S.A. Gülden, K. van der Moezel & U. Verhoeven, 19–49. Stuttgart: Franz Steiner.

Handel, Zev. In press. "The Cognitive Role of Semantic Classifiers in Modern Chinese Writing as Reflected in Neogram Creation." In *Seen Not Heard: Composition, Iconicity, and the Classifier Systems of Logosyllabic Scripts*. March 2–3, 2017 Proceedings. Oriental Institute Chicago.

Harel, Haleli. Forthcoming. *A Network of Lexical Borrowings in Egyptian Texts of the New*

Kingdom: Organizing Knowledge according to the Classifier System. Doctoral dissertation. Hebrew University, Jerusalem.

Hoch, James E. 1994. *Semitic Words in Egyptian Texts of the New Kingdom and Third Intermediate Period*. New Jersey: Princeton.

Imai, Mutsumi, and Henrik Saalbach. 2010. "Categories in Mind and Categories in Language: Do Classifier Categories Influence Conceptual Structures?" In *Words and the Mind: How Words Capture Human Experience*, edited by B. Malt and P. Wolff, 138–164. Oxford University Press.

Kammerzell, Frank. 1999. "Zur Interpretation einiger Beispiele graphemsprachlicher Varianz im Ägyptischen." *Göttinger Beiträge zur Sprachwissenschaft* 2: 61–97.

Kammerzell, Frank. 2015. "Egyptian Verb Classifiers." In *Proceedings of the Tenth International Congress of Egyptologists. Rhodes, 22–29 May 2008*, edited by P. Kousoulis & N. Lazaridis, 1395–1416. Leuven: Peeters.

Kemmerer, David. 2017. "Categories of Object Concepts Across Languages and Brains: The Relevance of Nominal Classification Systems to Cognitive Neuroscience." *Language, Cognition and Neuroscience* 32 (4): 401–424.

Kilarski, Marcin. 2013. *Nominal classification: A History of Its Study from the Classical Period to the Present*. John Benjamins Publishing Company.

Kilarski, Marcin. 2014. "The Place of Classifiers in the History of Linguistics." *Historiographia Linguistica* 41 (1): 33–78.

KRI= Kitchen, Kenneth A. 1969–1990. *Ramesside Inscriptions, Historical and Biographical*. 8 vols. Oxford: Blackwell.

Lefebvre, Gustave. 1955. *Grammaire de l'égyptien Classique*. Second edition, with Serge Sauneron. Cairo: Impr. de l'Institut français d'archólogie orientale.

LES= Alan. H. Gardiner. 1973. *Late-Egyptian Stories*. Bruxelles: Édition de la Fondation Égyptologique Reine Élisabeth.

Lincke, Eliese-Sophia. 2011. *Die Prinzipien der Klassifizierung im Altaegyptischen*. Wiesbaden: Harrassowitz.

Lincke, Eliese-Sophia. 2015. "Raumwissen in Hieroglyphen. Die sprachliche Kategorisierung des ägyptischen Ortes Sile (Tjaru) aus prototypentheoretischer Sicht." In *Raumwissen und Wissensräume. Beiträge des interdisziplinären Theorie-Workshops für Nachwuchswissenschaftler/innen*, edited by Kerstin P. Hofmann & Stefan Schreiber. = *eTopoi. Journal for Ancient Studies, Special Volume* 5: 127–155.

Lincke, Eliese-Sophia and Frank Kammerzell. 2012. "Egyptian Classifiers at the Interface of Lexical-Semantics and Pragmatics." In *Lexical Semantics in Ancient Egyptian*, edited by E. Grossman, S. Polis & J. Winand, 55–112. Hamburg: Widmaier.

Lincke, Eliese-Sophia and Silvia Kutscher. 2012. "Motivated Sign Formation in Hieroglyphic Egyptian and German Sign Language (DGS). Towards a Typology of Iconic Signs in Visual Linguistic Systems." In *Lexical Semantics in Ancient Egyptian*, edited by Eitan Grossman, Stéphane Polis & Jean Winand, 113–140. Hamburg: Widmaier.

Manassa, Colleen. 2012. “From Wool to Basketry: Materials, Contact Linguistics, and *tḫbs(t)* in Ancient Egyptian.” *Lingua Aegyptia* 20: 99–110.

Medinet Habu, Ramssess III = The Epigraphic Survey. 1930. *Earlier Historical Records of Ramses III*. Chicago: Oriental Institute.

Meeks, Dimitri. 2013. “Dictionnaire Hiéroglyphique, Inventaire Hiéroglyphique et Unicode.” *Document Numérique* 16 (3): 31–44.

Möller, Georg. 1909–1936. *Hieratische Palaographie*. 3 volumes. Leipzig.

Müller, Matthias. 2002. “Appendix.” In *Prophets, Lovers and Giraffes. Wor(l)d Classification in Ancient Egypt*, authored by Orly Goldwasser, 3*–46*. Wiesbaden: Harrassowitz.

Myers, James. 2019. *The Grammar of Chinese Characters: Productive Knowledge of Formal Patterns in an Orthographic System*. London: Routledge.

Naaman, Nadav. 2020. “Egyptian Centers and the Distribution of the Alphabet in the Levant.” *Tel Aviv: The Journal of the Institute of Archaeology of Tel Aviv University* 47 (1): 29–54.

Nyord, Rune. 2015. “Cognitive Linguistics.” In: *UCLA Encyclopedia of Egyptology*, J. Stauder-Porchet, A. Stauder and W. Wendrich (eds.). Los Angeles. http://digital2.library.ucla.edu/viewItem.do?ark=21198/zz002k44p6, Accessed January 30, 2020.

Payne, Annick. 2017. “Determination in the Anatolian Hieroglyphic Script of the Empire and Transitional Period.” *Altorientalische Forschungen* 44 (2): 221–234.

Peust, Carsten. 2012. “Review of *Die Prinzipien der Klassifizierung im Altägyptischen* by Eliese-Sophia Lincke, 2011.” *Lingua Aegyptia* 20: 303–308.

Polis, Stéphane. 2018. “The Functions and Toposyntax of Ancient Egyptian Hieroglyphs. Exploring the Iconicity and Spatiality of Pictorial Graphemes.” *Signata. Annales des sémiotiques/Annals of Semiotics* 9: 291–363.

Polis, Stéphane and Vincent Razanajao. 2016. “Ancient Egyptian texts in context. Towards a conceptual data model (The THOT data model—TDM).” *Bulletin of the Institute of Classical Studies* 59 (2): 24–41.

Polis, Stéphane and Serge Rosmorduc. 2015. “The Hieroglyphic Sign Functions: Suggestions for a Revised Taxonomy.” In *Fuzzy boundaries: Festschrift für Antonio Loprieno*, vol. 1, edited by H. Amstutz, A. Dorn, M. Müller, M. Ronsdorf, & S. Uljas, 149–174. Hamburg: Widmaier.

Polis, Stéphane, and Jean Winand. 2015. “Structuring the Lexicon.” In *Proceedings of the Tenth International Congress of Egyptologists. Rhodes, 22–29 May 2008*, edited by P. Kousoulis & N. Lazaridis, 1503–1512. Leuven: Peeters.

Pommerening, Tanja. 2017. “Classification in Ancient Egyptian Medical Formulae and its Role in Re-Discovering Comprehensive and Specific Concepts of Drugs and Effects.” In *Classification from Antiquity to Modern Times: Sources, Methods, and Theories from an Interdisciplinary Perspective*, edited by T. Pommerening & W. Bisang, 167–196. Berlin, Boston: De Gruyter.

Ranke, Herman. 1935. *Die altägyptischen Personennamen I–III*. Glckstadt/Hamburg: JJ Augustin.

Rude, Noel. 1986. "Graphemic Classifiers in Egyptian Hieroglyphics and Mesopotamian Cuneiform." In *Noun Classes and Categorization, Amsterdam/Philadelphie*, edited by C. Craig, 133–138. Amsterdam/ Philadelphia: John Benjamins.

RITA= Kitchen, Kenneth A. 1993–2014. *Ramesside Inscriptions Translated and Annotated: Translations*. 7 vols. Oxford: Blackwell; Chichester: Wiley-Blackwell.

Saalbach, Henrik and Mutsumi Imai. 2007. "Scope of Linguistic Influence: Does a Classifier System Alter Object Concepts?" *Journal of Experimental Psychology* 136 (3): 485–501.

Saalbach, Henrik, and Mutsumi Imai. 2012. "The Relation Between Linguistic Categories and Cognition: The Case of Numeral Classifiers." *Language and Cognitive Processes* 27 (3): 381–428.

Schneider, Thomas. 1992. *Asiatische Personennamen in ägyptischen: Quellen des Neuen Reiches*. Freiburg: Vandenhoeck & Ruprecht.

Schneider, Thomas. 2008. "Fremdwörter in der ägyptischen Militärsprache des Neuen Reiches und ein Bravourstück des Elitesoldaten (Papyrus Anastasi I 23, 2–7)." *Journal of the Society for the Study of Egyptian Antiquities* 35: 181–205.

Selz, Gebhard J., Colette Grinevald & Orly Goldwasser. 2018. "The Question of Sumerian Determinatives: Inventory, Classifier Analysis, and Comparison to Egyptian Classifiers." *Lingua Aegyptia* 25: 281–344.

Selz, Gebhard J. and Bo Zhang. In press. "Appositive Semantic Classification in Sumerian Cuneiform and the Implementation of *iClassifier*." *Ash-sharq: Bulletin of the Ancient Near East* 6, 27 pages.

Senft, Gunther. 2000. "What Do We Really Know About Nominal Classification Systems?" In *Systems of Nominal Classification*, edited by Gunter Senft, 12–49. Cambridge: Cambridge University Press.

Shalomi-Hen, Rachel. 2000. *Classifying the Divine: Determinatives and Categorization in CT 335 and BD 17*. Wiesbaden: Harrassowitz.

Shalomi-Hen, Rachel. 2006. *The Writing of Gods. The Evolution of Divine Classifiers in the Old Kingdom*. Wiesbaden: Harrassowitz.

Shalomi-Hen, Rachel. 2008. "The Bearded Woman and the Queen. The Formation and Transformation of Female Divine Classifiers." In *Sex and Gender in Ancient Egypt: 'Don your Wig for a Joyful Hour'* edited by C. Graves-Brown, 179–189. Swansea: Classical Press of Wales.

Te Velde, Herman. 1967. *Seth, God of Confusion: A Study of His Role in Egyptian Mythology and Religion*. Leiden, Brill.

Thuault, Simon. 2020. *La dissemblance graphémique à l'Ancien Empire. Essai de grammatologie cognitive*. Hamburg: Widmaier Verlag.

Van der Moezel, Kyra. 2018. "On signs, lists and standardisation," in: *Ägyptologische*

'Binsen'-Weisheiten III. Formen und Funktionen von Zeichenliste und Paläographie, edited by S.A. Gülden, K. Moezel, U. van der Verhoeven, 51–81. Stuttgart: Franz Steiner Verlag.

Werning, Daniel A. 2011. *Das Höhlenbuch. Textkrigrammatitische Edition und Textgrammatik*. 2 Vols. Wiesbaden: Harrassowitz.

Winand, Jean. 2016. "The Syntax-Semantics Interface in Earlier Egyptian: A Case Study in Verbs of Cognition." In *Coping with Obscurity: The Brown Workshop on Earlier Egyptian Grammar*, edited by J. Allen, M. Collier & A. Stauder, 109–139. Atlanta: Lockwood press.

Winand, Jean. 2019. "Did you say synonyms? The case of *pḥ* and *spr* in Late Egyptian." In *Festschrift für Hans-Werner Fischer-Elfert*, edited by M. Brose, F. Naether, D. Raue and T.S. Richter, 1235–1274. Berlin: de Gruyter.

Winand, Jean. 2021. "Horapollon ou la fin d'un parcours." In *Les Hieroglyphica d'Horapollon de l'Égypte antique à l'Europe moderne*, edited by J.-L. Fournet, 13–31. Paris: StudPAP 2.

Youssef, Ahmad. A. 1964. "Merenptah's Fourth Year Text at Amada." *Annales du Service des Antiquités de l'Égypte* 58: 273–280.

Zhang, Shi, and Bernd Schmitt. 1998. "Language-dependent classification: The mental representation of classifiers in cognition, memory, and ad evaluations." *Journal of Experimental Psychology Applied* 4 (4): 375–385.

Digital Projects Mentioned

"*Concepticon* Project's Semantic Fields-Concept Sets," Concepticon 2.2.0. Jena: Max Planck Institute for the Science of Human History, last accessed September 10, 2021, https://concepticon.clld.org/

"Demotic Word List Database," last accessed May 10, 2020, https://www.dwl.aegyptologie.lmu.de/index.php

"Determinatives Tab," Demotic Palaeographical Database Project (DPDP), last accessed September 10, 2021, http://129.206.5.162/beta/index.html

"Ramses Database," last accessed September 10, 2021, http://ramses.ulg.ac.be/

"Thot Sign List, Function type—Classifier," last accessed September 10, 2021, http://thotsignlist.org

"Thot Metadata Guidelines," last accessed September 10, 2021, http://thot.philo.ulg.ac.be/index.html

"TLA—Thesaurus Linguae Aegyptiae," last accessed September 10, 2021, http://aaew.bbaw.de/tla/servlet/TlaLogin

List of Figures—Credits

Figures are courtesy of the ©*iClassifier* (Goldwasser, Harel, Nikolaev) reuse guidelines, produced by Haleli Harel, and are based on the “Classifying the Other” dataset.

CHAPTER 7

Not Just *Another* Photogrammetry Report: Using Modern Technology to Help Solve an Ancient Riddle at Karnak Temple

Mark D. Janzen and Terrence J. Nichols

Abstract

For the past several decades, Egyptologists have debated the original authorship of the battle scenes on the western exterior wall of the Cour de la Cachette at Karnak Temple. A consensus has emerged that Merneptah was the original author of these scenes and texts. This study seeks to assist that ongoing conversation by analyzing the cartouches on the north end of the wall using modern technology, in particular photogrammetry and point clouds, in order to measure the depth of the cartouches and surrounding glyphs. Modern technology also allows us to gain a new perspective on the cartouches by creating a cross-section of the wall, further enhancing the ability to measure precisely. The results of these analyses demonstrate that there is insufficient depth for more than one usurpation of those cartouches giving further validity to the now widely accepted view that Merneptah's authorship was followed by the erasure of his names by Amenmesse before Seti II ultimately placed his name in the cartouches.

Keywords

Digital epigraphy – photogrammetry – point cloud – Merneptah – Ramesses II – Seti II – Amenmesse – Karnak temple – Cour de la Cachette – usurpation

1 Introduction

It is one task of archaeologists and Egyptologists to explore the riddles of ancient times, and there are indeed plenty of riddles despite significant leaps in understanding. It is the task of this study to pursue one such riddle and use modern technology to help solve it. A few years ago, the Karnak Great Hypostyle Hall Project was able to add the scenes on the western exterior wall

DOI:10.1163/9789004501294_008

of the so-called *Cour de la Cachette* (*Cachette* henceforth) to its epigraphic endeavors in consortium with the Tandy Institute for Archaeology,[1] now the Lanier Center for Archaeology.[2] At first glance, the scenes on this wall appear to be similar to those on the south wall of the Hypostyle Hall, so the scenes on the west wall are a natural extension of the larger project. Indeed, the two walls share a palimpsest of Ramesses II's Battle of Kadesh narrative.[3] At roughly the midpoint of the western wall of the *Cachette*, Ramesses II placed a copy of his peace treaty with the Hittites. Additionally, a bandeau text containing Ramesses' cartouches adorns the top of the wall. Thus, for decades Egyptologists assumed that Ramesses II was author of the scenes on the west wall of the *Cachette*.[4]

However, the key piece of evidence for ascribing authorship is *not* mere similarity to the style of reliefs from the south wall which is fairly subjective,[5] but rather the surviving cartouches on the west wall. These badly damaged, obviously usurped cartouches require a detailed examination that is not possible without a full epigraphic analysis. In other words, vital details about the erased glyphs are simply not observable from ground level or even with the naked eye. The intention of the present study is to examine the existent cartouches from the battle scenes of the west wall using modern digital epigraphic methods. The historical questions considered in this paper are 1) the provenance of the cartouches on the west wall regarding the authorship and identity of the pharaoh who originally inscribed the wall, and 2) the number of usurpations to these cartouches. The purpose of this study is to use photogrammetry to best determine the depth of the cartouches which will allow us to make a reasonable hypothesis regarding the number of successive usurpations. Before sharing the details of that investigation, brief surveys on the history of the late Nineteenth Dynasty and the surviving cartouches on the west wall are in order.

---

1 The scenes can be found in: PM II$_2$, 132 (491); Heinz 2001, 294, I.2.

2 The Tandy Institute for Archaeology was under the auspices of Southwestern Baptist Theological Seminary. As of 2020, the faculty, staff, and research of the Tandy, including this project, are now under the auspices of the new Lanier Center for Archaeology at Lipscomb University.

3 Egyptologists have long recognized this. See, Breasted 1903, pl. 7; Kuentz 1928, pl. 26; Schwaller de Lubicz 1999, 592 (fig. 51); Spalinger, 1985, 1–42.

4 For example, see Kitchen 1964, 47–70. Kitchen later corrected this and explained that he believed Merneptah to be the author of these scenes, see Kitchen 1999, 72–80.

5 In fact, close inspection of the details of the scenes on the west wall strongly suggests a different set of artisans. Various details, such as the arms of the enemies and the crenellations of the fortifications are executed with greater care on the west wall of the *Cachette* than on the south wall of the Hypostyle Hall. For a full analysis, see Brand 2011, 51–84. Brand's treatment

2 Historical Background and the Problem of Authorship

Following Ramesses II's unusually long reign, Merneptah ruled for only ten years, likely due to the fact that he was already old by ancient standards when he took the throne. His abbreviated reign did not allow him time to make major renovations at Karnak, in contrast to his illustrious predecessors. When it came time to celebrate his wars, as any Ramesside ruler should, most of the prime temple real estate had been decorated previously by Seti I and Ramesses II. This might explain why Merneptah chose the west wall of the *Cachette* despite the obvious presence of the Hittite Treaty. Recall that Ramesses II had begun to decorate this wall with a pictorial narrative of the Battle of Kadesh, a continuation of the same scenes on the south wall of the Hypostyle Hall. His artisans never finished this project, and only traces of it survive. Ramesses II abandoned it, likely due to the fact that the scenes were too large for the south wall and had to round the corner and continue onto the west wall, an artistic compromise that apparently did not sit well with the king. In other words, the incomplete Kadesh scenes were plastered over, probably because they offended Egyptian artistic sensibilities, namely "their love for symmetry."[6] Instead, Ramesses placed his famous Hittite Treaty on the west wall. Thus, in Merneptah's day the rest of the west wall appeared to be available for his use. All Merneptah had to do was carve his scenes around the Hittite Treaty and beneath the bandeau cartouches of Ramesses II.

The all-important cartouches in the battle scenes were subjected to erasure after the death of Merneptah because an interloper named Amenmesse ruled Thebes for a short period of time while Merneptah's legitimate heir, Seti II, ruled the north.[7] Ultimately, Seti II reunited Upper and Lower Egypt. In the meantime, someone, likely Amenmesse, hacked out the cartouches in the battle scenes on the west wall of the *Cachette*. This interloper did not attack the bandeau cartouche and Hittite Treaty of Ramesses II. When Seti II re-established control of Thebes, he placed his name in the badly damaged cartouches. There is no doubt his nomen and prenomen are visible in the cartouches (see Figure 7.2). Such is the wall as we see it today.

is by far the most definitive analysis of the authorship of these scenes and their authorship, using a host of data. To our knowledge, the debate appears to be largely settled as Brand's analysis has not been challenged by any further studies at this time.

6 Brand 2011, 54.

7 For more on Amenmesse see, Hopper 2010; Dodson 2010.

3 The Debate: Ramesses II or Merneptah?

A robust debate on the authorship of these scenes ensued after a series of articles by Frank Yurco, who believed the battle scenes on the west wall were the work of Merneptah.[8] Yurco's analysis was based on the surviving names in the cartouches, the proper place to start when ascribing authorship. Yet, Yurco's claims were controversial because he also connected the battle scenes to Merneptah's famous Victory Stele, meaning that he believed the west wall of the *Cachette* also contained a pictorial representation of early Israelites. Some scholars readily endorsed his claims (at times with modifications),[9] while others rejected Yurco's analysis, preferring to view Ramesses II as the author for all the scenes on the western wall.[10]

As Brand points out, this created an unfortunate scenario in which holding to Merneptah's authorship was connected to believing the Israelites were depicted on the wall, while those who rejected Yurco's Israelite theory continued to maintain that Ramesses II was the author of the scenes.[11] This is problematic because whatever importance one ascribes to the Israelite debate and/or the potential connection to Merneptah's Victory Stele, discussions on the authorship of the scenes on the wall should be based on what is visible on the wall itself before any other sources (tangential at best!) are consulted, nor does endorsing Merneptah's authorship *ipso facto* require one to connect the scenes the Israelites. It is perfectly reasonable to hold to Merneptah's authorship without endorsing Yurco's view regarding the theoretical depiction of the Israelites. Beyond these brief remarks, it is not our intention to address the Israelite question at this time, but rather to focus on the extant cartouches.

Unfortunately, as the debate escalated, there was no definitive, scientific epigraphic publication of the scenes which scholars on both sides of the debate could consult. Various arguments were put forth with the matter largely unsettled. Some scholars made attempts at drawing the reliefs, in particular the cartouches, but "the reliefs have not been subjected to a sufficiently thorough epigraphic and art historical analysis that might finally resolve the issue, espe-

8 Yurco published his findings in a series of articles stretching nearly two decades: Yurco 1978, 70; 1986, 189–215; 1990, 20–38; 1997, 28–53.

9 *RITANC* II, 72–80; Kitchen 1993, 21. Kitchen has previously ascribed the scenes to Ramesses II but was convinced by Yurco's epigraphic analysis. See also, Stager 1985, 56–64; Rainey 2001, 57–75, to name but a few.

10 Redford 1986, 188–200; Sourouzian 1989, 150; Iskander 2002, 316–329; Lurson 2003, 45–62.

11 Brand 2011, 51–52.

cially their relationship to those on the south wall of the Hypostyle Hall."[12] After a detailed examination of the west wall of *Cachette*, Brand concluded that Yurco was correct about Merneptah's authorship of the battle scenes despite Yurco erroneously believing to have found traces of the name of Amenmesse (more below).[13] Thus far, our own epigraphic analysis reinforces Brand's conclusion.

Several glyphs survive in the cartouches on the north end of the west wall (Figure 7.1 and Figure 7.2).[14] These undoubtedly contain only the names of Merneptah and Seti II. Yurco's detractors claimed that Ramesses II was the original author. Thus, they believe Merneptah usurped the cartouche of Ramesses II before Amenmesse usurped Merneptah's and then finally Seti II replaced Amenmesse's name (see Appendix, Figure 7.7 for the hieroglyphs of the names of these kings). There are substantial problems with this analysis.

Most problematic is the fact that in order to claim that Ramesses II was the original author of all the scenes on the west wall, one must believe that Merneptah usurped his cartouches. This is nearly impossible to accept on several grounds. The obvious implication of the name of Ramesses II surviving on both the bandeau text and the Hittite Treaty is that his name was never erased on this wall.[15] If, for example, Merneptah sought to usurp Ramesses II's battle scenes on west wall, one would expect him to also usurp the name of his father elsewhere on the wall. For that matter, if Ramesses II's name was usurped by Merneptah in the battle scenes on the west wall, there would be nothing stopping him from usurping Ramesses II's names elsewhere in the vicinity. One wonders why he did not usurp other monuments of his illustrious father. In the immediate context, if Merneptah had been so inclined there was nothing to stop him from also usurping the cartouches on the south wall of the Hypostyle Hall. Finally, Kitchen rightly notes that the scenes on the west wall do not date to the reign of Ramesses II because by this point in his reign, Ramesses II ceased celebrating his military victories in favor emphasizing his godhood.[16] A good example of this is the change in the epithet of Pi-Ramesses from "Great of Victories" to "Great *Ka* of Re-Horakhty."[17]

---

12 Brand 2011, 51.

13 Brand 2009, 29–48.

14 For a definitive treatment of the cartouches on block from the west wall that currently in the courtyard, see Brand 2011, 60–64.

15 Brand 2011, 56.

16 *RITANC* II, 74.

17 See Brand 2005, 31–36. The only war scenes of Ramesses II that dates to the period after the Hittite Treaty are those in the outer court of his temple at Derr. See Brand 2011, 55, n. 20.

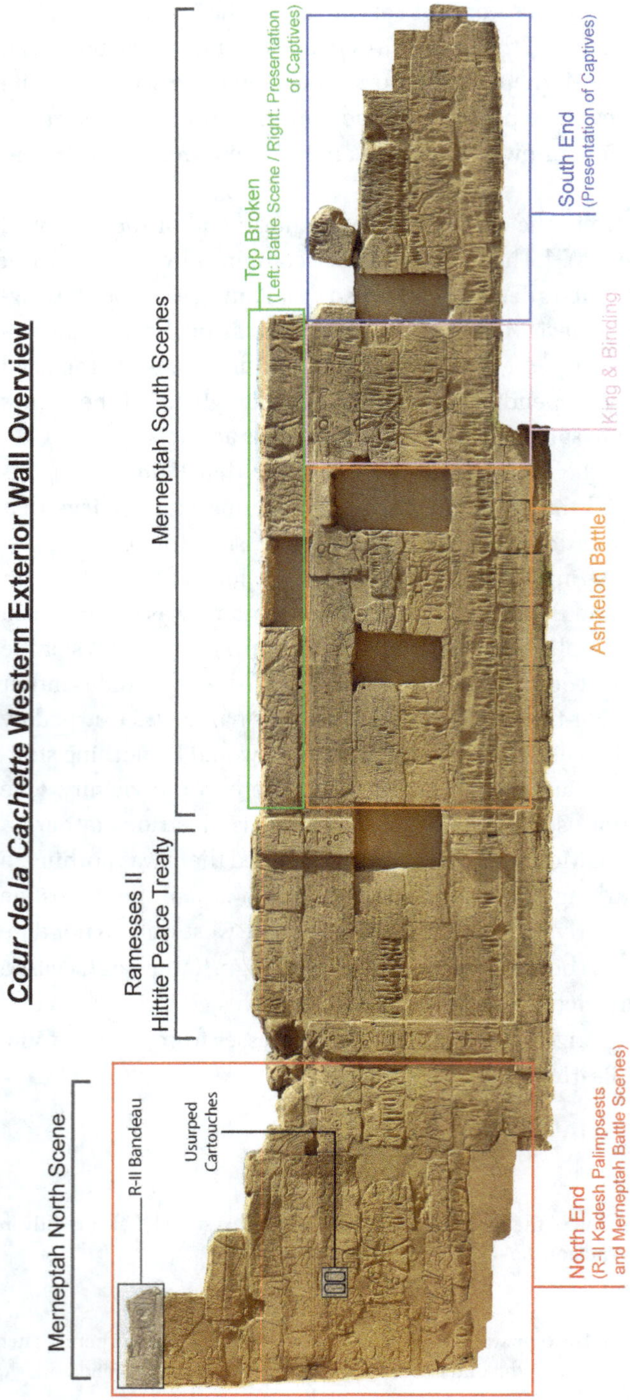

FIGURE 7.1 Detailed overview of the western wall

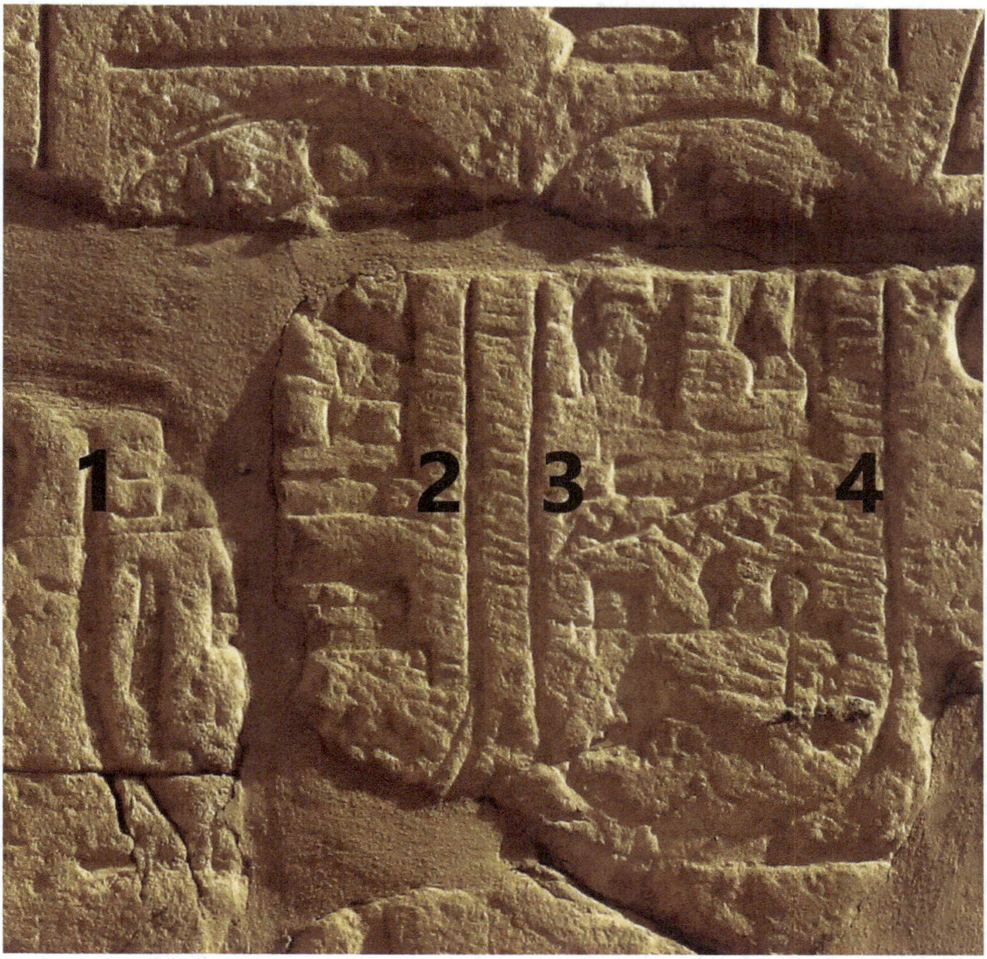

FIGURE 7.2 CDLC North End cartouches

While Yurco was correct in ascribing authorship of the scenes on the west wall of the *Cachette* to Merneptah, he erroneously believed to have found traces of the names of Amenmesse as well. Working on the wall the past few years, we found no traces of Amenmesse's name on the wall and certainly not in these cartouches, echoing the independent analysis of Brand.[18] A brief survey of the occurrences of the usurped cartouches of Seti II bolsters this conclusion.[19] Seti's preferred method of usurpation at Karnak was to chisel out the

18 Brand 2011, 55, n. 20.

19 For a more extensive treatment of this topic see, see Brand 2009, 29–48.

names of the previous kings by removing the stone matrix containing those names. He then polished the remaining stone, leaving a smooth depression into which his name was carved. In the few cases where traces of the usurped name can be found, they are Merneptah's each and every time.[20] Seti's methodology is important to understand because at the heart of the debate is the number of usurpations merely *possible* in the cartouches on the west wall of the *Cachette*. His thoroughness and the creation of a bowl-like depression leaves little room for additional usurpation.[21]

Depending on whether one believes Amenmesse carved his name or simply ran out of time, there would then be either three or four total names in the cartouches. Quite simply, a detailed examination reveals that no traces survive of either of the names of Ramesses II or Amenmesse, leaving only the readily observable names of Merneptah and Seti II. However, one could claim this was simply because the usurpers were so thorough. While we find this unlikely, we wanted another method to analyze the cartouches. Put another way, we wanted to come at the problem from a different *perspective*, which is where photogrammetry enters the discussion.

4 Digital Application: Photogrammetry

With the historiographical background and epigraphic problem enunciated, this paper turns its focus to digital technology and how its use allows us to solve the questions surrounding the cartouches on the west wall of the *Cachette*. To scrutinize the cartouches, the digital method chosen to address their provenance was photogrammetry.

Archaeology in general has witnessed a recent increase in the use of photogrammetry, especially with its capacity in 3D modelling. Presentations at national conferences like ARCE and ASOR demonstrate the use of photogrammetry by a number of archaeological projects. While photogrammetry is by

20 Brand 2011, 60. It should be noted that we are focusing on just one set of cartouches on the west wall of the *Cachette*. These are not as scooped out as most of the other erased Merneptah and Amenmesse cartouches at Karnak that Seti II re-inscribed. These more deeply carved out cartouches barely show traces of the initial names. Cf., Brand 2009, 23–24, figures 3–4.

21 Seti II does not appear to have used plaster. In other cases of re-inscribed cartouches with three different names where plaster was used, none were deeply scooped out. Examples include Horemheb/Ramesses I/Ramesses II on the Second Pylon and the large Ramesses IV/Ramesses VI cartouches on the lower parts of the columns in the hypostyle hall. For additional discussion, see Steele 1940; Revez and Brand 2015, 253–310.

no means colloquial, it does have popular appeal. The ability of archaeologists, Egyptologists, and historians to re-create a monument or object in 3D space rather than a simple picture or top plan has broadened the appeal of archaeology beyond specialists in the field. Yet the ability of photogrammetry to produce these models is a byproduct of the original purpose of the process. This paper carries the title of "Not Just Another Photogrammetry Report" as it will focus on using photogrammetry for its original purpose rather than the popular 3D model.

5 Origins of Photogrammetry

Photogrammetry is defined as "the science of extracting measurements from photographs."[22] The original purpose of photogrammetry was for measurement, seen in its original use in aerial photography for terrain mapping. The photogrammetric process begins with photographs, while the mathematical principles behind photogrammetry originate in projective geometry. The word 'photogrammetry' then is a blended word consisting of its three elements: Photo = light; -gram = something written/drawn; -metry = measurement. Modern photogrammetry began to develop in the 1840's with French engineers, and the term "photogrammetry" was first introduced by the German civil-engineer Albrecht Meydenbauer in 1867.[23] Photogrammetry in this era was concerned with combining individual terrain photos into a mosaic for a compilation of maps, now referred to as an orthomosaic photo. The shift from modern to contemporary photogrammetry started in the 1970's with the development of computers, culminating in the 1990's with soft-copy or digital processing of photographs.[24] Photography remains the primary and integral element for photogrammetry, and as such, must be defined, despite its obvious ubiquity. Photography is the "process of producing images on a sensitized surface *by the action of light* or other radiant energy."[25] Therefore, digital photogrammetry may be understood as a software program using algorithms that utilizes the 'action of light' recorded in a photograph to produce precise measure-

22 Walford 2017.

23 Grimm 2007, 54.

24 "History of Photogrammetry." Center for Photogrammetric Training 2008, 3–4. https://ibis.geog.ubc.ca/courses/geob373/lectures/Handouts/History_of_Photogrammetry.pdf. Accessed on 3/13/2019.

25 "Definition of Photography," Merriam-Webster.com, 2019, https://www.merriam-webster.com/dictionary/photography. Italics mine.

ments. Understanding the process of photogrammetry is necessary to see how it obtains precise measurements.

6 Process of Photogrammetry

Photogrammetry starts with a photograph. As photography has shifted to high-resolution, the precision of photogrammetry has also increased. The process of photogrammetry scans and analyzes a photo which then creates extracted measurements of each point in the photo known as a *point cloud*, whereby each 'point' represents one specific spot in the photo, while all the points collectively are called the 'cloud'. In the point cloud, extraneous points such as background or landscaping elements can be removed, leaving only the specified item. Each point of the item has a specific measurement on the XYZ matrix, where the X and Y axis denote width and height in two-dimension space, while the Z axis denotes depth for the third dimension. Once the process of photogrammetry is complete, every point in the point cloud has a specific measurement for XYZ matrix, which in turn allows for the creation of a 3D model as a secondary result. While photogrammetry primarily consists of the measurement of each point, 3D modelling has become the more prevalent aspect of the photogrammetric process. It is easy to understand why it is such, as datasets of XYZ matrix coordinates suitable for scholars in comparative studies lack the visual (read stimulating) appeal of a 3D digital reproduction or physical model. Nonetheless, photogrammetry produces a reliable and precise set of measurements. It is this primary function of the photogrammetric process that will be the focus of the following analysis of the cartouches on the west wall of the *Cachette*.

7 How It Was Done

In late November–early December 2016, the first field season of an epigraphic survey by the Tandy Institute for Archaeology at the west wall of the *Cachette* took place. The object of the first field season was to capture the entire western wall in a series of high-resolution photographs, for the purpose of using them in photogrammetry but also as a digital archive of the wall. The wall was captured using a Canon EOS 5D Mark III full-frame DSLR camera with a 50 mm lens shooting each picture at a resolution of 5760 × 3840 px. Nearly 400 overlapping photos were taken of the 25 m wall. During the second field season in December 2017, the focus turned to the inscriptions of the walls, and questions surrounding the two cartouches were further investigated, providing the impetus for this paper. A portion of 2016 photos from the north end of the west wall, including the clearly usurped cartouches and their immediate vicinity, were selected for this photogrammetry project. We imported these photos into the photogrammetry software Agisoft PhotoScan Pro v. 1.2.5. Multiple models within the north end were created for comparative purposes. Post-processing of the models was done using a second program CloudCompare v. 2.10.1 for direct and volumetric measurements, and we used Adobe Photoshop and Illustrator as a final edit to the models. The data provided by these models was then arranged into a database. The next section of the paper presents the data, as well as its interpretation and provides a solution to the historical problem investigated above.

8 Digital Solutions: Historical Problems

The historical questions considered in this paper are, 1) the provenance of the cartouches on the western wall regarding the authorship and identity of the pharaoh who originally inscribed the wall, and 2) the number of usurpations to these cartouches. The debate surrounding the cartouches has centered around the extrapolation, translation, and interpretation of the likely glyphs for the supposed pharaohs who inscribed or usurped each of their names on the western wall. Considering the implication that the cartouches were usurped in antiquity, only the depth of the inscription lines themselves was the focus. With the number of usurpations suggested by scholars, the depth of the cartouches and glyphs should give an indication how many usurpations or layers are likely: the more stone removed from the original surface, the deeper the surface of the cartouche face and inscribed lines should be.

We measured the depths of the inscription lines using photogrammetry due to its ability to provide precise measurements. The different depths of the cartouche lines and glyphs were compared to other inscriptions in the immediate vicinity of the cartouches within the same scene. The comparative depth of these inscribed lines will show that the inscription and remaining stone surface of the cartouches are not deep enough to allow for multiple usurpations.

9 Depth Data

Using the photogrammetric process to determine depth measurements, a number of datasets were prepared. The first dataset concentrates on the cartouche lines, the actual name-ring itself. The second set consists of glyphs within the cartouches and in the vicinity of the cartouches, in addition to some palimpsests of Ramesses II for comparison. The third dataset references back to the work of Frank Yurco enunciated in the historical background. This dataset compared Yurco's depth measurements noted from his field work in 1976–1977 to depth data collected by the Tandy in this publication. Each of these datasets will be presented as an average in the body of the paper, with the full data tables presented in the index. Each set will note any significant findings, with the interpretations and conclusions presented at the end of the section.

9.1 *Cartouche Name-Ring Depths*

The first dataset focuses on the inscribed name-ring lines making up the cartouche. The measurements below correspond to the vertical/straight side sections of the name-rings. In measuring these lines, it was discovered that these lines have a separate measurement for the inner and outer portions of the line and are included in the table below. Each line was measured for depth in 3 different spots, then averaged together. The inner and outer depth measurements were then averaged together to get an overall average depth. The outer ring line was measured to the original surface, while the inner lines were measured to the remaining surface. Additionally, the name-ring lines were measured for width. The abbreviated Table 7.1, shown below, shows only the average depth and width, while the full data table is included in Appendix, Table 7.6.

The initial depth measurements demonstrated that the interior name-ring line was shallower than the exterior. The interior ring depth is approximately 1 mm shallower than the exterior of the line. This corresponds to the reports of usurpations, as the interior stone surface of the cartouche would have been 'scooped' or shaved in preparation for the new name, particularly as discussed above regarding Seti II's usurpation.

It must be noted that the left cartouche's left ring-line (Line #1—see Fig. 7.2) is an outlier to the depth averages. Per the average, it is 2 mm deeper than the other name-ring lines. In addition, the average width for the left cartouche's left side is double the width of the other three inscribed name-ring lines, 1.4 cm vs 0.7 cm. While the reason for the left ring-line's larger averages in depth and

TABLE 7.1—ABBREVIATED

	Left cartouche			Right cartouche	
	1. Left side	*2. Right side*		*3. Left side*	*4. Right side*
			<Depth>		
Inner:	5.95	3.78		3.25	3.04
Outer:	6.76	4.45		3.17	4.67
Average:	6.36 mm	4.11 mm		3.21 mm	3.85 mm
			<Width>		
Average:	1.38 cm	5.79 mm		5.87 mm	6.59 mm

TABLE 7.2—ABBREVIATED

Left cartouche glyphs		Right cartouche glyphs	
𓂋: 6.19	𓁦: 4.5	—: 3.07	▯: 4.88
𓂋: 4.27	𓂋: 5.72	𓁢: 5.5	𓇌: 4.77
𓂋: 4.58		𓋹: 3.03	
Average:	5.05 mm	Average:	4.25 mm

width is unknown, this inscription is on a separate block, and it may be surmised that this block was subject to more exposure and weathering than the other blocks comprising the cartouche resulting in the larger measurements.

10.1 *Glyph Depths*

The second dataset to consider is the average depth of the glyphs inside the cartouches. It is important to note that in the analysis and data collection of the glyph depths, there were minimal assertions regarding the identification of the glyphs themselves, as it was only necessary to label the glyphs in order to construct and frame the data.[26] Each glyph was measured at minimum three times and then averaged, with some glyphs requiring more measurements due to their complexity. All depths were measured to the remaining surface. A full dataset is in Appendix, Table 7.7, with the abbreviated table provided below.

26 See the preceding sections and Table 7.2.

TABLE 7.3—ABBREVIATED

Glyphs in vicinity of cartouches		R-II Palimpsests (Averages)	
∆: 3.74	𓇌: 10.73	horse back: 9.84	soldier's shoulder: 4.05
𓊹 (line): 3.46	𓊹 (flag): 10.74		
=(under 𓂋): 4.87	¬: 4.51		
Average:	7.61 mm		

TABLE 7.4

Depth comparisons: ranges		
	Yurco 1976[a]	Tandy 2016
Merneptah—glyphs	9 mm	8.95 mm
Seti II—glyphs	7 mm—8.5 mm	6.84 mm—8.05 mm
Cartouche (inner surface)	1.5 mm—4 mm	2.31 mm—4.37 mm

a Yurco 2986, 196–197. Yurco collected his data in 1976, but published this article in 1986, per his own note on ibid., 196 as well as an asterisked note on ibid., 189.

This set also collected average depths of glyphs outside the cartouches but in the same area for comparison. Included are two Ramesses II palimpsests from the erased 'Battle of Kadesh' scene. Both sets were measured to the original surface. Appendix, Table 7.8 presents the full data with the abbreviated table here.

The glyphs outside of the cartouches have an average depth of about 7.5 mm. However, 2 out of 6 glyphs have a much deeper depth at nearly 11 mm. Taking these two glyphs out, the remaining average depth is 4.145, nearly the same depth of the right cartouche and comparable to the left.

10.2 *Depth Comparison*

The last set of data to consider is Yurco's depth measurements noted from his field work in 1976–1977. While the Tandy's data was collected via digital means, Yurco measured his depths manually in the field at the wall. It is worth noting that his measurements were collected in 1976 and the Tandy's in 2016, a 40-year difference. Both sets of depths were measured to original surface. This set serves as both a comparison but also as a corroboration towards the digital methodology of the Tandy contrasted with Yurco's manual measurements.

As it can be seen, Yurco's depth and the Tandy's are nearly identical, with only slight variations which can be accounted for both with the method of collection and the span of years between each set of measurements.

10.3 *Depth Findings*

In this paper, three sets of depth data were presented. First, the depth measurements for the cartouches indicate that the interior surface of the cartouche was approximately 1mm shallower than the original/exterior surface of the stone. This is significant in that it supports the claim of a one past usurpation where the inside surface of the cartouche was shaved to prepare one usurpation (likely Amenmesse's unfinished work; more below), but notably, is not deep enough accommodate multiple usurpations. Second, the depth of the cartouche name-rings and glyphs are between 3–5mm which are equal or shallower than the surrounding glyphs and inscriptions at between 4–10mm, and equal or shallower than the Ramesses palimpsests at 4–10mm. Lastly, the digital methodology for collection compares well with manual measurements. This seems to verify both datasets of measurements, but digital methodologies allow for additional measurements in a variety of contexts.

11 3D Model

The premise for this paper was that photogrammetry can be used to accurately measure depths rather than simply use it for the creation of a 3D model. While this paper demonstrated the accuracy of these measurements, in the process of consolidating the data, it was discovered that a 3D model presented a unique perspective not available in the field. The 3D model, as a digital creation, can also be dissected to expose unique perspectives, something one cannot do to the original monument. This is an innovative application for the use of photogrammetry.

11.1 *Cross Sections*

In setting up the cartouches for post-processing, the program CloudCompare allows the user to create cross-sections of a model. This was initially done to reveal certain angles to make it easier to measure depth. However, in doing so, it revealed a unique perspective of the cartouches not available before. As noted in the first depth measurements, the cartouches demonstrated that the interior was shallower than the exterior. Figure 7.3 shows the unique perspective of the

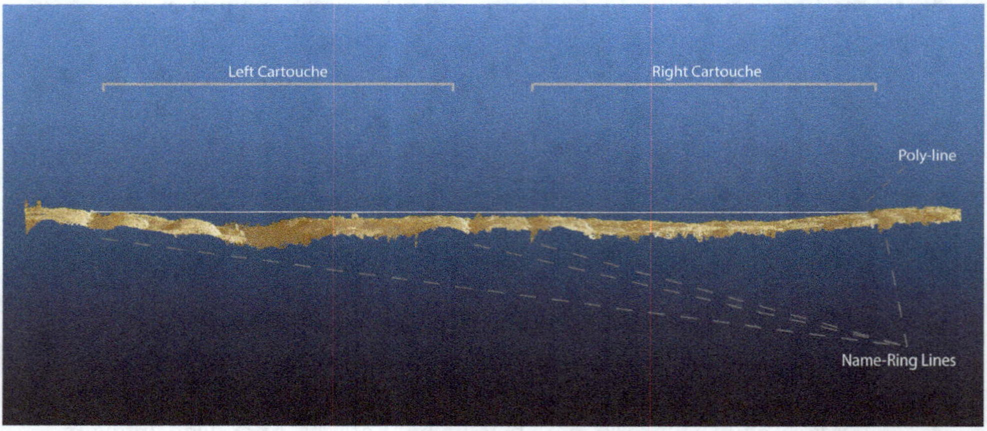

FIGURE 7.3 Cross section view of cartouches created using CloudCompare and edited in Illustrator

cross-section view of the cartouches where the 'shaving' effect of the usurpation is clearly seen. A white polyline was inserted to show the level of the original surface, while the name-ring lines were marked to show the edges of the cartouches. The 'scooping' effect of usurpations is notable in the right cartouche. Please note that the left cartouche is damaged, and the deep cut in its center is the break from one block to the next, shown in Figure 7.3 but the cartouche does begin to show the same scooping effect. Although quite damaged, it was the left cartouche that provided a significant key to unlocking the historical problems. The top portion of the left name ring shows clear evidence of re-carving, see Figure 7.4.

The re-carving of the line created an artificial 'ledge', which allowed a way to measure depths recognizing at least one usurpation of the cartouche. The re-carved line's 'ledge' also created a surface to match depths across the cartouche. In Figures 7.5 and 7.6, white polylines have been added to show the connection of the ledge to corresponding points inside the cartouche. Another data table included below shows the depths of these re-carved lines and their overall depth.

FIGURE 7.4 A view of the 2 cartouches with an arrow pointing to the re-carved lines

FIGURE 7.5 Level perspective of the left cartouche with re-carved top lines. Created using PhotoScan

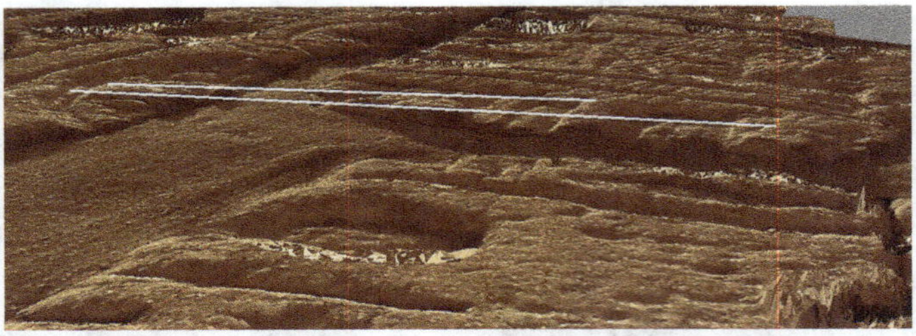

FIGURE 7.6 Another perspective of the left cartouche with inserted polylines. Created using PhotoScan

TABLE 7.5

Left cartouche: re-carved top						
Top Line Depth	3.81	4.2	4.04	5.43	Average:	4.37
Bottom Line Depth	3.51	2.59	4.03	3.64	Average:	3.4425

12 Conclusion

This paper set out to find a solution to historiographical and epigraphic problems regarding the cartouches of the West Wall of the *Cour de la Cachette*. The historical questions considered in this paper were the provenance of the cartouches on the west wall regarding the authorship and identity of the pharaoh who originally inscribed the wall, and any usurpations to these cartouches. The chosen digital application to do this was photogrammetry, which was used to measure depth. The comparative depth of the inscribed lines showed that the inscriptions and stone surface of the cartouches are not deep enough to allow for multiple usurpations.

While the glyphs were measured to an average depth between 4.25–5.05 mm, there is not enough depth to the original surface to support more than 2 layers of glyphs. The interior surface of the cartouche is only 1 mm shallower than the surrounding original surface. In addition, the deepest inscription of the cartouche is only 8.95 mm deep. Subtracting the 1 mm of removed stone material off the inner surface of the cartouches, this leaves approximately 8 mm of depth. Based on the findings that the glyphs average approximately 4 mm in depth, this signifies only two stacked layers of glyphs could exist. The 3D model and cross section demonstrate the perspective that only 1 usurpation is likely, based on the amount on material removed. In short, there simply is not enough room for more than one usurpation. The data shows that both the amount of stone material removed and the average depth of the inscriptions are not deep enough to support more than 2 layers of glyphs: the original layer and one usurpation. The depth data solves one of the historical problems: the number of usurpations to these cartouches. This allows us to focus on the second historical problem, the provenance of the original author.

The results of the photogrammetry bolster the analysis offered above that the battle scenes on the west wall of the *Cachette* were originally commissioned by Merneptah and usurped only one time. Because Amenmesse had good reasons to deny the validity of his rival Seti II, son and rightful successor of Merneptah, he is the obvious candidate for the perpetrator of the erasure of the names. However, he did not have time (apparently) to carve his own names, as Seti II reunited Egypt and placed his own names in the erased cartouches.

13 Other Concluding Thoughts

Photogrammetry is not a replacement to epigraphy. Faint inscribed lines may still be invisible to the camera lens yet are visible to naked eye, which is why the field work of epigraphy is still so important. Rather, photogrammetry presents a way to examine the monuments in lieu of being in the field and without physically touching the wall, such as for measurements. In addition, it represents a method to dissect a monument or space in ways that cannot be done to the actual monument, creating cross-sections or finding other unique perspectives not available in a real-world setting. This is an innovative application for the use of photogrammetry. In short, digital and "in field" epigraphy need each other! This paper was not "just another photogrammetry report", because it applied the original purpose of photogrammetry in new ways that complement traditional epigraphy.

Appendix

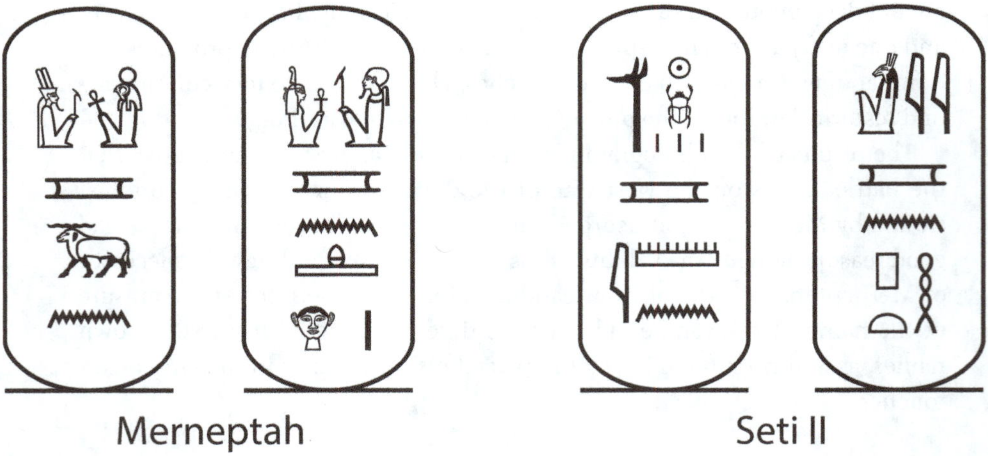

FIGURE 7.7 Cartouches of the North End scene in the CDLC

FIGURE 7.8 CDLC North End Cartouches as they appear on the wall today

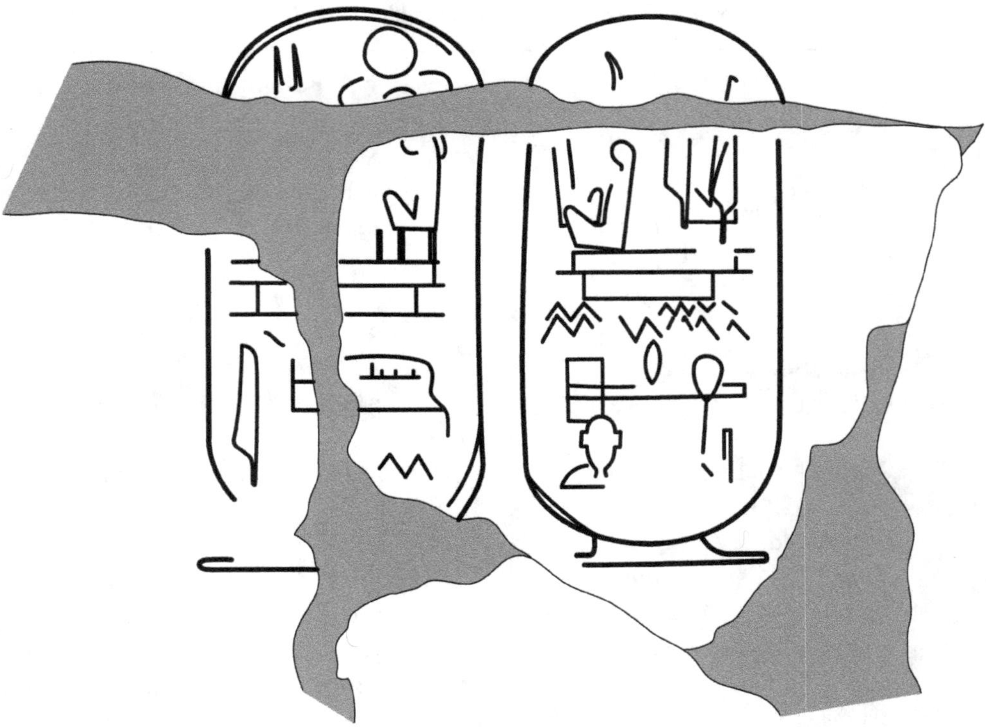

FIGURE 7.9 Initial draft version of the epigraphic drawings of the cartouches

TABLE 7.6 Cartouche name ring line depth and width

Left cartouche		*Line:*	Right cartouche	
		Depth		
1. Left Side			3. Right Side	
*Outer**	*Inner***		*Inner***	*Outer**
6.36	.58		2.78	4.91
6.93	.25		3.03	5.18
6.99	.03		3.32	3.91
Averages:			Averages:	
6.76	5.953333		3.0433333	å4.666667
Overall Average:	6.356667		Overall Average:	3.855
2. Right Side			4. Left Side	
*Inner***	*Outer**		*Inner***	*Outer**
3.37	4.37		3.1	3.52
3.77	4.46		3.45	2.77
4.19	4.51		3.21	3.22
Averages:			Averages:	
3.776667	4.446667		3.17	3.253333
Overall Average:	4.111667		Overall Average:	3.21167
		Width		
1. Left Side			3. Right Side	
1.1**cm**	.2 cm			6.24
1.7 cm	.5 cm			6.59
	.4 cm			6.94
Average:	.38 cm		Average:	6.59
2. Right Side			4. Left Side	
	6.59			5.32
	5.25			5.67
	5.53			6.61
Average:	5.79		Average:	5.87

* Outer line depths measured from original surface.
** Inner line depths measured from remaining surface.

TABLE 7.7 Cartouche glyphs depths with notes. Depths taken from remaining surface

Left cartouche		**Notes:**	**Right cartouche**		**Notes:**
𓏠		𓏠 = Right side only was measured; remainder too damaged	𓎛		𓎛 = deeper in corners, shallow on curves
	5.84			3.1	
	5.21			2.7	
	6.12			2.69	
Average:	5.723333			3.66	
			Average:	3.0375	
𓊃		𓊃 = double-stacked glyph, lower is deeper but damaged with rounding/sloping edges	𓈖		𓈖 = damaged heavily on R side, taken primarily from L side
along top	7.19	R edge		2.96	
	6.98	R edge		3.11	
	4.4	L upper		3.16	
Average:	6.19		Average:	3.076667	
top-to-bottom	4.15		𓊪		𓊪 = bottom of glyph is damaged (averages 0.5 mm or less)
	4.39			4.13	
	4.26			5.21	
Average:	4.266667			5.04	
				5.14	
on bottom	5.24		Average:	4.88	
	3.93				
	4.07		𓇌		𓇌 = depths taken from both reed leaves
	5.08			5.22	(drops to 2 mm average in reed stems due to damage)
Average	4.58			4.57	
				3.3	
				4.81	

TABLE 7.7 Cartouche glyphs depths with notes. Depths taken from remaining surface (*cont.*)

Left cartouche	Notes:	Right cartouche	Notes:
Seated figure		5.07	
5.55		4.92	
5.0		5.53	
4.99		Average: 4.774286	
5.69			
4.24		**Seated figure**	
2.93		5.36	
3.1		5.33	
Average: 4.5		4.8	
		6.16	
		6.15	
		6.59	
		4.23	
		Average: 5.517143	

TABLE 7.8 Comparative glyph depths near cartouches. Depths taken to original surface

Comparative Glyphs

	6.96
	2.72
	2.62
	2.21
	2.21
	2.96
	4.91
	5.3
Average:	3.73625
 **(2 parts)**	
line	4.19
	2.72
	4.36
	2.58
Average:	3.4625
flag	10.4
	12
	9.83
Average:	10.74333
R-II Palimpsests	
Horse back and leg	
	9.19
	10.5
Average:	9.845
Soldier shoulder & arm (top of wall near missing blocks)	
	4.75
	3.49
	3.93
Average:	4.056667

	10.9
	8.96
	8.98
	16.2
	8.61
Average:	10.73

	4.45
	4.43
	4.67
Average:	4.516667
= (under above left cartouche)	
	4.6
	5.42
	4.15
	4.95
	4.86
	5.1
	5.06
Average:	4.877143

Bibliography

Brand, Peter J. 2011. "The Date of the War Scenes on the South Wall of the Great Hypostyle Hall and the West Wall of the Cour de La Cachette at Karnak and the History of the Late Nineteenth Dynasty." In *Ramesside Studies in Honour of K.A. Kitchen*, edited by M. Collier and S. Snape: 51–84. London: Rutherford Press.

Brand, Peter J. 2009. "Usurped Cartouches of Merenptah at Karnak and Luxor," in *Causing His Name to Live: Studies in Egyptian History and Epigraphy in Memory of William J. Murnane*, 29–48. Leiden; Boston, MA: Culture and History of the Ancient Near East 37, Brill.

Brand, Peter J. 2005. "Ideology and Politics of the Early Ramesside Kings (13th Century BC): A Historical Approach," in *Prozesse des Wandels in historischen Spannungsfeldern Nordostafrikas/Westasiens: Akten zum 2. Symposiumdes SFB 295 Mainz, 15.10.–17.10.2001*, ed. Walter Bisang et al, 31–36. Mainz: Verlag.

Breasted, James Henry. 1903. *The Battle of Kadesh. A Study in the Earliest Known Military Strategy by James Henry Breasted*. Chicago: University of Chicago Press.

De Reu, Jeroen, Philippe De Smedt, Davy Herremans, Marc Van Meirvenne, Pieter Laloo, and Wim De Clercq. 2014. "On Introducing an Image-Based 3D Reconstruction Method in Archaeological Excavation Practice." *Journal of Archaeological Science* 41: 251–262.

Dodson, Aiden. 2010. *Poisoned Legacy: The Fall of the Nineteenth Egyptian Dynasty*. Cairo: American University in Cairo Press.

Grimm, Albrecht. 2007. "The Origin of the Term Photogrammetry." Photogrammetric Week 2007. Stuttgart, Germany: International Society for Photogrammetry and Remote Sensing-ISPRS: 53–60. https://www.isprs.org/society/history/Grimm-The-Origin-of-the-Term-Photogrammetry.pdf (accessed 02-25-2022).

Heinz, Susanna Constanze. 2001. *Die Feldzugsdarstellungen Des Neuen Reiches: Eine Bildanalyse*. Wien: Verlag Der Osterreichischen Akademie Der Wissenschaften.

"History of Photogrammetry." 2008. Center for Photogrammetric Training: 1–36. https://ibis.geog.ubc.ca/courses/geob373/lectures/Handouts/History_of_Photogrammetry.pdf (accessed 02-25-2022).

Hopper, Roy. 2010. *The Monuments of Amenmesse and Seti II: A Historical Inquiry*. PhD Dissertation, University of Memphis.

Iskander, Sameh. 2002. "The Reign of Merenptah." PhD Thesis, New York University.

Kitchen, Kenneth A. 1964. "Some New Light on the Asiatic Wars of Ramesses II." *The Journal of Egyptian Archaeology* 50 (1): 47–70.

Kitchen, Kenneth A. 1993. "L'Egypte Ancienne et l'Ancien Testament: Aperçus Nouveaux." *Bulletin de la Société Française d'Égyptologie:* 12, 21.

Kitchen, Kenneth A. 1999. *Ramesside Inscriptions: Translated and Annotated Notes and Comments. Vol. 2*. Oxford: Blackwell.

Kuentz, Charles. 1928. *La Battaille de Qadech*. Memories de l'Institut Français d'Archeologie Orientale 55. Cairo: Institut Français d'Archeologie Orientale.

Lurson, Benot. 2003 "Israël sous Merenptah ou le Sort de l'Ennemi dans l'Égypte Ancienne." *Estrangers et Exclus dans Le Monde Biblique: Colloque International a l'Université Catholique de l'Quest. Angers. Les 20 et 21 Février 2002*: 45–62. Anger: Université Catholique.

Rainey, Anson F. 2001. "Israel in Merenptah's Inscription and Reliefs." *Israel Exploration Journal* 51: 57–75.

Redford, Donald B. 1986. "The Ashkelon Relief at Karnak and the Israel Stela," *Israel Exploration Journal* 36: 188–200.

Remondino, Fabio, and Sabry El-Hakim. 2006. "Image-Based 3D Modelling: A Review." *The Photogrammetric Record* 21 (115): 269–291. https://doi.org/10.1111/j.1477-9730.2006.00383.x (accessed 02-25-2022).

Revez, Jean and Peter J. Brand. 2015. "The Notion of Prime Space in the Layout of the Column

Decoration in the Great Hypostyle Hall at Karnak." Pages 253–310 in *Cahiers de Karnak* 15. Centre franco-égyptien d'étude des temples de Karnak. Cairo.

Sapirstein, Phillip. 2016. "Accurate Measurement with Photogrammetry at Large Sites" *Journal of Archaeological Science* 66: 137–145.

Schwaller De Lubicz, R.A., Georges De Miré, Valentine De Miré, and Lucie Lamy. 1999. *The Temples of Karnak*. Rochester, Vt.: Inner Traditions.

Sourouzian, Hourig. 1989. *Les Monuments du roi Merenptah*. Sonderschrift / Deutsches Archäologisches Institut Kairo 22: Mainz.

Spralinger, A.J. 1985. "Notes on the Reliefs of the Battle of Kadesh." In *Perspectives on the Battle of Kadesh*, edited by Hans Goedicke et al, 1–42. Baltimore: Halgo.

Stager, Lawrence E. 1985. "Merenptah, Israel, and the Sea Peoples: New Light on an Old Relief." *Eretz–Israel* 18: 56–64.

Steele, Keith C. 1940. *The Coregency of Ramses II with Seti I and the Date of the Great Hypostyle Hall at Karnak*. Studies in Ancient Oriental Civilization 19. Chicago: University of Chicago Press.

Walford, Alan. 2017. "What is photogrammetry?" http://www.photogrammetry.com/ (accessissed 02-25-2022).

Yurco, Frank J. 1978. "Merenptah's Palestinian Campaign." *Journal of the Society for the Study of Egyptian Antiquities:* 8,70.

Yurco, Frank J. 1986. "Merenptah's Canaanite Campaign." *Journal of the American Research Center in Egypt* 23: 189–215.

Yurco, Frank J. 1990. "3,200-Year-Old Picture of Israelites Found in Egypt." *Biblical Archaeology Review* 16, no. 5: 20–38.

Yurco, Frank J. 1997. "Merenptah's Canaanite Campaign and Israel's Origins." In *Exodus: The Egyptian Evidence*, edited by E.S. Frerichs and L.H. Lesko et al, 28–53. Winona Lake, IN: Eisenbrauns.

CHAPTER 8

The 3D Digital Documentation of Shaft K24 in Saqqara

Matthias Lang, Ramadan Hussein, Philippe Kluge

Abstract

In our contribution we discuss in detail the three-dimensional digital documentation strategy of the intact burial chambers of Shaft K24 of the recently discovered mummification workshop complex by the Saqqara Saite Tombs Project (SSTP). This 30 m deep shaft served as a communal burial site. In six tombs dug into the walls of the shaft at different depths, a large number of significant archaeological materials were found in situ. After an introduction to the project, we discuss the different 3D technologies, which we used to document the site and the excavation process, highlighting their advantages and disadvantages. Besides the technical description, a main focus of the paper will be the influence of 3D documentation on the research and conservation processes.

Keywords

Ancient Egypt – 3D-documentaion – Saqqara – Terrestrial Laser-scanning – Image-Based-Modelling

1 Introduction

The digital three-dimensional documentation of archaeological sites, features and artifacts has become a standard approach in the past decade, replacing traditional analog methods almost completely. The digital methods are more efficient and accurate. They also make possible to produce a digital copy of a physical reality, whereas the conventional methods are always a derivation of the three-dimensionality of the actual shape of a place, space or object. Such a derivation always loses something and can never represent the shape and appearance of reality to such a degree of accuracy. Moreover, a drawing, sketch or plan is always interpretative and depends on the experience and the specific research interest of the draughtsman.[1] To a certain extent, this also applies to photography, since the photographer must decide on a shooting distance or angle, which again depends on a specific interest of the researcher. In theory, digital 3D methods are capable of capturing a scaleable and neutral copy of an excavated feature, whose shape is independent of the experience and interest of the researcher, whereby, of course, the decision to remove a layer is already an interpretative act that is very much dependent on the factors mentioned and this statement refers only to the pure process of documentation. From this digital facsimile of the site, all views, plans and illustrations can then be driven easily at any scale. In fact, digital recording in this case is a preservation by documentation, in which fragile objects and features are transformed from their physical forms into digital format. In addition, through digital recording of the excavation processes, archaeologists harvest large corpora of digitally born datasets that would offer present and future scholars unlimited research potentials.

In practice, however, this type of ideal documentation cannot be expected because no technology has yet been able to capture all the details of a site at all necessary scales in one single step. In the reality of fieldwork, researchers and digital specialists must decide beforehand what is to be recorded and at what level of detail, and which technology is best suited for this purpose. In most cases it is necessary to combine different technologies and methods to meet the requirements of archaeological research in the best possible way.[2] All 3D data can be integrated into globally valid coordinate systems and combined in this way to produce ever newer and demand-oriented derivatives of reality that serve a specific research interest. The term copy does not really apply either, since such a copy can only preserve the geometry and coloration of a space or object. All the other senses we use in the perception of an archaeological site have so far been difficult or impossible to preserve virtually. And of course, even the best facsimile remains a copy and cannot preserve or replace a physical site or object even with the most sophisticated visualization or printing techniques.[3]

However, there is no doubt that there is no available approach at this stage that would allow us to copy and preserve such a large amount of archaeologically relevant information. Digital documentation is proven to be of utmost importance in the recording of fragile artifacts, organic materials, and human

1 Morgan and Wright 2018.
2 Forte 2014; Siebke et al. 2018.
3 McCoy 2020, 196; Forte 2014.

remains in their varied contexts. Indeed, human remains and archaeological objects made of organic, degradable and corroded materials are the most endangered finds, because they are prone to loss and disintegration once exposed to environmental agents of deterioration during excavation.

A problem that has received little attention so far is that 3D information is rarely integrated as such in the research process. Rather, it is usually used to derive two-dimensional plans and views from the three-dimensional data sets. This is not surprising to us, since archaeologists are used to and experienced in conducting their research with such derivatives for their printed books and articles, which have been for long the only possibility of knowledge transfer and presentation. Working with 3D data is moreover new to many archaeologists and often requires the use of complex software that must first be learned. Also, hardware and software are often missing to work efficiently with the recorded datasets. Therefore, specialists are usually required to extract information from the complex 3D data to present it to the researchers as conventional products like pictures or plans. Used in this way, the 3D models are only an intermediate product for creating traditional visualizations and as such have no value of their own. Another decisive reason for the limited distribution of 3D models is certainly the lack of possibilities to publish them in standardized and sustainable formats in combination with conventional publications. As Olson already pointed out, we run the risk of simply imitating traditional methods digitally and making them faster and more accurate, but not realizing the analytical potential of 3D methods.[4]

When we work directly with 3D data on a computer screen, we must always bear in mind that this is already a two-dimensional derivation that has lost some essential information, such as the impression of the size of the object or space. With the technologies of Virtual Reality (VR), however, possibilities have been available for years now to experience contexts and objects in their real size and their actual environment. So far, the use of such tools has mostly been limited to the dissemination of research results to the public and is hardly reflected in research or university education.[5] Moreover, archaeological VR applications focus more on the reconstruction and immersive experience of the past and less on the current appearance of a site after or even during excavation.[6] An exception to this is the Çatalhöyük project, which for decades has served as a testing ground for various digital technologies in a large-scale archaeological

4 Olson and Placchetti 2015.
5 Bekele et al. 2018; Hageneuer 2020; Kevin Kee 2014.
6 Holter and Schwesinger 2020; Forte 2014.

excavation.[7] Already in 2009 the project started to document the excavation process in 3D in order to analyze the data in VR environments.[8] The project has clearly demonstrated the potential of such an approach and it can be seen as a methodological blueprint for the digital recording and analysis of a site in the different phases of its investigation.[9] However, it also shows the enormous technical effort and knowledge that is necessary for this. The technical and human resources used are only available to a fraction of the archaeological field projects, and a direct transfer to another project is therefore usually out of the question.

In our contribution we want to discuss how we have developed a digital and three-dimensional documentation strategy for the tombs in Shaft K24 in Saqqara with significantly less use of personnel and hardware, which nevertheless meets the requirements of the project and offers the researchers a direct scientific benefit, which could not have been achieved with conventional methods. We will focus specially on how we integrated the results directly into the research and conservation processes and how they have been influenced by it.

2 The Saqqara Saite Tombs Project (SSTP)

The Saqqara Saite Tombs Project (SSTP) of the University of Tübingen received two rounds of funding from the German Research Foundation from 2016–2019 and 2020–2023. It began as essentially a second round of excavation and documentation of the Saite-Persian tombs (Dynasties 26 and 27, ca. 664–404 BC) located to the south and east of the pyramid of King *Wenis* of Dynasty 5 (ca. 2345–2315 BC) at Saqqara.[10] Since one of the main goals of the SSTP is to produce exact facsimiles of the texts of these tombs, we discussed the advantages of the employment of 3D terrestrial laserscanning (TLS) and image-based modelling (IBM) to obtain rectified and high-resolution images of the texts on the vaulted ceilings of the burial chambers. We also decided to employ a confluence of digital technologies in the mapping of the site's subterranean and aboveground structures. The digital documentation strategy for the tombs

7 Berggren et al. 2015.

8 Forte 2014.

9 Lercari et al. 2018.

10 These are the tombs of Tjaninanihbu, Psamtek, Padinist, Padinit, and Hekaemsaf, see: Barsanti and Maspero 1900b; Barsanti and Maspero 1900a; Bresciani, Giangeri-Silvis, and Pernigotti 1977.

was only recently described and discussed by the authors.[11] The combination of laser scanning and image-based modeling has proven to be an efficient documentation method for the tombs investigated up to now. The integrated approach offered the possibility to record every feature in the desired scale and resolution. As already mentioned, the bigger part of the features recorded so far were excavated decades ago and the 3D models always represent the final stage of the excavation process. The process itself is no longer reproducible by using the models because the information was not recorded at the time. This reduction to the last phase of the excavation process applies not only to the shaft tombs in Saqqara that we investigated, but to the same extent to all other previous 3D documentation projects in Egyptian tombs such as those of *Seti, Nefertari* or *Tutankhamun*.[12]

The situation is quite different in the newly discovered Shaft K24. Here, for the first time, it was possible to record all phases of the excavation of an untouched Egyptian tomb complex in 3D and to integrate the results directly in the research and conservation processes.

Recognizing the advantages and research potential of digital documentation, we decided to digitally record the excavation processes of the hallways and burial chambers in Shaft K 24 of the Saite mummification workshop complex at Saqqara. Shaft K 24 is spatially and functionally associated with two embalming facilities, namely a subterranean embalming room and a tent of purification, i.e., a structure called *ibu*. The shaft is located in the middle of the *ibu*-structure and measures 3 m. × 3.50 m. It reaches down to a depth of 30 m and served as the communal burial shaft of the mummification workshop complex (Fig. 8.2). It has six tombs cut into its walls at different depths. Some of them are simple loculi with one or two mummies (Tombs 1 and 4), a large room with multiple burials (Tomb 5), a complex of niches arranged along corridors (Tomb 2) or a complex of burial chambers laid out around hallways (Tombs 3 and 6) (Fig. 8.3).

Tomb 6 is cut into the north wall of Shaft K 24 at a depth of 30 m. It consists of two hallways on a north-south axis and six burial chambers. These burial chambers are arranged in pairs around the two hallways: one pair on the west (K24 W1-W2), another on the east (K24 E1-E2), and a third on the north (K24 N1-N2). These burial chambers yielded diverse and significant archaeological finds, including 17 badly decayed human mummies, 19 calcite and pottery canopic jars, 4 limestone sarcophagi, 10 badly thermo-disintegrated and decayed wooden coffins, thousands of faience *shawabti* figurines, a dozen

11 Lang et al. 2020.

12 Lowe 2018; Factum Arte 2009.

miniature marl clay and faience embalming cups, small symbolic sundried mud boat models, and a gilded silver mummy mask. Only the adoption and implementation of 3D digital technologies allowed us to precisely record these fragile human remains and degraded artifacts in their original archaeological contexts.

3 3D Documentation of the Archaeological Excavation Process

3D technologies are becoming an increasingly essential tool in the toolbox of field archaeology. Due to their accuracy and efficacy, they replace or have already completely replaced conventional methods.[13] This development is mostly driven by the availability of efficient, robust, and easy-to-use software-environments based on image-based algorithms like Structure-From-Motion and Multi-Stereo-View.[14] This Image-Based-Modelling (IBM) is comparatively inexpensive and its basic features are easy to learn by an archaeologist and do not require extensive specialist knowledge.[15] Furthermore, no special equipment is required. A standard digital camera for recording the necessary overlapping image data sets, a Total Station or DGPS for locating and scaling the results and a computer for calculating the models. These are available equipment in almost every project. The necessary software is in most cases inexpensive for scientific use and the manufacturers all provide free trial versions that allow a low-threshold entry into IBM.

Terrestrial Laser Scanning (TLS) on the contrary is comparatively expensive and the necessary equipment is hardly affordable for most archaeological projects. A laser scanner, powerful computers and storage systems and the required special software easily reach the one hundred-thousand-dollar threshold. Furthermore, the skills of a specialist are necessary. This applies less to the scanning itself than to the processing of the data in special software packages, whose extensive functionalities are difficult to master.

These observations easily explain IBM's great success in archaeology, while TLS has always been a highly specialized application, available only to a few projects, and could not establish itself as the standard documentation method for archaeology due to cost reasons and complex data processing. In the following we want to discuss why we nevertheless combined both methods for the documentation of the burial chambers in shaft K24.

---

13 Doneus, M. et al. 2011; Reu et al. 2014; Reu et al. 2013; Galeazzi 2016.

14 Verhoeven, G. et al. 2013.

15 Aicardi et al. 2018; Douglass, Lin, and Chodoronek 2015; José Luis et al. 2019.

4 3D Documentation Strategy of Shaft K24

The image-based approach has obviously some clear advantages over the laser scanner, which have been particularly effective in narrow spaces. The scanner must always be placed on a tripod, which is often a challenge, especially within complex archaeological contexts, and requires extreme caution from the operator. As the scanner is not able to capture the area under the device, a part of the space to be captured is always in the so-called scan shadow, which always requires the device to be placed several times in order to capture the whole feature.

Since the scanner uses an active sensor, it offers the advantage that it can be used in the dark without additional light sources to capture highly accurate 3D information.[16] However, this only applies to the geometric information, so that we could only use this advantage in the corridors and shafts whose surfaces do not show archaeologically relevant color information. If the scan is to be provided with color information, this must be recorded in a second step with the scanners built-in camera system or an external setup. As the color information undoubtedly represents a decisive component of the most features and artefacts, these must be illuminated. As a result, the data acquisition with the scanner is a complex, physically demanding task and the use of the heavy equipment with the large tripod requires extreme caution by the operating crew.

In spite of these obvious disadvantages, laser scanning offers some benefits over the image-based methods. For example, several hundred individual scans can be combined in high accuracy in a semi or even fully automated process to create an integrated model of the entire site. This has made, in the case of the SSTP, the complex interrelationships of the underground and above-ground features visible (Fig. 8.2). Image-based models, on the other hand, usually have to be connected manually, and never reach the accuracy of the registered scans.[17] Furthermore, they are not scaled, and control points measured with a total station or a measuring tape are necessary to scale the object to its actual size. On the contrary, the point-clouds recorded by the laser-scanner are always scaled in a metric system and no additional information to derive accurate measurements is required.[18] They are therefore perfectly suited to derive highly accurate plans and views that cannot be achieved with any other technology

16 Historic England 2018.
17 Kersten, Mechelke, and Maziull 2015.
18 Reu et al. 2014.

at this level of accuracy, precision, and speed. In this manner, the scientists involved can access detailed plans and sections shortly after the recording process.

The result of the TLS process is always a dense point-cloud, which is only conditionally suitable to represent reality, since our physical environment does not consist of a multitude of points distributed in space, but of solid geometric objects with a closed surface. For the plans and sections, this becomes visible only at a very small scale and can therefore be neglected. In the case of photorealistic derivations or direct use of the objects in digital 3D environments, however, a strong alienation effect occurs, which makes the models look very far removed from reality.[19] Using Poisson Surface Reconstruction, the point-clouds can be meshed together, but this process requires the point-normals, which indicate the orientation of the surface.[20] While IBM calculates them automatically, TLS requires them to be derived from the point-clouds, such a step may lead to an incorrect reconstruction. In addition, the common tools for registering point-clouds are not able to calculate the meshes and another specific—in the most cases very expensive—software is required. Texturing of the meshes based on the images for the point-cloud coloring or on the color of the point-cloud itself is also possible, but the results are rarely satisfactory—especially in comparison to the models resulting from the image-based methods (Fig. 8.1).

Without doubt, the scanner with its resolution of 6.3 mm at a distance of 10 m is not suitable to capture all features in sufficient detail. In order to record all relevant information in a reasonable resolution, it would be necessary to apply a whole series of different scanner systems that have the respective suitable resolution. This approach is not very practical, for it requires extensive and expensive equipment with special hardware and software that produces a variety of different data formats.[21] In contrast, the achievable resolution of the IBM is not limited by the used camera, but is defined by the resolution and size of the camera sensor, the selected focal length and the distance from the object to be recorded. This makes the procedure highly variable and can easily be adapted to the diverging requirements by changing the basic parameters like the lens or the distance from the object. In this way, the IBM approach is suitable for both large rooms and small objects.[22]

In contrast to the data sets acquired by laser scanning, the data for the IBM can be directly meshed and textured with the high-resolution images used

19 Olson 2016.
20 Kazhdan, Bolitho, and Hoppe 2006.
21 Siebke et al. 2018.
22 Historic England 2017.

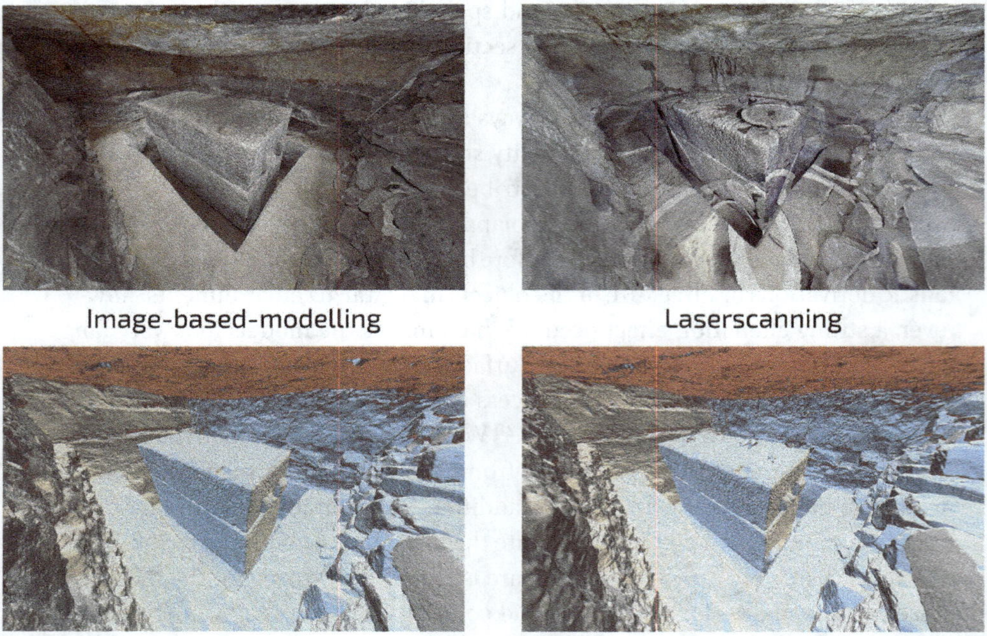

FIGURE 8.1 Comparison of textures and meshes from IBM and TLS of chamber K24 E1. The top two shots show the textured models and the bottom ones only the meshes.

for reconstruction in a combined and seamless workflow in the same software.[23] Thus the models generated with IBM reproduce reality much more accurately than the colored point-clouds recorded with the laser scanner as already described above (Fig. 8.1).

In contrast to those benefits, the approach has some clear downsides compared to the TLS. Although image-based methods are able to combine several hundred digital pictures in order to derive a 3D model, the processing of the data is time consuming and can easily take up several hours or even days related to the size of the image-set and the available computing resources.[24] Therefore, it will always take some time to control the results of the data acquisition before the excavation process can proceed and previously documented features can be removed.[25] Meanwhile, the data recorded by the laser-scanner are available directly after, or even during, the scanning process. The control of the generated models is of high significance, because not every acquisition is successful on

23 Reu et al. 2013; Verhoeven, G. et al. 2013; Galeazzi 2016; Davies; Davis et al. 2017.
24 Doneus, M. et al. 2011.
25 Olson and Placchetti 2015.

the first attempt. This can be caused by a lack of overlapping of images, incorrect exposure, blurring due to too long exposure times or the lack of depth of field. During the recording process these errors are easily overlooked and only become visible during or after processing. Especially structures with only a few visible features like smooth walls or highly reflective objects are often difficult to capture and, in many cases, require alternative capture strategies.[26] All this easily leads to incomplete or highly interpolated inaccurate models. In some cases, the model generation will only partially succeed or even fail. However, in this recording method, if the archaeological feature removed during excavation and before checking of the obtained records, the resulting models would be incomplete at best, and the context will be lost in worst cases.[27] For this reason, adding more images iteratively to faulty models is a common procedure, but this is only possible if the recorded feature is left untouched for the time it takes to create a satisfactory model. Particularly in an excavation with a very tight schedule, such as the one carried out in K24 with a large number of specialists with different tasks involved in a very confined space. This requirement could only be met in the rarest of cases and the recording of the data for the IBM usually had to be done in one attempt, so as not to interrupt the ongoing excavation and conservation processes. In this case, it is essential that the documentation team has extensive experience in recording image data sets for the IBM.

Furthermore, the number of images to be processed simultaneously is limited by the computing power of the available hardware. Today, a few thousand photographs can be combined into a model in a single process, but this requires extremely powerful computers and the computing process still takes several days. The resulting models usually cannot be displayed and used in their full resolution and the reduction to a fraction of the original size is essential in order to make it possible to work with such a model. This is, however, one of the great strengths of the TLS, which allows the registration of several hundred scans in a manageable time in an integrated, usable model in its full resolution.

As we have shown, both methods have advantages and drawbacks. We have therefore decided to combine both technologies in order to record all relevant information in a suitable scale, resolution and accuracy.

26 Galeazzi 2016.

27 Olson and Placchetti 2015.

5 Laserscanning of Shaft K24

In 2017 and 2018, the already excavated shafts as well as all surface structures and features were completely recorded with 240 individual laser scans, which have been merged into a single model of the whole site. Shaft K24 was individually scanned in order to integrate it into the general model. We used a Leica P40 scanner, which proved to be an excellent and efficient tool, especially on the surface with its range of 270 meters.[28] A full dome 360° scan with full range can be obtained in less than two minutes with a resolution of 6.3 millimeters at a distance of ten meters. Plans and sections in a scale of 1:20 could thus be derived easily. We chose the resolution as a suitable compromise between speed, resolution and data-size, as we used an image-based approach for the more detailed models. With a higher resolution, the size of the data and the scanning-time increases significantly and so does the time needed for registration and post-processing. Less satisfactory, however, is the built-in camera of the scanner, which is necessary to colorize the point-cloud. With this camera, even under the best lighting conditions, the acquisition of an image set takes up to eight minutes and due to the use of a very small image sensor, the quality of the images is not sufficient for our purposes. The duration of scanning and image acquisition is of immense importance, especially on a busy site like Saqqara, where it is extremely difficult to keep workers, archaeologists and tourists away from the scanning area. Therefore, we have decided to use an iSTAR 360 panoramic camera, which collects a fifty-megapixel HDR data set within a few seconds, depending on the light conditions. The iSTAR panoramic camera sped up the process significantly, compared to the scanner's built-in camera, and the undesirable capturing of the images of passersby and other individuals was consequently reduced. Despite the high resolution of the iSTAR 360 camera, the quality of the images is not entirely satisfactory. In comparison with a standard DSLR, the main problems are the unreliable white balance and the problematic behavior of the camera in backlight or sidelight, which very quickly leads to blurred and low-contrast images. Since the images are processed directly in Leica Cyclone, subsequent adjustment is hardly possible. Particularly when taking images in confined spaces, it quickly becomes apparent that the unrecorded area under the camera is larger than the shadow of the scanner. This means that in each scan an uncolored ring remains around the location of the scanner and camera, which must be cut out manually in a time-consuming process.

28 Walsh 2015.

Also, in the narrow shafts, corridors and chambers, the limitations and problems of this approach became evident. The Leica P40 laser scanner weighs almost 13 kilograms and must be levelled on a tripod for each scan. It was often very difficult to find a suitable place to set up the tripod in the narrow chambers without disturbing the features and artifacts. An even bigger problem was again the large scan shadow. Especially in the very narrow burial chambers it was not always possible to place the device in such a way that the feature could be completely captured (Fig. 8.1). Equally complex was the acquisition of the image data set for the coloration of the point-cloud. For this purpose, the features had to be illuminated with hand-held LED lights that had to be placed behind the camera as evenly as possible. This proved to be very difficult due to the wide angle of the panorama camera and required almost artistic and physical skills from the operators in order not to step on artifacts distributed on the ground.

The post-processing of the data has been carried out in Leica Cyclone, Leica Cyclone REGISTER 360 and Autodesk Recap. First, we combined the scans with the images collected with the iStar-camera in a semi-automated process and exported a colorized version of every scan-station in Leica Cyclone. Second, we used the fully automated registration process, based on an Iterative-Closest-Point-algorithm (ICP) implemented in Leica Cyclone 360 REGISTER, to merge the single scans.[29] The use of two different tools was necessary because the merging of panorama and scan only worked in Cyclone and the automated registration of scans was only available in Cyclone 360 REGISTER. The georeferencing of features and structures was carried out using black and white targets measured with the Total Station, which was stationed in the UTM/WGS84 coordinate system used by the project, one that has been used for decades in Saqqara as the basis for all surveying work in order to link the maps of all archaeological projects.[30] In a final step, the merged data set was exported into a standardized data format (E57) and imported into Autodesk Recap to clean them. The resulting Recap data set could be directly opened in Autodesk AutoCAD or PointCab to derive highly accurate plans and sections (Fig. 8.2). The data exportation to a standardized open data format guarantees their long-time usability and makes the data independent from Leica's proprietary, expensive, unstable, and not very user-friendly in-house software.

For each archaeological context we created a completely new model after each phase of the excavation and then connected them together. This model

29 Holz et al. 2015; Besl and McKay 1992.

30 Tavares 2011.

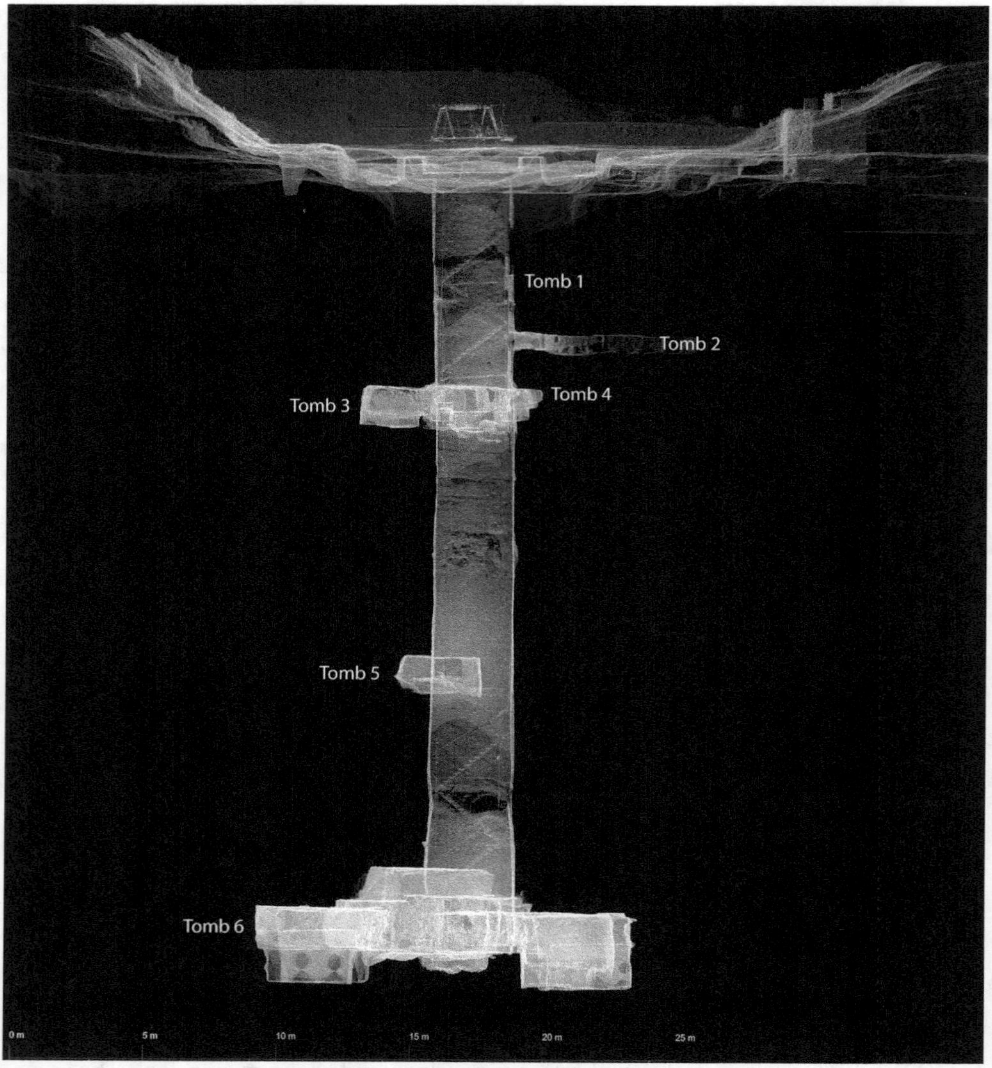

FIGURE 8.2 Shaft K24 connected with the features on the surface by TLS

of shaft K24 was then integrated into the model of the entire site (Fig. 8.2). In Autodesk Recap, the individual rooms and contexts can thus be switched on and off at all stages, making both the chronological sequence of the research process and the complex spatial relationships of the underground and above-ground features visible. Although the procedure has proven to be less than ideal, the chambers, corridors and shafts could usually be surveyed in a short time and the resulting plans and sections could be made available to archaeologists and conservators a few hours later (Fig. 8.3).

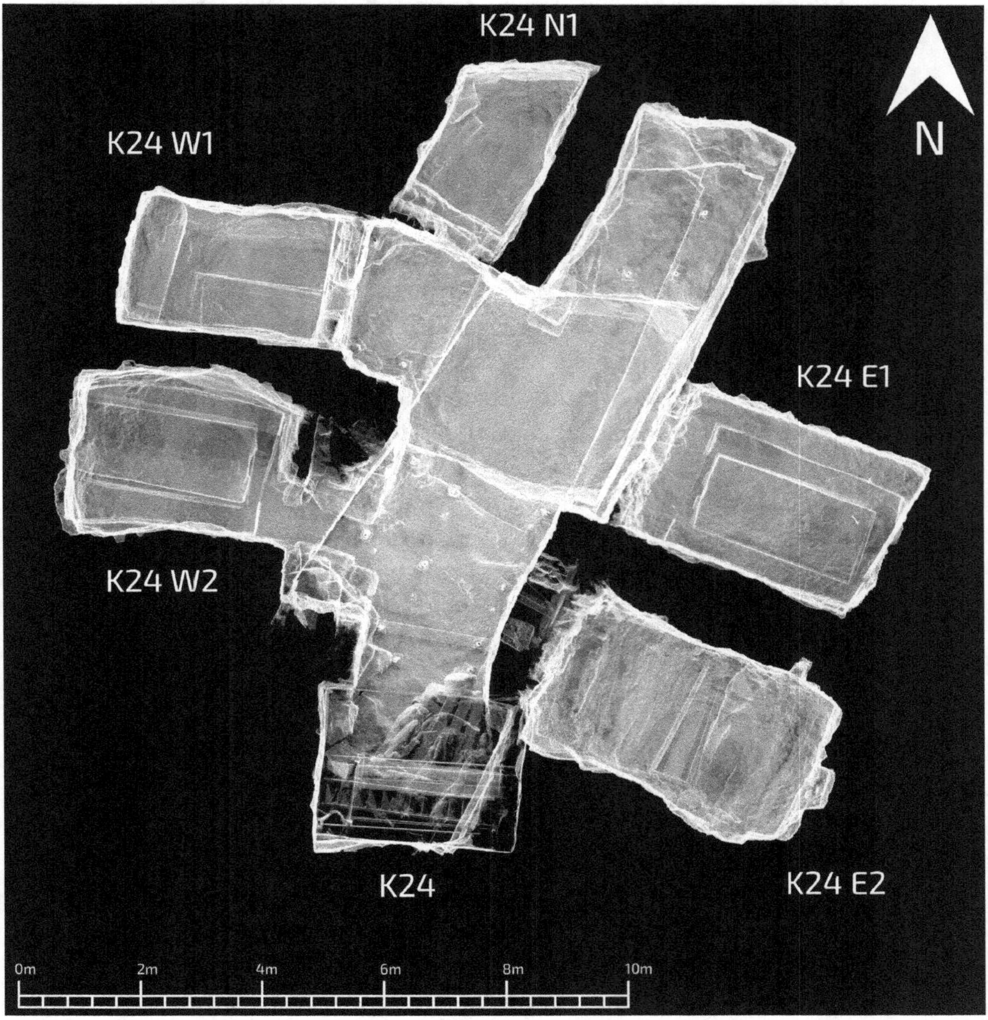

FIGURE 8.3 TLS generated floor plan of the chambers

6 Image-Based 3D-Documentation

To obtain photorealistic virtual copies of the contexts and artifacts, we supplemented the laser scanning with an image-based approach based on Structure-From-Motion and the Multi-Stereo-View algorithms. We chose the Agisoft Metashape Pro for the whole process, from the orientation of the images to the texturing of the meshed model. Metashape is the most widespread IBM tool in archaeological contexts, due to its functionality, usability, stability and very moderate pricing compared to the competitors on the market.

As the burial chambers of Tomb 6 of Shaft K24 are all located deep underground, it was necessary to illuminate them artificially in order to obtain a well exposed image data-set for the 3D-reconstruction. In earlier campaigns, we used a handheld LED-light for the documentation of the Saite Persian Tombs. We tried to place it parallel to the camera sensor in order to obtain even lightning. Especially in narrow spaces, it was not always possible to shoot every image under the same lighting conditions. This led to uneven textures, which are also extremely difficult to adjust in the post processing. Even with a strong LED-lamp, the shutter speeds were so slow that it was not possible to hold the camera by hand. Therefore, a sturdy tripod and a remote shutter release were necessary. Despite the use of a Nikon D750 full format camera, increasing the ISO value led to image noise that became overly visible in the results. The process proved to be time consuming and unsatisfactory as it was not always possible to position the tripod and the LED well in the confined spaces. Therefore, we decided to experiment with a camera-mounted flash. Compared to the LED-light and tripod setup, the hand-held approach accelerated the process drastically and the results were much more evenly illuminated and color-controlled. Therefore, we decided to change our workflow and used the camera-mounted flash for all areas without natural light. The flash was directed upwards at an angle of 45° and equipped with a diffuser to illuminate the object to be photographed as completely as possible and to prevent too strong shadows. The procedure proved to be extremely flexible in order to capture all details of the complex of burial chambers. For example, the floor of chamber K24 E2 was completely covered with the remains of wooden coffins, mummies and grave goods, making it impossible to work without disturbing the context. We improvised in order to obtain a complete record of the situation; we attached the camera to a pole with mounted flash and used a smartphone as display and a remote shutter release.[31]

To scale and georeference the models, we placed small markers in the scene and measured them with the total station in the same grid as the laser scans. Since we always left all markers in place, it was possible to scale and locate each phase of the excavation accurately with the same set of control points.

The next step was to import all images into Agisoft Metashape Pro after some minor adjustments in Adobe Photoshop Lightroom to generate the 3D models. This semi-automated process consisted of six consecutive steps, (1) the image orientation and (2) sparse point-cloud generation, (3) dense 3D point-cloud generation, (4) meshing of the dense point-cloud, (5) texture mapping and (6)

31 José Luis et al. 2019.

ortho-image generation.[32] In addition to the pictures, the coordinates of the control-points can be imported into the software to combine them manually with the markers visible on the images in a semi-automated process in order to scale and georeference the model and its derivatives.

In the next step, the necessary ortho views were derived from each model and the achieved Ground Sampling Distance (GSD) was less than 0.25 mm in all recorded burial chambers. It was usually possible to completely generate the models of one or two chambers or other features after the field work in order to make the results available to the researchers the following day and to check whether a sufficiently overlapping image data set was recorded or whether further images had to be taken to fill gaps in the model. If this was necessary, it could be done directly as a first step in the morning before continuing with the archaeological work. New models of the chambers were also created using IBM after each phase of the excavation (Fig. 8.4)

Due to the considerable flexibility of the method, it was also possible to record a large number of the artifacts from the tombs in addition to the features. These were first brought to the depot and could then be photographed there under good lighting conditions on a rotating plate. We focused on the artifacts that were particularly fragile, such as gilded silver mummy mask from chamber K24 W2 (Fig. 8.5), and on the objects made of organic materials that are difficult to conserve, such as such as the inscribed, yet badly decayed, wooden coffin of Tadihor in chamber K24 E1.

Many of the objects from the chambers were in a poor condition that they could not be removed without destroying them. This was particularly true for the wooden boxes and coffins and the mummies. As an example, we show the mummies in the opened sarcophagi in chambers K24 W1 and K24 W2 (Fig. 8.6). These objects were documented in situ before they were removed from the chambers. Here a resolution between 0.1 and 0.2 mm could be achieved for all objects.

32 Since the basics of Image Based Modelling have been widely discussed in recent years, we will only refer to further literature at this point: Verhoeven, G. et al. 2013; Reu et al. 2013; Reu et al. 2014; Historic England 2017; Aicardi et al. 2018; Kersten, Mechelke, and Maziull 2015; Howland, Kuester, and Levy 2014; Galeazzi 2016; Zachar, Horňák, and Novaković 2017.

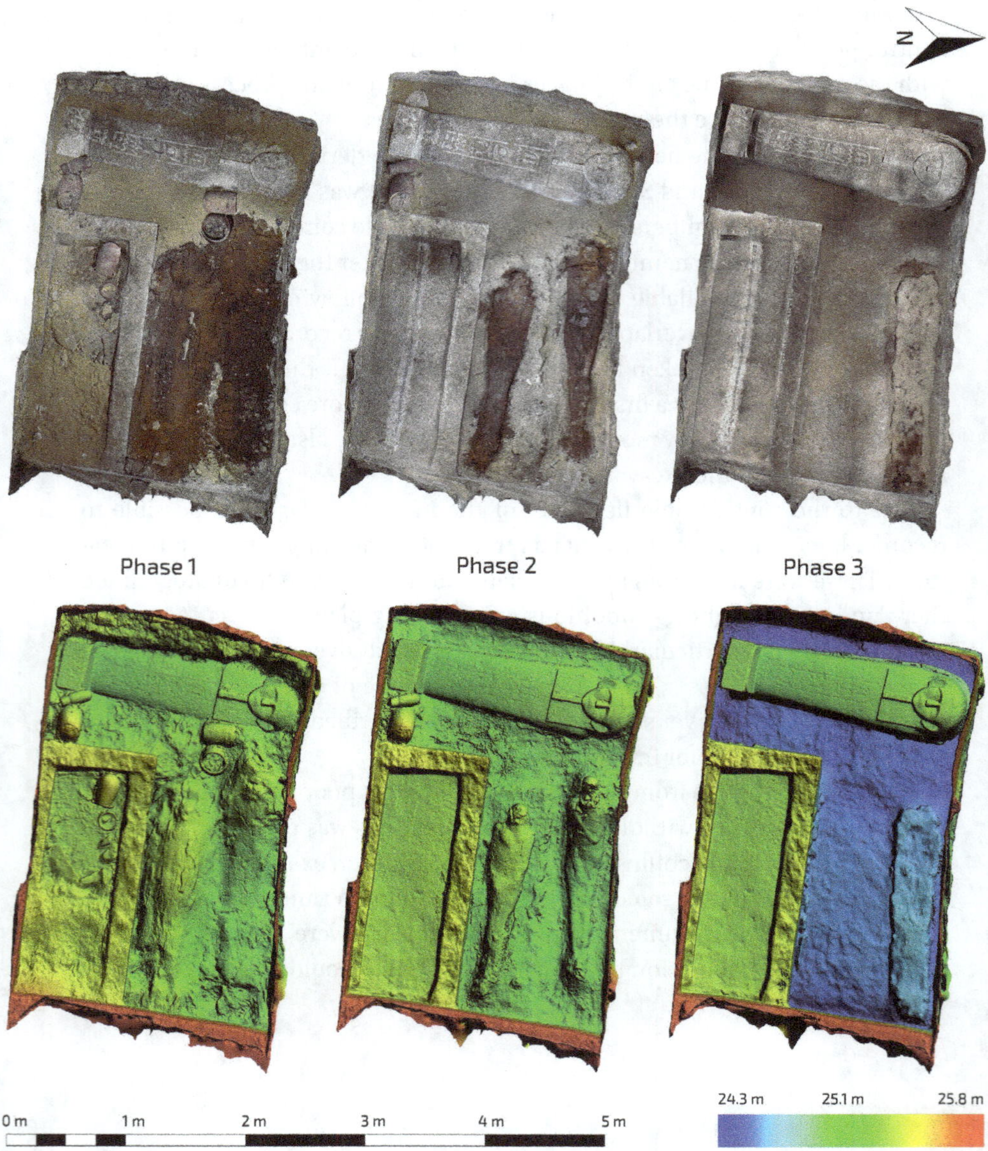

FIGURE 8.4 Three different phases of excavation of chamber K24 W1

FIGURE 8.5 Rendering of the silver-golden mask from K24 W2

FIGURE 8.6 Rendering of the open sarcophagi in the chambers K24 W1 and K24 W2

7 Data Management

One of the problems we have encountered in recent years is the amount of data produced. More than four terabytes of raw and processed data were accumulated, representing records of irrecoverable and unique features and structures that need to be preserved for the future.[33] In the last two decades countless attempts have been made to develop new standards and infrastructures to save and secure data. We have decided to follow the guidelines of good practice published by the English Archaeology Data Service (ADS), which fully cover our requirements.[34] This applies in particular to the extremely detailed metadata schema, which allows a meaningful description of the data in technical and in domain-specific aspects. First of all, we transformed all data to open and sustainable formats like TIFF for the imagery and E57 and OBJ for the 3D-data. All data will be described with metadata according to the guidelines for depositors and stored in the research-data-portal FDAT provided by the University of Tübingen.[35]

8 Results and Experiences

The hybrid approach we chose, to combine TLS and IBM, has proven to be effective for the project presented here, as both methods complement each other perfectly. With the TLS we were able to survey the chambers, shafts and hallways in a short time and the resulting plans and sections could be made available to the archaeologists and conservators a few hours later in order to support their decision-making process at work. All data could be integrated directly into the overall model of the site and thus the virtual copy becomes more and more condensed, revealing the extremely complex spatial relationships of the features below and above ground more clearly. Our project is not the first to visualize a part or even the whole necropolis in an integrated 3D model.[36] However, unlike other projects, our model is not based on the reconstruction of the past ritual landscape, but rather on highly accurate measurements that create

33 Koller, Frischer, and Humphreys 2009; Richards-Rissetto and Schwerin 2017; Lowe 2018; Niven and Richards 2017.

34 Archaeology Data Service 2016.

35 eScience-Center 2018.

36 Sullivan 2020. For a summary of the problems of reconstructing past archaeological landscapes see in particular. Der Manuelian 2013–2013.

a visualization as neutral as possible,[37] whose purpose is not to interpretively reconstruct, but to virtually copy the current state of the site. There can be a multitude of versions of this actual state on a timeline, documenting changes through scientific or illegal excavation, destruction and decay, but also through restoration and new discoveries.

We used IBM and TLS to created different versions of the same archaeological contexts due to the discussed limitations of these two technologies. It is hoped that the rapid technical development of 3D technologies will make it possible in the future to use only one method consistently. However, our approach of combining IBM and TLS has proven to be successful and has enabled us to record all relevant information in a target-oriented accuracy and resolution. Due to the highly accurate georeferencing of all models, we are always able to map the results from TLS and IBM to each other. In the achieved accuracy, both data sets can be regarded as equivalent, a comparison of both methods in chamber E1 shows a mean error of 4 mm. With Reality Capture, which has been available for some time now, it is even possible to process both data types together, thus combining the advantages of both technologies. However, due to the small measurement error, it did not seem necessary to adapt the workflow we use.[38]

Although laser scanning did not prove to be ideal for fully capturing the narrow spaces in K24, it did offer the decisive advantage of the fast availability of the results in form of plans and views. The further progress of the work was then dictated and coordinated on the basis of these derivatives from the 3D models. In addition, the scans made it possible to quickly decide where to support the ceilings of the chambers and corridors in order to prevent them from collapsing after removing walls and debris. Moreover, the scans make it possible to always view the new discoveries directly in the overall spatial and temporal context of the site and not as isolated phenomena.

The IBM-based models, on the other hand, are used by researchers to study the individual features and artifacts. They allow the chambers to be viewed repeatedly from all sides and from all distances under perfect lighting conditions without having to carefully move around the extremely fragile objects in the dark rooms. Orthoimages show the researchers the rooms from a bird's eye view, which makes many of the complex micro-spatial relationships between the individual burials and the objects surrounding them understandable. This applies in particular to the wooden objects and their painted plasterwork. Their

37 See Ch. 1, from the present volume, for "neutrality" in digital representations.

38 Luhmann et al. 2019.

poor condition made it impossible in almost all cases to remove them from the chambers without destroying them. We were surprised that the various specialists in archaeology, conservation, geology and epigraphy, in addition to the plans and views, increasingly demanded a 3D PDF that would allow easy access to the 3D information.

As Polig and Llobera have already pointed out, archaeological research and its results depend heavily on discovering patterns and identifying relationships and connections.[39] The recognizability of these, in turn, depends very much on the type of visualization available to understand these connections. Frischer also emphasizes the importance of data visualization, referring to Colin Ware, who lists five points with which visualization can support the process of understanding and interpreting.[40]

1. It may facilitate the cognition of large amounts of data
2. It can promote the perception of unanticipated emergent properties
3. It sometimes highlights problems in data quality
4. It makes clear the relationship of large-and small-scale features
5. It helps us to formulate hypotheses

The integration of all information in an integrated, neutral and scale-independent 3D visualization thus represents a major advantage over other, conventional visualization methods when it comes to recognizing patterns and correlations and interpreting them, since it fully meets the points that Ware has established. Of course, this integrated overall model represents an ideal solution that cannot be implemented in this way because the virtual research environment required for this purpose does not exist. The quantity, complexity and semantics of the data-set far exceed the capabilities of such a software at present. In the future, however, increasingly powerful software environments will make it possible to connect and visualize such complex data sets semantically correctly in order to recognize still unknown patterns and relationships. It is therefore crucial and our duty that we describe our data carefully and transparently with metadata and store them in sustainable data formats to allow their later use in superordinate virtual systems.

9 Perspectives

As we discussed in the introduction, we see the derivatives from the 3D models only as an intermediate step towards virtual realities that allow direct immer-

39 Polig 2017; Llobera 2011.
40 Frischer 2009; Ware 2004.

sive research in the digital copies of the site. Initial experiments with Unreal engine 4 show that the underground chambers with their clear spatial boundaries are particularly suitable for this purpose. The chambers are exclusively determined by the archaeological context and can be derived directly from the 3D models, adding further elements such as a sky or a sound stage is not necessary here to facilitate the creation of an immersive environment. Also, the artificial lights used during the excavation could be modelled and integrated into the scene very easily. Although this creates a fascinating opportunity to visualize the individual steps of the excavation in an immersive virtual copy of the tombs, we believe it is still too early to attempt to integrate the approach into the research process. In our opinion, the effort to prepare the individual models for integration into the VR environment is still too big and the available head mounted displays are simply too impractical for longer use.[41] Moreover, the possibilities to interact with the virtual world are still very limited and only possible with hand-held controllers that do not allow an intuitive and natural interaction with features and objects. The potential of the technology is undoubted and the rapid technical development of such technologies will solve the problems addressed, thus allowing the researcher to visit and study the site immersively in its various phases of excavation.

We already consider such environments to be ideally suited to make the results of our research tangible and understandable for everyone, while the ever-changing excavation itself is only accessible to a few specialists. VR makes it possible to experience the excavation and to virtually look over the scientists' shoulders as they work. The artefacts from the tombs can thus be viewed directly in their original context and not as an isolated object in a museum showcase. In particular, the usefulness of such environments in the education of students cannot be overestimated, as they can offer students a direct experience of the site and the artefacts that can be enriched with any amount of additional information.[42] Last but not least, it contributes to the democratization of the study of archaeology, as excursions and field schools are only open to a privileged few who can afford the high costs travelling the world. It is therefore essential that we make our data available under free and open licenses in standardized data formats and not, as is the case with most publications, behind paywalls that exclude non-privileged scholars and students. For this reason, it is equally essential that we as scientists are able to master and apply the necessary technologies ourselves so that we do not have to rely on the help of a few

41 Cassidy et al. 2019; Bekele et al. 2018.

42 Kevin Kee 2014.

specialists and commercial companies. Therefore, we have to integrate the procedures and technologies discussed in this paper into the study of archaeology in order to enable future colleagues to assess and apply them independently.[43]

References

Aicardi, Irene, Filiberto Chiabrando, Andrea Maria Lingua, and Francesca Noardo. 2018. "Recent Trends in Cultural Heritage 3D Survey: The Photogrammetric Computer Vision Approach." *Journal of Cultural Heritage* 32:257–266. https://doi.org/10.1016/j.culher.2017.11.006 (accessed 02-25-2022).

Archaeology Data Service. 2016. "Guides to Good Practice." Accessed September 19, 2018. http://guides.archaeologydataservice.ac.uk/g2gp/3d_Toc (accessed 02-25-2022).

Barsanti, Alexandre, and Gaston Maspero. 1900a. "Fouilles Autour De La Pyramide D'ounas (1899–1900). II. Les Tombeaux De Psammétique Et De Setariban. Les Inscriptions De La Chambre De Psammétique." *Annales du Service des Antiquités de l'Égypte* 1: 161–184

Barsanti, Alexandre, and Gaston Maspero. 1900b. "Fouilles Autour De La Pyramide D'ounas (1899–1900). IV. Le Tombeau De Pétènisis." *Annales du Service des Antiquités de l'Égypte* 1: 230–259.

Bekele, Mafkereseb Kassahun, Roberto Pierdicca, Emanuele Frontoni, Eva Savina Malinverni, and James Gain. 2018. "A Survey of Augmented, Virtual, and Mixed Reality for Cultural Heritage." *J. Comput. Cult. Herit.* 11 (2): 1–36. https://doi.org/10.1145/3145534 (accessed 02-25-2022).

Berggren, Åsa, Nicolo Dell'Unto, Maurizio Forte, Scott Haddow, Ian Hodder, Justine Issavi, Nicola Lercari, Camilla Mazzucato, Allison Mickel, and James S. Taylor. 2015. "Revisiting Reflexive Archaeology at Çatalhöyük: Integrating Digital and 3D Technologies at the Trowel's Edge." *Antiquity* 89 (344): 433–448. https://doi.org/10.15184/aqy.2014.43 (accessed 02-25-2022).

Besl, Paul J., and N.D. McKay. 1992. "A Method for Registration of 3-D Shapes." *IEEE Trans. on Pattern Analysis and Machine Intelligence* 14 (2): 239–256. https://doi.org/10.1109/34.121791 (accessed 02-25-2022).

Bresciani, Edda, Maria Paola Giangeri-Silvis, and Sergio Pernigotti. 1977. *La Tomba Di Ciennehebu, Capo Della Flotta Del Re.* Tombe d'eta saitica a Saqqara 1. Pisa: Giardini.

Cassidy, Brendan, Gavin Sim, David Wayne Robinson, and Devlin Gandy. 2019. "A Virtual Reality Platform for Analyzing Remote Archaeological Sites." *Interacting with Computers* 31 (2): 167–176. https://doi.org/10.1093/iwc/iwz011 (accessed 02-25-2022).

43 Olson 2016.

Davies, Hugh E.H. "Design and Construction of Roman Roads in Britain: Ph.D." University of Reading.

Davis, Annabelle, David Belton, Petra Helmholz, Paul Bourke, and Jo McDonald. 2017. "Pilbara Rock Art: Laser Scanning, Photogrammetry and 3D Photographic Reconstruction as Heritage Management Tools." *Herit Sci* 5 (1): 48. https://doi.org/10.1186/s40494-017-0140-7 (accessed 02-25-2022).

Der Manuelian, Peter. 2013. "Giza 3D: Digital Archaeology and Scholarly Access to the Giza Pyramids: The Giza Project at Harvard University." In Proceedings of *DigitalHeritage 2013 [Digital Heritage International Congress], Marseilles, France, October 28—November 1, 2013*, vol. 2, 727–734. https://dash.harvard.edu/handle/1/12560998 (accessed 02-25-2022).

Doneus, M., G. Verhoeven, M. Fera, Ch. Briese, M. Kucera, and W. Neubauer. 2011. "From Deposit to Point Cloud—a Study of Low-Cost Computer Vision Approaches for the Straightforward Documentation of Archaeological Excavations." *Geoinformatics FCE CTU* 6: 81–88. https://doi.org/10.14311/gi.6.11 (accessed 02-25-2022).

Douglass, Matthew, Sam Lin, and Michael Chodoronek. 2015. "The Application of 3D Photogrammetry for in-Field Documentation of Archaeological Features." *Adv. archaeol. pract.* 3 (02): 136–152. https://doi.org/10.7183/2326-3768.3.2.136 (accessed 02-25-2022).

eScience-Center. 2018. "Forschungsdatenarchiv (FDAT)." Accessed September 19, 2018. https://fdat.escience.uni-tuebingen.de/portal/#/start (accessed 01-20-2023).

Factum Arte. 2009. "Factum Arte's Work in the Tombs of Tutankhamun, Nefertari and Seti 1." Unpublished manuscript, last modified July 07, 2019. http://www.factum-arte.com/resources/files/ff/publications_PDF/Tutankhamun_Report_may2009.pdf (accessed 02-25-2022).

Forte, Maurizio. 2014. "3D Archaeology." *Journal of Eastern Mediterranean Archaeology & Heritage Studies* 2 (1): 1. https://doi.org/10.5325/jeasmedarcherstu.2.1.0001 (accessed 02-25-2022).

Frischer, Bernard. 2009. "Introduction." In *Beyond Illustration: 2D and 3D Digital Technologies as Tools for Discovery in Archaeology*, edited by Bernard Frischer, v–xxiv. BAR International series 1805. Oxford: Hadrian Books.

Galeazzi, Fabrizio. 2016. "Towards the Definition of Best 3D Practices in Archaeology: Assessing 3D Documentation Techniques for Intra-Site Data Recording." *Journal of Cultural Heritage* 17:159–169. https://doi.org/10.1016/j.culher.2015.07.005 (accessed 02-25-2022).

Hageneuer, Sebastian, ed. 2020. *Communicating the Past in the Digital Age: Proceedings of the International Conference on Digital Methods in Teaching and Learning in Archaeology (12th–13th October 2018):* Ubiquity Press.

Historic England. 2017. *Photogrammetric Applications for Cultural Heritage: Guidance for Good Practice*. Swindon.

Historic England. 2018. *3D Laser Scanning for Heritage: Advice and Guidance on the Use of Laser Scanning in Archaeology and Architecture*. Swindon.

Holter, Erika, and Sebastian Schwesinger. 2020. "Modelling and Simulation to Teach (Classical) Archaeology: Integrating New Media into the Curriculum." In *Communicating the Past in the Digital Age: Proceedings of the International Conference on Digital Methods in Teaching and Learning in Archaeology (12th–13th October 2018)*, edited by Sebastian Hageneuer: 167–177: Ubiquity Press.

Holz, Dirk, Alexandru E. Ichim, Federico Tombari, Radu B. Rusu, and Sven Behnke. 2015. "Registration with the Point Cloud Library: A Modular Framework for Aligning in 3-D." *IEEE Robotics Automation Magazine* 22 (4): 110–124. https://doi.org/10.1109/MRA.2015.2432331 (accessed 02-25-2022).

Howland, Matthew D., Falko Kuester, and Thomas E. Levy. 2014. "Structure from Motion: Twenty-First Century Field Recording with 3D Technology." *Near Eastern Archaeology* 77 (3): 187. https://doi.org/10.5615/neareastarch.77.3.0187 (accessed 02-25-2022).

José Luis, Pérez-García, Mozas-Calvache Antonio Tomás, Barba-Colmenero Vicente, and Jiménez-Serrano Alejandro. 2019. "Photogrammetric Studies of Inaccessible Sites in Archaeology: Case Study of Burial Chambers in Qubbet El-Hawa (Aswan, Egypt)." *Journal of Archaeological Science* 102:1–10. https://doi.org/10.1016/j.jas.2018.12.008 (accessed 02-25-2022).

Kazhdan, Michael, Matthew Bolitho, and Hugues Hoppe. 2006. "Poisson Surface Reconstruction." In *Proceedings of the Fourth Eurographics Symposium on Geometry Processing*: 61–70. SGP '06. Goslar, DEU: Eurographics Association.

Kersten, Thomas., Klaus. Mechelke, and Lena Maziull. 2015. "3D Model of Al Zubarah Fortress in Qatar—Terrestrial Laser Scanning Vs. Dense Image Matching." *Int. Arch. Photogramm. Remote Sens. Spatial Inf. Sci.* XL-5/W4:1–8. https://doi.org/10.5194/isprsarchives-XL-5-W4-1-2015 (accessed 02-25-2022).

Kevin Kee, editor, ed. 2014. *Pastplay: Teaching and Learning History with Technology*. s.l. University of Michigan Press.

Koller, David, Bernard Frischer, and Greg Humphreys. 2009. "Research Challenges for Digital Archives of 3D Cultural Heritage Models." *Journal on Computing and Cultural Heritage* 2 (3): 1–17. https://doi.org/10.1145/1658346.1658347 (accessed 02-25-2022).

Lang, Matthias, Ramadan Hussein, Benjamin Glissmann, and Philippe Kluge. 2020. "Digital Documentation of the Saite Tombs in Saqqara." in press. *Studies in Digital Heritage*.

Lercari, Nicola, Emmanuel Shiferaw, Maurizio Forte, and Regis Kopper. 2018. "Immersive Visualization and Curation of Archaeological Heritage Data: Çatalhöyük and the Dig@IT App." *J Archaeol Method Theory* 25 (2): 368–392. https://doi.org/10.1007/s10816-017-9340-4 (accessed 02-25-2022).

Llobera, Marcos. 2011. "Archaeological Visualization: Towards an Archaeological Information Science (AISc)." *J Archaeol Method Theory* 18 (3): 193–223. https://doi.org/10.1007/s10816-010-9098-4 (accessed 02-25-2022).

Lowe, Adam. 2018. *Scanning Seti: The Re-Generation of a Pharaonic Tomb: 200 Years in the Life of a Tomb*. Accessed September 19, 2018. http://www.factum-arte.com/resources/files/ff/articles/seti_basel_36.pdf (accessed 02-25-2022).

Luhmann, T., M. Chizhova, D. Gorkovchuk, H. Hastedt, N. Chachava, and N. Lekveishvili. 2019. "Combination of Terrestrial Laserscanning, UAV and Close-Range Photogrammetry For 3D Reconstruction of Complex Churches in Georgia." *Int. Arch. Photogramm. Remote Sens. Spatial Inf. Sci.* XLII-2/W11:753–761. https://doi.org/10.5194/isprs-archives-XLII-2-W11-753-2019 (accessed 02-25-2022).

McCoy, Mark D. 2020. *Maps for Time Travelers: How Archaeologists Use Technology to Bring Us Closer to the Past*. Oakland: University of California Press.

Morgan, Colleen, and Holly Wright. 2018. "Pencils and Pixels: Drawing and Digital Media in Archaeological Field Recording." *Journal of Field Archaeology* 43 (2): 136–151. https://doi.org/10.1080/00934690.2018.1428488 (accessed 02-25-2022).

Niven, Kieron, and Julian D. Richards. 2017. "The Storage and Long-Term Preservation of 3D Data." In *Human Remains: Another Dimension: The Application of Imaging to the Study of Human Remains*, edited by Tim Thompson and David Errickson: 175–184. Place of publication not identified, s.l. Elsevier Ltd.

Olson, Brandon Richard. 2016. "The Things We Can Do with Pictures: Image-Based Modeling and Archaeology." In *Mobilizing the Past for a Digital Future: The Potential of Digital Archaeology*, edited by Erin W. Averett, Jody M. Gordon, and Derek B. Counts. Version 1.1 (updated November 5, 2016): 237–250.

Olson, Brandon Richard, and Ryan A. Placchetti. 2015. "A Discussion of the Analytical Benefits of Image Based Modeling in Archaeology." In *Visions of Substance: 3D Imaging in Mediterranean Archaeology*, edited by Brandon R. Olson, William R. Caraher, and Sebastian Heath: 17–26. Grand Forks: The Digital Press at The University of North Dakota.

Polig, Martina. 2017. "3D GIS for Building Archeology—Combining Old and New Data in a Three-Dimensional Information System in the Case Study of Lund Cathedral." *Studies in Digital Heritage* 1 (2): 225–238. https://doi.org/10.14434/sdh.v1i2.23253 (accessed 02-25-2022).

Reu, Jeroen de, Gertjan Plets, Geert Verhoeven, Philippe de Smedt, Machteld Bats, Bart Cherretté, Wouter de Maeyer et al. 2013. "Towards a Three-Dimensional Cost-Effective Registration of the Archaeological Heritage." *Journal of Archaeological Science* 40 (2): 1108–1121. https://doi.org/10.1016/j.jas.2012.08.040 (accessed 02-25-2022).

Reu, Jeroen de, Philippe de Smedt, Davy Herremans, Marc van Meirvenne, Pieter Laloo, and Wim de Clercq. 2014. "On Introducing an Image-Based 3D Reconstruction Method in Archaeological Excavation Practice." *Journal of Archaeological Science* 41:251–262. https://doi.org/10.1016/j.jas.2013.08.020 (accessed 02-25-2022).

Richards-Rissetto, Heather, and Jennifer von Schwerin. 2017. "A Catch 22 of 3D Data Sustainability: Lessons in 3D Archaeological Data Management & Accessibility." *Digital Applications in Archaeology and Cultural Heritage* 6:38–48. https://doi.org/10.1016/j.daach.2017.04.005 (accessed 02-25-2022).

Siebke, Inga, Lorenzo Campana, Marianne Ramstein, Anja Furtwängler, Albert Hafner, and Sandra Lösch. 2018. "The Application of Different 3D-Scan-Systems and Photogrammetry at an Excavation—A Neolithic Dolmen from Switzerland." *Digital Applications in Archaeology and Cultural Heritage* 10:e00078. https://doi.org/10.1016/j.daach.2018.e00078 (accessed 02-25-2022).

Sullivan, Elaine A. 2020. *Constructing the Sacred: Visibility and Ritual Landscape at the Egyptian Necropolis of Saqqara*. Stanford: Stanford University Press.

Tavares, Ana. 2011. "Coordinate Systems and Archaeological Grids Used at Giza." In *Giza Plateau Mapping Project: Season ... Preliminary Report*, edited by Mark Lehner: 203–216. Giza occasional papers 5. Boston: Ancient Egypt Research Assoc.

Verhoeven, Geert, Christopher Sevara, Wilfried Karel, Camillo Ressl, Michael Doneus, and Christian Briese. 2013. "Undistorting the Past: New Techniques for Orthorectification of Archaeological Aerial Frame Imagery." In *Good Practice in Archaeological Diagnostics: Non-Invasive Survey of Complex Archaeological Sites*, edited by Cristina Corsi:31–67. Natural Science in Archaeology. Cham, Heidelberg: Springer.

Walsh, Gregory. 2015. *Leica Scanstation: White Paper*. Heerbrugg.

Ware, Colin. 2004. *Information Visualization: Perception for Design*. Amsterdam: Elsevier.

Zachar, Jan, Milan Horňák, and Predrag Novaković, eds. 2017. *3D Digital Recording of Archaeological, Architectural and Artistic Heritage*. CONPRA series vol. 1. Ljubljana.

CHAPTER 9

Digital Archaeology and Ancient Egypt: Reflections on the Results of the 2017 El-Hibeh Digital Archaeology Project

Jean Li, with contributions by Jimmy Tran and Devin Ostrom

Abstract

In 2016–2017, the el-Hibeh Project began a digital initiative with members from Toronto Metropolitan University (formerly Ryerson University). Over the course of a year and half, the project created a Virtual Reality rendering of the temple of Amen, and deployed robotic explorations of the labyrinth of tunnels that riddle the site during a short season in summer 2017. The qualified success of the project has led to considerable reflection on the issue of sustainability of digital enterprises. I ponder the aims and goals of digital archaeology in scholarly research and public outreach through discussion of early and current digital projects, such as Çatalhüyük's Okapi Island and the CRANE project focused on the Orontes Watershed, as well as examine contemporary commercial enterprises, such as Lithodomos. In doing so, whilst thinking about the next steps for the el-Hibeh digital project, I wrestle with the question, "if archaeology is by nature material, how do we reconcile the virtual?"

Keywords

El-Hibeh – Virtual Reality – Augmented Reality – robot – digital archaeology

1 Introduction

The goal of archaeology is simple: to contribute to knowledge of the past. Since the transformation of archaeology from antiquarian collection to a scholarly discipline, practitioners have wrestled with issues of evolving methodologies, theories, and philosophies of a practice that destroys its evidence as it reveals it. Today, archaeology is at a crossroads. As a discipline that engages explicitly with the past, its practitioners are also products of the present, and our present is a digital age. Information abounds: produced, collated and easily accessible via smartphones, tablets and computers, stored on drives and in Clouds. As we become ever more digital, so too are our scholarly concerns.

Digital Archaeology is an evolving discipline.[1] Although there are variations in definitions, Digital Archaeology can be summed up as the "use of computerized—especially internet connected and portable—tools and systems aimed at facilitating the documentation, interpretation and publication of material culture."[2] Thus defined, this approach has resulted in the pronouncement that "we are *all* digital archaeologists."[3]

Morgan and Eve's often repeated call to arms[4] seems indicative of the zeitgeist of archaeological research. Projects embrace digital technology and proponents highlight the streamlined, systematized, and efficient workflow that results from adopting digital archaeology and going paperless in the field. Publications are populated with terms such as "efficiency," "workflow," "wire frames," and "born digital," characterizing not only the push to "go digital," but an assertion that archaeology is, in fact, undergoing a paradigm shift.[5] Digital archaeology is presented as the panacea that will cure our disconnection between present static and past dynamics. Digital technologies offer a way to represent the real world in a compact and efficient package: allow one to count, do statistics, evaluate and measure, efficiently model and simulate real world processes, and make possible the creation of virtual worlds and their transmissions to a wider audience.[6] In this new world, technology is king.

The approaches and applications in digital archaeology are wide ranging, reflected in the multiplicity of terminology used by archaeologists. Under the umbrella term of "Digital Archaeology" operate approaches such as Virtual Archaeology,[7] 3D archaeology,[8] cyber-archaeology, and tele-immersive archaeology.[9] These terms are also reflective of the epistemological perspectives of their practitioners and the evolution of ideas of digital archaeology.

Virtual archaeology, one of the earliest terms used, was first coined by Paul Reilly and focused on the replication of the original archaeological data, that is, simulation.[10] In the decades since Reilly's conceptualization of a system of vir-

---

1 Evans and Daly 2006, 1–3.
2 Averett et al. 2016, 3.
3 Morgan and Eve 2012, 523.
4 For examples see, Averett et al. 2016, 11; Costopoulos 2016, 1–3; Walker 2014, 217–235.
5 See Averett et al. 2016, 5; Roosevelt et al. 2015, 325–346.
6 Evans and Daly 2006, 9.
7 Reilly 1991, 132–139.
8 Forte, Maurizio. 2014a.
9 Kruillo and Forte 2012; Forte, Maurizio, 2014b; Lanjouw 2016.
10 Reilly 1991, 133.

tual replication, projects such as Catalhöyük, Pompeii, Digital Karnak, and Giza 3D have given us beautiful films and 3D models that visualize the past. But as archaeologists produce increasingly vibrant pictures interpreting the past, critics are also calling for more reflections on what it means to do archaeology in a digital age.[11]

In the decades after Reilly's conception of virtual archaeology, scholars became increasingly frustrated with the seemingly lack of epistemological progress and hermeneutic critique, criticizing the models produced as "static without any interrelation with human activities or social behaviours."[12] This focus on the end product led Maurizio Forte to discard "Virtual Archaeology," turning instead to practicing "Cyberarchaeology." Cyberarchaeology does not translate analog data into digital as in virtual archaeology, instead the data is "born digital," and therefore erases the "top-down" or "bottom-up" phases of interpretation, integrating the collection, documentation and interpretation of archaeological data.[13] Lastly, in order to further blur the linear trajectory of collection and interpretation, Forte also introduced a model of "Teleimmersive archaeology" in which archaeologists work simultaneously with the digital data in the material and virtual world through a Virtual Reality environment.[14] Given the wide range of approaches informed by a variety of theoretical underpinnings, this was a complicated arena to navigate in my initial venture into digital archaeology. What follows is a summary of the inception, progress, results, and reflections on a digital archaeology collaboration that took place between 2016 and 2017 from the perspective of a non-technician, archaeologist end-user.

In an attempt to be a "digital archaeologist," I teamed up with a number of faculty, staff, and student members at Toronto Metropolitan University (formerly Ryerson University),[15] to initiate a collaboration focused on the site of el-Hibeh, in middle Egypt. Founded at the beginning of the Third Intermediate Period (c. 1069–664 BCE), the provincial town of el-Hibeh boasts a once-impressive town wall enclosing a substantial settlement. When control of Egypt was divided between the priesthood of Amun centered at the ancient southern city of Thebes, and the nominal kings of the Twenty-first Dynasty in the north,

11 Walker 2014; Huggett 2015a; Huggett 2015b; Caraher 2016; and Morgan 2016.

12 Forte 2014b, 117.

13 Forte 2014b, 117–119.

14 Kruillo and Forte 2012.

15 In 2022, Ryerson University formally changed its name to Toronto Metropolitan University. However, since the activities of the 2016–2017 period were concluded under the University's former name, "Ryerson" will be used for the remainder of the present chapter.

FIGURE 9.1 El-Hibeh Egypt, extensive looting

el-Hibeh marked the northern political boundary of the territory controlled by the priesthood of Amun.[16] It is an important site for our understanding of Egyptian history.

Since 2001, under the auspices of the Ministry of State of Antiquities, the University of California, Berkeley project, directed by Professor Carol Redmount (Near Eastern Studies, University of California, Berkeley), has undertaken conservation and excavation activities at el-Hibeh. However, since 2011 the site has undergone systematic and sustained looting (Fig. 9.1). The long hiatus of archaeological excavation alongside the changed archaeological landscape called for a reconsideration of archaeological research, heritage preservation and raising awareness. This is where the potential benefits of digital archaeology may be tested. How do we apply current technology to the archaeological exploration and heritage management in the archaeology of Hibeh?

In this digital archaeology project, I teamed up with a number of colleagues and students at Ryerson University. Dr. Michael Carter (Industry Director of

16 Wenke 1984.

Master in Digital Media Program) connected me with Professor Alexander Ferworn (Graduate Program Director, Computer Science) who brought with him his (then) graduate students and colleagues, especially Dr. Jimmy Tran (Computer Science) and Devin Ostrom (Technical Officer, Mechanical and Industrial Engineering), who were responsible for the design and construction of a number of Unmanned Ground Vehicles (UGV), or robots. An additional UGV was designed and built by Professor Rob Blain (Program Coordinator, Multimedia Design and Development at Humber College). Mr. Namir Ahmed (Digital Media Experience Coordinator, Ryerson University Library) offered additional assistance and UVG design. The Virtual Reality temple at el-Hibeh was the creation of Professor Kristian Howald (Computer Animation, Sheridan College).[17]

In discussions with colleagues at Ryerson, ideas flew fast and furious. We can build robots, or unmanned ground vehicles (UGV), to explore the looted shafts and tunnels; we can do photogrammetry and 3D map the tunnels, we can make VR models, first of the temple and then perhaps the entire site. In the end we decided on two directions as a pilot project: the building and deployment of robots to explore the site *in situ*, especially the looted contexts to assess safety and potential for further exploration, and the rendering of a VR model of the small temple of Amun to test new ways of fostering wider engagement and education. The results of this initiative have pushed me to consider practicing digital archaeology as collaboration, design and practice, efficiency, and the epistemological issues and questions of technological use, fetishization, and fashion.

2 Robotic Deployment: Technology, Design and Practice:

One of the main objectives of the July 2017 field season, then, was to field-test the use of robots in assessing the extent of looting, the types looting, and ascertain areas of interest for future archaeological research. What follows is a summary of activities as well as preliminary recommendations for future design modifications and protocols from my perspective as a non-technician, archaeologist operator.

17 This initiative was generously funded by the Office of the Provost, The Deans of the Faculties of Science and Arts, as well as with a variety of financial, infrastructure and staff support from the Yeates School of Graduate Studies, the Library, and the Office of the Vice President of Research and Innovation.

FIGURE 9.2 Field testing Indy II

In July 2017, Dr. Michael Carter and Mr. Namir Ahmed entered Egypt with one robot, Indy II (Fig. 9.2).[18] Due to unforeseen circumstances that delayed the opening of the season Ahmed and Carter were only able to operate the robot on July 12, 2017 in a desert field test. Between July 15th and July 25th, I was the main operator for the robot. Ms. Brooke Norton and Ms. Kea Johnston (PhD candidates, University of California, Berkeley) assisted in the robotic operation and exploration.

During our short, two-week season, we were able to deploy the robot a total of four days, several times each day. Over this period we explored a few different contexts; some highlights including a shaft that led to a tunnel (Southwest Road

18 Indy I and II were designed by Dr. Jimmy Tran and Mr. Devin Ostrom. For technical specifications regarding the robot design and fabrication, please see Tran's and Ostrom's discussion in the Addendum.

FIGURE 9.3 North Cemetery 08, Burial Cave with visible disturbed material culture

Side 09), a burial cave (North Cemetery 08), and a very long tunnel (Southwest Road Side 10). One of the most successful runs of the robot was the burial cave, North Cemetery 08 (NC 08).

NC 08 was our first deployment because from its wide entry we were able to see visible material culture, while at the same time it was sufficiently deep that we could not easily determine the dimensions of the cave and its contents (Fig. 9.3). The robot navigated well over sharp limestone debris of varying sizes, and in the course of the exploration, we were able to view and determine that there were about 14 roughly hewn sarcophagi stacked in the large main chamber of the cave. All were disturbed and there were no visible artifacts.

In the course of remote exploration of NC 08, the robot's cameras helped archaeologists locate additional offshoot passages, either looting tunnels or ancient tunnels reused by looters. In the northeast area of the main chamber, about 5.5 meters (18 feet) from the burial cave entrance we located an opening. This opening was too tall for the robot to navigate, and subsequent, *in situ* explorations by team members found a chamber dominated by two monumental sarcophagi. Another passage heading in a southeast direction in the main chamber also led to an opening that led to another chamber. Interestingly,

FIGURE 9.4 Chamber in North Cemetery 08 with blocked-off wall

the recorded footage showed what appears to be a blocked off wall, confirmed later by in-person exploration (Fig. 9.4). These are contexts for potential future explorations.

Another contexts we explored were looting tunnels carved into the limestone bedrock. It appears, as already noted by Wenke, the entire site of el-Hibeh is riddled with tunnels, and in many different periods people dug into the limestone foundation to inter their dead.[19] Some of these tunnels were quite extensive. The greatest distance we were able to send the robot was in Southwest Roadside 10, a tunnel that ran about 27 meters, or 88 feet (Fig. 9.5) before reaching a further opening into which the robot was unable to navigate. Along the way, we found a burlap bag and cigarette stubs attesting to recent looting activities.

2017's "deployment" of the robot was the end of a long process of development. Months of discussions about design and preparation, rushing to finish and test prototypes were necessary stages of the process. The collaborative process was not without challenges in and out of the field.

19 Wenke 1984, 12.

FIGURE 9.5 Southwest Roadside 10, 27 meter long tunnel

Digital Archaeology is necessarily a collaborative endeavor, involving scholars and technicians outside the traditional humanities fields. The greatest challenge I encountered was the issue of "learning to speak each other's languages." Although Egyptology, and the Computer Science, and Digital Media partners shared a desire to explore issues of cultural heritage preservation using technology, our differing disciplines with respective disciplinary foci resulted in difficulties. As an Egyptologist, I was primarily interested in application of the robot, while my colleagues necessarily focused on the design and construction, which nonetheless required substantive knowledge of field conditions and archaeological practice in Egyptology. Moreover, because of the nature of el-Hibeh as a Third Intermediate Period site, and its extensive looting and our previous inability to get on site, we did not have a lot of what would seem to be basic information.

It was undoubtedly frustrating for my Computer Science colleagues, especially Dr. Tran and Prof. Blain, to hear, "we don't know," in response to seemingly basic questions of, "how big would the shaft opening be?" "How deep and far would the tunnels extend?" "What kind of conditions would the robot encounter?" Although field archaeologists are good at description, and pictures supposedly are worth a thousand words, there is no substitute for the real-life experience of being on site.

FIGURE 9.6 Two-person team operating Indy II

We did anticipate certain conditions that determined the design of the robot. We anticipated having to lower the robot into shafts, and that the site location and subterranean context necessarily precluded the use of wireless or Bluetooth technology as a form of control and navigation. As a result, the prototype robots were wheeled and tethered. We had two prototypes: One—which we eventually took to Egypt—was the robot with foam core wheels, the other was build low and compact with tracks.

Indy II, the wheeled prototype we took to Egypt, was designed with a tether (150 feet length) that was also an Ethernet AV cable—in that sense it functioned as a physical tie and information feed. This also meant that it was designed so that should it become stuck or lodged, with an experienced operator, one can potentially drive it out. The foam core wheels were also designed to navigate the rough terrain with limited damage.

Because of the tether that was attached to the robot and the spool, which was attached to the monitor, a two-member team was necessary (Fig. 9.6). One would be the driver and the other would be in charge of the tether. The small screen and poor visibility of the monitor (especially under the bright summer sun) also meant that often times it was not possible for both people to see the video feed. Therefore, the driver had the task of driving, giving instructions to loosen or tighten the tether, as well as observing the environment. It was difficult to concentrate on anything other than driving the robot, especially when trying to navigate it over difficult terrain. Communications between team members were essential, and the driver spoke observations aloud for other team members to record.

The 2017 field project demonstrated some of the capabilities and limitations of the robotic exploration and collaboration. As we explored different areas of the *tell* we further observed that the robot was not equipped to work in the way it was originally designed. The navigation cameras recorded some great footage, but the quality was not reflected in the real-time monitoring feed, resulting in some difficulties for the operators when navigating the robot over the rough terrain. The robot had tendency to get entangled in its tether, overset or become trapped, situations that have as much to do with environment and our piloting inexperience as well as the design. In the end, this combination of factors resulted in an almost 100 % rate of human retrieval of the robot (https://doi.org/10.6084/m9.figshare.21896892),[20] which not only resulted in archaeologists crawling through some quite claustrophobic conditions, but also my own questions about the supposed efficiency of technology.

Other issues included troubleshooting by non-technical personnel. The first day included hotwiring (Fig. 9.7) the robot assisted by phone calls to Toronto to one of the robot's designers using WhatsApp. Additional troubleshooting and maintenance of the robot in and out of the field was necessary as we continued to use it. Mechanical maintenance was complicated by the many handcrafted elements in the structure of the robot. The amount of time doing repairs by non-specialists calls into question the exhortation of efficiency in digital archaeology.

Overall, Indy II successfully carried out its intended scouting and reconnaissance function. We felt cautiously comfortable entering into some very confined contexts because we had been able to observe the environment beforehand via the robot's cameras. However, robot retrieval was undertaken with the understanding that the conditions were potentially dangerous. With additional refinement the robotic exploration will prove a valuable tool for investigations and assuring the safety of archaeologists.

20 See short video.

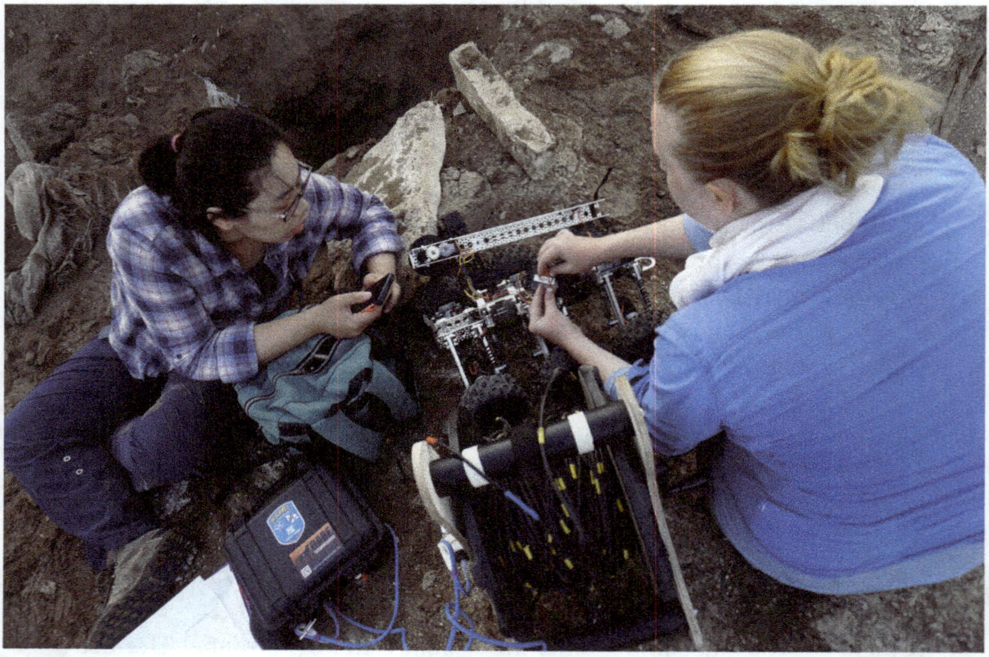

FIGURE 9.7 International tech support and hotwiring the robot

3 Virtual Reality: Knowledge Production and Dissemination

Around the same time the robot project was initiated, the Master in Digital Media Program at Ryerson University also assisted the el-Hibeh project in creating an initial Virtual Reality experience focused on the temple of Amun at el-Hibeh (Fig. 9.8).[21] This work was the Major Research Project of Professor Kristian Howald.[22] Howald's aims explored not just the utility of the technology, but also examined the potentials of created digital environments to convince, and be consumed by the observers to result in new knowledge.[23]

21 Due to time constraints, we were unable to add color or the decorative program on to the temple façades.

22 Professor Kristian Howald is a faculty member in Post Graduate Computer Animation Program in the Faculty of Animation, Art and Design at Sheridan College.

23 Professor Howald undertook this project as his Major Research Project as part of the fulfillment of his degree in the Master in Digital Media Program. He discusses the process in Howald 2017, 9. For more information on the process of creation, readers may consult Howald's blog: https://elhibeh.blog/, accessed 03-22-2022.

FIGURE 9.8 Façade of Virtual Reality model of the small temple to Amun, Great-of-Roarings at el-Hibeh

The basic foundations of this digital model were formed from analog information. When Hermann Ranke excavated el-Hibeh from 1913–1914,[24] he took a number of high-quality photographs whose glass plate negatives are currently in the University of Heidelberg collection. In recent years, Heidelberg has made the images from el-Hibeh accessible online.[25] At the inception of this project I studied and printed out the images from heidlCON and in a very analog fashion, mapped the images onto a floor plan of Ranke and Breith's (Ranke's project architect) published architectural drawing, which Howald then used as a starting point for mapping images onto his digital model (Fig. 9.9).

Overall the creation of the temple took four months and utilized more than seven different software programs to produce the Virtual Reality experience.[26] Autodesk Maya[27] was the main modeling tool used to build all the inanimate objects, Autodesk Sketchbook[28] to create 2D patterns for textures of objects,

24 Ranke 1926.

25 The object and multimedia database of Heidelberg University (heidlCON): https://heidicon.ub.uni-heidelberg.de/search, accessed 03-22-2022, using the terms, "Grabung El-Hibe".

26 Howald, "El-Hibeh," 19–20.

27 For information on Autodesk Maya, see: https://www.autodesk.com/products/maya/overview?AID=10282382&PID=3662453&SID=725X1342Xd1857051e5b5ac3f2c77fb393d5b534f&mktvar002=afc_us_deeplink&cjevent=f276e9e153f511ea827002de0a240611&affname=3662453_10282382, accessed 03-22-2022.

28 https://sketchbook.com/, accessed 03-22-2022.

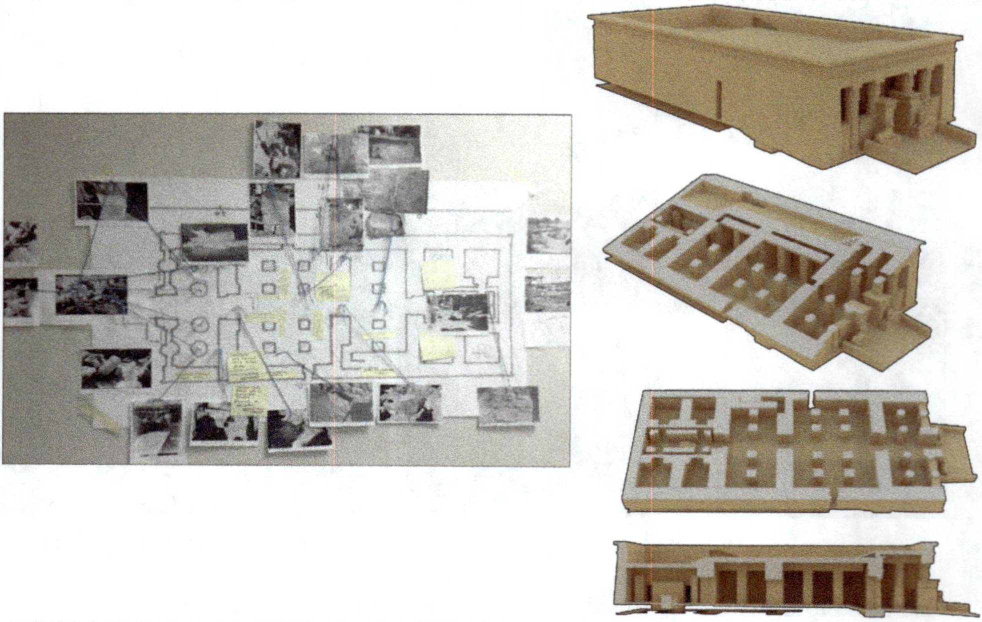

FIGURE 9.9 Analog to digital plan of the temple

Substance Painter[29] for detailing objects with real world properties, and Zbrush used to create the avatars.[30]

The process of creation was full of challenges, resulting in the creation of two models. The first preliminary construction was unsatisfactory mostly due to Howald's lack of knowledge of Egyptian architecture, and resulted in an unbalanced model. In the second iteration, Howald refined his methodology by breaking up the model into individual components based on their position in the temple. He used Photoshop to create comparative images between Ranke's photographs and published illustrations, which also proved challenging. As archaeologists we can appreciate the high resolution of the glass plate negatives of Ranke's photographs, but for Howald, there was a lot of occluded information due to the angles of the photographs, forcing him to rely more heavily on the published architectural illustrations.[31] In order to provide a sense

29 https://www.substance3d.com/education/, accessed 03-22-2022.

30 Zbrush (https://pixologic.com/, accessed 03-22-2022) was originally developed for the fashion industry but is now regarded by the 3D community as the best organic modeling package for creating realistic clothing and cloth, Howald, Personal Communication, February 2020.

31 Howald 2017, 22.

FIGURE 9.10 Avatar used to provide spatial awareness and scale

of spatial awareness for the VR user, Howald introduced inhabitants into the temple.[32] The original intent was to create a stylized anthropomorphic figure, but in the desire to not distract the observer with real looking people, the deliberately sketchy avatars became distractingly off-putting. As a compromise Howald created more realistic looking figures with lightly grey skins to avoid confusion with humans (Fig. 9.10).[33]

The entire temple and contents were then imported into the Unreal software engine, to run the Virtual Reality experience. To add to the aura of authenticity, Howald included some audio that I had recorded in the environment near el-Hibeh during July 2017. Users navigated and experienced the Virtual Temple through HTC Vive (hardware).[34]

In the Virtual Reality project, we encountered some of the same issues of interdisciplinary collaboration as already mentioned in the robot project. There were many times when I was at a loss for answers to questions posed by

32 Howald 2017, 28.
33 Howald 2017, 28.
34 Howald 2017, 36.

FIGURE 9.11 Questions from Howald

Howald, such as additional sources of lighting in the temple, what would have been around and under the entrance ramps, or what kind of wool would be depicted on criosphinxes (Fig. 9.11)? These conversations between myself and Howald about construction techniques, and the questions posed by Howald often led me to much deeper reflections about what kind of information and knowledge Egyptologists privilege contrasted with those needed by digital craftsmen. I've had to think about translating my knowledge and understanding of Egypt, el-Hibeh and priorities in field research to an audience that is skilled and proficient in what they do, but had little background or disciplinary knowledge. It is in the process of reconstruction that we realize not just the gaps in knowledge, but how much we engage in a process of "suspension of disbelief." The ancient Egypt that we created with traditional scholarship methods "work" because they did not require us to run simulation that confront us to examine the limits of our scholarly understandings.[35] Moreover, my own lack

35 I would like to thank the anonymous reviewer who raised this point.

of technical knowledge of and practice with the various software programs proved frustrating for me, exacerbated by a desire to learn but a lack of time and consistent access to the various technologies. The challenges associated with interdisciplinary research, and the issues of communication and efficiency will necessarily result in archaeologists, or Egyptologists in our case, become ever more technical—the scholar-technician.

Other challenges include economics and funding, and more importantly, a real consideration of the goals of digital and technological applications in research. Some of the tools are costly and prohibitively so (see discussions below and in Tran and Ostrom) in many cases. Moreover, whilst interest in applying digital technology to archaeology may be the way to funding, the next issue that arises is the perception of goals. In this particular project I found that the technology often took center stage in terms of interests by students, university administration (funding), and collaborators, privileging technology as the goal rather than a means to a goal. I found myself questioning whether we were using technology in the production of knowledge or had knowledge become the servant of technological production? What exactly did we hope to achieve with this Virtual Reality experience?

At the conclusion of the project Howald reflected on whether his hard work resulted in a "unique, authentic experience." Of particular interest to Howald were participants' behaviors within the virtual environment. He noted that participants navigated the virtual environment as if the constructs were real, for example, walking around objects they saw. Participants also wanted to interact with objects—to reach out to touch the digital models—but the illusion would be shattered at this point as their "hands" passed through the virtual object.[36] This disconnection between the senses inhibited an experience that would presumably produce new knowledge and a glimpse into the "ancient Egyptian mind." If one of the aims of this initiative is to disseminate knowledge, and attract attention, then certainly the Virtual Reality project has achieved one aim of digital archaeology. As a tool of imagination and attraction, Virtual Reality seems a way to reach a wider audience, but I can't help by wonder: How do we reconcile the verisimilitude of Virtual Reality with our desires for the tangible, or material?

36 This phenomenon is commonly called the Swayze Effect, that is, the feeling of having no tangible engagement with an embodied experience in Virtual Reality. See, http://www.vrglossary.org/glossary/swayze-effect/, accessed 03-22-2022.

4 Digital Archaeology at El-Hibeh: Aims, Results and More Questions

From a non-digital specialist and archaeologist's perspective, I found this Digital Archaeology collaboration a success not just in terms of results, but because it proved to be a thought-provoking exercise. At the end of this project I find myself with more questions that need answers before moving forward: Are the ways by which archaeologists producing knowledge really undergoing a paradigm shift or merely a methodological shift? Is the adoption of ever newer and more powerful tools and technology contributing to the epistemology of archaeological theory or are we in danger of succumbing to technological fetishism?[37] What does the increased reliance on technology mean for the theory of archaeological practice? And what does a field practice of digital archaeology involve?

I started out this project suffering from a bit of technological fetishism; that is, an idea that anything is possible given the right technology. The issue of human retrieval of the robot, while anticipated, was a valuable lesson in theory and practice. However, the limitations of technology may be that of ontology rather than capability. The robot engaged in reconnaissance, and the recorded images and video were of sufficiently high quality that it allowed us to get better sense of some of the conditions of the looting tunnels (for example the blocked off walls). The review of footage merely stimulated more curiosity and desire to "see for oneself." Both the robot and the VR model achieved what they needed to do, but did they go beyond instruments that mediated experience and understanding of the human archaeologist? This is an especially important question to ask of Virtual Reality.

The allure of Virtual Reality is undeniable. The digital craftsman translates archaeological and architectural information into a medium that immediately appeals to a wide range of audiences. And when a user puts on the Oculus Rift, they are "transported" to an environment that is at once immersive and disconcerting. There is often visual and cognitive dissonance involved—we instinctively reach out and try and touch, or crouch down to get a closer look at an interesting detail—but what we see in the end has been pre-interpreted for us. The experience then becomes one of reception rather than interaction.

Is the goal of virtual archaeology to create such a real environment that we lose ourselves in it and yet somehow will allow us to access the ancient mind and elicit the interest of numerous stakeholders? Does this created world, a product of technical skill, artistic eye, and scholarship, actually reveal anything

37 Huggett 2004.

of the past? In many ways, the VR environment becomes another work of art to be admired and perceived, but not grasped. And more importantly, perhaps admired briefly and then forgotten (sadly, I have never had a student ask for a second walkthrough of the temple after the initial VR session). If that is the case, then is Virtual Reality a useful approach to engage a wide and sustained audience who would be invested in preserving el-Hibeh, the original aim of the project?[38] What are the alternatives?

5 Engagement and Sustainability

The issue of sustainability in digital archaeology, certainly in terms of Virtual Reality and 3D models and, to a lesser extent, other forms of digital archaeology, seems tied to the engagement of the audience. For Virtual Reality simulations and 3D replicas of artifacts, the perception of "authenticity" is integral to successful engagement. Howald's product is a work of art, yet clearly not a "replication" of the temple, which leads to questions of authenticity and purpose. We can all differentiate between an original and a replica and react differently to each. But what is *the* quality that a replica, simulation, model, or visualization must possess to increase its efficacy and believability? Walter Benjamin used the term "aura," that is, an indefinable quality that makes something unique, which lends power and authority to an object or piece or art when analyzing authenticity and significance of an object.[39] Stuart Jeffery suggested that the significance of an object is defined by narrative mediated by the materiality of the object, but because digital objects lack materiality, they would lack aura and reduce efficacy of engagement.[40] Ruth Tringham suggested a key component of a successful digital media project is multisensorial interactivity rather than just reception and perception.[41]

Several past and current projects have tried or are addressing engagement and interactivity as avenues of sustainability.[42] A pioneering project was Berkeley's virtual translation of their archaeological excavation of the East Mound

38 Of course a major component to sustainability is the issue of economics. The software employed by Kris Howald was funded by the various university and college institutions, and often with an educational discount involved. However, with licensing and subscriptions, a single project will have to budget at least $5000 for software alone.

39 Benjamin 1968.

40 Jeffrey 2015.

41 Tringham 2013; Tringham 2016.

42 The projects mentioned here constitute a very limited selection of different approaches taken by past and ongoing digital projects.

at Catalhöyük to Second Life,[43] an online 3D user-generated world in which users, called "residents," live a virtual life.[44] First created in 2006, the Berkeley project, named OKAPI (Open Knowledge And the Public Interest Research Group) Island, was eventually defunct in 2012.[45] During its lifespan, OKAPI Island practiced open-access and interactivity; visitors had agency to do whatever they wanted on site, including go into houses, interact with the archaeologists, and reflecting real life, vandalize and abuse. In one incident, virtual squatters built a floating parking lot in the airspace above OKAPI Island, compelling the project to restrict access and limit the range activities allowed by visitors.[46] By 2010, the project became increasingly unsustainable—one of the reasons being that Second Life had raised rent prices.[47] But perhaps the most telling part in the death of Okapi Island is Tringham and Ashley's observation that maintenance "required the presence of us archaeologists to welcome visitors, to create a feeling of an inhabited space."[48] Even in the virtual people craved the real.

Recently, one of the nine research projects that comprise the CRANE (Computational Research on the Ancient Near East) project is exploring another facet of user-driven engagement through gamification, in hopes that there will be sustained public interest, without the investment of an archaeologist present at all times.[49] This particular initiative includes projects focused on 3D modeling of the Orontes Watershed in Minecraft, another interactive online world. The goal is to have visitors traverse through the environment and interact with avatars and Non-Player Characters (such as Supiliulium) with educational storylines to teach players about Iron Age Orontes Valley. With this "archaeogaming" model, the project hopes for a result of increased public outreach and sustained interest; that is to keep people coming back over and over again.

---

43 Morgan 2009.

44 https://secondlife.com/, accessed 03-22-2022.

45 Tringham 2013, n. 1.

46 Morgan 2009, 478.

47 To get a sense of the economic costs of maintaining a presence in Second Life: in 2008, OKAPI Island cost $1800 per year for land use. In 2010 Linden Labs, which owns Second Life, discontinued the academic discount and doubled rent prices resulting in the eventual abandoning of the project due to economic constraints in 2012. However, between 2010–2012, Okapi Island had already begun to fall apart. See Tringham and Ashley 2015, 21–40.

48 Tringham and Ashley 2015, 33.

49 Batiuk 2019. For additional information on the project: https://www.crane.utoronto.ca/3d-modeling.html, accessed 03-22-2022.

Unlike our el-Hibeh temple, neither Second Life nor Minecraft aim for photorealism or a hyperreal representation of the past. This is one of the approaches researchers are taking—that of an explicitly unreal representation of the past, focusing instead on narrative to drive interest and continued visits. So here we have two very important aspects to consider in digital reconstructions: representation and interaction. Realistic representation appears to sacrifice interactivity and interactivity lacks the patina of authenticity. Is the real or perceptions of the real tied to the material? Often we perceive authenticity as synonymous with "real." So then where do we go from here? Is Augmented Reality the way forward?

Augmented Reality is a means of a mediated reality, in which elements of the real world is combined with the virtual.[50] Often this involves using mobile devices or Oculus Rift at archaeological sites. The commercial potential is evident, and a number of companies and organizations have been exploring the application of Augmented reality. One such commercial enterprise is Lithodomos.[51] Lithodomos' staff consists of archaeologist-technicians who create VR experiences for primarily the tourism industry. Tourists visit the sites, put on their Oculus Rift and are transported into a version of the past life of the site. Here we see the connection of the material and the virtual. In terms of broad scope engagement, Augmented Reality may be one of the more successful approaches in engagement and learning—nothing replaces the "real thing"—that is the material.[52]

If the equation is that sustainability = maximum number of people engaged, perhaps all of this complexity and exploration simply means that, as researchers, we need to think in terms of scale—of tiers and stages—in our applications that fit with the clearly defined aims of any respective project. GlobalXplorer, initiated by Sarah Parcak, is an example of a more restrained and currently achievable approach to public engagement and sustainability. GlobalXplorer crowd sourced information—people went to look at tiles to find potential looting and archaeological sites, which were then followed up by archaeological investigation.[53]

---

50 Tringham 2016, slide 18.

51 Young 2019. For more information: https://www.lithodomosvr.com/, accessed 03-22-2022.

52 Of course, there are numerous critiques that can be applied to this commercial approach, including the absence of the voices of different groups and that the narrative presented is a singular vision of the client/tour company. Moreover, such a business/education model faces an uncertain future in a post-pandemic world.

53 https://www.globalxplorer.org/, accessed 03-22-2022, although there does not appear to be a next project.

At the end of this initial foray into the various approaches that exist under the aegis of Digital Archaeology, I have few insights. The potentials of digital technology are undeniable, but as scholars move forward in creating these various virtual worlds and representations, we need to do so with a clear understanding of their effects and sustainability. Lastly, the two sides of the el-Hibeh digital archaeology project, the robot and the VR, have consistently reinforced for me the importance of the material in our understanding of the past. Be it driving the robot over rough terrain or feeling the absence of a real environment as I "walk" through the virtual reconstruction of the temple of Amun, I can't help but think that if sustainability requires widespread engagement of a variety of audiences, then sustainability hinges upon achieving the ever-elusive goal of archaeology, digital or traditional: to make the intangible, tangible.

6 Addendum: Exploration Robot for El-Hibeh Site

Jimmy Tran and Devin Ostrom

6.1 *Introduction*

We discuss our approach for the interdisciplinary use of robotic technology for the purposes of archaeological reconnaissance. Specifically, we will discuss the development of the Unmanned Ground Vehicles (UGVs) used during the 2017 el-Hibeh (http://www.hibeh.org/) field survey.

The use of UGVs in Archaeological study is no longer novel. It has been done in Richardson's Giza pyramid exploration.[54] Their solutions were specific to pyramids, which have rigid structures and well-understood environments. The goal of the el-Hibeh project was to use the UGV to enter tunnels dug by looters and to scout ahead to collect video data. The hope was that this data would allow us to assess the level of looting or damage that may occurred but more importantly it would tell us if it is safe for people to access the tunnels.

The initial plan for the project was to purchase commercially-off-the-shelf (COTS) solutions. This was quickly abandoned due to their exorbitant cost. The most challenging part about the problem was that there was little information known about the environment that the robot had to explore. It would have been wasteful to purchase an expensive robot that may not work for an unknown scenario. In the next few sections, we will discuss how we eventually arrived at the design requirements for the UGV, our approach at the design and development of the UGV, and finally the lesson learned after the robot was used on site.

54 Richardson et al., 2013.

6.2 *Design Requirements*

The first part of solving any engineering problem is to clearly define it. Unfortunately, in the context of this specific project it was particularly difficult because of the lack of access to the archaeological site for the last few years and not since the looting tunnels were created. Many assumptions had to be made.

The first issue was the size of the tunnels. The tunnels were dug by people, possibly children. From this information, the conjecture is that the tunnels were large enough for at least a small person to fit through, which was approximated to be about 0.5 m to 0.7 m in diameter. Most parts of the tunnel would be pitch black.

The following information was unknown: the overall depth of the tunnel (how far the robot must travel), the configurations of the tunnels (such as any twist or turns or angle of inclinations), and the type of terrain or obstacles that maybe inside the tunnel. Since these were unknown, the worst conditions had to be assumed and anticipated. This included the possibility of rocks of various sizes, sand covering most surfaces, and even garbage debris. In the next sections, we will describe how our team handled these unknown challenges in our design process.

The system was built in Canada and sent to Egypt to be used by non-robotics expert users. The team that built the robot in Canada would not be able to be in Egypt. The system must be user-friendly such that users with limited or no prior experience with operating a robot can use it. It also has to be user-serviceable because any maintenance or reparation would have to be performed by non-expert users.

In this context our goal was to design a user friendly, user serviceable UGV that could traverse distances in small tunnels of unknown configuration, broadcast and record video, and be easily retrievable from the site. We call the UGV system Indy.

6.3 *Design Approach*

Separate from the design requirements, the project itself presented other constraints including a specific budget and a deliverable date. These ultimately impacted the final design of the UGV. The team was given a small budget, and a timeline of 4 weeks to order all the parts, build the robot, test it, and train the users on the system. In approaching the design, we broke the project into three main categories: terrain scaling capability, communications and control, and durability and maintainability.

6.4 *Terrain Scaling*

The terrain that was assumed in the design requirements dictated the size of the robot to be smaller than 0.5 m inches in width and height. The robot could be long but then it would not be able to turn around sharp corners in the tunnels. Although the robot had to be small to fit through the tunnel, it could not be too small because it still needed to be able to climb over rough, uneven terrain.

After approximating size of the robot, the next decision was the mode of locomotion. Track vehicles are well suited for uneven terrain since many areas of track could make contact with the ground giving the robot more traction. However, developing a track robot is a time consuming and costly approach because there are many inherent problems. The main issue with tracks is that rocks and debris can get into them, causing the track to come off unless there are careful tensioning and designs to prevent this.

The final design utilized a six-wheel drive system in which each wheel had independent suspension and was powered by a motor (Figure 9.12). The independent suspension system combined with a six-wheel drive allowed the robot to have multiple contacts with many uneven surfaces in various configurations as shown in Figure 9.12. Having power applied directly to each wheel allowed us to rapidly build a prototype.

For the prototype we used a COTS barebones frame kit called the 6WD Mantis (Figure 9.13) from Servocity.com.[55] The kit came with a high-quality aluminum frame, independent suspension system, and six DC motors with planetary gearbox and wheels. From this kit, we built the electronics, power system, and pan and tilt cameras system ourselves. The shocks on the original kit were plastic base and thus not robust enough for our specific needs, so we replaced them with aluminum oil-filled shocks (Figure 9.14).

55 https://www.servocity.com/6wd-mantis, accessed 03-22-2022.

FIGURE 9.12 Independent suspension

FIGURE 9.13 6WD Mantis robot frame kit

FIGURE 9.14 Indy Robot on the rocks navigating well due to aluminum oil filled shocks

6.5 *Communications and Control*

The driving force behind the design of the UGV system was the ability to maintain communication and control of the robot. Given that the environment was underground, with possible twist and turns, wireless communication was not possible. With the added fact that the robot may have to be retrieved, a tether needed to be attached to it. The tether also allowed the robot to be lowered into shafts and pulled up at the end of its deployment. Therefore, the optimal way to communicate with the robot and to control it was through a wired connection.

Physics determined that the robot must have a certain mass to pull a tether of a certain mass. No matter how much torque the motors could apply, the robot could not pull anything heavier than itself. Therefore, the size of the robot greatly determined the mass of the robot, which in turn, determined how long of a tether it could pull.

An Ethernet cable was chosen to be the tether of the robot as it is readily available at low cost, made to transfer data, and strong enough to pull at least 20 kg of mass. The final mass of the robot was approximately 8 kg, which

FIGURE 9.15 UGV system that includes robot, tether management system and OCU

through testing and calculation, allowed it to easily pull about 100 m of Ethernet cable. The cable was solely used for communications because the robot was powered with an on-board battery.

The communication scheme was simple. The robot transmitted video from its onboard camera to the Operator Control Unit (OCU) via an analog signal and received commands from the OCU via digital serial data (Figure 9.15). When a signal is relayed through a wire, the signal degrades as the length increases. At 100 m, this was sure to cause problems. To overcome this issue, we used a tether interface system called a Fathom-S boards from Seascape Subsea B.V.[56] These boards were designed specifically for Ethernet cables and are rated to work at over 300 m.

The robot was equipped with two cameras, but only one video feed could be transmitted at a given time. The front camera had a pan and tilt system, while the back camera was fixed. Both of these cameras were equipped with arrays of infrared (IR) light emitting diodes (LEDs) to allow visibility. The robot was also equipped with a light sensor to automatically adjust the IR LEDs in bright or lit conditions. With the pan and tilt capability, the front camera was designed so that the user could see and explore the tunnels remotely. The back camera allowed the user to drive backwards and to see and follow the tether

56 https://www.seascapesubsea.com/product/thether-interface/, accessed 03-22-2022.

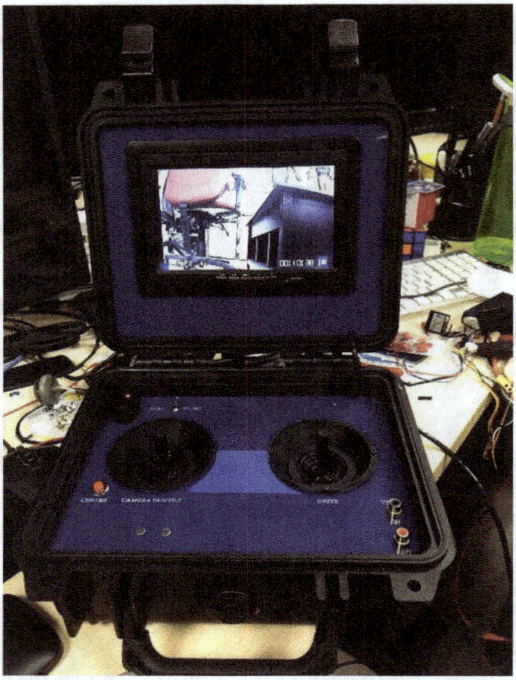

FIGURE 9.16 Indy's OCU

to get back out. In an enclosed space, it can be difficult to keep track of where you are. Therefore the tether was the perfect navigation system for returning the robot back to the user.

Managing a 100 m tether during operation was extremely cumbersome. It is difficult to spool and unspool the tether without any knots or tangle at each deployment. To save time during deployment and setup, we created a tether management system. Anyone who has watered their lawn knows that a spool makes it a lot easier with long hoses. What people may not realize is at the end of the spool where the water line is attached, there has to be a connection that allows it to freely twist around as the spool is rolled or unrolled. Our system consisted of a special designed tether spool that connected the OCU to the robot. The spool also had a built-in a slip ring (a mechanism that allows the Ethernet connection to rotate freely while still hold the connections and not continuously twist the cable). This feature allowed the robot to be driven while the spool unrolled or rolled the tether.

The OCU was a simple design (Figure 9.16). It had an LCD screen that display the video feed. There was a built-in video recorder that allowed the user to save the data onto a flash card. There were two joysticks, one to control the drive direction of the robot, the other to control the pan and tilt of the front camera. The drive control was proportional meaning that the further out from the center of the sticks, the higher the power applied to the motors. Although there were six motors, the robot was controlled like a tank with differential steering. An algorithm limited the sudden acceleration of the robot in a confined space. There were also additional control switches that allowed users to re-center the camera and alternate between the front and back camera. The OCU was built in a rugged Pelican case that had its own power unit.

6.6 *Durability and Maintainability*

Since the team building the robot was not be able to be on site and there were many unknowns about the environment, durability and maintainability were highly desired but difficult to achieve—especially with the time and budget constraints. It was decided early on that there would be two duplicate systems. If one robot failed, another could be swapped in (we applied the same logic to the OCU or tether management system). Using this strategy, the entire UGV system was modular allowing changes to be made simply. Most of the electronics systems had keyed quick disconnect systems that allowed components to be easily swapped.

6.7 *Lessons Learned*

The two-robots strategy did not work as only one system was allowed into Egypt. Fortunately, when it came to malfunctions, there was only one time where a power wire was severed inside the robot. Other failures came in the form of wheels that did not do well on a thin layer of sand on a hard surface on a steep incline. This scenario is difficult for most UGVs and was expected. The robot was also too tall at some places in the tunnels. The biggest problem was that the robot got stuck every time. One crucial feature that was proposed in the initial design but was not implemented, due to time constraints, was a self-righting feature. With a better understanding of the challenges, we believe that a second version would do much better.

Bibliography

Averett, Erin Walcek, Jody Michael Gordon, and Derek B. Counts. 2016. *Mobilizing the Past for a Digital Future. The Potential of Digital Archaeology*. Grand Forks: the Digital Press @ The University of North Dakota.

Batiuk, Stephen. 2019. "Archaeogaming?? The Computational Research on the Ancient Near East (CRANE) project's use of Video Game Systems for the Exploration of Archaeological Data." Public Lecture 16 January 2019.

Benjamin, Walter. 1968. "The Work of Art in the Age of Mechanical Reproduction." In *Illuminations*, edited by Hannah Arendt, transl. Harry Zohn, 217–251. New York: Schocken Books.

Caraher, William. 2016. "Slow Archaeology: Technology, Efficiency and Archaeological Work." In *Mobilizing the Past for a Digital Future*, edited by Erin Walcek Averett, Jody Michael Gordon and Derek B. Counts, 421–441. Grand Forks: the Digital Press @ The University of North Dakota.

Evans, Thomas L. and Patrick Daly. 2006. *Digital Archaeology: Bridging Method and Theory*. Abington; New York: Routledge.

Forte, Maurizio. 2014a. "3D Archaeology: New Perspectives and Challenges—the Example of Çatalhöyük." *Journal of Eastern Mediterranean Archaeology and Heritage Studies* 2(1): 1–29.

Forte, Maurizio. 2014b. "Virtual Reality, Cyberarchaeology and Teleimmersive Archaeology." In *3D Recording and Modelling in Archaeology and Cultural Heritage, Theory and Best Practices*, BAR International Series 2598, edited by Fabio Remondino, and Stefano Campana, 115–134. Oxford: Archeopress.

Howald, Kristian. 2017. *El-Hibeh: Reconstructing the Past in Virtual Reality*. MA Thesis, Ryerson University.

Huggett, Jeremy. 2004. "Archaeology and the New Technological Festishism." *Archeologia e calcalatori* 15: 81–92.

Huggett, Jeremy. 2015a. "Challenging Digital Archaeology." *Open Archaeology* 1: 79–85.

Huggett, Jeremy. 2015b. "A Manifesto for an Introspective Digital Archaeology." *Open Archaeology* 1: 86–95.

Jeffrey, Stuart. 2015. "Challenging Heritage Visualisation: Beauty, Aura and Democratisation," *Open Archaeology* 1: 144–152.

Kansa, Eric. 2106. "Click Here to Save the Past." In *Mobilizing the Past for a Digital Future: The Potential of Digital Archaeology*, edited by Erin Walcek Averett, Jody Michael Gordon and Derek B. Counts, 443–472. Grand Forks: the Digital Press at The University of North Dakota.

Kersel, Morag M. 2016. "Response: Living a Semi-digital Kinda Life." edited by Erin Walcek Averett, Jody Michael Gordon and Derek B. Counts, 475–492. Grand Forks: the Digital Press @ The University of North Dakota.

Kruillo, Gregorij and Maruizio Forte. 2012. "Telearch-integrated visual simulation environment for collaborative virtual archaeology." *Mediterranean Archaeology and Archaeometry* 12(2): 1–10.

Landjouw, Tijm, 2016. "Discussing the obvious or defending the contested: still discussing the 'scientific value' of 3D applications in archaeology?" In *The Three Dimensions of Archaeology, Proceedings of the XVII UISPP World Congress (1–7 September 2014, Burgos Spain) Volume 7/Sessions A4b and A12*, ed. Hans Kamermans, Wieke de

Neef, Chiara Piccoli, Axel Posluschny and Roberto Scopigno, 1–11. Oxford: Archaeopress Publishing ltd.

Morgan, Colleen and Stuart Eve, 2012. "DIY and Digital Archaeology: What Are You Doing to Participate?" *World Archaeology* 44(4): 521–537.

Morgan, Colleen. 2009. "(Re)Building Çatalhöyük: Changing Virtual Reality in Archaeology." *Archaeologies: Journal of the World Archaeological Congress* 5(3): 468–487.

Morgan, Colleen. 2016. "Analog to Digital: Transitions in Theory and Practice in Archaeological Photography at Çatahöyük." *Internet Archaeology* 42: 1–24.

Ranke, H. 1926. *Koptische Friedhöfe bei Karâra und der Amontempel Scheschonks I bei El Hibe: Bericht über die Badischen Grabungen in Ägypten in den Wintern 1913 und 1914.* De Gruyter; Auflage.

Reilly, Paul. 1991. "Towards a virtual archaeology." In *CAA 90 Computer Applications and Quantative Methods in Archaeology 1990*, edited by Sebastian Rahtz and Kris Lockyear, 132–139. Oxford: Tempus Reparatum.

Richardson, Robert, Shaun Whitehead, T.C. Ng, Zahi Hawass, Andrew Pickering, Stephen Rhodes, Ron Grieve, Adrian Hildred, Arjun Nagendran, Jason Liu, William Mayfield, Mehdi Tayoubi, and Richard Breitner. 2013. "The 'Djedi' Robot Exploration of the Southern Shaft of the Queen's Chamber in the Great Pyramid of Giza, Egypt." *Journal of Field Robotics* 30(3): 323–348.

Roosevelt, Christopher, Peter Cobb, Emanuel Moss, Brandon R. Olson, and Sinan Ünlüsoy, 2015. Excavation is ~~Destruction~~ Digitization: Advances in Archaeological Practice. *Journal of Field Archaeology* 40 (3): 325–346

Tringham, Ruth. 2016. "PROWESS-ING THE PAST: CONSIDERING THE AUDIENCE." Presentation at Conference on Modeling Culture: 3D Archaeology and the Future of the Past, UC Santa Cruz, April 1, 2016. https://www.academia.edu/27033910/PROWESS-ING_THE_PAST_CONSIDERING_THE_AUDIENCE._Presentation_at_Conference_on_Modeling_Culture_3D_Archaeology_and_the_Future_of_the_Past_UC_Santa_Cruz_April_1_2016, accessed 03-22-2022.

Tringham, Ruth. 2015. "Becoming Archaeological," *Journal of Contemporary Archaeology* 2(1): 21–40.

Tringham, Ruth. 2013. "A Sense of Touch—the Full-Body Experience—in the Past and Present of Çatalhöyük, Turkey." In *Making Senses of the Past, Toward a Sensory Archaeology*, edited by Jo Day, 177–195. Carbondale and Edwardsville: Southern Illinois University Press, 2013.

Tringham, Ruth and Michael Ashley. 2015. "Becoming Archaeological," *Journal of Contemporary Archaeology* 2(1): 29–41.

Walker, Dominic. 2014. "Antisocial Media in Archaeology?" *Archaeological Dialogues* 21(2): 217–235.

Wenke, Robert J. 1984. *Archaeological Investigations at el-Hibeh 1980: Preliminary Report,* American Research Center in Egypt Reports 9, 7. Malibu: Undena Publications.

Young, Simon J. 2019. "Virtual Reality: a Powerful Technology that Enables us to Step into Lost Environments of the Ancient World." Public presentation 11 January 2019.

CHAPTER 10

Digitizing and Annotating Ancient Egyptian Coffins: The Book of the Dead in 3D

Rita Lucarelli and Mark-Jan Nederhof

Abstract

3D visualizations of heritage objects such as ancient Egyptian coffins can be better used for general and specialistic studies if they also provide annotations. This paper presents the system of annotations developed for the "Book of the Dead in 3D Project," which applies photogrammetry and digital annotations to coffins and sarcophagi produced in the 1st millennium BCE. The annotated models of the project include the transcription, translation and transliteration of the magical texts inscribed on the coffins, which can be interactively read by the user while navigating the 3D model.

Keywords

3D visualizations – photogrammetry – ancient Egyptian coffins – Book of the Dead – 3D annotations

1 Introduction: Cultural Heritage Digitization and 3D Modelling through Photogrammetry

Since the time the so-called "digital revolution" started, in the late 1950's, the application of digital technologies to the study and preservation of cultural heritage has been progressing very quickly. In particular, the digitization of artifacts through 3D visualizations and the advancement of software for photogrammetry have provided new tools and inspired new approaches for the study of different categories of artifacts and their contextualization within a museum environment.

In Egyptology, where the study of material culture is a main part of the discipline, photogrammetry techniques are being employed for virtual museum exhibitions and for disseminating research projects focusing on particular archaeological sites[1] or on other tangible cultural heritage, from sculpture to pottery and funerary objects.

When documenting an object in 3D and choosing which technique to use, the object's size plays a central role. While 3D scanning technology is especially effective on small size objects, photogrammetry works well on artifacts of bigger size since it results in a textured 3D model of high-resolution, obtained through photos taken with one or more cameras. The 3D model becomes then a new medium itself and can be navigated on a digital viewer through which one can analyze it in detail. A particular category of objects that is proving to be ideal for photogrammetry, within the ancient Egyptian material culture, are the coffins and sarcophagi used to protect the mummified bodies of the deceased and found in tombs throughout the long span of the Pharaonic history. Coffins are also among the most studied objects of the ancient Egyptian funerary culture because of the incredible amount of information they can provide on the ancient Egyptian funerary customs, burial practices and beliefs.[2]

In particular, the anthropoid wooden coffins and sarcophagi produced during the 1st millennium BCE are rich in textual and iconographic information on their lids and exteriors, since they are completely covered in offering formulas, magical spells and images for the protection of the deceased in the netherworld. Their 3D visualizations offer an optimal point of study for scholars and wider public interested in the ancient Egyptian religion and funerary culture and literature, besides providing a user-friendly reconstruction of the object, which can be flipped around in space, and is easier to observe and study in comparison to the original kept in a museum behind glass or in a storage room that is difficult to access.

1.1 *The Book of the Dead in 3D*

In comparison to other media of funerary magic—from 2D objects such as papyri and inscribed tomb walls to 3D items such as amulets, statuettes, stelae and other burial equipment—coffins stand out for literally enfolding and surrounding the body of the deceased, which needs to remain intact in order to ensure a successful journey in the netherworld and the final transfiguration into an *akh*, a blessed, effective spirit.

"Coffin magic" applies to the variety of spells and magical iconography selected for the coffins since the Middle Kingdom, with the so-called Coffin

1 See for instance the "Giza Project" of Harvard University discussed by Nick Picardo in this volume.

2 Among the most recent studies on coffins from different periods, see: Strudwick and Dawson 2019; Taylor and Vandenbeusch 2018.

Texts, and then evolving in the New Kingdom and later on anthropoid coffins and stone sarcophagi whose exteriors are abundantly covered in spells, so that we can talk about "textualization" of coffins.[3]

Although there is a good number of projects and studies on ancient Egyptian coffins and a high number of them is being used for 3D modeling, they mostly focus on stylistic and typological studies rather than on their textualization. When the text analysis of a coffin or sarcophagus is included through a facsimile in printed publications, it becomes difficult to show its peculiar distribution on its medium—the coffin—especially when such a text is not copied in an homogeneous block but rather scattered all over the coffin surface and complemented by images.[4] This is where photogrammetry and 3D visualizations of coffins, including annotations to provide information on the origin, archaeological context, iconography, and the text's transcription, transliteration and translation, become useful in order to reconstruct a story around the object, which can be presented in museums and used by scholars for research.[5]

The *Book of the Dead in 3D* Project applies photogrammetry and digital annotations to coffins and sarcophagi produced in the 1st millennium BCE which are still unpublished and mostly kept in minor Egyptian collections housed in museums with limited gallery space. Producing annotated models of these artifacts facilitates their dissemination and study online through a digital viewer (see below) and allows the user to interact with the object, understand its context and magical function through the annotations.

2 Coffins in 3D

The project team[6] has realized, at present, eighteen 3D models (sixteen wooden coffins and two stone sarcophagi), which are accessible on the project website (https://3dcoffins.berkeley.edu, accessed 03-23-2022) where the digital viewer

3 On the textualization of coffins and coffin magic, see Elias 1993, 323.

4 See the example of the coffin of Irethoreru: https://3dcoffins.berkeley.edu/coffins/san-diego-museum-man-20010630001, accessed 03-21-2022.

5 On the "biography of the object" realized through 3D technology, see C. Greco 2019. https://www.int-arch-photogramm-remote-sens-spatial-inf-sci.net/XLII-2-W11/5/2019/, accessed 03-21-2022.

6 At the time when this article is being written, the team is composed by the two authors of this article, three graduate students of UC Berkeley (Kea Johnston, Jessica Johnson and Jason Silvestri) and one independent scholar (Matt Whealton).

is currently embedded as well.[7] Each item on the website has a general introduction providing its whereabouts and history of acquisition, if known.

Six items, presently digitized in 3D, are kept in the Phoebe A. Hearst Museum of Anthropology (PAHMA) at UC Berkeley, which is where the project's idea itself was born; the PAHMA is a perfect example of a museum housing a very interesting, large and almost completely unpublished collection of ancient Egyptian artifacts, most of which have to be kept in storage rooms due to the limited gallery space. For the same reason, PAHMA has been recently renewing its digital database and portal,[8] through which one can access the vast majority of the museum objects. In the case of the coffins studied for the Book of the Dead in 3D, the Hearst Museum portal provides the 3D models together with the photos and other information related to each coffin.[9] In this way, the tridimensional visualizations replace, for the museum visitor, the experience of seeing the original in the museum gallery and add a more direct interaction with the object, which can then be seen and analyzed through a variety of data, from 2D to 3D. The application of photogrammetry to large size artifacts like coffins therefore allows an optimal dissemination of ancient material culture's productions through a digital experience, eventually valorizing the object itself, which would be otherwise "forgotten" in the museum's storage room.[10]

Even if on exhibition in the museum gallery space, though, a coffin can hardly be analyzed in full because of its size, which does not make it easy to look at it from each side. Very often in museums, coffins are placed standing against a wall or lying horizontally, in both cases hiding the bottom part, which is often also decorated, from view. Showing a coffin's interior can be difficult too; 3D visualizations solve this sort of issue by providing a digital replica that can be observed on each side flipped around and zoomed in to look at details that often cannot be seen behind glass when the coffin is on exhibition in a museum gallery. This is, for instance, the case of a carved cedar wooden coffin housed at the Legion of Honor in San Francisco, the coffin of Irethoriru,[11] proba-

---

7 All the 3D models are also available on the project's page on sketchfab (https://sketchfab.com/bookofthedead3d, accessed 03-21-2022).

8 https://portal.hearstmuseum.berkeley.edu/, accessed 03-21-2022.

9 See for instance the page for the sarcophagus lid of Psamtik (PAHMA 5–522): https://portal.hearstmuseum.berkeley.edu/catalog/d5eb08d9-69b5-4121-8548-d5a88a9521f2, accessed 03-21-2022.

10 On the relationship between digital and material culture and the use of 3D techniques for the study of cultural heritage, see Rossi 2019. https://www.int-arch-photogramm-remote-sens-spatial-inf-sci.net/XLII-2-W11/1025/2019/, accessed 03-21-2022.

11 https://3dcoffins.berkeley.edu/coffins/fine-arts-museums-san-francisco-20022a-b, accessed 03-21-2022.

FIGURE 10.1 The Coffin of Irethoriru (FAMFSF 2022.2a–b)
HTTP://WWW.LIVELYFOUNDATION.ORG/WORDPRESS/?P=1883, ACCESSED 03-21-2022

bly dating to the Thirtieth Dynasty or early Ptolemaic Period on a stylistic basis, standing in the gallery spaces behind glass (fig. 10.1). Its 3D model realized for the project allows instead the viewer to analyze more carefully the beautifully carved hieroglyphs covering the lid, which were once also painted in color and reproduce a variant of Spell 72 of the Book of the Dead.[12] Traces of blue-green color and of white cement used to probably set glass inlays into the central inscription can be noted while zooming in on the 3D model (fig. 10.2/3D model for the online version), which is also useful for studying the text.[13]

Since ancient Egyptian coffins are spread out not only in museums housing extensive Egyptian collections but also in museum institutions where sometimes they are the only ancient Egyptian object as part of an antiquity collec-

12 On Spell 72 and its frequent use on coffins and sarcophagi of the 1st millennium BCE, see Elias 1993.

13 A preliminary translation of this text has been made by the project's team and will be published in the form of annotations in the near future.

FIGURE 10.2 3D model of the coffin of Irethoriru
SCREENSHOT FROM THE SKETCHFAB PAGE OF THE PROJECT: HTTPS://SKETCHFAB.COM/3D-MODELS/FAMSF-20022A-B-IRETHORIRU-E1544D6E19B441F1BBE18444A6221B85, ACCESSED 03-21-2022

tion, realizing 3D models online is an optimal tool for letting these "isolated" items be known by the Egyptological community. This is the case, for instance, of a coffin-ensemble (outer and inner coffin) of the Late Period housed at the Washington State Historical Society in Tacoma (WA), which belongs to a priest from Akhmim called Ankh-Wennenefer. Both the outer and inner coffin (figs. 10.3, 4/3D models for the online version) of this individual present relevant texts and illustrations taken from the repertoire of the magical spells of the Book of the Dead and other funerary texts that show the creativity of the coffin-makers and artists of the Late Period in adapting texts and illustrations taken from longer spell variants on papyrus and adapting them on the reduced tridimensional surface of the coffin. The 3D models of this coffin ensemble can therefore be used to study these two pieces, which have been rarely mentioned in Egyptological studies until now.[14]

14 These coffins are under study by Kea Johnston and will be published in her dissertation on

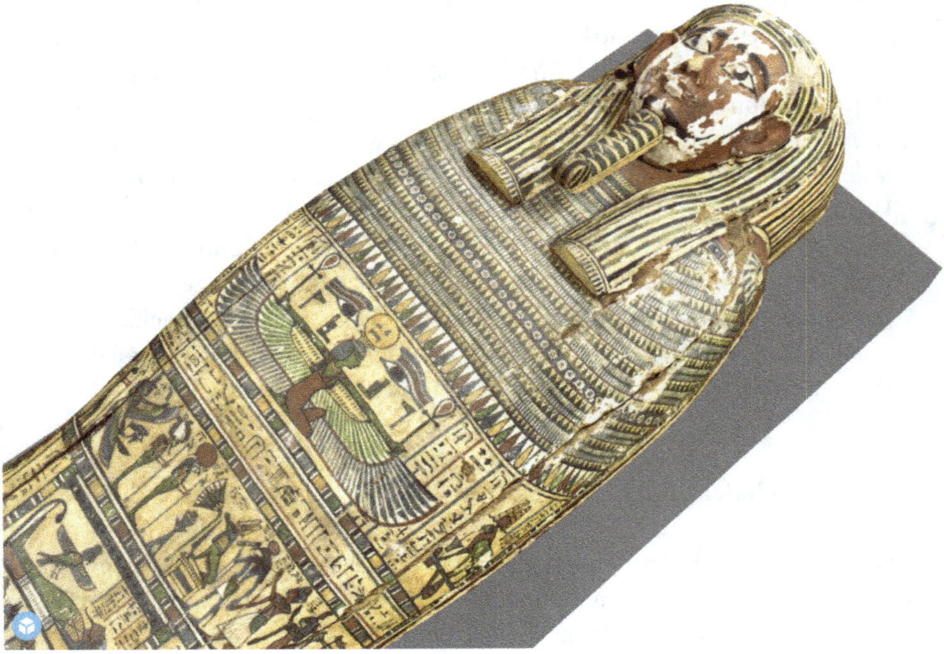

FIGURE 10.3 3D model of the inner coffin of Ankh-Wennenefer (Washington State Historical Society, TA)
SCREENSHOT FROM THE SKETCHFAB PAGE OF THE PROJECT: HTTPS://SKETCHFAB.COM/3D-MODELS/WSHS-INNER-COFFIN-OF-ANKH-WENENNEFER-CDE92142987B45E3A3136F41CABF4DEF, ACCESSED 03-21-2022

Next to the items mentioned below and housed in the PAHMA, the Legion of Honor in San Francisco and the Washington State Historical Society in Tacoma, the project's team has been able to work with other museum institutions such as the Global Museum of San Francisco State University,[15] the

coffins from Akhmim of the 1st millennium BCE. For more details on this coffin-ensemble, see https://sketchfab.com/3d-models/wshs-outer-coffin-of-ankh-wenennefer-f9d1b53bbbce4e0c9f0ce32e1951937b, accessed 03-23-2022, and https://3dcoffins.berkeley.edu/coffins/ankh-wenennefer-outer-coffin-and-lid-wshs, accessed 03-21-2022.

15 The Global Museum houses the so-called Sutro Collection, which includes the painted wooden coffin of the Third Intermediate Period belonging to an anonymous woman (https://3dcoffins.berkeley.edu/coffins/sfsu-global-museum-10411042, accessed 03-21-2022) and the coffin-ensemble (outer and middle coffins, cartonnage) of Nespaperrenub (22nd Dynasty, Thebes; https://3dcoffins.berkeley.edu/coffins/sfsu-global-museum-10512, accessed 03-21-2022). The publication of Nespaperrenub's coffin ensemble is in preparation through a joint study of Lissette Jemenez, Kea Johnston and Rita Lucarelli.

Chrysler Museum of Art in Norfolk (VA),[16] the Museum of Man in San Diego,[17] the Rosicrucian Museum in San Jose,[18] and the Utah Museum of Fine Arts.[19]

Museums like those participating in this project and which are in possession of cultural heritage from the Middle East and Egypt, are also deeply involved in the mission of decolonizing their collections and reconsidering best practices for conserving and displaying the objects while at the same time providing a service for the community and functioning as cultural hubs for the latter.[20] Digital contents and 3D reconstructions like those offered by the Book of the Dead in 3D project for ancient Egyptian coffins are useful in helping museums (both institutional and university-related) to contextualize its objects, to reach remote audiences and to tailor contents according to the specific objects' typology and function, complementing the information already available on the museums' websites and virtual exhibitions, in some cases.

3 Technical Notes

Our framework for developing annotated models is built around the software package PhilologEg.[21] For manipulating 3D graphics, it makes use of libGDX,[22] a Java library that was originally designed for computer games. Displaying a model in the resulting webpage is realized by means of the three.js library.[23]

16 The stone sarcophagus of Psamtik-Seneb (26th Dynasty) housed at the Chrysler museum will be published in full by Matt Whealton. A preliminary translation of the text on the lid is available on the project website through annotations: https://3dcoffins.berkeley.edu/coffins/chrysler-museum-art-712254

17 The Museum of Man houses a painted wooden coffin of the Late Period of uncertain origin but that we believe to be dated to the Late Period and originally coming from Middle Egypt: https://3dcoffins.berkeley.edu/coffins/san-diego-museum-man-20010630001, accessed 03-21-2022.

18 Among others, the Rosicrucian Museum has acquired the painted wooden coffin of Tadinanefer, probably from El-Hibeh and dated to the Saite Period or First Persian Period: https://3dcoffins.berkeley.edu/coffins/rosicrucian-egyptian-museum-rc-2186, accessed 03-21-2022.

19 The Utah Museum of Art is in possession, since 2008, of a Late Period wooden coffin painted in yellow on black, belonging to Padiusir: https://3dcoffins.berkeley.edu/coffins/utah-museum-fine-arts-2008171, accessed 03-21-2022.

20 On the museum as a cultural hub, see Simpson, Fukuno and Minami 2019.

21 https://mjn.host.cs.st-andrews.ac.uk/egyptian/align/, accessed 03-21-2022.

22 https://libgdx.badlogicgames.com/, accessed 03-21-2022.

23 https://threejs.org/, accessed 03-21-2022.

FIGURE 10.4 outer coffin of Ankh-Wennenefer (Washington State Historical Society, TA) SCREENSHOT FROM THE SKETCHFAB PAGE OF THE PROJECT HTTPS://SKETCHFAB.COM/3D-MODELS/WSHS-OUTER-COFFIN-OF-ANKH-WENENNEFER-F9D1B53BBBCE4E0C9F0CE32E1951937B, ACCESSED 03-21-2022

Our framework for 3D models builds on an earlier framework for photographs, which was used to annotate three stelae.[24] In the default view, a stela is displayed in the left half of the browser window, which offers the possibility to zoom and navigate. On the right, there are several tabbed pages, which each display different kinds of information, concerning, for example, the iconography, the translation of the text, or the lexicon. One can click on parts of such a page, which navigates the photograph on the left, to display and highlight

24 Angela McDonald and Mark-Jan Nederhof 2014, 2015; Gabor Toth, Michael McClain and Mark-Jan Nederhof 2016.

the relevant area. Conversely, if one clicks on an area of the photograph, then the page on the right scrolls to the relevant part, which is also highlighted. For example, in the case of a tabbed page containing a lexicon, a click on the occurrence of a word in the photograph makes that the relevant lemma in the lexicon is highlighted, and conversely, a click on a lemma makes the photograph navigate to its first occurrence in the text (a second click navigates to the second occurrence, etc.).

This framework for ancient Egyptian stelae was in turn inspired by an Adobe Flash application that presents the annotation of a Mayan stela.[25] On the left side there is a line drawing of the stela, in which pairs of signs are enclosed in rectangles. One can click within such a rectangle to display the corresponding transliteration and translation on the right side. One can also play an audio recording of the transliteration. In addition, there are explanations related to Mayan writing and culture. Next to the line drawing, a high-resolution photograph can be displayed. Regrettably, Adobe stopped supporting Flash in January 2021, and most modern browsers block all Flash content.

4 Annotation

Annotating the text on a coffin follows a methodology whereby different tiers, such as transcription, transliteration, translation, lexical analysis, and so on, are represented in individual files.[26] When the webpage is constructed, the tiers are combined through automatic alignment, resulting in interlinear text, as well as a customized lexicon.

Transcription of the hieroglyphic text is encoded in RES.[27] Where possible, this is converted to Unicode in the resulting webpage, but we find that Unicode is at present too limited for faithful encoding of typical inscriptions; a particular issue is the lack of a primitive for rotation of signs by 90°. Our fallback option is to render RES directly on an HTML canvas.

One tier of annotation contains the 3D model itself. It pairs areas on the model to occurrences of signs. For each sign occurrence, its area is delimited by a polygon of points on the surface of the object.[28] Moreover, the direction

25 Mark Van Stone and Barbara MacLeod 2008.
26 Nederhof 2009.
27 Nederhof 2002.
28 Upon rendering, the 3D points are projected to 2D points, to give flat polygons. We find that this suffices for our purposes. More advanced approaches to annotation of 3D models exist. Cf. Ponchio, Callieri, Dellepiane and Scopigno 2020.

from which the area is best seen, and possibly a desired rotation of the camera, are part of the annotation. This provides necessary information to the web application, to find a suitable camera position from which the text can be conveniently read. Moreover, this information may be used to suppress display of polygons that are currently 'behind' the object, or more precisely, where the angle between the direction from which the area is preferably seen and the current direction of the camera is greater than 180°.

Where a word consists of multiple signs, the polygons from the individual signs are automatically joined in the process of automatic alignment. This results in a larger polygon delimiting the word occurrence; the preferred camera directions and rotations from the individual signs are averaged. Further polygons may be created on the surface of the 3D model. These can be used to link areas that are of e.g. iconographical interest to parts of the currently displayed tabbed page.

5 Navigation

Interaction with virtual 3D environments has been studied for the purpose of various applications and as a problem in its own right.[29] Central issues are navigation within the environment, and control to achieve navigation and to manipulate objects.

Coffins with text pose particular problems to navigation, as a result of different pieces of text appearing with different orientations. Text on the front of a coffin is typically such that the "animal signs" face either the middle of the front of the coffin, or one of the sides. Conceptually, such text can be read if the coffin is lying on its back, and the reader stands at its feet and bends over the coffin. Text on the feet however is sometimes upside-down relative to the text on the front, and such text can be read more conveniently if the reader stands near the head, and once more bends over the coffin (fig. 10.5).

Text on the sides may have the same orientation as text on the front, with animal signs facing the front of the coffin, but here the reader would need to stand near the sides and tilt their head by half a turn. In other cases, text on the sides has a different orientation, with animal signs facing the top of the head. Conceptually, such text can be read if the reader stands at one of the sides, then without needing to tilt their head.

29 LaViola, Kruijff, McMahan, Bowman,. Poupyrev 2017.

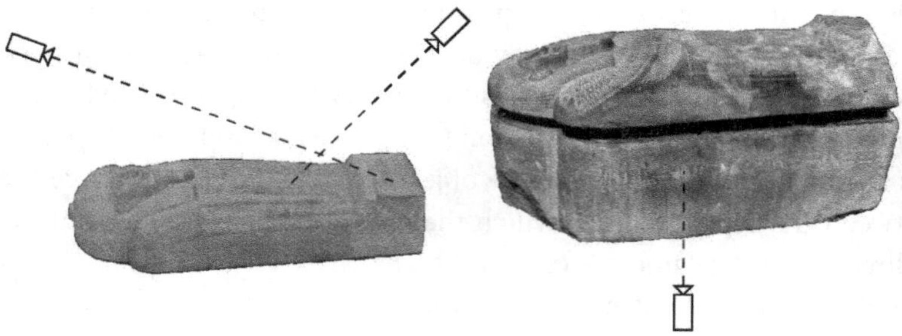

FIGURE 10.5 Orbit control

The challenge is to find an intuitive form of navigation allowing different pieces of text on the coffin to be read, without the user getting disoriented. We have chosen for orbit-control as the main form of navigation, assuming the coffin to be lying in its back. More precisely, the camera is located on a sphere at a fixed distance from the geometric center of the coffin, as if it is moving in orbits around the object. Horizontal drags of the mouse (or slides on a touchscreen) are translated to changes in azimuthal angle (eastward or westward) and vertical drags are translated to changes in polar angle (northward or southward). Where appropriate, for example if the underside of the object has not been scanned, the polar angle is constrained to be between 0° and 90° (between north pole and equator), as there would be no purpose in inspecting the model from underneath.

The top of the view always points towards the conceptual north pole (polar angle 0°), to avoid the user getting disoriented. Focusing in on details of the object is by decreasing the field of view (by mouse scrolls, key presses or moving two fingers away from each other on a touchscreen), rather than by moving the camera towards the center of the object. The reason we have not chosen for the latter is that if the camera moves through the surface of the object, then the user gets disoriented by the resulting empty view. Furthermore, moving the camera very close to the surface of an object tends to result in an unnaturally distorted view. The smaller the field of view, the smaller the changes in camera position per unit of distance travelled by mouse drags.

By default, the camera is directed at the geometric center of the object, with the camera on the sphere at a fixed distance from that center. In order to conveniently read text when having zoomed in, the camera can also be moved laterally away from its usual position on the sphere, by pressing arrow keys (or by moving two fingers in the same direction on a touchscreen). This is illustrated in fig. 10.6.

The lateral displacement of the camera relative to its usual position on the sphere is preserved upon small changes in that position through orbit control.

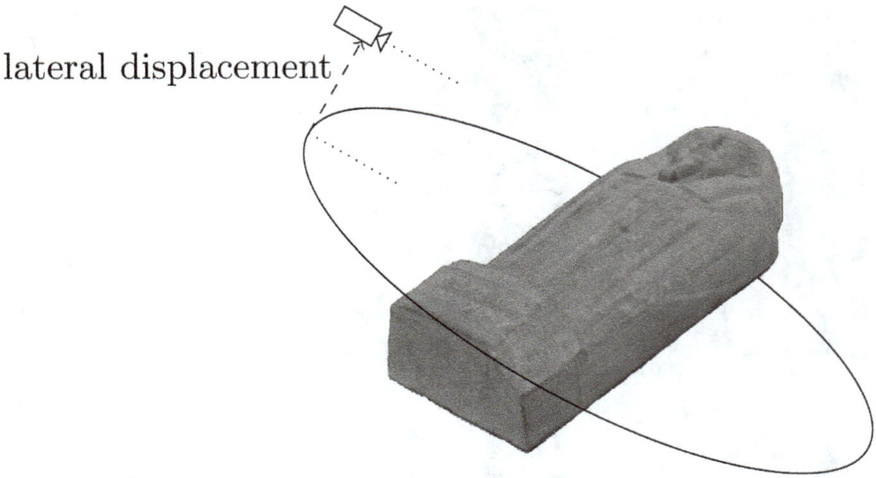

FIGURE 10.6 Displacement

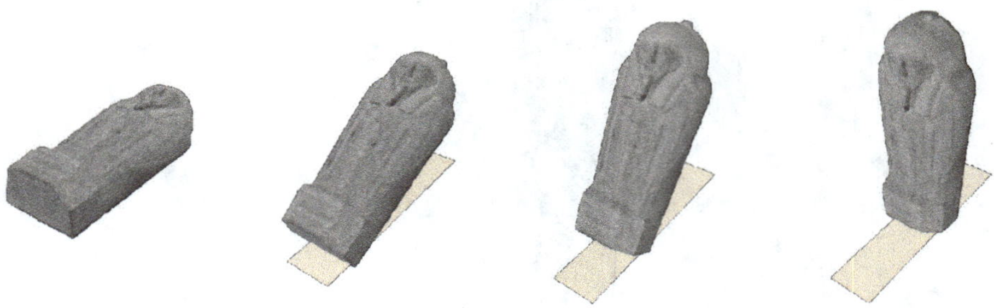

FIGURE 10.7 Tilt

However, upon larger changes, the lateral displacement proportionally decays. This keeps the user from getting disoriented if the camera looks past the object, showing an empty view, or shows a side of the object at a sharp angle. The translation of arrow keys to changes in lateral displacement is once more affected by the field of view: the smaller the field of view, the smaller the corresponding displacement; one arrow key press corresponds to roughly a quarter of the current view.

None of the above however would allow the user to conveniently read text on the sides of a coffin with animal signs facing the middle of the front. For this reason, the object can conceptually be tilted up (or down), by increments of 30° (by key presses or swipes on a touchscreen). A "floor" then appears beneath the object (fig. 10.7).

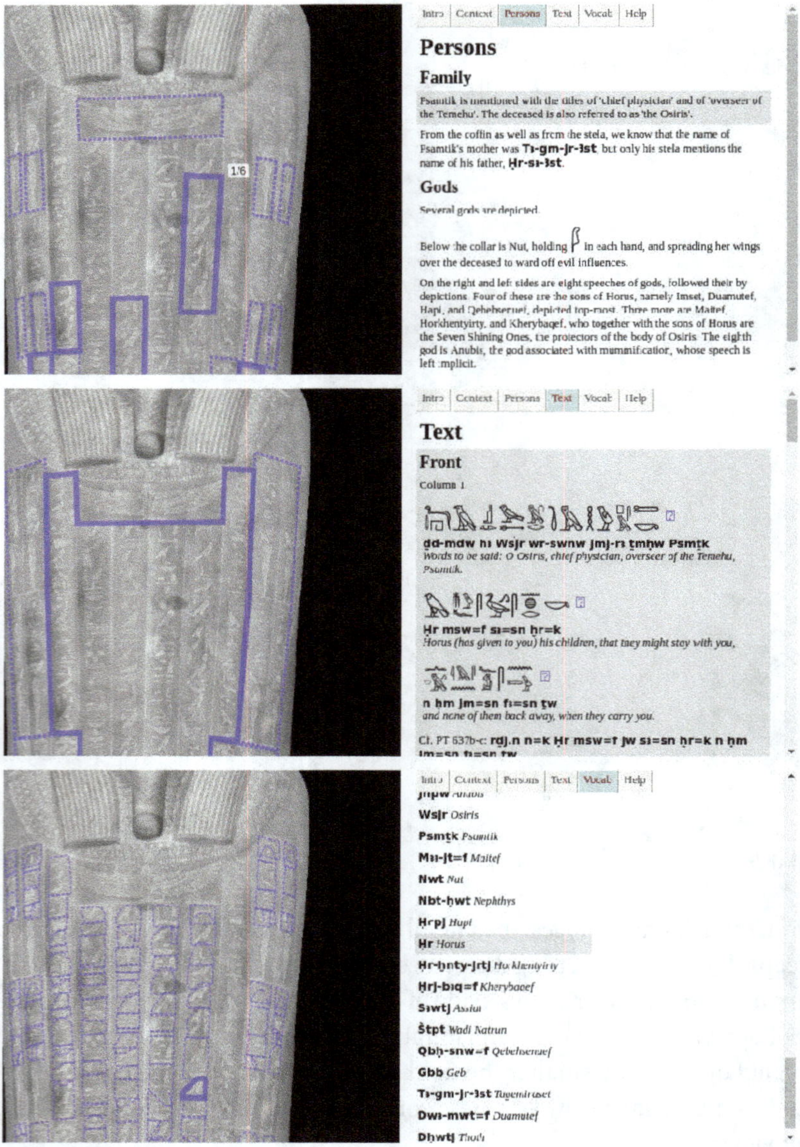

FIGURE 10.8 Interactivity

6 Interactivity

There is interaction between the model, in the left half of the screen, and the current tabbed page, in the right half. Clicking on an area of the model results in highlighting of relevant text in the tabbed page. Conversely, clicking on text in a tabbed page results in navigation of the model, and highlighting of the relevant area. This navigation makes use of the preferred direction from which the area is to be viewed, which was determined during annotation of the model. From this, the application computes the optimal combination of azimuthal and polar angles and lateral displacement, taking into account any tilting of the object. The position of the camera then gradually changes to that view. Switching between tabbed pages does not in itself change the camera position (fig. 10.8). All this allows exploration of the object with a considerable amount of flexibility and continuity.

Bibliography

Elias, J.P. 1993. *Coffin inscription in Egypt after the New Kingdom: a study of text production and use in elite mortuary preparation*. Ann Arbor: UMI.

C. Greco. 2019. "The Biography of Objects," in: *The International Archives of the Photogrammetry, Remote Sensing and Spatial Information Science*, XLII–2. https://www.int-arch-photogramm-remote-sens-spatial-inf-sci.net/XLII-2-W11/5/2019/, accessed 03-21-2022.

LaViola, Joseph J., Ernst Kruijff, Ryan P. McMahan, Doug Bowman, Ivan P. Poupyrev. 2017. *3D User Interfaces: Theory and Practice*, Second Edition. City? Addison-Wesley Professional.

McDonald, Angela and Mark-Jan Nederhof. 2014. *The stela of Iy*. Dundee: McManus art gallery and museum. https://mjn.host.cs.st-andrews.ac.uk/egyptian/stelae/iy-dundee/, accessed 03-21-2022.

McDonald, Angela. Mark-Jan Nederhof and Campbell Price. 2015. *The stela of Ramose. City:* Manchester Museum. https://mjn.host.cs.st-andrews.ac.uk/egyptian/stelae/ramose-manchester/, accessed 03-21-2022.

Nederhof, Mark-Jan. 2009. "Automatic Creation of Interlinear Text for Philological Purposes". ATALA (*Traitement Automatique des Langues*), 50(2): 237–255. Publisher?

Nederhof, Mark-Jan. 2002. "A revised encoding scheme for hieroglyphic". In *Proceedings of the 14th Table Ronde Informatique et Égyptologie*. Pisa: publisher. On CD-ROM.

Ponchio, F., M. Callieri, M. Dellepiane and R. Scopigno. 2020. "Effective Annotations Over 3D Models". *Computer Graphics Forum*, 39(1): 89–105. City: Publisher.

Rossi, C. 2019. "Aristotle's Mirror: Combining Digital and Material Culture", *The Interna-*

tional Archives of the Photogrammetry, Remote Sensing and Spatial Information Science, XLII–2. City: Publisher. https://www.int-arch-photogramm-remote-sens-spatial-inf-sci.net/XLII-2-W11/1025/2019/, accessed 03-21-2022.

Simpson, Andrew, Akiko Fukuno and Hiroshi Minami. 2019. "University museums and collections as Cultural Hubs: The Future of Tradition", *University Museum and Collections Journal*, 11, 9–12; on best practices in museums with Middle Eastern collections, see Emberling, Petit eds, *Museums and the Ancient Middle East Curatorial Practices and Audiences,*

Strudwick, H. and J. Dawson (eds). 2019. *Ancient Egyptian coffins: past—present—future*. Oxford: Oxbow Books

Taylor, J.H. and Vandenbeusch, M. (eds.) 2018. *Ancient Egyptian coffins: craft traditions and functionality*. British Museum Publications on Egypt and Sudan 4. Leuven: Peeters.

Toth, Gabor, Michael McClain and Mark-Jan Nederhof. 2016. *The stela of Qema-mar* (Zagreb Archaeological Museum), http://egypt-grammar.rutgers.edu/qemamar/, accessed 03-21-2022.

Van Stone, Mark, and Barbara MacLeod. 2008. *Decode Stela 3*, https://www.pbs.org/wgbh/nova/mayacode/translate.html, accessed 03-21-2022.

CHAPTER 11

Photogrammetry and Face Carvings: Exploring the 'Face' of the Egyptian Anthropoid Coffins by 3D Modeling

Stefania Mainieri

Abstract

This paper aims to illustrate the initial results of a new methodological approach to the study of anthropoid coffins. We analyze the coffins using a new methodological approach that will support and complement traditional analytical methods used for these finds. The project is based on the analysis of "facial features" carved on the lid and the so-called "minor attributes." 3D modeling provides us with an excellent record of surface morphology.

Keywords

photogrammetry – coffins – sarcophagi – mummy – facial analysis – portraits

1 Introduction

This paper will explain a new methodological approach for studying Ancient Egyptian coffins. The new methodology is based on analyses of the facial features of the coffin lid masks and other elements using 3D models created by photogrammetry: an extremely versatile, simple and economical technique that records accurate textures based on photographic shots.[1]

Starting with decontextualized Late Period coffins—on the occasion of a collaboration with the Vatican Coffin Project (VCP)[2]—a pilot project on a new

1 Fiorini 2012, 213–227; Remondino and Campana 2014; Russo, Remondino and Guidi 2011, 169–198.

2 Amenta 2014, 483–499; Amenta and Guichard 2017; Weiss 2018. The *Vatican Coffin Project* (*VCP*) is an International Team Project, set up by the Egyptian Department of the Musei Vaticani, directed by Alessia Amenta, in collaboration with the Diagnostic Laboratory for

class of artifacts was started in October 2018: the so-called "yellow coffins." The conference in Bloomington on "Ancient Egypt and New Technology" was the occasion for the presentation of this use of photogrammetry and an explanation of interesting aspects revealed and preliminary considerations born within the *VCP*, to be analyzed in depth from February 2021. In the months following the conference, in fact, a well-defined, specific and exact methodology was established, and the project has been awarded a Marie Skłodowska-Curie (IF) Global Fellowship (within the H2020 program), with the title "Faces Revealed Project. Investigating the faces of "yellow coffins" through photogrammetry. A comparison of 3D models and digital twins for a new analysis of the manufacture, painting techniques and re-use" (H2020-MSCA-GF 2019: 895130).[3]

2 Starting Point: The Late Period Coffins

This methodology was conceived during the researcher's Ph.D. studies at the University of Naples "L'Orientale," (UniOr), while studying two bivalve coffins

Conservation and Restoration of the Musei Vaticani, directed by Ulderico Santamaria, to study the 'yellow coffins' of the Third Intermediate Period. Other partners are: Rijksmuseum van Oudheden in Leiden, Musée du Louvre in Paris, Museo Egizio di Torino, Centre de Recherche et de Restauration des Musées de France (C2RMF) in Paris, Centro Conservazione e Restauro La Venaria Reale in Turin and Xylodata in Paris. Kathlyn M. Cooney (UCLA University, Los Angeles) is collaborating in the project for the study of the reuse of coffins of the 21st Dynasty.

3 This prestigious achievement will allow the researcher to lead—as Principal Investigator—the Project, through a two-year research/training path at the University of California Los Angeles, USA (Host) with a reintegration time at the Egyptian Museum of Turin (Beneficiary) during the third year. The project will be coordinated by Christian Greco, Director of the Museo Egizio di Torino, and Kathlyn Cooney, Professor of Egyptology at UCLA. Corinna Rossi, Professor of Egyptology, will host the researcher for a secondment at the Politecnico di Milano University. This research has been made possible thank to the suggestion, support and collaboration of several scholars. I would like to thank Alessia Amenta (Curator, Ancient Egypt and Near Eastern Antiquities Dept., Musei Vaticani, Città del Vaticano), Angela Bosco (CISA—University of Naples "L'Orientale", Italy), Kathlyn M. Cooney (Professor of Egyptian Art and Architecture and Chair of the Department of Near Eastern Languages and Cultures at UCLA, California), Andrea D'Andrea (CISA—University of Naples "L'Orientale", Italy), Christian Greco (Director of the Museo Egizio, Turin, Italy), Hélène Guichard (Curator, Ancient Egypt Antiquities Musée du Louvre, Paris), Rita Lucarelli (Associate Professor of Egyptology, UC Berkeley) Rosanna Pirelli (Professor of Egyptology, University of Naples "L'Orientale", Italy), Rosario Valentini (CISA—University of Naples "L'Orientale", Italy), and Lara Weiss (Curator, Ancient Egypt Antiquities, Rijksmuseum van Oudheden, Leiden).

in the Archeological Museum of Naples (MANN),[4] sold by Camillo Borgia in 1814 and repainted during the 19th century.[5] The chronology of coffins is usually based above all on the figurative apparatus and the paleography. Therefore, the current condition of both lids did not allow us to understand and reconstruct their cultural and chronological context. For this reason, a different kind of analysis was used. At first, the structure and morphology were considered, and then other specific details were investigated, above all the face. Features such as eyes, nose and eyebrows are important elements for the typology and classification of statues,[6] but had never been taken into consideration for coffins, perhaps because they are secondary elements—the so called "minor attributes"[7]—and because it is difficult to clearly discern them under the superimposed pictorial decoration. The presence of pictorial layers, in fact, makes the correct and objective perception of the sculpture of the face—the only carved element in anthropoid coffins with hands—less clear. However, I found photogrammetry, created in 2015 in collaboration with Centro Interdipartimentale Servizi per l'Archeologia (CISA) at the UniOr for documenting the condition of the coffins in the MANN before the restoration, to be of great help for this type of analysis. Along with its value in designing, documenting and monitoring the objects, 3D models could also be useful for research and study because they give us an excellent record of surface morphology. The monochrome solid model, curvature shading and shadows—important for conveying shape—all allow us to discover the modeling of the face under the pictorial layer and can reveal fine surface details that are not always discernible in color (fig. 11.1).

This possibility of "reading" the object and analyzing in greater detail the sculpted "face" on the lids was fundamental for my research. In the case of one of the repainted Borgia coffins (the coffin of Tchahapyemiu, inv. gen. nos. 2340, 2345), for example, photogrammetric analysis made clear all the "invisible" sculpted physiognomic traits: eyes, eyebrows, the line of the nose and mouth. All these traits, especially the eyebrows, represent features commonly observed in stone specimens, dated precisely to the 25th/26th Dynasty (Perdu 2012). This chronology was later confirmed by the reconstruction of the original texts and

4 Anonymous coffin, inv. gen. nos. 2342, 2346, and coffin of Tchahapyimu, inv. gen. nos. 2340, 2345.

5 Mainieri 2016, I, 338–347, cat. nos. 15.3–4, 423–469; AA.VV. 2016, 120–121, n. 7; Mainieri 2019, 63–71.

6 Bryan, 1987, 3–20; id. 2010, 913–943; id. 2015, 375–396; Laboury 2010, 1–18; Perdu 2012; Sorouzian 1999, 55–74.

7 Taylor 2003, 102.

FIGURE 11.1 Comparison of the model textured (above) and model solid (below) of the coffins at the MANN. From the right: anonymous inner coffin lid, inv. no. 2346 (Saqqara (?), 24th–30th Dyn.); inner coffin lid of Tchahapyemiu, inv. no. 2345, (Menfi (?), 25th–26th Dyn.); Coffin of Ankhhapy, inv. no. 114.113 (Akhmim, early Ptolemaic Period)
COURTESY OF THE MINISTERO DEI BENI E DELLE ATTIVITÀ CULTURALI E DEL TURISMO—MUSEO ARCHEOLOGICO NAZIONALE DI NAPOLI (© MANN)

decoration, thanks to the accurate descriptions and the "photographic" copy of the hieroglyphs made by Zoëga in 1789 (NKS 357b fol. III, I), which allowed me to combine the morphological data with the paleographic and stylistic ones, thus corroborating the proposed chronological identification and linking the coffin to a specific typology from the north.[8]

3 The Pilot Project: "Yellow Coffins" and 3D Models—First Results

The collaboration with the *Vatican Coffin Project* at the end of 2018 led to the shifting of the project to a new class of artifacts: the "yellow coffins" appearing in Thebes at the end of the New Kingdom and used for more than a millennium,

8 Mainieri *in press.*

peaking during the 21st Dynasty (ca. 1069–945 BCE).[9] These coffins are characterized by a marked sense of *horror vacui*. In the Third Intermediate Period (ca. 1069–664 BCE), the pictorial and textual tradition of the tomb walls found its way onto the coffins. We are therefore witnessing what van Walsem calls the "architectonization" of the coffin, which becomes a small universe, at the center of which is the deceased themselves, who becomes the source of their own regeneration and rebirth.[10] The development of these artifacts has been extensively studied,[11] thanks to the coffins found in the 19th century in the two Theban cachettes,[12] now stored in diverse museums around the world.[13] In 1988, Niwiński identified different typologies with a precise chronology based on layout, style and images.[14] Nevertheless, the numerous stylistic and iconographic variables, as well as the various techniques of execution and the widespread reuse of coffins during the Third Intermediate Period, continue to place these artifacts at the center of a lively debate. While earlier scholars focused almost exclusively on layout, iconography and texts,[15] recent years have seen a shift of attention towards production, materials and painting techniques.[16] The Pilot Project seeks to close the existing gap in our knowledge by focusing specifically

9 Niwiński 1988; Taylor 1989; van Walsem R. 1997.

10 Van Walsem 2015, 390–397.

11 Aston 2009; Niwiński 1988; Sousa, 2018b; Taylor 1989; id. 2001, 164–181; id. 2006, 263–291; van Walsem 1997.

12 In 1881 at Deir el-Bahari the first *cachette* (TT 320) was found, with royal coffins and mummies and 14 'yellow coffins' of the Priests of Amun during the 21st Dynasty (ca.1069–945 BCE). Ten years later (1891), an inviolate burial chamber was found in the northeast corner of the Hatschepsut's temple (ca. 1473–1458 BCE), with 250 coffins dating back to the 21st Dynasty, later called Bab el-Gasus.

13 Part of the Bab el-Gasus coffins was retained for the Giza Museum and another part was divided into XVII lots and sent in 1893 to countries involved in political and diplomatic operations in Egypt: Lot I (France), Lot II (Austria), Lot III (Turkey), Lot IV (United Kingdom), Lot V (Italy), Lot VI (Russia), Lot VII (Germany), Lot VIII (Portugal), Lot IX (Switzerland), Lot X (USA), Lot XI (Netherlands), Lot XII (Greece), Lot XIII (Spain), Lot XIV (Sweden-Norway), Lot XV (Belgium), Lot XVI (Denmark), Lot XVII (Città del Vaticano). Originally 17 museums profited from Khedive's gift, but subsequently the coffins were reallocated and today at least 35 museums are known to house objects from Bab el-Gasus. 'The gate of the Priests Project', directed by Rogério Sousa (University of Lisbon) and involving the Egyptian Dept. of the Musei Vaticani, the University of Leiden and the University of California, Los Angeles, has as its main purpose recreated/reconstructed the original integrity of the Egyptian tomb of the priests of Amun (Bab el-Gasus) dating from the 21st Dynasty and found undisturbed since antiquity, see Sousa et al. (eds) 2021.

14 Niwiński 1988.

15 E.g., Niwiński 1988.

16 Amenta 2014, 483–499; Amenta and Guichard 2017; Dawson and Strudwick 2016; Dawson and Strudwick 2018; Weiss 2018.

on the material aspects of the Egyptian coffins together with their symbolic and cultural value. The reason for this is to understand whether the construction and painting/decorative techniques can also be functional elements in creating a typology.[17] An important development in burial practices was the subsequent reuse of "yellow coffins" during the 21st Dynasty which had economic and social implications during what was a period of scarcity.[18] In recent years, the question of the identity of ateliers and the reuse of the objects have been the subject of specific projects characterized by a combination of different but interconnected skills and competences from diverse disciplines: Egyptology, Diagnostic and Conservation.[19] Thanks to such collaborations, a protocol of scientific analyses and a specific methodology have been devised that aim at studying the composition of the various materials and deepening our understanding of the practices lying behind ancient reuse, through the use of the latest technologies available in our cultural heritage.

The Pilot Project fits into this new line of research and started from a small group of 13 artifacts coming from the Bab el-Gasus Cache, a coherent *corpus* for dating, provenance and commissioning:

1) Lot XVII at the Musei Vaticani (MV) donated by the Egyptian Government to Pope Leo XIII: outer coffin and mummy board of Ikhy, MV 250353.1–3;[20] inner coffin of Takhybiat, MV 25015.2.1–2;[21] two anonymous inner coffins, MV 25016.2.1–2[22] and MV 51515;[23] and two anonymous mummy boards, MV 25020[24] and MV 25022.[25]
2) Lot XI at Rijksmuseum van Oudheden in Leiden (RMO): anonymous inner coffin (RMO F 93/10.4); coffin set of Nesytanebtawy, outer and inner coffin and mummy board (RMO F 93/10.2a–c); coffin set of Tjenetpenherunefer, inner coffin and mummy board (RMO F 93/10.3a–b).[26]

The first step was to create a 3D model of the upper part of each coffin lid. The photos were taken with Nikon D5200 that allowed the acquisition of photos with very high resolution, in order to produce a final result with many accu-

17 Amenta 2014, 483–499.
18 Cooney 2007; id. 2011, 3–44; id. 2014, 45–66; id. 2017, 101–112. id. 2018a, 295–322; id. 2018b, 96–108.
19 Amenta and Guichard 2017; Dawson and Strudwick 2018; Weiss 2018.
20 Cooney 2014, 48, fig. 11; id. 2017, 101–112; Gasse 1996, 81–97.
21 Du Quesne 1998, 615–617; Gasse 1996, 98–109.
22 Gasse 1996, 13–23; Grenier 1993, 23.
23 Gasse 1996, 129–130.
24 Gasse 1996, 132–134.
25 Gasse 1996, 135–137.
26 Weiss 2018.

rate details, using Color Checker (for the color) and a metric reference (for the size). The settings of the shots were calibrated in relation to the features of the objects but, above all, to the light and shadows. The coffins, in fact, are stored in both the exhibition rooms and storerooms, so the light conditions changed in relation to this; moreover, some of them are exhibited in a vertical position, others in very dark rooms or in non-optimal light conditions and often they cannot be moved.[27]

Once the 3D model had been created and the monochrome solid processed with Agisoft Metashape program (1.4.0.5076 version), in order to compare the facial geometry and the underlying features with the decoration, the model solid of each coffin was compared with its 3D model textured. Unexpected information emerged, different from previous data obtained for Late Period coffins, but equally important because more details can be added to the information that we already have on these coffins. We have to consider this information differently from the Late Period coffins, where the paint was directly applied onto the wood or a very thin layer of plaster—so the 3D model "reflects" the sculpted wooden facial features. In the "yellow coffins," the painted layer is applied over generally two layers of plaster (differing in composition and granulometry). Moreover, the plaster does not lie on the carved wooden surface in a uniform manner. Often, we can find a large amount of plaster used to help create a curved surface and to give a better shape to the head, face and upper part of the trunk. So we have to emphasize that the 3D model reveals more of the surface level of plaster modeling in the "yellow coffins."

The first result was to observe that the faces seemed to be different from each other. This might suggest that there is no "standardization" (fig. 11.2): they can be rounded or square, sometimes not symmetrical or well proportioned (MV 51515; MV 25020); in some cases the eyes are at different levels (RMO F 93/10.2c), the mouth is out of position regarding the central axis, which can create a distortion of the two halves of the face (RMO F 93/10.4; MV 51515), or individual features are not placed in the correct relationship to each other (MV 25015.2.1); in one case, the jaw seems to display prognathism (MV 25022); with all the specimens, the eyebrows follow the extension of the line of the nose that continues from the curve of the eyebrows: this curve, however, can be straight and slightly tilted at the ends, rounded or arched; the nose can be thin and proportionate, large at the base or very big; the mouth can be narrow and fleshy, or thin and straight.

---

27 In general, a value ISO 800 was preferred, with an aperture between f/7 and f/10 and lens of 24 mm in order to guarantee a good compromise between depth of field and brightness. The distance from the object was 1 meter maximum.

FIGURE 11.2 The 'yellow coffins' (Lot XVII and Lot XI) from the Bab el-Gasus *cache*
COURTESY OF THE *VATICAN COFFIN PROJECT* (© MUSEI VATICANI) AND OF THE NATIONAL MUSEUM OF ANTIQUITES, LEIDEN (© NATIONAL MUSEUM OF ANTIQUITES, LEIDEN

Nevertheless, starting from a simple autoptical observation of the form of the face, it is possible to arrange the finds into four groups:

Group I round face (full cheeks; small, round chin; narrow, wide forehead)
Group II square face (wide forehead; square, protruding jaw)
Group III triangular face (high cheekbones; wide forehead; pointed chin)
Group IV oval face (wide forehead; shallower jaw; protruding chin).

Owner		Inv. n.	Date	Group I	Group II	Group III	Group IV
Takhybiat	IC	MV 25015.2.1–2	Late 21st Dyn.				X
Anonymous	IC	MV 25016.2.1–2	Late 21st Dyn.	X			
Anonymous	IC	MV 51515	Late 21st Dyn.	X			
Anonymous	MB	MV 25022	Late 21st Dyn.		X		
Anonymous	MB	MV 25020	Late 21st Dyn.			X	
Ikhy	MB	MV 25035.3.2	Late 21st Dyn.			X	
Ikhy	OC	MV 25035.3.1–3	Late 21st Dyn.	X			
Nesytanebtawy	OC	RMO F 93/10.2a	Late 21st Dyn.	X			
Nesytanebtawy (?)	IC	RMO F 93/10.2b	Late 21st Dyn.	X			
Nesytanebtawy (?)	MB	RMO F 93/10.2c	Late 21st Dyn.				X
Tjenetpenherunefer	IC	RMO F 93/10.3a	Early 21st Dyn.			X	
Tjenetpenherunefer	MB	RMO F 93/10.3b	Early 21st Dyn.			X	
Anonymous	IC	RMO F 93/10.4	Late 21st Dyn.			X	

Certainly other elements could be different or could be linked to another group: in some cases, there may be some minute details which certain masks belonging to different groups have in common (e.g. mummy boards MV 25035.3.2 and MV 25020), as well as masks in the same group but belonging to a different set having strong correspondence in other facial features (e.g. the inner coffin of Takhybiat, MV 25015.2.1 and the mummy board of Nesytanebtawy, RMO F 93/10.2c having the same rendering of lips and mouth).

Another element emerged during this preliminary observation regarding the eyes. In most cases the eyes appeared as a rounded shape where the upper part seems to define the eyelids, while the lower part is the area for the iris and cosmetic lines; sometimes they were painted on without following the sculpted line (e.g. MV 25020 and MV 51515), a characteristic often found on objects dating from a successive period (Sousa 2018a, 48). Moreover, in some cases the eyes were not sculpted in detail on the wood but were modelled with plaster in order to create more or less three-dimensionality. Good examples come from the inner coffin lid of a Takhybiat (MV 25015.2.1) and the anonymous mummy board MV 25022, both in the Musei Vaticani.

Maybe the wooden mask of Takhybiat was adapted at a later date or at a different stage of production, to make the face smaller (fig. 11.3). The wooden mask is 15.3 cm long, and the new one (modelled with plaster on wood) is 2.2 cm smaller. The poor condition of the coffin allowed us also to observe the wooden

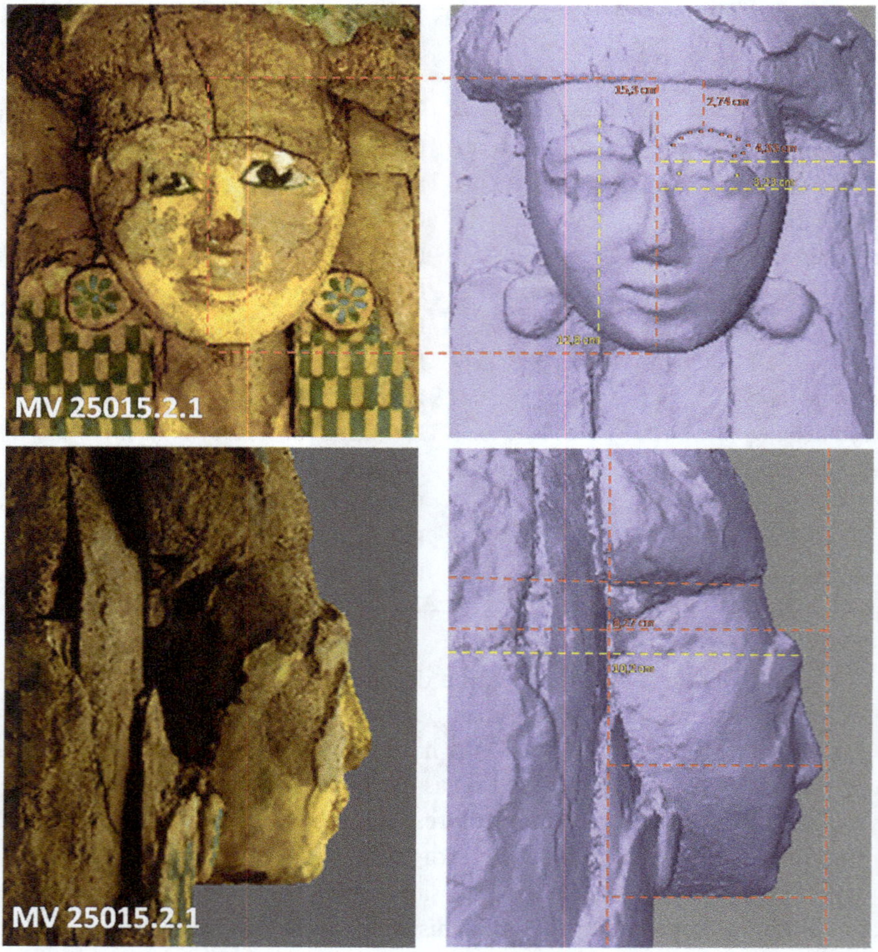

FIGURE 11.3 Particular of the inner coffin lid of Takhybiat in the Musei Vaticani
COURTESY OF THE *VATICAN COFFIN PROJECT* (© MUSEI VATICANI)

base and see one of the carved eyes on it. The eye starts 2.74 cm under the hairline. It is a very basic eye, narrow and long (around 4.33 cm between outer and inner corner of the eye fissure), simply carved onto the wood without any detail and three-dimensionality, but possibly only carved to indicate the position where the eye was supposed to be placed. The "new" eye, instead, was created by modeling the plaster (1.93 cm thick). It is aligned with the carved eye, but smaller and painted on a crudely made, slight bulge for three-dimensionality.

Conversely, in the anonymous mummy board (MV 25022), the model solid made it clear that the eyes are almost flat. The lines of the eyes and eyebrows

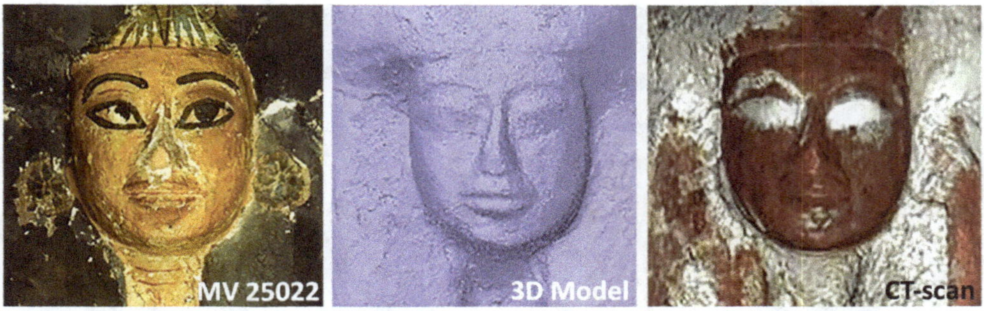

FIGURE 11.4 Comparison of the model textured, model solid and CT-scan (cardiovascular filter) of the anonymous mummy board in the Musei Vaticani
COURTESY OF THE VATICAN COFFIN PROJECT (© MUSEI VATICANI)

are distinguishable under the paint but different from the forms observed in the other artifacts. The eyes are narrow and elongated and seem to have the form of an almond; both eyelids have the same curvature but are extended upward. Moreover, the face is crudely carved, and the paint is not refined, as if the painting was applied in a hurry. The observation of the Computer Tomography (CT) scan images in Volume Rendering Technique (VRT) carried out on the artefact within the VCP[28] shows that much plaster was used to shape the eyes, and we can observe the same form in the photogrammetry (fig. 11.4).

One of the most important results of these preliminary findings regards the coffin set belonging to the Chantress of Amun, Ikhy (MV. 2035.3.1–3), in particular her mummy board. Studied for a long time, both iconographically and textually, its physiognomic traits had never been taken into consideration. The face could be part of group III but its square-shaped, striking features are very different from the features on the other Bab el-Gasus coffins in the Musei Vaticani and even from the features on its own outer coffin.

If we observe in detail the painted faces of both artifacts, we can recognize the same painting style and the same colors, which might be indicative of the same workshop or craftsman, or that they were painted at the same time. Both faces are painted yellow; the eyes, with a black iris painted on a white background, are large with cosmetic lines painted in black, as well as the eyebrows, which are slightly tilted at the ends; the lips are outlined in red with a shadowy effect created by small circles painted in the corners of the mouth. However, this uniformity in the painted features cannot be observed in the sculpted ones. The 3D monochrome solid, in fact, immediately reveals that the sculpted face

28 Amenta 2018, 323–335.

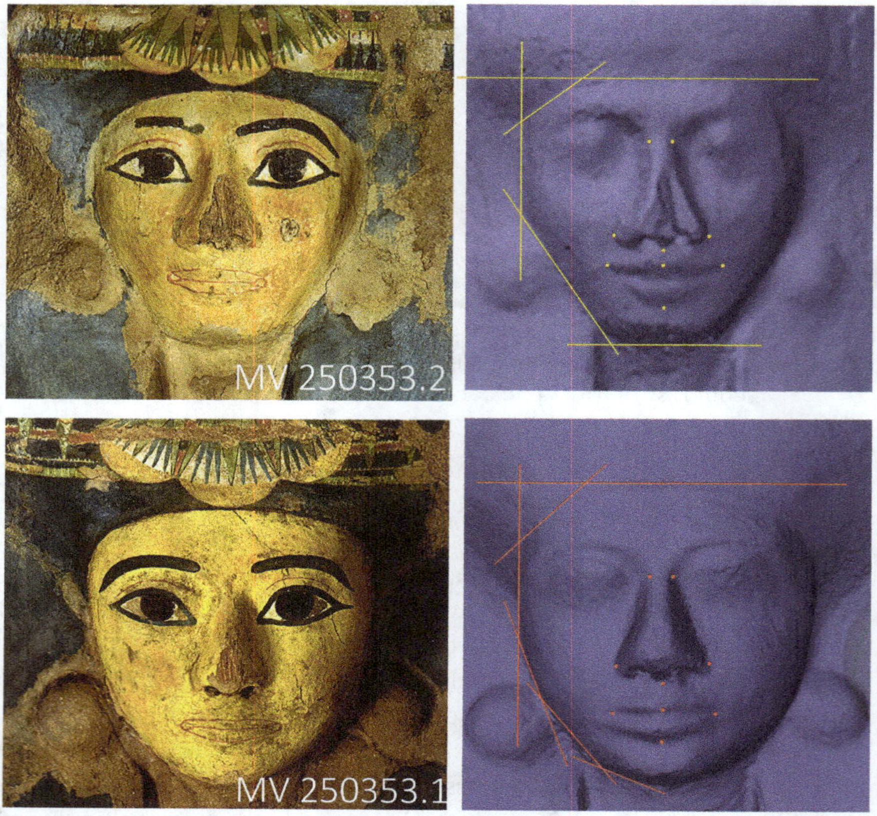

FIGURE 11.5 Comparison of model textured and model solid of the mummy board of Ikhy (above) and its outer coffin (below)
COURTESY OF *THE VATICAN COFFIN PROJECT* (© MUSEI VATICANI)

and physiognomic features of the mask on the mummy board are totally different from the face on the outer coffin (fig. 11.5). The mummy board has a more square-shaped face with striking features, such as high cheekbones and a square chin; the nose is long, very large in the lower part, with flared nostrils; the mouth is closed, forming a hint of a smile with well-defined, large, fleshy lips. Conversely, the face of the outer coffin presents fine features with rounded cheekbones, chubby cheeks and a slightly pointed chin; the nose is small and well-proportioned with "regular" nostrils, and the mouth, closed too, is linear and fleshy but less so than the mummy board. Why they are so different in form? Moreover, along with these so different facial features, we note that the mummy board presents a modification of some sculpted elements with paint, possibly made to unify the style of decoration both in the mummy board and the outer coffin. This is evident in the mouth: the paint does not reflect the

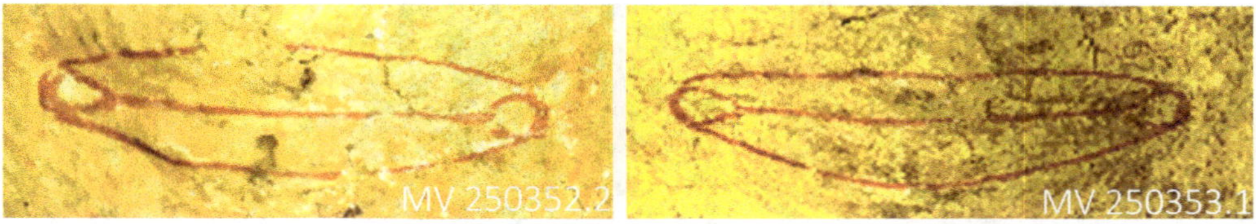

FIGURE 11.6 Particular of the painted lips on the mummy board of Ikhy (on the left) and on its outer coffin
COURTESY OF THE VATICAN COFFIN PROJECT (© MUSEI VATICANI)

underlying features, but changes them, from a hint of a smile and a mouth with well-defined, large, fleshy lips to a smaller, closed, unsmiling mouth (fig. 11.6).

4 *Faces Revealed Project* and the New Methodology

To sum up, the investigation of this small group of artifacts and the comparison of the 3D models solid with their models textured show us that sometimes the faces seem to be painted without following the underlying sculpted/ modelled traits. Moreover, in some cases we have a large layer of plaster that completely changes the carved wooden traits, thus remodeling some facial features. This evidence has never been considered before and therefore has much potential to add to both the recent projects concerning the "yellow coffins" and the more traditional analytical approaches to these objects. In any case, these results consequently pose a series of interesting questions that are also the main focus of the "Faces Revealed" project, in other words to understand:

a) How important the paint is when compared with the modelled/sculpted masks, alongside how faithfully it reproduces them
b) Whether the different physiognomic modelled/sculpted traits and proportions of the faces can be linked to different workshops and/or can reflect the stylistic features of a certain period (as also proposed for the statues)
c) How important the plaster is in the modeling of the original sculpted traits in creating forms and three-dimensionality, and whether its thickness can be linked to the economy of the workshop or different location or linked to the reuse of the masks/coffins. We have to consider, in fact, that around 53.82 % of coffins dated to this period provide circumstantial evidence of reuse.[29] The presence of excess plaster on the coffin arms,

29 Cooney 2014, 48; id. 2017, 101–112; Prestipino 2017, 397–406.

FIGURE 11.7 Example of marks and polylines to individuate the correspondence between paint and modelled traits on the outer coffin of Ikhy
COURTESY OF THE VATICAN COFFIN PROJECT (© MUSEI VATICANI)

for example, is suggestive of reuse, because plaster was a common way to cover over the older style of modelled forearms and elbows. Could the same be the case for the faces?

In order to attempt to answer the questions that emerged during this pilot project, the analysis was extended to all "yellow coffin" (included "stola coffins") from the 19th to the beginning of the 22nd Dynasty and coming from other archaeological sites for a broader investigation, and a project with a well-defined, specific and exact methodology was developed in three main steps:

1) Analysis of the faces (fig. 11.7). After the creation of the 3D model, for each coffin the following processes will be carried out: a) conducting an autoptic comparison of the 3D models solid and their models textured to understand whether or not the pictorial layers correspond to the underlayer; facial features will be drawn with polylines on the Metashape program and then inspected on the 3D model for resemblance; b) Virtually eliminating the paint to identify landmarks on the solid model for measurement.

FIGURE 11.8 Example of the morphometric approach to facial proportion and measurements on the outer coffin of Ikhy
COURTESY OF THE VATICAN COFFIN PROJECT (© MUSEI VATICANI)

2) Morphometric approach to facial proportion, measurements, and grouping. The facial features will be measured in order to understand the proportions of the faces. Each face will be inserted into a grid, which divides it into symmetrical halves and where vertical and horizontal distances will be the kinds of variables mostly related to four of the major facial components (fig. 11.8). Physiognomic marks will be identified and measured. Faces will be grouped together for their similarities and the local distance along the predominant modes of variation for different facial components will be computed to understand whether there are any correlations between face carvings.[30] If correlations are discovered, they may

30 These measurements will be done on an ad hoc basis with a pre-calibrated and fixed grid that will be adapted for each face. During the project, in collaboration with the UCLA and Dipartimento di Architettura, Ingegneria delle Costruzioni e Ambiente Costruito at the Politecnico di Milano, it is my intention also testing face recognition software for a faster and more precise investigation.

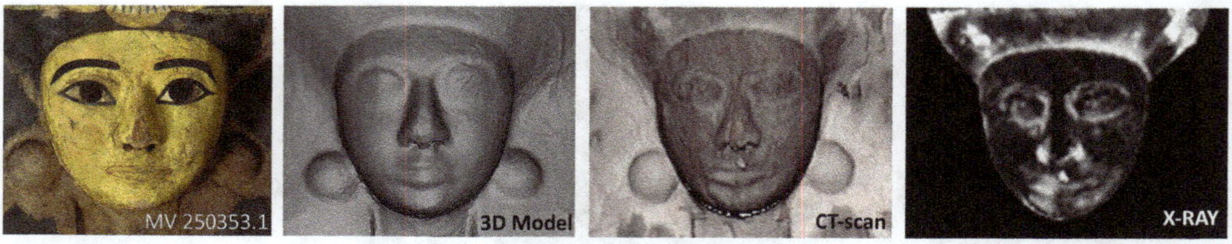

FIGURE 11.9 Comparison of the 3D model textured, 3D model solid, CT-scan and X-Ray of the outer coffin of Ikhy
COURTESY OF THE VATICAN COFFIN PROJECT (© MUSEI VATICANI)

indicate how many and which workshops were used in crafting masks and whether the proportions could be linked to a specific workshop or to a precise historical period.

3) Database and data connection. The entire corpus of the selected coffins will be managed in a relational database including all the newly acquired data, which will be organized into different tables according to a relational structure. During this phase the Faces Revealed database will be made accessible to the Vatican Coffin Project team. The collaboration of the researcher with the VCP is a unique opportunity to combine data acquired from the most recent and ongoing research on "yellow coffins" in Europe, such as that gained from the last 10 years by the VCP, with that obtained from the *Faces Revealed Project*. Such a collaboration will allow an integration of the data with results obtained through other scientific analyses and a comparison of the photogrammetry methodology with the latest technologies that the international team have been using. Comparing 3D models textured and solid ones with CT-scans and X-rays (fig. 11.9) will allow an analysis of the thickness and density of the plaster and give us a plethora of technical information.[31]

In conclusion, the Faces Revealed project takes its lead from innovative research trends, which see the combination of different but interconnected skills and competences, and which seeks to contribute to the study of the coffins by developing a new and efficient methodology based on an extremely fast, simple, cost-effective and portable technique allowing the acquisition of a non-invasive partial stratigraphy. This technology will provide further insights into the manufacture, production, workshops and maybe ancient reuse of the "yellow coffins" and could also help to create a new way of classifying these coffins, considering not only their iconographic and textual *apparatus* but all the various characteristics of these artifacts. The innovative project focuses on two

31 Amenta 2018, 323–335.

areas never considered before: the physiognomic traits of the coffin faces virtually removing the pictorial layer, and the impact and importance of the plaster at first in "modeled," as well as in "remodeled" sculpted wooden masks. Moreover, combining all the data (iconography, texts, layout, morphology, and technical aspects) will allow us to isolate comparative models and outline a new, more precise classification of "yellow coffins" and their provenances. Interpretation will have an impact on our scholarly knowledge of the ancient Egyptian coffin and of the funerary religion as a whole, because coffins represent "a social hieroglyph," "one of the enshrined material foundations for Egyptian society"[32] that we are trying to discover.

Bibliography

AA.VV. 2016. *Guida alla Collezione Egiziana del MANN*. Napoli: National Archaeological Museum of Naples

Amenta, Alessia. 2014. "The Vatican Coffin Project." In *Thebes in the First Millennium BC*, edited by Elena Pischikova, Julia Budka and Kenneth Griffin, 483–499. Cambridge: Cambridge Scholars Publishing.

Amenta, Alessia. 2018. "New results from the CT scanning of a coffin." In *Ancient Egyptian Coffins: Craft, Traditions and Functionality*, edited by John. H. Taylor and Marie Vandenbeusch, 323–335. British Museum Publications on Egypt and Sudan 4. Leuven: Peeters.

Amenta, Alessia and Hélène, Guichard (eds). 2017. *Proceedings of the First Vatican Coffin Conference (Vatican Museums, 19–22 June 2013)*. Città del Vaticano: Edizioni Musei Vaticani.

Aston, David A. 2009. *Burial Assemblages of Dynasty 21–25. Chronology—Typology—Developments*, Vienna: Verlag der Österreichischen Akademie der Wissenschaften.

Bryan, Betsy M. 1987. "Portrait Sculpture of Tuthmose IV." *Journal of the American Research Center in Egypt*, 24, 3–20.

Bryan, Betsy M. 2010. "New Kingdom Sculpture," in *Companion to Ancient Egypt*, edited by Alan B. Lloyd, 913–943. New York: John Wiley & Sons.

Bryan, Betsy M. 2015. "Portraiture." In *Companion to Ancient Egypt*, edited by Melinda Hartwig, 375–396. New York: John Wiley & Sons.

Cooney, Kathlyn M. 2007. *The Cost of Death: The Social and Economic Value of Ancient Egyptian Funerary Art in the Ramesside Period*, Leiden: Nederlands Instituut voor het Nabije Oosten.

32 Meskell 2004, 130; see also Cooney 2007.

Cooney, Kathlyn M. 2011. "Changing Burial Practices at the End of the New Kingdom: Defensive Adaptations in Tomb Commissions, Coffin Commissions, Coffin Decoration, and Mummification." *Journal of the American Research Center in Egypt* 47: 3–44.

Cooney, Kathlyn M. 2014. "Ancient Egyptian funerary arts as social documents: Social place, reuse, and working towards a new typology of 21st Dynasty coffins." In *Body, cosmos, and eternity: New research trends in the iconography and symbolism of ancient Egyptian coffins*, edited by Rogério Sousa, 45–66. Archaeopress Egyptology 3, Oxford: Archaeopress.

Cooney, Kathlyn M. 2017. "Coffin reuse: Ritual materialism in the context of scarcity." In *Proceedings of the First Vatican Coffin Conference (Musei Vaticani, 19–22 June 2013)*, edited by Alessia Amenta and Hélène Guichard, vol. 1, 101–112. Città del Vaticano: Edizioni Musei Vaticani.

Cooney, Kathlyn M. 2018a. "Coffin reuse in Dinasty 21: A case study of the coffins in the British Museum." In *Ancient Egyptian Coffins: Craft, Traditions and Functionality*, edited by John. H. Taylor and Marie Vandenbeusch, 295–322. British Museum Publications on Egypt and Sudan 4. Leuven: Peeters Publishers.

Cooney Kathlyn M. 2018b. "Patterns of coffin reuse from Dynasties 19 to 22." In *Ancient Egyptian Coffins: Past—Present—Future*, edited by Julie Dawson and Helene Strudwick, 96–108. Oxford: Oxbow Books.

Daressy, Georges. 1900. "Les sepoltures des prêtres d'Ammon à Deir el-Bahari." *Annales du Service des Antiquités de l'Égypte* 1: 141–148.

Dawson, Julie and Helene Strudwick (eds.). 2016. *Death on the Nile: Uncovering the Afterlife of Ancient Egypt*. London: GILES.

Dawson, Julie and Helene Strudwick (eds.). 2018. *Ancient Egyptian Coffins: Past—Present—Future*. Oxford: Oxbow Books.

Du Quesne, Terence. 1998. "Seth and jackals." In *Egyptian religion: The last thousand year. Studies dedicated to the memory of Jan Quaegebeur*, edited by Willy Clarisse, Antoon Schoors, and Harco Willems, 613–628. Orientalia Lovaniensia Analecta 84–85. Leuven: Uitgeverij Peeters.

Fiorini, Andrea. 2013. "Tablet pc, fotogrammetria e pdf 3d: strumenti per documentare l'archeologia." *Archeologia e Calcolatori* 23: 213–227.

Gasse, Annie. 1996. *Les sarcophages de la Troisième Période Intermédiare du Museo Gregoriano Egizio*. Monumenti Musei e Gallerie Pontificie. Museo Gregoriano Egizio *Aegyptiaca Gregoriana* Vol. III. Città del Vaticano: Musees Vatican.

Grenier, Jean-Claude. 1993. *Museo Gregoriano Egizio (Guide Cataloghi Musei Vaticani 2)*. Rome: "L'erma" di Bretschneider.

Laboury, Dimitri 2010. "Portrait Versus Ideal Image." *UCLA Encyclopedia of Egyptology*, 1(1), 1–18.

Lipinska, Jadwiga. 1993–1994. "Bab el-Gusus: Cache-tomb of the priests & priestesses of Amen." *KEMET* 4: 48–60.

Mainieri, Stefania. 2016. *La sezione egiziana del Museo Archeologico di Napoli* (MANN). *Storia, documenti d'archivio e reperti inediti delle collezioni di Stefano Borgia e Giuseppe Picchianti*, 4 vols. PhD diss., University of Naples "L'Orientale."

Mainieri, Stefania. 2019. "An example of 'dangerous' nineteenth-century restoration work at the Museo Archeologico Nazionale di Napoli (MANN)." In *Anthropology of forgery. A multidisciplinary approach to the study of archaeological fakes* (2017), edited by Monica Baggio, Elisa Bernard, Monica Salvadori and Luca Zamparo, 63–71. Padova: Padova University Press.

Mainieri, Stefania. In press. "The Borgia coffins at MANN: a contribution to the study of Late Period coffins." In *Proceedings of the First Vatican Coffin Conference* (*Vatican Museums, 6–9 June 2017*). Città del Vaticano: Edizioni Musei Vatican.

Marucchi, Orazio. 1896. "Sessione I. 30 gennaio 1896." In *Dissertazioni della Pontificia accademia Romena di Archeologia*, series 2, Tomo 6. Rome: Nella Stamperia de Romanis.

Marucchi, Orazio. 1899. *Il Museo Egizio Vaticano descritto ed illustrato*. Rome: V. Salviucci.

Meskell, Lynn. 2004. *Object Worlds in Ancient Egypt. Material biographies Past and Present*. Oxford-New York: Routledge.

Niwiński, Andrzei. 1988. *21st Dynasty coffins from Thebes: Chronological and typological studies*. Theben 5. Mainz am Rhein: Mainz Philipp von Zabern.

NKS 357b fol. III, I. *Catalogo dei monumenti egiziani esistenti in Velletri nel Museo Borgiano composto ed ordinato dal Sig. Giorgio Zoega dotto danese nel mese di Ottobre 1784*. Det Kongelige Bibliotek Copenhagen, Zoegas Papirer.

Perdu, Oliver. 2012. *Les statues privées de la fin de l'Égypte pharaonique* (*1069 av. J.-C.–395 apr. H.-C.*). Paris: Diffusion de Boccard.

Prestipino, Giovanna. 2017. "The Vatican Coffin Project: Observations on the construction techniques of Third Intermediate Period coffins from the Musei Vaticani." In *Proceedings of the First Vatican Coffin Conference* (*Musei Vaticani, 19–22 June 2013*), edited by Alessia Amenta and Hélène Guichard, vol. II, 397–406, Città del Vaticano: Edizioni Musei Vaticani.

Remondino, Fabio and Stefano Campana (eds). 2014. *3D Recording and Modelling in Archaeology and Cultural Heritage. Theory and Best Practices*. BAR International Series 2598. Oxford: BAR publishing.

Russo, Michele, Remondino, Fabio and Gabriele, Guidi. 2011. "Principali tecniche e strumenti per il rilievo tridimensionale in ambito archeologico." *Archeologia e Calcolatori* 22: 169–198.

Sorouzian, Hourig. 1999. "A bust of Amenophis II at the Kimbell Art Museum." *Journal of the American Research Center in Egypt*, 28, 55–74.

Sousa, Rogério. 2018a. *Gleaming Coffins. Iconography and Symbolism in Theban Coffin Decoration* (*21st Dynasty*), vol. I, Coimbra: Coimbra Imprensa da Universidade de Coimbra.

Sousa, Rogério. 2018b. "The Tomb of the Priests of Amun at Thebes: The History of the Find." In *The Coffins of the Priests of Amun, Egyptian coffins from the 21st Dynasty in the collection of the National Museum of Antiquities in Leiden*, edited by Lara Weiss, Alessia Amenta and Christian Greco, 21–33. Leiden: Sidestone Press.

Sousa, Rogério, Amenta, Alessia and Kathlyn M. Cooney (eds). 2021. *Bab El-Gasus in Context: Rediscovering the Tomb of the priests of Amun*. Rome and Bristol: "L'Erma" di Bretschneider.

Taylor, John H. 1989. *Egyptian Coffins*. Shire Egyptology 11. Princes Risborough: Shire.

Taylor, John H. 2001. "Patterns of colouring on ancient Egyptian coffins from the New Kingdom to the XXVI dynasty, an overview." In *Colour and Painting in Ancient Egypt*, edited by William V. Davies, 164–181. London: British Museum Press.

Taylor, John H. 2003. "Theban coffins from the Twenty-second to Twenty-sixth Dynasty: dating and synthesis of development." In *The Theban Necropolis. Past Present and Future*, edited by Nigel Strudwick and John H. Taylor, 95–121. London: The British Museum Press.

Taylor, John H. 2006. "The coffin of Padiashaikhet." In *Egyptian Art in the Nicholson Museum, Sydney*, edited by Karin N. Sowada and Boyo G. Ockinga, 263–291. Sydney: Mediterranean archaeology.

Van Walsem, René. 1997. *The Coffin of Djedmonthuiufankh in the National Museum of Antiquities at Leiden*. Leiden: Nederlands Inst. voor het Nabije Oosten.

Van Walsem, René. 2015. "Dalla fossa nella sabbia al sarcofago a tomba-tempio: 'trasformazione architettonica', iconografia e testi nei sarcofagi del I millennio a.C." In *Egitto Splendore millenario. La Collezione di Leiden a Bologna*, edited by Daniela Picchi and Paola Giovetti, 390–397. Milano: Skira.

Weiss, Lara (ed.). 2018. *The Coffins of the Priests of Amun, Egyptian coffins from the 21st Dynasty in the collection of the National Museum of Antiquities in Leiden*. Leiden: Sidestone Press.

CHAPTER 12

VÉgA (Vocabulaire de l'Égyptien Ancien): A New Definition of a Dictionary

Anaïs Martin

Abstract

VÉgA* (Vocabulaire de l'Égyptien Ancien, Vocabulary of Ancient Egyptian) is an online digital dictionary developed in collaboration between Egyptologists, designers and computer engineers specialized in the creation of digital interfaces. The first goal was to create a platform where we can model and present the evolving knowledge of Ancient Egyptian by bringing together and cross-checking words, references, their hieroglyphic forms, as well as creating links to a collection of records related to the texts in question. Since 2017, VÉgA has been accessible using a simple web browser. There are no other installation requirements. Its interface is tailor-made and user-friendly to address the public, whether they be an amateur, professional, student or linguist with needs from the simple translation to a complete lexicographical study. However, after two years of platform development, the outcome ended being radically different from the initial idea. The freedom offered by digital technology to create what we need, the way we want it, pushed the team to rethink how we used a dictionary and to model a new methodology along the way, especially taking advantage of the opportunity to open the tool to scientific and even non-scientific users.

Keywords

lexicography – dictionary – digital publication – database – open source – crowd source

Digital technology has widely spread through all aspects of our lives and it is only natural that its amazing potential should be put in the service of research.

* This project is supported by LabEx ARCHIMEDE from "Investissement d'Avenir" program ANR-11-LABX-0032–01.

However, if academics can be skilled users, to create proficient software remains the business of digital professionals, pushing new ways of handling the hard documentation at our disposal. But digital technology is not an end in itself either, only a means to facilitate the work of the discipline that embraces it, especially if we are not afraid to revise our research practices and introduce Egyptology in the field of the Digital Humanities.

VÉgA is an online digital dictionary of ancient Egyptian, translated into French, English, German and Arabic.[1] It has been developed as a technological program within the LabEx ARCHIMEDE (Laboratory of Excellence for Archeology and History of the Mediterranean and ancient Egypt), supported by the University Paul-Valéry Montpellier 3 and the CNRS, in collaboration with the private company Intactile DESIGN, specialized in the creation of digital interfaces.[2]

At the University of Montpellier, the study of ancient Egyptian Texts is one of the research team's main areas and the need for a global lexicographical tool has always been felt. But challenging the Academy of Berlin's *Wörterbuch der ägyptischen Sprache* is a task not to be lightly undertaken. Begun in 1897, printed between 1926 and 1931 for the five main volumes, the Berlin Dictionary contains approximately 16,000 headwords and is still today the largest and most complete printed dictionary of ancient Egyptian, even though the knowledge it presents was essentially accumulated before World War I. Since then, new words have been discovered, and new analyses have changed the translation of lemmas or even ruled out some of them that are now deemed nonexistent, which constitutes a rather big flaw in the present use of the dictionary. But already in 1945, A.H. Gardiner was somewhat critical about this "masterpiece" of ancient Egyptian lexicography and the sentiment of final achievement given by the publishers, arguing that "it behooves all students of Egyptian philology to consider what steps could be taken to supply the crying need for a dictionary adequate to the present position of our knowledge."[3]

Since then, an everlasting debate has been going on how to make a "good" dictionary, exhaustive and up to date.[4] In the end, a consensus was made on what such a tool should offer:

1 It uses a graph database with HTML5 API allowing access to the tool via only a web browser. At the time of publication of this article, the software is on its version V3.1.0.

2 For the history of the project and the collaboration with our different partners, see our website: http://vega-vocabulaire-egyptien-ancien.fr, accessed 03-23-2022.

3 Gardiner 1947, XIII.

4 See the various conferences held on the matter, the last workshop to date being the one in Liège in 2009 on "Lexical Semantics in Ancient Egyptian" (Grossman et al. 2012).

- every known word of the ancient Egyptian lexicon, with translations
- every primary source reference, in context, to distinguish the different usages and meanings, while allowing the scholar to check for himself
- a commentary with bibliography on the ongoing discussions on lexemes when necessary
- a lexical analysis, with semantical and morphological data, along with links between lexemes formed on the same root or of the same lexical field

But as anyone who has undergone the process of such a publication can attest, on the scale of the whole lexicon, the volume of data to collect and analyze is such that the need for exhaustivity seems to be in perpetual conflict with the wish for constant updating, and even more so when you have to fix it in a printed book. To counter some of these issues, each new dictionary or specialized lexicon chose to focus only on one aspect of the language (a period, a type of documentation ...). However, though it can prove very useful when working on a small corpus of documents, the global information is too widely spread on a multitude of media, not always easy to handle. The idea of a real solution had already been germinating in the mind of A.H. Gardiner in his preface of the *Ancient Egyptian Onomastica*: "would it not be better for a number of special dictionaries or vocabularies to be started in different places, and then ultimately combined into a great general dictionary which would constantly refer back to these smaller but fuller ones? [...] All these special vocabularies would advance simultaneously in different places and would go into greater detail as regards their own particular fields than could the comprehensive dictionary to be eventually superimposed upon them as a copying-stone."[5]

Ultimately, the decision on how to process this kind of data is a matter of choices, and in its ambition to be a global platform for ancient Egyptian vocabulary, VÉgA is no exception. It is however fundamental that these choices should be well-grounded and part of an all-encompassing approach on the development of the project (and its associated tool) over the long term. With that in mind, and right from the start in 2008, the digital tool came as a sensible choice for its flexibility, the opportunity to stock and update as much data as needed, and the benefit of having access to all the documentation in one place. But as with any new device, we must learn new ways to adapt. And in order to develop something effective we needed the help of specialists in computer engineering and design, hence our collaboration with the private company intactile DESIGN, elaborating and realizing digital interfaces for specialist sectors taking into account the position of the user, and with whom we

5 Gardiner 1947, XX.

were able to fashion a tailor-made tool through co-design work and Agile software development meetings.[6] Their expertise in knowledge modeling allowed the Egyptologists to implement new ways of conceiving and organizing lexicographical data centered specifically on the user, regardless of their status and their particular needs.

1 Using Design for Research

As a proper digital tool, VÉgA was never perceived as a mere database of ancient Egyptian vocabulary where data is compiled in a way only pertinent to the subject itself and its intrinsic patterns. Its priority should instead be to facilitate the task of the user, putting him at the core of the process. The Egyptologists had then to embrace some of the design philosophy and seek the help of specialists in research *by* design,[7] integrating a multidisciplinary team of designers and computer engineers exploring usages together with the targeted users.

At first, they spent some time observing the Egyptologists in their daily routine of library consultation, collecting data and checking of references. They managed to decompose the process in multiple steps and summarize them in technical terms, understanding the different issues needed to be dealt with in this specific field. During co-design meetings the different components of the tool were then established one by one, using *low-tech* devices: storyboards and paper prototypes (Figure 12.1). It allowed for a quick development of ideas and testing of the components, whilst micro-suitability tests on the digital items were constantly carried out. It took four years of back-and-forth collaboration and three different prototypes (with various iterations) until the V1.0 of the software was delivered in December 2014.[8] The merits of this methodology are most highlighted by the end product being so radically different from the initial idea, as the successive versions of the storyboards and prototypes made the functionalities more and more precise.

From the start, the team knew the dictionary would be handled by all types of users, from amateurs and students to professionals. Thus, the focus was put on how to organize and display the data, and how to browse it most practically. The software was designed with a didactic intention, centered on the experience of research using tailor-made and user-friendly desktop and notes.

6 Chauveau 2015, 82–87.
7 On the methodology, cf. Jacques et al. 2006, 321–334.
8 Cassier et al. 2017.

The most noteworthy change is the user not being constrained anymore by the rigidity of traditional structures such as alphabetical order, which is discarded in favor of the most appropriate search method, be it via search field directly—typed as transliteration,[9] translation or ID number—or through links activated between headwords, grouping variants or lexemes starting with different letter but sharing a common root (Figure 12.2). In reverse translation, words with similar meanings can also be reunited in a single list of results (Figure 12.3).

FIGURE 12.1 The team is working on a paper prototype during a co-design meeting.

9 In a "search as you type" manner: for each typed character, the number of results decreases.

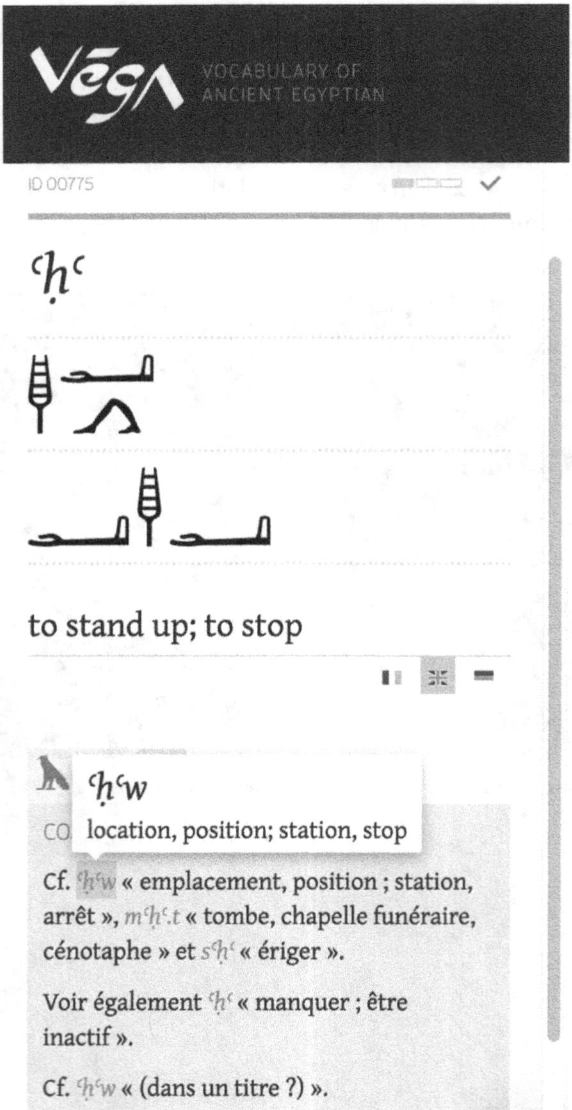

FIGURE 12.2
Snapshot of the VÉgA ID 775 note. In the "Comments" tab, three links are made towards words constructed on the root *ʿḥʿ*: *ʿḥʿw* "location, position; station, stop" (VÉgA ID 3560), *mʿḥʿ.t* "tomb, funerary chapel, cenotaph" (VÉgA ID 4896), and *sʿḥʿ* "to erect, to set up". The last one is still in orange, meaning the headword is not yet recorded in the database and the link is not active.

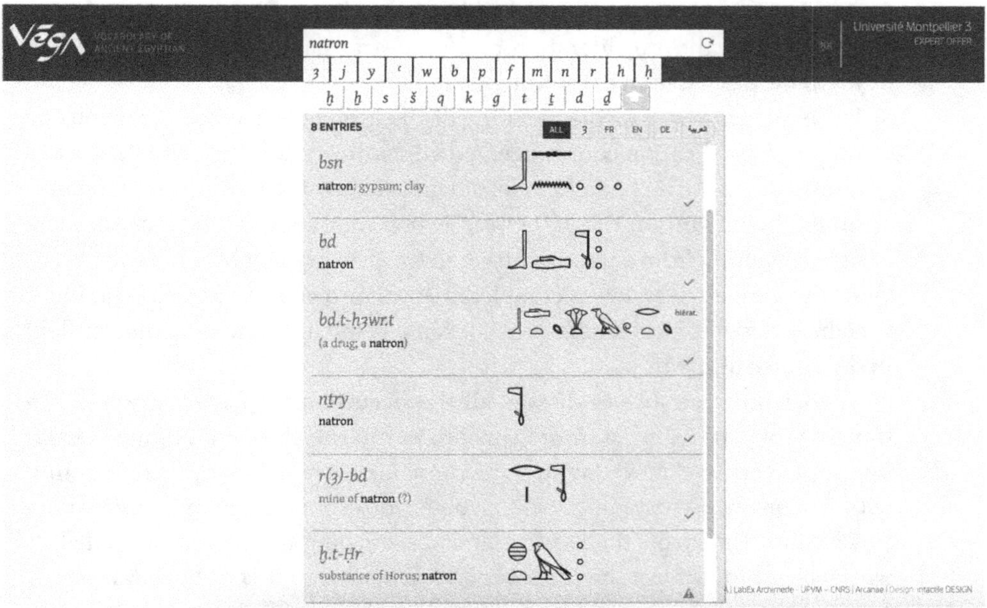

FIGURE 12.3 Snapshot of the VÉgA desktop with the list of results for the word "natron" in the search field (without filter). The list shows all the entries containing this word in the translation.

2 Collecting Data

Among the first choices faced by the development team was the type of raw material to be used. The obvious and most academic one would have been primary texts, analyzed and referenced as on the *Wörterbuch* Zettel.[10] In the process, it would have been necessary to check a list of dictionaries and other lexicographical tools to compare. This first step was subsequently skipped in favor of a list of 21 dictionaries and lexicons of reference, as the majority of known sources are published and already referenced. The goal was simply to assess the state of ancient Egyptian vocabulary at present, to establish its perimeter and inventory all the known words. Each one is then attributed a status of validity based on the accuracy of the data at our disposal to offer a translation:

– **Valid:** the existence and meaning of the lexeme have been verified and validated

10 Available online in the *Digital Slip Archive* of the *Thesaurus Linguae Aegyptiae* (http://aaew.bbaw.de/tla, accessed 03-23-2022).

- **Problematic**: the existence of the lexeme has been validated but the hieroglyphic form, the transliteration, and/or the translation are problematic. The matter is discussed in a "Comments" tab
- **False**: the existence or the meaning of the lexeme is deemed erroneous by the VÉgA publication committee. A distinction is made between *false entry* and *false word*. The first one refers to an existing word misread or misinterpreted by an author; the *false word* is one that does not exist at all, most often resulting from a wrong cutting in the primary text by an author.

Once this general overview is completed, it will provide a strong base for more specific lexicographical studies, each lexeme being linked to its semantic field in the course of the process.

VÉgA is not yet able to display all the occurrences of each word—even though it is planned for a future version. But in the case of a unique attestation (or of very few ones), mentioning the source is mandatory, in conjunction with its correct hieroglyphic spelling, date, corresponding bibliography and a small commentary on the context and the ongoing discussion if needed. It was decided to not use the term *hapax legomenon* as it is not always easy to distinguish a true new word from a simple variant or nuance of a known lexeme.

3 Classifying and Display

Designers and computer engineers helped the Egyptologists structure all the needed data to be formally added to the database, in order to clearly present each element without clouding the consultation of others. With that in mind, a note is made of different layers, from the basic translation to more "encyclopedic" features focusing on the real meaning of the word, along with semantic considerations, ethnological details, and specific bibliography.

A first unit displays the basic information required with the headword in transliteration,[11] its main hieroglyphic forms and translations in the four languages of the academic area (French, English, German and Arabic) (Figure 12.4). The translations presented are generally, for now, the ones given by the dictionaries and glossaries of reference. However, they can be updated according to recent studies and publications mentioned in the bibliography. On top of this

11 The transliteration conventions adopted are those established in Grandet and Mathieu 2003.

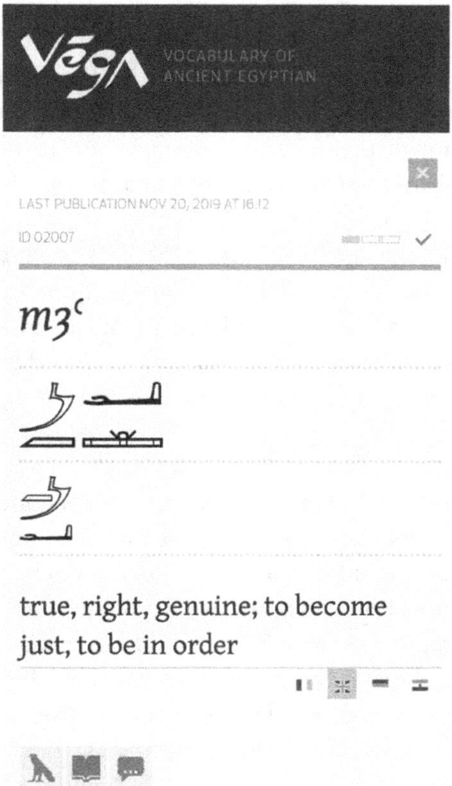

FIGURE 12.4
Snapshot of a note in VÉgA in its basic display, along with essential data: transliteration, main hieroglyphic forms, and translations (available in French, English, German and Arabic). It is completed at the top with the unique ID number and progress indicators of the lexicographical study and status of validity.

are the unique ID number of the note, an indicator of the status of validity[12] and one on the progress of the lexicographical study, which is determined by three grades:

- 1st grade: indicates a note as produced by the VÉgA publication committee;
- 2nd grade: indicates a note is still available for consultation, but has been assigned to a contributor in charge of the complete lexicographical study;
- 3rd grade: indicates an exhaustive lexicographical note, signed by the contributing author. All information about the corresponding word has been recorded and updated.

A timeline can also be featured in the current state, at least when the lexeme is known by one, or very few, attestation(s). It is however a mandatory feature of exhaustive lexicographical notes.

12 See Grandet and Mathieu 2003.

Under this first unit, additional detailed data is divided into three different tabs:

1) **Catalog of hieroglyphic forms**
   There are displayed the most usual graphic spellings[13] and the most atypical ones. Those are not systematically dated, nor identified by their sources in the current state, though this feature is planned for in the future. The mention "hiérat." is also added in the case of a hieratic source.
2) **Dictionaries and glossaries**
   It regroups all the occurrences of the headword in the dictionaries and glossaries of reference, covering all the stages of the language, from ancient Egyptian to Coptic.[14] In the end, the "Other references" are an arbitrary selection of book glossaries covering as many subjects as possible. Not all are always checked systematically, but at least the most meaningful to the subject.
3) **Comments**
   In this section can be found all the information needed to understand the meaning of a word: grammatical class (when it can be misleading), etymology, primary sources, relevant and/or specific bibliography, characteristic nuances and metaphorical uses, noteworthy expressions and constructions, and more importantly clickable links toward other headwords with similar roots or lexical fields and research leads. This tab is automatically displayed if the lexeme is deemed problematic or false.

Except for the "Comments" tab, which is an open field, all features are standardized so that any data can be processed by the database and brought up more easily. And thanks to the flexibility of the digital medium, the structure can be changed at will, for example, if we notice that a certain type of information becomes recurrent despite not being expected beforehand. For instance, in a future version, the ability to indicate the grammatical class in a more systematic way would be welcome, as we observed it could be misleading in some cases or could generate mistakes.

4 A New Dawn of Digital Publication

For the academics used to the inertia of the scientific practice of collecting data, analyzing, publishing and waiting to be corrected in a very public man-

13 The hieroglyphics that currently exist on VÉgA have been generated using the JSesh software, created by S. Rosmorduc.

14 The titles are presented in an abridged version and the complete bibliographical reference appears in a rollover, by passing the cursor on the abbreviation.

ner, which can be a very long process, the flexibility of the digital technology is without a doubt one of its best assets. In our case, it allows the team to proceed in stages, in short and long terms, and to give access to results (almost) in real time, without lingering over finished work for months—or years—until the whole project is completed. For now, VÉgA is still in its early stage, which will only be completed once the perimeter of the ancient Egyptian lexicon is finalized, with a complete review of all lexicographical data scattered across many dictionaries, lexicons, and punctual publications. This new methodology has its pros and cons, as a quicker pace may allow more room for errors at first, which however can be rectified very quickly as well, with minor inconvenience. But at any rate, it should not be an obstacle as long as we acknowledge the limits of the tool at each step.

Thus, VÉgA is designed to be in constant evolution, updated permanently with a minimum amount of latency between the discovery of a new piece of information, its validation by the scientific committee and its publication. It implies that everything in a dictionary entry can be modified at any time, from the transliteration of the headword to the translations and remarks.

As an example, the case of the word  "pyramid". In all dictionaries and glossaries—except the digital ones—the lexeme is read as *mr*. So when the note VÉgA ID 4227 was first created, it was given just *mr* as a headword. Yet, an article of J. Fr. Quack[15] showed that the reading of the first sign was uncertain as it could also be red *mḥ*, hence a reading *mḥr* "pyramid". This piece of information emerged only during the survey of words beginning with *mḥ*, some months later, but as soon as it did, it was included in the previous note, together with a comment discussing the new problem of a word thought to be well known just by looking into paper dictionaries.

This example is one of the more extreme situations the team may be faced with, along with cases of phonemic alternation in initial position, where the same word can be recorded under different letters.[16] Most of the time, however, revisions concern only the addition of new references or spell checking. Nevertheless, it raises the issue of the preservation of the data and the need for archives to record the modifications and the different stages a note can go through.

Preservation of digital data is a very current preoccupation going far beyond the scope of our academic field and some institutions are fully dedicated to this endeavor on a very large scale. Developing a digital tool means being aware of

15 Quack 2003.

16 Very frequent in the case of semiconsonants *ꜣ*/ *j* / *w*, but also recurrent with the palatals *ẖ* / *š*, the dentals *d* / *t*, and the bilabials *b* / *p*.

this question, even if we are not specialists, to ensure its ability to endure and connect with other platforms. It concerns mostly the technical aspect of physical storage, data formats, hardware and software facing digital obsolescence. But if the sustainability of the database is an obvious issue in our development process, it does not concern immediately the user requiring assurance of the accuracy and consistency of the data provided by VÉgA. To that end, each iteration of a note should be archived in the database. But this feature is not implemented yet, as the team judged the revisions to be still too frequent and would continue to be until the completion of the lexicon perimeter. It would generate too many archives, neither manageable nor useful in the current state. It is nonetheless planned for a future version of the software, as we established that data integrity is a requirement of digital preservation. In the meantime, and as for any online reference, it is necessary to add the date of consultation when citing a VÉgA note, along with its ID number.

5 What Next?

Over the next few years, the first stage of outlining the ancient Egyptian vocabulary will come to an end, and VÉgA will be ready to widen its horizon and reach its full potential by becoming a real research platform in lexicography, by implementing new features in one hand and with the help of the international scientific community in the other hand.

At the beginning of the program, notes were modeled to be exhaustive, showing all known occurrences of a lexeme sorted by nuances with translations, dates, a full catalog of hieroglyphic forms and an extended commentary on its etymology and meaning. However, a true lexicographical study is a very long process and at the moment the team cannot afford to spend so much time on a single word. As an example, it took a full year to a scientific collaborator to solely collect all the attestations of *ꜣpd* "bird, poultry". As a result, there are only a few complete notes available at the moment. In the future, the opportunity to make a complete note shall then be given to the community, VÉgA providing a platform to publish quickly all lexicographical data under the supervision of an international scientific committee. And contrary to present notes being the product of the whole team, a complete note shall also be signed by its author, as any academic work would. Ideally, when envisioning each note as complete, it would result in every single piece of ancient Egyptian texts to be available in the database. In the end, this could resolve one of the biggest flaws of the tool compared to other digital dictionaries at the moment: the absence of textual resources and bodies of texts.

But the collaboration with individual scientists is not the only one to be considered. Many digital projects are developed worldwide, each reaching to the same conclusion: to create a database and just hope for it to be used is not enough anymore. It has to be conceived in a global perspective of interconnections between different databases handling different aspects. With that in mind, a first connection has been established with the *Karnak* project,[17] aiming to organize and make available textual documentation from the temples of Karnak.[18] Whenever a word is identified from a monument in Karnak, a link can be made toward their database *via* the KIU[19] number, providing a high-definition photograph of the inscription and all the information associated with it (Figure 12.5). It seems indisputable that for an effective digital tool in the future, the more connections between the different existing databases the better. Although it raises another issue concerning the use of a common terminology, which cannot be addressed yet considering the diversity of methods, each one with its own merits.

In the digital world, given the endless possibilities, it can be hard not to rush towards the most striking features but to ponder each step in order to give a database strong foundations. It does not prevent the team to plan for the long term and to have many ideas for future developments and new features, taking also into consideration feedback and acknowledging the needs of the community. For example, besides the ongoing reflection on how to model and present the evolving knowledge of ancient Egyptian vocabulary, another feature regularly asked for is the implementation of interactive hieroglyphs,[20] enabling the search by glyphs in the database. But on a larger scale, these glyphs should be provided by a separate catalog where they would be indexed and supplemented with metadata, supplying a whole new range of semiotic information on phonograms, logograms, and classifiers, inseparable from the comprehension of the ancient Egyptian language.

6 Conclusion

The idea behind VÉgA, besides updating knowledge on the ancient Egyptian vocabulary, is to create a tool, specifically a digital one, enhancing the work

17 CNRS—LabEx ARCHIMEDE, ANR-11-LABX-0032-01, "Investissement d'Avenir" program, USR 3172—CFEETK / UMR 5140, Équipe ENiM.

18 http://sith.huma-num.fr, accessed 03-23-2022.

19 For "Karnak: Identifiant Unique".

20 Encoded with the Manuel-de-Codage conventions.

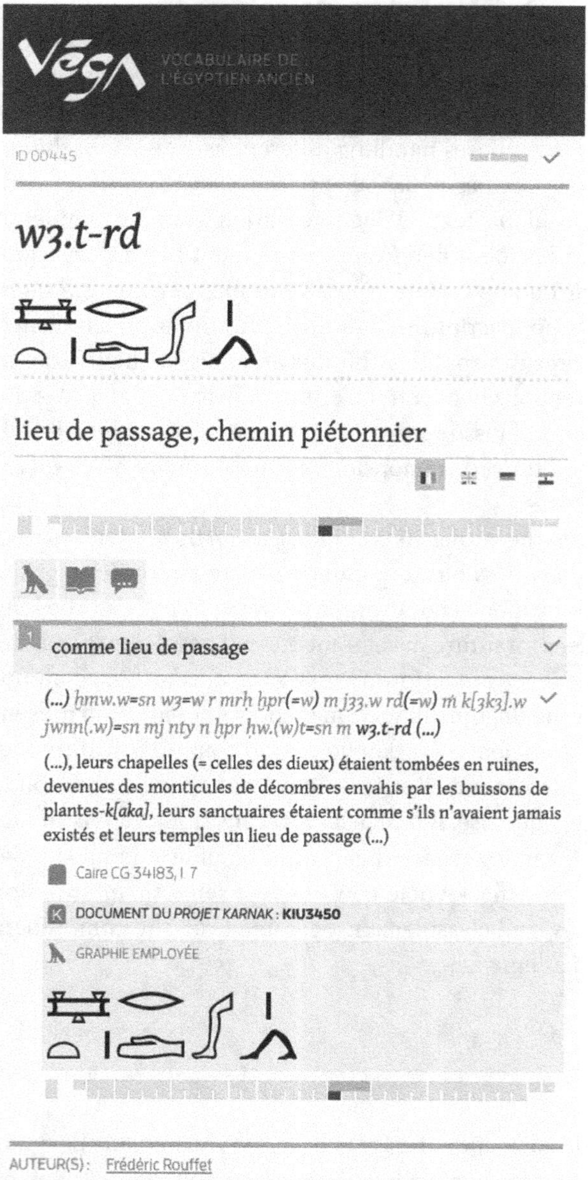

FIGURE 12.5 Snapshot of the VÉgA ID 445 note. The word is only attested on the Restauration stela, so the author made a complete note with a translation of the concerned text and signed it. Under the mention of the source, a link "Document du Projet Karnak" is activated toward their database.

capacity of Egyptologists and giving access to the language to a wider range of people. By bringing on board designers and computer engineers, the Egyptologists wanted to imagine new ways to handle old data.

For now, VÉgA is still in its infancy and cannot assume the title of a true lexicographical tool yet. And being not yet complete means it still misses some

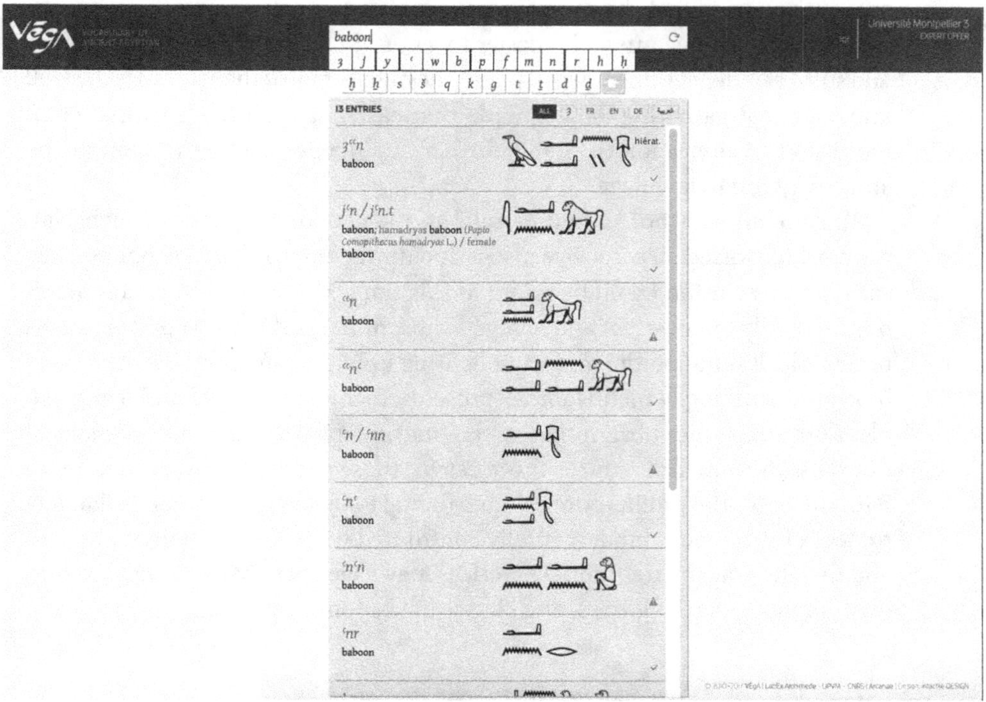

FIGURE 12.6 Snapshot of the VÉgA desktop with the list of results for the word "baboon" in the search field (without filter). The list shows all the entries containing this word in the translation, and at least eight of them seem to be variants of the same word, may it start by the letter *ꜣ*, *j* or *ꜥ*.

important information, which can change the content of a note at any moment. But the more notes are recorded, the further the team go in linking lexemes between each other, and the more some patterns in the linguistic structure become apparent. When confronting the dictionaries of J. Kahl,[21] R. Hannig,[22] L.H. Lesko[23] and P. Wilson,[24] all focusing on different periods, and trying to fit them under a single headword, one can begin to grasp what 3,000 years of linguistic evolution entails, especially when adding references to Demotic and Coptic dictionaries. For instance, when searching for the word "baboon" in ancient Egyptian, thirteen entries are displayed at the moment, but eight concern the same word with its variants, not necessarily beginning by the same letter (Figure 12.6). It proves the lexeme has a wide range of evolution depend-

21 Kahl 2002–2004.
22 Hannig 2003 and 2006.
23 Lesko 2002–2004.
24 Wilson 1997.

ing on the period and the region, most significantly concerning the semiconsonants. Thus, it becomes a lot easier to ascertain lexical fields, roots, prefixes and suffixes, alternation between phonemes, etc. Some of them are already well known, but if gathering data is made that easier, it is all the scientific community VÉgA hopes for, to enable further and deeper comprehension of the ancient Egyptian language.

To conclude, a word should be said on the economic aspect of such platform, which raised an extensive discussion during the conference, but no truly satisfactory solution could be found at this time. Matter-of-factly, it oscillates between fully open access and charged subscriptions. If the first option seems preferable, it relies on the project benefiting from a secure fund over the years to keep it running, which is under present circumstances less and less feasible. Conversely, the subscription system makes a project more autonomous, if not self-sufficient, but reduces the access to those willing to pay, even more so with the potential proliferation of such digital tools. What is certain is that any project of this type is fundamentally confronted with the question of finances and that, here again, traditional practices may have to evolve with the development of new operating ways, at all levels of academic research.

Bibliography

Cassier, Charlène, Nicolas Chauveau, Magali Massiera, and Frédéric Rouffet. 2017. "VÉgA, une plateforme numérique de recherches lexicographiques pour la connaissance et la diffusion de l'égyptien ancien", in *Systèmes d'organisation des connaissances et humanités numériques. Actes du 10e colloque ISKO France 2015*, edited by Emmanuelle Chevry Pébayle: 274–288. Paris.

Chauveau, Nicolas. 2015. "Le design numérique au service de la recherche en SHS: une étude de cas du projet VÉgA, Vocabulaire de l'Égyptien Ancien." *Sciences du Design* 2/2, 2015: 82–87.

Claes, Wouter, and Ellen van Keer. 2014. "Les ressources numériques pour l'égyptologie." *BiOr* 71, no. 3/4: 297–306.

Erman, Adolph, and Hermann Grapow. 1935–1959. *Wörterbuch der ägyptischen Sprache im Auftrage der Deutschen Akademien* (5 vol.). Berlin, Leipzig.

Gardiner, Alan H. 1947. *Ancient Egyptian Onomastica* I, Oxford.

Grandet, Pierre, and Bernard Mathieu. 2003. *Cours d'égyptien hiéroglyphique* (nouvelle édition revue et augmentée, 5e tirage). Paris: Khéops

Grossman, Eitan, Stéphane Polis and Jean Winand, ed. 2012. *Lexical Semantics in Ancient Egyptian, LingAeg StudMon* 9. Hamburg: Widmaier.

Grunert, Stefan, and Ingelore Hafemann, ed. 1999. *Textcorpus und Wörterbuch. Aspekte zur ägyptischen Lexikographie, PdÄ* 14. Leiden, Boston, Köln: Brill.

Hannig, Rainer. 2003. *Ägyptisches Wörterbuch* I. *Altes Reich und Erste Zwischenzeit* (*HannLex 4*), *KAW* 98. Mainz am Rhein: Philipp von Zabern.

Hannig, Rainer. 2006. *Ägyptisches Wörterbuch* II. *Mittleres Reich und Zweite Zwischenzeit* (*HannLex 5*), *KAW* 112 (2 vol.). Mainz am Rhein: Philipp von Zabern.

Jacques, Emmanuelle, Sandra Solinski, Claire Ollagon, and Yves Rinato. 2006. "La conception numérique, entre espace intime et monstration. À la recherche des intelligences collectives", in *Intelligence collective. Rencontres 2006*, edited by Jean-Michel Penalva, 321–334. Paris.

Kahl, Jochem. 2002–2004. *Frühägyptisches Wörterbuch unter Mitarbeit von Markus Bretschneider und Barbara Kneißler* (3 vol.). Wiesbaden.

Lesko, Leonard H. 2002–2004. *A Dictionary of Late Egyptian* (2 vol.), 2nd edition. Berkeley.

Quack, Joachim Fr. 2003. "Zum Lautwert von Gardiner Sign-List U 23." *LingAeg* 11: 113–116.

Wilson, Penelope. 1997. *A Ptolemaic Lexicon. A Lexicographical Study of the Texts in the Temple of Edfu*, *OLA* 78. Leuven: Peeters.

CHAPTER 13

The Egyptian Road Most Taken: Mapping the Least Cost Path Routes from the Nile to the Red Sea Coast

Morgan E. Moroney

Abstract

This paper applies Geographic Information Systems (GIS) to satellite data to examine the movements of royal expedition routes through Egypt's Eastern Desert during the Middle Kingdom. In this paper I employ Least Cost Path analysis, a GIS tool that analyzes the landscape and pinpoints the most cost-effective route for a traveler. This paper first overviews the use of GIS in the Digital Humanities, before examining evidence for historical expedition routes during Egypt's Middle Kingdom and analyzing possible routes through the Eastern Desert.

Keywords

GIS – Least Cost Path analysis – Middle Kingdom Egypt – Eastern Desert – expedition routes – Digital Humanities

1 Introduction

From the earliest days of the ancient Egyptian state, goods and people, Egyptians and foreigners, were on the move—into, out of, and around Egypt. Trade expeditions, army campaigns, ritual processions, religious pilgrimages, neighborly visits, and kingly journeys—these were taken on foot, donkey, horse, cart, sledge, or boat.[1] The main "highway" for the Egyptians was the Nile. From Aswan to the Mediterranean, the Egyptian Nile brought life and transportation ease as it flowed north through the Eastern and Western Deserts of Egypt. Egypt's geography is central to its understanding as a civilization. Maps of Egypt, the Nile and its Delta, the surrounding deserts and seas, are ubiquitous.

1 Many thanks to Professor Michael Harrower for his support with this project.

Almost every modern publication dealing with Egypt features a map. However visually informative these maps often are, they are not always based on precise mapping techniques and their accuracy cannot be entirely reliable. They can be more general drawings than measured certainties. More precise mapping should be pursued to answer more specific questions, particularly when thinking of archaeological questions related to mobility and landscapes.

In the pursuit of better maps and data, GIS and satellite imaging have entered Egyptology.[2] Utilizing GIS to better understand how and why ancient peoples might have moved across their landscapes, while applied in archaeology more generally, has been less utilized for ancient Egypt. But Egypt's rugged terrain—particularly its deserts—and the ancient Egyptians' ability and desire to move across these difficult landscapes—is a rich ground for study. With this in mind, this paper applies GIS to satellite data in order to examine the movements of royal expedition routes through Egypt's Eastern Desert during the Middle Kingdom (ca. 2055–1650 BCE). More specifically, my research employs Least Cost Path analysis, a GIS tool that produces the most cost-effective route for a traveler over a landscape. In this context, 'cost-effective' means the easiest route for an individual to walk from point A to point B across a terrain with various slopes.

During the Middle Kingdom, approximately 15–20 known state-run expeditions journeyed from the banks of the Nile across the desert to the Red Sea Port of Mersa Gawasis, ancient *sꜣww*.[3] This paper utilizes Least Cost Path analysis to map possible paths these missions may have taken through the desert. Before going into my methodology for studying these human pathways using Least Cost Path analysis, I will first give a brief and general introduction to the use of GIS within the Digital Humanities. I will then summarize the archaeological and textual evidence for these Middle Kingdom expeditions, and their movements and routes through the Eastern Desert.

2 GIS and the Digital Humanities

GIS, or Geographic Information Systems, examines spatial locations and organizes them into maps and 3D models.[4] Its application within the Digital Humanities has become more popular of late, through a mix of quantitative

---

2 See, for example, Förster and Riemer 2013; Parcak 2009; 2019.

3 Bard and Fattovich 2018, 176–177.

4 Esri, "What is GIS?" accessed September 22, 2020, https://www.esri.com/en-us/what-is-gis/overview.

and qualitative analyses.[5] GIS can be employed on a macro-scale, such as the examination of satellite data as seen in this paper's study, or on much smaller, micro-scales with the mapping of a building, room, or even single object in an archaeological site. GIS can be used to combine archaeological and textual data. Paper maps can be georeferenced onto modern coordinate systems to help better understand how ancient sites correspond to physical locations in the modern world. Places mentioned in historical and literary works can also be mapped with GIS to gain a fuller understanding of these documents and their historical times.[6] GIS can help capture a moment in time and space. Its analysis tools allow various methods to examine how peoples, ancient and contemporary, moved and interacted with both small spaces and vast landscapes. GIS has opened up a myriad of new opportunities for the ways humanities data and studies can be further enriched through spatial analyses.

3 Middle Kingdom Seafaring Expeditions and Mersa Gawasis:

Ideologically, the Egyptian king was the chief actor for the gods on earth, operating on their behalf by means of rituals and offerings to preserve Maat and ensure creation and life continued. Through this role the king was, at least in name, the sole and high priest, while his other main responsibility was as political and military leader.[7] Kings' access to, control over, and presentation of natural resources was also central to the ideology of kingship. The king was expected to provide materials to enrich the cult and mortuary temples. Desirable, rare, and imported riches also enabled them to assert political legitimacy.[8]

Ostensibly, the king also controlled all who moved across the lands of Egypt, but in actuality, royal authority over the periphery was usually limited to specific and significant sites. During the focus of our study, the Middle Kingdom, the state attempted to monitor the movement of peoples and goods across the Egyptian border to varying degrees. Kings built huge fortresses at the southern border with Nubia, and forts across the northeastern periphery of the Delta called the Way of Horus.

During this time, the state also directed large missions to exploit natural resources in Egypt and in its neighboring lands. Within Egypt, precious gemstones like amethyst were mined, and granite and limestone blocks were

5 Altaweel 2017.
6 McHaffie, Sungsoon, and Follett 2018, 11–45.
7 O'Connor and Silverman 1994, XX–XXI.
8 Bard and Fattovich 2018, 1–7.

quarried and reformed into statues, obelisks, and temples. To retrieve specific resources from far-flung lands, kings sent seafaring military expeditions to Byblos for Lebanese wood, to the Sinai for turquoise, and south to Punt for incense, ebony, and ivory.[9]

The land of Punt, thought to include parts of modern-day Eritrea, Djibouti, Ethiopia, eastern Sudan, and/or Yemen, was accessed by both land and sea.[10] During the Middle Kingdom, the powerful Kushite kingdom of Kerma ruled Nubia to the south, making land travel difficult. This formidable barrier needed to be bypassed, and so to reach Punt and Bia-Punt (the "mine" of Punt), Middle Kingdom kings launched nautical expeditions from the Red Sea port of the aforementioned Mersa Gawasis.[11] Under order of the king, these sea voyages procured precious resources such as myrrh, ebony, resin, and electrum, objects important to religious rituals intimately tied to the role of kingship.

The Red Sea harbor of Mersa/Wadi Gawasis is located at approximately 34°2'8.25"E and 26°33'4.03"N and sits at the eastern edge of the Wadi Qena system. It is about 2 km south of the eastern edge of the Wadi Gasus and about 26 km north of the Roman and Islamic port of Qadim al-Quesir, ancient Myos Hormos.[12]

Mersa Gawasis, or ancient *sꜣww*, was primarily in use during the Middle Kingdom, but there is also evidence of Old Kingdom and early Eighteenth Dynasty occupation.[13] George W. Murray discovered the site in the 1920s and misidentified it as the Ptolemaic-Roman port of Philoteras/Aenum.[14] Abdel Monem A.H. Sayed worked at the site in the 1970s and correctly postulated that it was the Twelfth Dynasty port of *sꜣww*.[15]

Most recently, Bard and Fattovich, worked at the site from 2001–2011.[16] These excavations uncovered stelae erected by expedition leaders and participants, as well as ritual structures, hearths, anchors, Kerma, Canaanite, and Yemeni ceramic ware, ropes, wooden boxes labeled "Punt," and the remains of wooden sailing ships stored in caves.[17] These ships were 30 meters long and manufactured with imported Lebanese wood such as cedar and pine.[18] They were kept

9 Bard and Fattovich 2018, 5–7.
10 Bradbury 1988, 127–156; Meeks 2003, 53–80; Bard and Fattovich 2018, 156–191.
11 Bard and Fattovich 2007; Bard and Fattovich 2015, 4–10; Bard and Fattovich 2018.
12 Bard and Fattovich 1977, 140–178; Whitcomb and Johnson 1982; Peacock and Blue 2006.
13 Bard and Fattovich 2007; 2015; 2018.
14 Murray 1925, 138–150; Bard and Fattovich 2018, 18.
15 Sayed, 1997.
16 Bard and Fattovich 2007; 2015; 2018.
17 Bard and Fattovich 2007; 2015; 2018.
18 Ward 2012, 53–64.

FIGURE 13.1 Relief images with products of Punt, including ebony, from Hatshepsut's temple at Deir el-Bahri

FIGURE 13.2 Map of the study area

in the caves to be reused for some later journey that never occurred. Based on the inscriptional and archaeological evidence at the site, approximately 12 to 20 Middle Kingdom expeditions were launched from Mersa Gawasis.[19]

On average, the Eastern Desert extends from the Nile Delta in the north to Sudan in the south and stretches about 225 km east to west. Mersa Gawasis is about 170–180 km east from the closest site on the Nile, Coptos. Coptos or Qift, ancient *gbtw*, is located on the Qena Bend, the area on the Egyptian Nile closest to the Red Sea. Coptos was a very ancient site and a main cult center for the god Min, who was believed to protect the desert, its resources, and those who traveled through it. Two large Eastern Desert wadi systems begin at or near Coptos and then branch off into smaller wadis: Wadi Hammamat runs southeast, and Wadi Qena runs northeast. Based on inscriptions, Coptos was the primary launching site in the Nile Valley for expeditions to Mersa Gawasis, and for quarrying missions into the Eastern Desert to the Wadi Hammamat.[20] At the Wadi Hammamat, greywacke stone for statues was quarried, and the site is famous for its hundreds of inscriptions dating from the Predynastic through modern times.[21] An inscription dating to the reign of Pepi I (ca. 2321–2287 BCE) is one

19 Bard and Fattovich, 2015; 2018.
20 Couyat and Montet 1912, 81–84; Sayed 1977, 169–173.
21 Couyat and Montet 1912; Goyon 1957.

of the earliest attestations of Min, lord of Coptos, and Eastern Desert mining and quarrying expeditions might have been launched from Coptos as early as the Predynastic period.[22]

4 Textual Evidence

Egyptian military and trade campaigns, mining parties, herders, and various individuals have traversed (on foot and on hoof) the wadis and hills of the Eastern and Western Deserts of Egypt since Predynastic times. However, Egyptians did not explicitly record exactly where and how they traveled, although there are clues. Some of the main sources we have regarding expeditions include tomb autobiographies of expedition leaders or participants, and inscriptions left along routes and at destinations.[23]

In addition to inscriptions, recent excavations along the Red Sea coast confirm a prime motive for desert travel: access to the Red Sea and its sailing routes to the Sinai and to the land of Punt in the south. Mersa Gawasis is one of three known and recently excavated pharaonic harbors. Wadi el-Jarf, discovered in 2008, was used during the reigns of Old Kingdom kings Sneferu and his successor Khufu for expeditions to the Sinai.[24] Ayn Soukhna, discovered in 1999, became the other main launch site to the Sinai in the Old Kingdom, most likely replacing Wadi el-Jarf at the end of Khufu's reign.[25]

Previous attempts to understand and map the movements of these expeditions from Coptos to Mersa Gawasis have been based primarily on inscriptional evidence, as well as information collected from archaeological survey. The first relevant text dates to the Eleventh Dynasty and the reign of Mentuhotep III (ca. 2004–1992 BCE).

The text, Wadi Hammamat 114, was inscribed at the greywacke stone quarry of the Wadi Hammamat.[26] According to the text, some 3000 men under the direction of the royal Steward Henu marched across the desert to the Red Sea. He was one who "made the path as a river" (*jr.n mṯn m jtrw*) and the "desert as a field edge" (*tꜣ-dšr.t m ꜥꜣd n sḫ.t*). Henu boasted of establishing a route from Cop-

22 Couyat and Montet 1912, 59–60.

23 Couyat and Montet 1912; Goyon 1957.

24 El-Raziq, Castel, and Tallet 2017, 19–21.

25 Tallet and Marouard 2014, 4. See also, Pierre Tallet and Gregory Marouard, "The harbor facilities of King Khufu on the Red Sea shore: The Wadi al-Jarf/Tell Ras Budran system," *JARCE* 52 (2016): 135–177.

26 Couyat and Montet 1912, 81–84.

FIGURE 13.3 Inscriptions from the Wadi Hammamat. Henu's inscription, Wadi Hammamat 114, is in the bottom left.

tos to the Red Sea through unknown regions called Idahet and Yaheteb, digging wells and keeping his men provisioned. The route back to the Nile after they returned from Punt was noted to be through another unidentified region called *wꜣg*, and then through the Wadi Hammamat to quarry stone, before heading back to the Nile Valley.[27] This text does not name Mersa Gawasis, but it is possible that these men used the port. However, Mentuhotep III is not attested at Mersa Gawasis.

Sayed uncovered the next relevant text, the Antefiker stela, at Mersa Gawasis in the 70s. This stela dates to the reign of Senwosret I (ca. 1956–1911 BCE), slightly later than Henu. In this inscription, 3756 men are listed as being part of the expedition to the "Mine of Punt" that began at the "dockyard" (*wḫr.t*) of Coptos.[28] These two texts detail how Punt-bound expeditions on the Red Sea

27 Couyat and Montet 1912, 81–84.
28 Sayed 1977, 169–173.

began at Coptos. According to Antefiker, at Coptos workers built sailing ships of imported wood. These boats were then disassembled, and the ship parts were transported overland across the desert wadis to the Red Sea. On the shore of the Great Green the ships were reassembled and launched.[29] No archaeological evidence for these shipbuilding facilities has yet been uncovered at Coptos or Qena.[30]

These, and other similar expedition monumental inscriptions, describe how thousands of men were part of these state-run trips. The accuracy of such large numbers has been debated, and based on the size of the boats discovered in the caves near Mersa Gawasis, far fewer than 3000 individuals could have participated in the actual sailing journeys to Punt.[31] There is also no archaeological evidence at Mersa Gawasis that the site could have or did support thousands of men, even temporarily.[32] The Mersa Gawasis excavators do propose that these expeditions, however large they were, were also connected to mining, quarrying, and other activities in the Wadi Hammamat and elsewhere, and men enlisted to carry the timber across the desert would then go on to participate in other manual labor elsewhere in the Eastern Desert.[33] Henu's text mentions the sea voyage coupled with quarrying, which supports this idea. These large numbers were potentially more ideological than truthful, commemorative for the gods and those who come after, rather than the actual personnel count.[34] Also regarding desert routes, the Mersa Gawasis excavators suggest that the presence of two stelae found in the Wadi Gasus, about 7 km west of Mersa Gawasis, confirm that this wadi was used as part of the route to the coast.[35] But again these texts only offer hints. The exact expedition route(s) across the complicated Eastern Desert terrain remain(s) undetermined.

5 Survey and Maps

Data collected from survey provides another source of evidence for these expeditions and their routes. Previous survey of the Eastern Desert has tended to

29 Zazzaro and Calcagno 2012 87–104; Bard and Fattovich 2015; Herbert and Berlin 2003; Creasman and Doyle 2010, 14–30.
30 Bard and Fattovich 2018, 181; Herbert and Berlin 2003.
31 See, for example, Köpp 2013, 107–132.
32 Bard and Fattovich 2018.
33 Bard and Fattovich 2018.
34 Bard and Fattovich 2018, 68; Creasman and Doyle 2010, 15; Parkinson 2002, 86; Bloxam 2006, 286.
35 Bard and Fattovich 2018, 182–183.

focus on the collection of inscriptions and the mapping of quarries, mines, wells, and ancient sites. A comprehensive and systematic survey has not been conducted, and data collection has tended to focus on later time periods, particularly the Roman-era, as well as on modern geomorphic and hydrological studies. South of Mersa Gawasis, excavators at Quseir al-Qadim have conducted a survey near the ancient Roman port, while a survey was conducted from Coptos to Quseir al-Qadim by the University of Michigan's survey in the 1990s.[36] From 1998–1999, under the direction of David Rohl, the Eastern Desert Survey (EDS) project surveyed the Central Eastern Desert to map inscriptions collected by Henry Winkler in the 1930s in an area south of the Wadi Hammamat, but this excluded the northern and eastern Wadi el-Atwani branches, the areas closest to theorized routes.[37] From 1989–1993 Klemm and Klemm surveyed for gold mines between the 28° and 22° parallel North in a wide area of the Eastern Desert.[38] North of Mersa Gawasis, the site of Mons Claudianus has been excavated.[39] Further south, the University of Michigan and Asyut University surveyed between the Nile Valley and the Roman Port of Berenike.[40]

More recent and relevant, the University of Naples "L'Orientale" with an Italian and Egyptian team under the direction of Irene Bargantini, has been conducting a survey in the central Eastern Desert slightly north and west of Mersa Gawasis and south of the Wadi Hammamat and the Qena region outside of Thebes.[41] Using information collected from the "L'Orientale" surveys, ancient texts, previous surveys, and general geological surveys, many of which took place around the turn of the century and the 1950s,[42] Andrea Manzo created models in GIS of possible routes.[43] Many of the mapped ancient sites date to the Roman era, a period with more available data, but almost 2000 years after Mersa Gawasis went out of known use.[44] This work is helpful, however, there is still a lack of systematic survey of the region between Coptos and Mersa Gawasis.

36 Cuvigny 2003; Herbert and Berlin 2003.

37 Rothe, Miller, and Rapp 2008; Winkler 1938.

38 Klemm and Klemm 2013; Bragantini, Pirelli, and Università degli Studi di Napoli "L'Orientale" (UNO) 2013, 47–156.

39 Peacock and Maxfield 1997.

40 Sidebotham and Gates-Foster 2019.

41 Bragantini et al. 2013; Bragantini, Pirelli, and Università degli Studi di Napoli "L'Orientale" (UNO) 2015, 165–177.

42 Barron and Hume 1902; Nessim 1954; Ossman and Sidebothman 2000, 7–30.

43 Bard, Fattovich, and Manzo 2013, 533–556; Bard and Fattovich 2018.

44 Bragantini et al. 2013; 2015; Bard, Fattovich, and Manzo 2013.

Water was a critical resource for these missions, and thus water access is an important factor in Manzo's maps. Based on Henu's and other inscriptions as well as excavated donkey mandibles from Mersa Gawasis, we know that donkeys were a critical part of these expeditions: donkeys, like humans need water.[45] Water was carried, but wells were a prime resource for desert travelers. Henu notes his party dug some 16 wells, and some ancient wells are marked in the desert with inscriptions, even if they have disappeared from the modern landscape, which many have.

Manzo mapped his proposed routes along possible pathways near wells located at a distance of less than 80 km from one another,[46] this being based on an established assumption that donkeys can cover a maximum 100–120 km in two days without water.[47] This distance is shorter than the 170–180 km trek to the coast from the Qena Bend. Manzo's maps also consider known natural resources, quarries, and mines, assuming that these expeditions were also connected to mining and quarrying. Manzo and the excavators did not run GIS models based on topographical slope, surface, or time, nor Least Cost Path analysis specifically.

Many important factors have been taken into account to map these possible routes to Mersa Gawasis. However, because survey of the area has been limited, there is a good chance that sites, wells, and inscriptions left from Middle Kingdom expeditions have been overlooked and thus are not part of Manzo's original GIS route models. Also, from flash floods and rock collapse, ancient (and modern) water sources disappeared from the landscape naturally, and wells could easily have been dug and not recorded or inscribed and then lost to the archaeological record. Most of the known wells also are mainly Late Period and/or Roman or later sites. In addition, the ancient pathways themselves disappear from the surface.

6 Least Cost Path Analysis and Methodology

Satellite imagery and GIS have proved effective tools for studying the Egyptian landscape. Sarah Parcak's remote sensing work to detect unknown monuments and to track looting is a prime example.[48] Bubenzer and Bolten employed Digital Elevation Models, or DEMs, in conjunction with historical sources and

45 Bard and Fattovich 2015.
46 Bard, Fattovich, and Manzo 2013; Bard and Fattovich 2018, 183.
47 Wainwright 1935, 259–261; Bard, Fattovich, and Manzo 2013.
48 Parcak 2009; 2019.

ground-truthing, to determine and map ancient caravan routes in the Western Desert of Egypt.[49] DEMS are digital models of the earth's surface captured from above the earth. They store the measurement of terrain elevation data. Our specific context of Middle Kingdom Eastern Desert expedition routes provides a rich opportunity for Least Cost Path analysis, particularly when combined with the available recorded data.

It is important to state that Least Cost Path analyses are theoretical, in that they approximate a human's potential travel decisions based solely on the landscape and their presumed knowledge of that landscape. Despite this, Least Cost Path analysis is a precise way to map a landscape, and possible human movements across it. Least Cost Path analysis has proven to be a useful tool for archaeologists. Utilizing DEMS, including Least Cost Path analyses, Phillips and Leckman re-created prehistoric trails in the desert of south-central New Mexico that have since disappeared.[50] On a smaller scale, Least Cost Path analysis has proven informative about the changing movements of urban peoples over time at the site of Kerkenes Dağ in Turkey under Scott Branting's direction.[51] Harrower and D'Andrea mapped the Least Cost Path routes from Aksum to Adulis on the Red Sea in the highlands of Ethiopia.[52]

Utilizing ArcMap, I created Least Cost Paths through the Eastern Desert wadi systems to Mersa Gawasis. These are based on topographical data and time, meaning: What is the fastest and most cost-efficient route across the desert? Would Least Cost Path analysis produce similar maps to those created based on wells and inscriptions? For my analysis, I generated routes to Mersa Gawasis from Coptos, where we know from inscriptions that some expeditions began. I also ran routes from Qena, a site in the Nile Valley about 25 km north of Coptos, which was used in later periods. The city of Qena sits at the mouth of the Wadi Qena and might also have been a launching place for some earlier expeditions. This site is directly across the Nile from the Temple of Dendera, an important religious site dating back to at least the Old Kingdom. The Romans later used Qena, as well as Coptos, as Nile Valley bases to launch mining expeditions into the desert and sailing missions to the port of Quesir al-Qadim, about 26 km south of Mersa Gawasis. Quesir al-Qadim was a Roman port that sits at the eastern mouth of the Wadi Hammamat system. The Oriental Institute excavated the site in the 1980s, and Southampton worked there in the 1990s.[53] It was a

49 Bubenzer and Bolten 2013, 61–76.
50 Phillips and Leckman 2012, 46–56.
51 Phillips and Leckman 2012, 46–56.
52 Harrower and D'Andrea 2014, 513–541.
53 Burke 2005; Peacock and Blue 2006.

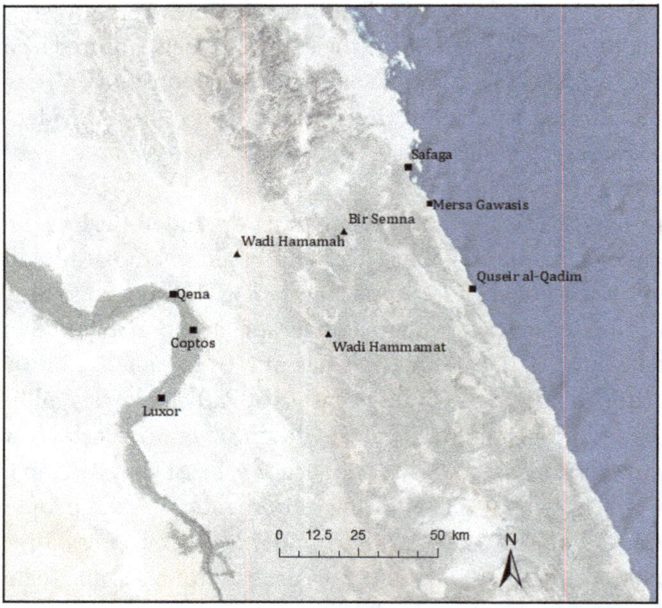

FIGURE 13.4 Map of the study area, including Coptos, Qena, Mersa Gawasis, and Quesir al-Qadim

significant Roman and Islamic port, but there is no evidence that it overlapped in use with Mersa Gawasis.[54] Manzo's maps include routes in the desert beginning from Qena and Coptos, and it seemed productive to see if and how the Least Cost Path outputs changed based on the origin at the Nile. I also mapped a path to Quesir al-Qadim from Coptos to see if it could also help me determine whether the Wadi Hammamat was actually an easier and faster topographical distance to traverse compared to the northern wadi systems, particularly the Wadi Qena system.

For these models, Least Cost Path analysis was run in ArcMap (10.5). I created a geographical study area created based on 6 ASTER Global Digital Elevation Models (DEMs) images, each with a resolution of 30 meters by 30 meters. These were downloaded from United States Geological Survey (USGS) Earth Explorer and mosaicked into a single raster. This same resolution has proven effective in other Least Cost Path archaeological studies.[55]

The study area includes the sites of Mersa Gawasis, Coptos, Qena, and Quesir al-Qadim. This area includes two UTM zones, zones 35 and 36, so I created

54 Burke 2005.

55 See, for example: Harrower and D'Andrea 2015.

a unique projected coordinate system based on a general African coordinate system with a central meridian at 30° to ensure correct analysis of the topography. In ArcGIS, the cost paths were implemented using spatial analysis tools. From the DEM, a least-cost distance raster and a backlink raster were created. Least Cost Path measures the accumulated cost within each cell of a raster and creates a path to the border cell that has the least accumulative cost, thus the easiest to navigate. The DEM study area was the input cost (slope) raster for the Least Cost Path analysis from Coptos and then from Qena, and from that the backlinks were created and Mersa Gawasis and then Quesir al-Qadim were set as the destination for each respective path.

In order to calculate the time-cost of moving up and down different slopes in the landscape, I used the Tobler's hiking function.[56] This function quantifies the time cost of moving over terrain of differing slope [from -XX to +xx degrees] in km/hr with an estimated average travel time of 5.037 km/hr.

7 Results and Analysis

For these models, I produced three different Least Cost Path routes, none of which overlapped. The first route was from Coptos to Mersa Gawasis. The most cost and time-efficient route resulted in a central path starting in the southern part of Wadi Qena system. This ran east roughly through the Wadi Atallah el-Mur, the Wadi Abiyad, and then to the coast.

I next mapped the Least Cost Path route from Coptos to Quesir al-Qadim, which produced a path that followed a southern route directly through the Wadi Hammamat. This route then ran east through the Wadi Um Arat and the Wadi Seyala to the coastal Roman site.

Lastly, the Least Cost Path from Qena to Mersa Gawasis produced the most northern route, through the Wadi Qena system. This route went north through part of the Wadi Hammah to the Wadi Garya, the Plain of Markh, and through the Wadi Abu Shia. The route hit the sea at the mouth of the Wadi Gasus, slightly north of Mersa Gawasis, and then ran south along the coast to the ancient harbor.

These three paths represent the fastest and the most cost-efficient routes for walking along various slopes. The estimated length of time needed to traverse each route does not vary greatly. However, the origin on the Nile and the destination did change the time each relative journey would take. The route walking

56 Tobler 1993.

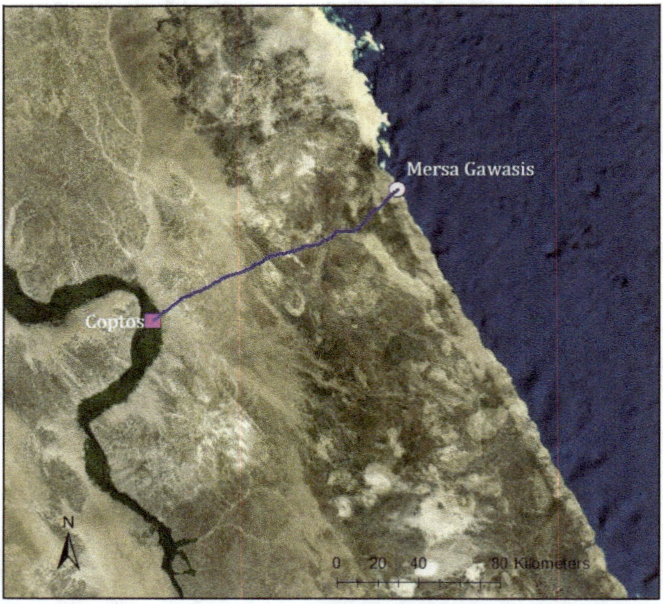

FIGURE 13.5 The Least Cost Path route from Coptos to Mersa Gawasis

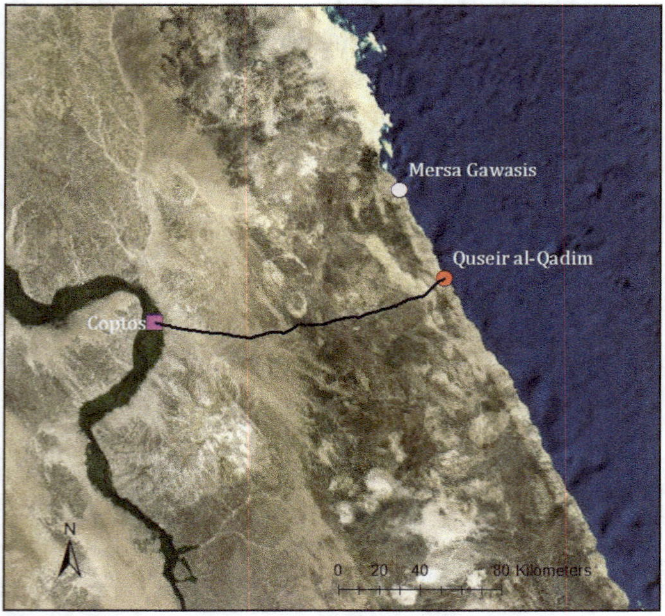

FIGURE 13.6 The Least Cost Path route from Coptos to Quesir al-Qadim

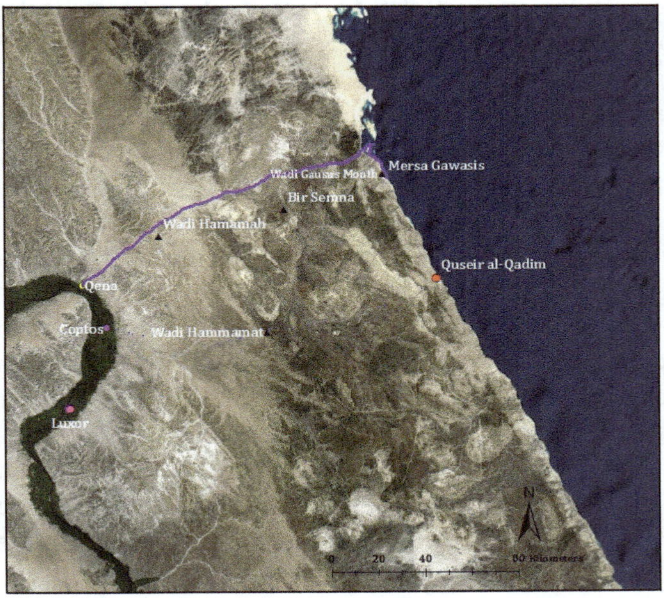

FIGURE 13.7 The Least Cost Path route from Qena to Mersa Gawasis

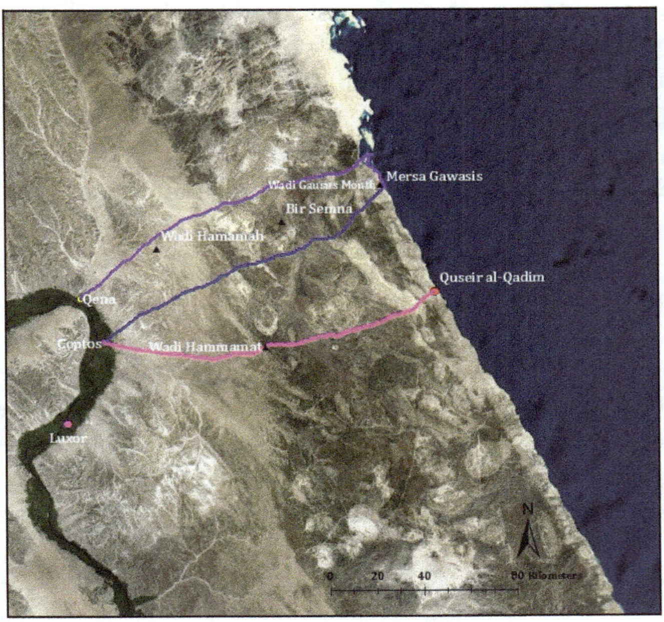

FIGURE 13.8 Map with the three Least Cost Path routes

from Coptos to Mersa Gawasis was calculated to take about 31.9 hours; Coptos to Quesir was about 33.04 hours; and Qena to Mersa Gawasis was about 29.8 hours. These times are calculated again based on Tobler's function, which does not consider rest periods. To have more useful and accurate data, these rate and time calculations should be further broken down to consider necessary periods of rest, which I considered.

There have been calculations for the average distance modern humans and donkeys are able to walk in a day.[57] Köpp compares various estimated rates of desert travel in Egypt, compiling evidence from texts and practical studies.[58] Förster estimated a distance of 25–40 km/day for donkey caravans to travel to the Dakhla Oasis during the Old Kingdom and First Intermediate Period.[59] In general, both humans and donkeys can average about 25–40 km walking per day, including rests and loads. The individuals in the Mersa Gawasis expeditions were most certainly on foot except perhaps for the expedition leader, and the donkeys would have been carrying loads—water, ship timber, etc. These three Least Cost Path routes would each have taken a little over 5 days if the estimated rate with rest periods was calculated at the slowest of 25 km/day.

Although all three of these Least Cost Path routes would have taken similar amounts of time, based on this topographical analysis, the path from Qena to Mersa Gawasis appears to have been the fastest. Although we only have inscriptional evidence documenting departures from Coptos—and only two extant examples at that—one might wonder if Qena was also used to launch these missions in the Middle Kingdom. Indeed, the Romans were known to use Qena.[60] Starting a journey through the Wadi Qena system proved faster from both Qena and Coptos to the coast than through the Wadi Hammamat.

8 Discussion

The Mersa Gawasis excavators suggest two possible paths to the harbor site based on known inscriptional evidence and mining sites: "A northern route along the Wadi Hammamh, Wadi Abu Jarida, Wadi Safaga, Wadi Simna and Wadi Saqi to the coast, and a southern route along the Wadi Hammamat and Wadi Qush".[61] These two routes are very similar to the Least Cost Path analysis

57 Köpp 2013.
58 Köpp 2013.
59 Förster 2013, 381–390.
60 Rothe, Miller, and Rapp 2008; Burke 2005.
61 Bard and Fattovich 2018, 183.

that I produced from Qena to Mersa Gawasis and from Coptos to Quesir al-Qadim. However, no "central" route was mentioned that is similar to my Least Cost Path analysis results of the route from Coptos to Mersa Gawasis, suggesting further systematic investigation of this area of the Eastern Desert is worthwhile.

The excavators' northern route and my Least Cost Path northern route both exit at the coast through the Wadi Gasus and then head south. The Wadi Gasus is about 2 km north of Mersa Gawasis. The harbor of Mersa Gawasis was a lagoon, and its protected setting would have allowed ships to dock while being protected from the winds.[62] The harbor was not a permanent site, and the excavated evidence suggests it was used only on a temporary basis for these expeditions. The closest natural spring is 7 km away in the Wadi Gasus system, which may account for Mersa Gawasis's temporary use.[63]

Middle Kingdom royal expeditions were not relatively frequent events, another possible reason why Mersa Gawasis was only temporarily occupied. These journeys took place every 15–20 years, based on inscriptional information.[64] The "central" Least Cost Path produced from Coptos to Mersa Gawais might not necessarily present the closest paths to these specific expedition routes, only in that these expeditions were grand events undertaken for many reasons, convenience not necessarily being the primary objective. The king and expedition leaders may not have cared about the most cost-efficient route, rather taking troops and goods through the more familiar Wadi Hammamat, Wadi Hammamh, and/or where water sources were known.

One must consider, however, the lack of inscriptional mentions of *sꜣww* in the many inscriptions of the Wadi Hammamat.[65] Was a route through this wadi then in fact used for these expeditions after Henu? Middle Kingdom kings are especially well represented in the Wadi Hammamat inscriptions, as well as in stelae and inscriptions at Mersa Gawasis and its vicinity.[66] Farout and, later, Gasse have reflected on the intertextuality between contemporary royal inscriptions at Mersa Gawasis and the Wadi Hammamat, particularly during the reign of Senwosret I, but also some later Twelfth Dynasty kings.[67] These connections are made primarily based on the titles of oarsmen and fishermen, perhaps also emphasizing an overlap in these missions.[68] The Eleventh

62 Bard and Fattovich 2018, 183.
63 Bard and Fattovich 2018, 61.
64 Bard and Fattovich 2018, 176–177.
65 Couyat and Montet 1912; Gasse 2012, 133–144.
66 Bard and Fattovich 2018.
67 Farout 2006 43–52; Gasse 2012.
68 Gasse 2012.

Dynasty Henu and his team, one of the earliest known expeditions of the Middle Kingdom, may have been more familiar with the Wadi Hammamat, where his inscription is found. Again, he described how he made the desert navigable and dug new wells, presumably for this and future expeditions. Perhaps later in the Middle Kingdom some Nile Valley dwellers would have been more familiar with the landscape of the Eastern Desert, applying accumulative knowledge of other paths and relative speed outcomes and desired resources, and thus other routes aside from the Wadi Hammamat were chosen. The excavators of Mersa Gawasis discuss how the Eastern Desert was very much a policed, monitored environment in the Twelfth Dynasty, with individuals constantly on the move.[69] There is a mining site close to Mersa Gawasis (known today as the "Roman fort") that is an example of a much more frequented site than Mersa Gawasis, and it was in use the same time as the harbor.[70] The fact that my Least Cost Path model from Coptos to Mersa Gawasis does not run through the Wadi Hammamat intriguingly matches up with the lack of textual evidence for travel between these two localities.

Again, Mersa Gawasis (*sꜣww*) is not named in the numerous inscriptions from the Wadi Hammamat. The southern Least Coast Path route from Coptos to Quesir al-Qadim traveled through the Wadi Hammamat. This frequented site had wells and inscriptions, and was quarried from Predynastic times, onward. Despite not being as cost-efficient or speedy, the Wadi Hammamat could have been utilized in order to procure natural resources on the way to and from the coast, or because it was the traditional route. However, if the Wadi Hammamat was used, one might assume there would be more inscriptional evidence there regarding these expeditions. We also know that ships were stored at the coast for future trips after the completion of a journey, so ships were not being carried across the desert on every trip. It is possible that chosen routes varied because of this. However, these GIS generated routes travel near known mines, wells, and sites, suggesting topography might have been a factor in decision making as well. If large ship beams were being carried, the Least Cost Path would logically have been desirable when considering such a load. And in places where they do not match, further investigation of the area might yield significant results. In addition to these expeditions, other individuals of varying classes and occupations were crossing the Eastern Desert for numerous reasons. These accurately mapped paths also help give a better understanding to how people in general

69 Bard, Fattovich, and Manzo 2013.
70 Bard and Fattovich 2018.

traveled over the landscape and thus yield further insight into Eastern Desert travel in general.

My Least Cost Path analysis presents somewhat overlapping but also alternative routes to those maps made based on wells, sites, and inscriptions. The Least Cost Path models display a more measured reality of ancient movements, particularly compared to maps with drawn paths between mines, wells, and inscriptions. These known data points can sometimes be many kilometers apart, and Least Cost Paths is an accurate measurement of terrain within meters. However, when combined, both methods together augment the accuracy of predicting ancient humans' movements and route realities. The possibility of two or more different routes for these expeditions is very plausible; the reasons for various routes could have and probably did change, depending on the mission. The fact that the Least Coast Path models produced a third route from Coptos to the coast further emphasizes the need for ground truthing and more archaeological survey in this central part of the desert.

Topography, including slope and surface terrain, is a major factor that can affect human behavior, and it is worthwhile to consider these models. Combined with Manzo's maps and other data,[71] Least Cost Path analysis provides deeper insight into routes. It also includes another major criterion—namely the difficulty of navigating certain areas of the landscape—and examines how these factors can also influence human choice while traveling over land. These models also offer guidance for locating future sites and water sources which can only be further deduced with eventual ground truthing, presenting alternative options that should be explored. There are many factors, beyond topography—and wells and inscriptions—as to why ancient peoples moved as they did. These influences need to be combined for a fuller understanding of how and why ancient people traversed their landscape.

Bibliography

Abd El-Raziq, Mahmoud, George Castel, and Pierre Tallet. 2017. "Ayn Sukhna and the Red Sea: A Preliminary Report on the Excavations of the Site 2001–2006." *Abgadiyat* 12 (1): 19–21.

Altaweel, Mark. "GIS and Digital Humanities." GIS Lounge. Last modified July 10, 2017. https://www.gislounge.com/gis-digital-humanities/.

Bard, Kathryn A. and Rodolfo Fattovich. 2007. *Harbor of the Pharaohs to the Land*

71 See also, Cooper 2020.

of Punt: Archaeological Investigations at Mersa/Wadi Gawasis, Egypt, 2001–2005. Naples: University of Naples "L'Orientale."

Bard, Kathryn A. and Rodolfo Fattovich. 2015. "Mersa/Wadi Gawasis and Ancient Maritime Trade in the Red Sea." *Near Eastern Archaeology* 78 (1): 4–10.

Bard, Kathryn A. and Rodolfo Fattovich. 2018. *Seafaring Expeditions to Punt in the Middle Kingdom: Excavations at Mersa/Wadi Gawasis, Egypt*. Leiden; Boston: Brill.

Bard, Kathryn A., Rodolfo Fattovich, and Andrea Manzo. 2013. "The Ancient Harbor at Mersa/Wadi Gawasis and How to Get There: New Evidence of Pharaonic Seafaring Expeditions in the Red Sea." In *Desert Road Archaeology in Ancient Egypt and Beyond*, edited by Frank Förster and Heiko Riemer, 533–556. Köln: Heinrich-Barth-Institut.

Barron, Thomas and W.F. Hume. 1902. *Topography and Geology of the Eastern Desert of Egypt, Central Portion*. Cairo: National Print Dept.

Bloxam, Elizabeth. 2006. "Miners and Mistresses: Middle Kingdom Mining on the Margins." *Journal of Social Archaeology* 6: 277–303.

Borojevic, Ksenija and Rebecca Mountain. 2011. "The Ropes of Pharaohs: The Source of Cordage from 'Rope Cave' at Mersa/Wadi Gawasis Revisited." *Journal of the American Research Center in Egypt* 47: 131–141.

Bradbury, Louise. 1988. "Reflections on Traveling to 'Gods' Land' and Punt in the Middle Kingdom." *Journal of the American Research Center in Egypt* 25: 127–156.

Bragantini, Irene, Rosanna Pirelli, and Università degli Studi di Napoli "L'Orientale" (UNO). 2013. "The Archaeological Mission of 'L'Orientale' in the Central-Eastern Desert of Egypt." *Newsletter di Archeologia CISA* 4: 47–156.

Bragantini, Irene, Rosanna Pirelli, and Università degli Studi di Napoli "L'Orientale" (UNO). 2015. "The Preliminary Report on the Second Season of the Italian Archaeological Mission in the Eastern Desert of Egypt." *Newsletter di Archeologia CISA* 6: 165–177.

Branting, Scott. 2012. "Seven Solutions for Seven Problems with Least Cost Pathways." In *Least Cost Analysis of Social Landscapes*, edited by Devin White and Sarah L. Surface-Evans, 209–224. Salt Lake City: University of Utah Press.

Breyer, von Francis. 2015. *Punt: Die Suche nach dem "Gottesland"*. Boston: Brill.

Bubenzer, Olaf and Andreas Bolten. 2013. "Top down: New satellite data and ground-truth data as base for a reconstruction of ancient caravan routes. Examples from the Western Desert of Egypt." In *Desert Road Archaeology in Ancient Egypt and Beyond*, edited by Frank Förster and Heiko Riemer, 61–76. Köln: Heinrich-Barth Institut.

Burke, Katherine Strange. 2005. "*Quesir al-Qadim*" *Oriental Institute Annual Report* (*2003–2004*). Chicago: Oriental Institute.

Cooper, Julien. 2020. Toponym on the Periphery: Placenames of the Eastern Desert, Red Sea, and South Sinai in the Egyptian Documents from the Early Dynastic until the End of the New Kingdom. *Problem der Ägyptologie* 39. Boston: Brill.

Couyat, J., and Pierre Montet. 1912. *Les inscriptions hiéroglyphiques et hiératiques du Ouâdi Hammâmât*. Cairo: Imprimerie de l'Institut Français d'Archéologie Orientale.

Creasman, Pearce Paul and Noreen Doyle. 2010. "Overland Boat Transportation During the Pharaonic Period: Archaeology and Iconography." *Journal of Ancient Egyptian Interconnections* 2 (3): 14–30.

Creasman, Pearce Paul. 2014. "Hatshepsut and the Politics of Punt." *The African Archaeological Review* 31 (3): 395–405.

Cuvigny, Hélène ed. 2003. *La route de Myos Hormos: l'armée romaine dans le désert oriental d'Egypte*. Le Caire: Institut Français d'Archéologie Orientale.

Esri. "What is GIS?" Accessed September 22, 2020. https://www.esri.com/en-us/what-is-gis/overview

Farout, D. 2006. "Des expéditions en mer Rouge au début la XII dynastie." *Égypte Afrique & Orient* 41: 43–52.

Förster, Frank. 2013. "Beyond Dakhla: The Abu Ballas Trail in the Libyan Desert (SW Egypt)." In *Desert Road Archaeology in Ancient Egypt and Beyond*, edited by Frank Förster and Heiko Riemer, 381–390. Köln: Heinrich-Barth-Institut.

Förster, Frank and Heiko Riemer, eds. 2013. *Desert Road Archaeology in Ancient Egypt and Beyond*. Köln: Heinrich-Barth-Institut.

Gasse, Annie. 2012. "Wadi Hammamat and the Sea from the Origins to the End of the New Kingdom." In *The Red Sea in Pharaonic Times: Recent Discoveries along the Red Sea Coast. Proceedings of the Colloquium held in Cairo / Ayn Soukhna 11th–12th January 2009*, edited by Pierre Tallet and El-Sayed Mahfouz, 133–144. Cairo: Institut Français d'Archéologie Orientale.

Goyon, Georges. 1957. *Nouvelles inscriptions repestres du Wadi Hammamat*. Paris Adrien-Maisonneuve.

Harrower, Michael and Catherine D'Andrea. 2014. "Landscapes of State Formation: Geospatial Analysis of Aksumite Settlement Patterns (Ethiopia)," *The African Archaeological Review* 31 (3): 513–541.

Herbert, S.C. and A. Berlin. 2003. *Excavations at Coptos (Qift) in Upper Egypt, 1987–1992*. Portsmouth: Journal of Roman Archaeology.

Klemm, R., D.D. Klemm. 2013. *Gold and Gold Mining in Ancient Egypt and Nubia: Geoarchaeology of the Ancient Gold Mining in the Egyptian and Sudanese Eastern Deserts*. Berlin—Heidelberg: Springer.

Köpp, Heidi. 2013. "Desert travel and transport in ancient Egypt: An overview based on epigraphic, pictorial and archaeological evidence." In *Desert Road Archaeology in Ancient Egypt and Beyond*, edited by Frank Förster and Heiko Riemer, 107–132. Köln: Heinrich-Barth-Institut.

Manzo, Andrea. 2011. "Punt in Egypt and beyond: comments on the impact of maritime activities of the 12th Dynasty in the Red Sea on Egyptian crafts with some historical and ideological thoughts." *Ägypten und Levante* 21: 71–85.

Manzo, Andrea. 2017. "BiAw Pwnt in the archaeological record. Preliminary results and perspective of research." In *Cultural and Linguistic Transition explored. Proceedings of the ATrA closing workshop Trieste, May 25–26, 2016*, edited by Ilaria Micheli, 87–108. Trieste: EUT Edizioni Università di Trieste.

Meeks, Dimitri. 2003. "Locating Punt." In *Mysterious Lands*, edited by David B. O'Connor, Stephen Quirke, 53–80. London: Institute of Archaeology, University College London.

McHaffie, Patrick, Sungsoon Hwang, Cassie Follett. 2018. "GIS and Digital Humanities." In *GIS: An Introduction to Mapping Technologies*, edited by Patrick McHaffie, 11–45. Boca Raton: CRC Press.

Murray, George W. 1925. "The Roman Roads and Stations in the Eastern Desert of Egypt." *Journal of Egyptian Archaeology* 11: 138–150

Nessim, Shafik. 1954. *Report on the Prospecting Work Carried Out in Wadi Hammamat area-South Eastern Desert, 1953–1954*. Cairo.

O'Connor David, and David P. Silverman, eds. 1994. *Ancient Egyptian Kingship*. Leiden; New York: E.J. Brill.

Ossman, R.A.K. and S.E. Sidebothman. 2000. "Geomorphology and Archaeology of the Central Eastern Desert of Egypt." *Sahara* 12: 7–30.

Parcak, Sarah. 2009. *Satellite Remote Sensing for Archaeology*. New York: Routledge.

Parcak, Sarah. 2019. *Archaeology from Space: how the future shapes our past*. New York: Henry Holt and Company.

Parkinson, R.B. 2002. *Poetry and Culture in Middle Kingdom Egypt. A Dark Side to Perfection*. London and New York: Continuum.

Peacock, D.P.S., and V.A. Maxfield. 1997. *Mons Claudianus: Survey and Excavation, 1987–1993*. Cairo: Institut Français d'Archéologie Orientale.

Peacock, D.P.S. and L. Blue, eds. 2006. *Myos Hormos-Quseir Al-Qadim: Roman and Islamic ports on the Red Sea*. Oxford: Oxbow Books.

Phillips, Shaun M. and Phillip O. Leckman. 2012. "Wandering the Desert: Least Cost Path Modeling for Water Transport Trails in the Jornada Mogllon Region, Fort Bliss, South-Central New Mexico." In *Least Cost Analysis of Social Landscapes*, edited by Devin White and Sarah L. Surface-Evans, 46–56. Salt Lake City: University of Utah Press.

Rothe, Russell D., William K. Miller, and George (Rip) Rapp. 2008. *Pharaonic Inscriptions from the Southern Eastern Desert of Egypt*. Winona Like, Indiana: Eisenbrauns.

Sayed, Abdel Monem A.H. 1977. "The Discovery of the Site of the 12th Dynasty Port at Wadi Gawasis on the Red Sea Shore." *RE 29*: 140–178.

Sidebotham, Steven E. and Jennifer Gates-Foster. 2019. *The Archaeological Survey of the Desert Roads between Berenike and the Nile Valley: Expeditions by the University of Michigan and the University of Delaware to the Eastern Desert of Egypt, 1987–2015*. Boston: American Schools of Oriental Research.

Tallet, Pierre and Gregory Marouard. 2014. "The Harbour of Khufu: on the Red Sea Coast at Wadi al-Jarf, Egypt." *Near Eastern Archaeology* 77 (1): 4–14.

Tallet, Pierre, and El-Sayed Mahfouz, eds. 2012. *The Red Sea in Pharaonic Times: Recent Discoveries along the Red Sea Coast. Proceedings of the Colloquium held in Cairo / Ayn Soukhna 11th–12th January 2009*. Cairo: Institut Français d'Archéologie Orientale.

Tallet, Pierre. 2013. "The Wadi el-Jarf site: A harbor of Khufu on the Red Sea." *Journal of Ancient Egyptian Interconnections* 5 (1): 76–84.

Tobler, Waldo. 1993. "Three presentations on geographical analysis and modeling: Non-isotropic geographic modeling; speculations on the geometry of geography; global spatial analysis." *National center for geographic information and analysis*, 93–91.

Wainwright, G.A. 1935. "Review: C.G. Seligman. 'Egypt and Negro Africa: A Study in Divine Kingship,' London 1934." *Journal of Egyptian Archaeology* 21: 259–261.

Ward, Cheryl. 2006. "Boat-building and its social context in early Egypt: interpretations from the First Dynasty boat-grave cemetery at Abydos." *Antiquity* 80, (307): 118–129.

Ward, Cheryl. 2012. "Ancient Egyptian Seafaring Ships: Archaeological and Experimental Evidence." In *The Red Sea in Pharaonic Times: Recent Discoveries along the Red Sea Coast. Proceedings of the Colloquium held in Cairo / Ayn Soukhna 11th–12th January 2009*, edited by Pierre Tallet and El-Sayed Mahfouz, 53–64. Cairo: Institut Français d'Archéologie Orientale.

Whealey, David, and Mark Gillings. 2002. *Spatial Technology and Archaeology: The Archaeological Applications of GIS*. New York: Taylor and Francis.

Whitcomb, Donald S. and Janet H. Johnson. 1982. *Quseir al-Qadim 1980: Preliminary Report*. Malibu: Undena.

White, Devin A. and Sarah L. Surface-Evans, eds. 2012. *Least Cost Analysis of Social Landscapes: Archaeological Case Studies*. Salt Lake City: The University of Utah Press.

Winkler, Hans A. 1938. *Rock-Drawings of Southern Upper Egypt*. London: The Egypt Exploration Society, H. Milford Oxford University Press.

Zazzaro, Chiara and Claire Calcagno. 2012. "Ship Components from Mersa Gawasis Recent Finds and their Archaeological Context." In *The Red Sea in Pharaonic Times: Recent Discoveries along the Red Sea Coast. Proceedings of the Colloquium held in Cairo / Ayn Soukhna 11th–12th January 2009*, edited by Pierre Tallet and El-Sayed Mahfouz, 87–104. Cairo: Institut Français D'Archéologie Orientale

CHAPTER 14

Secondary Epigraphy in Egypt: A Case for a Research Infrastructure

Hana Navratilova

Abstract

Project concept "SEE"—Secondary Epigraphy in Egypt—aims at developing a strategy for a research e-infrastructure concerned with secondary epigraphy in Egypt.[1] It also promotes the articulation of research infrastructures as an organic part of the research landscape of Egyptology. This is a brief introduction of the SEE model, which is a research- and capacity-building digital resource that would enable access to complex information about secondary epigraphy in Egypt, enabling us to reconstruct the history of the sites through analysis of secondary epigraphy.

Keywords

Secondary Epigraphy in Egypt – research infrastructures – digitization – cyberinfrastructures

1 The contribution draws from a number of project planning discussions and grant-writing sessions at the University of Reading and the King's College, London. A number of colleagues (listed further in a random order, but all having had a decisive input) contributed to the development of the idea of a secondary epigraphy infrastructure. I would like to thank Ian Rutherford, Rachel Mairs, Kathryn Piquette, Paul Caton, Pamela Mellen, Emma Aston, Samantha Sherry, Charlotte Johnson and the University of Reading research development team. In addition, input and ideas for this concept were gained over a number of years, going back to an early idea of a more limited graffiti database presented in 2004. A great impetus was provided by the *Informatique et Égyptologie* meetings in Oxford in 2006, and in Vienna in 2008, as many other colleagues contributed by being willing discussants. I have learned a lot about research infrastructures during my time (2008–2011) at the *Topographical Bibliography*, at the Griffith Institute, Oxford, under the supervision of Jaromir Malek and working alongside the "TopBib" team. Inconsistencies and mistakes in this paper remain, of course, my own responsibility.

 | DOI:10.1163/9789004501294_015

1 What Is a Research Infrastructure?

"'Research infrastructure' means facilities, resources and related services that are used by the scientific community to conduct top-level research in their respective fields and covers major scientific equipment or sets of instruments; knowledge-based resources such as collections, archives or structures for scientific information; enabling Information and Communications Technology-based infrastructures such as Grid, computing, software and communication, or any other entity of a unique nature essential to achieve excellence in research. Such infrastructures may be 'single-sited' or 'distributed' (an organized network of resources)."[2] The concept of research e-infrastructures, or cyberinfrastructures[3] is based on expanding the research infrastructures with digital means, offering an enhanced research resource with greater accessibility, but also coming with its own set of challenges, particularly interoperability, technology updates, and sustainability,[4] but also a distance of the researcher from their physical resource.[5]

There is a debate concerned with the roles of research infrastructures—do they provide data, or can they also set research agendas and shape research experience? It has been argued that they can do both.[6] The research epistemology and methods are influenced by digital research infrastructures and in turn contribute to their formation. Digital humanities infrastructures often present also a "spatial/visual" turn,[7] especially where the potential of spatial humanities, such as a "deep map" is being explored.[8]

Anderson, in her insightful outline on research infrastructures, proposed "that we need to view infrastructure as a material and experiential presence that is embedded in the practices and experience of research, which builds on and enhances that which already exists, that unites scholars with archivists,

2 A definition promoted by RISCAPE (European Research Infrastructures in the International Landscape), https://blogs.helsinki.fi/riscape-project/what-is-a-research-infrastructure.

3 Cf. Anderson 2013, 4.

4 On an example of a technology & equipment infrastructure that had faced challenges, Foka et al. 2018, with further references. In case of digital datasets and specially designed applications, the software interoperability issues, financial sustainability, data migration and memory demands represent further challenges.

5 Barker et al. 2012, 185.

6 Anderson 2013.

7 Barker et al. 2012, 185.

8 Bodenhamer, Harris and Corrigan 2013, see further in more detail Bodenhamer, Harris and Corrigan 2015.

librarians, and museum curators, and that also finds a place for the amateur."[9] In this incarnation, cyberinfrastructures may be particularly adaptive for the tasks of encompassing the wide historical span of records in humanities—from legacy to newest born-digital data. This aspect makes them particularly useful for archaeology and epigraphy.[10] The other aspect is enhancing accessibility—a diaspora of academic data across institutions and countries can become reachable, leveling somewhat the inequalities of access to hardcopy libraries and archives.[11]

2 Why a Research Infrastructure on "Secondary Epigraphy"?

Secondary epigraphy—also known as graffiti—represents a visual and written communication that reflects different social worlds of the past, from the elites to the people marginalized in other records. Secondary epigraphy[12] is a large corpus of texts and figures of non-monumental character, widespread in, but not limited to, the ancient world. Egypt is particularly rich in secondary epigraphy, and its ancient corpus from 3000 BC to 400 AD is multilingual and multicultural, as indeed are the subsequent corpora until the present day.

Graffiti, i.e. texts and figures added to man-made structures and in anthropogenic landscapes, are a widely known but underestimated resource, although they illustrate cultural patterns and personal agency expressed by a large part of past populations, elite and non-elite alike.[13] They represent a unique access to the history of humanity—and the research of their "siblings," the rock inscriptions and art, is already using digital technology with increasing quality and quantity of results,[14] and enabling a qualitative change in mapping and interpreting this vast resource.[15]

9 Anderson 2013, 4.

10 There are many examples of good practice around—a large site-specific case study is represented by the Giza Archives online—the address being, as of November 2019, http://www.gizapyramids.org/. Now http://giza.fas.harvard.edu/

11 As noted, e.g., by Quirke 2013 on a particular example of materials for history of Egyptology.

12 Ragazzoli 2018.

13 Keegan 2014, compare Darnell et al. 2002, and also Grajetzki 2008.

14 See Wilding Brown 2017, and for visualization Urcia et al. 2018.

15 On the social and cultural history potential see most recently Brown 2017, and Polkowski 2016, with references.

Within the discussions on the multilingual corpus of secondary epigraphy,[16] we have identified fragmented research efforts, and a compartmentalization of study of secondary epigraphy at Egyptian archaeological sites, particularly where diachronic development is concerned.[17] Many ancient sites had a role in local communities for generations after the demise of or change in their original function, and this role was attested in graffiti. Can we emphasize the complexity of historical continuity and change by harnessing the capability of digital infrastructures to map secondary epigraphy?

The SEE project proposes that we can. Its intended result, a gazetteer on Egyptian graffiti locating and linking available graffiti information, with links to other research infrastructures, should provide information concerning site, monument, date, script, and language of the graffiti. This is essential for a contextual study of secondary epigraphy advocated explicitly at least since 1976.[18] The infrastructure would not be limited to one type of writing, one language, or one historical period, but would pool data on all secondary epigraphy features present at a site, and information about their spatial and geographical location.[19]

3 Egyptology, DH and Research Infrastructures

The present volume showcases a number of approaches to and uses of digitization in Egyptology as of 2019/2020.[20] Digital photography, multispectral pho-

16 I am grateful for the input of Rachel Mairs and Ian Rutherford in particular, on the multilingual context.

17 As noted by Rutherford 2003.

18 Caminos 1976.

19 There is an awareness of the challenges posed to spatial humanities, especially when using historical and ambiguous data, or when dealing with a changing landscape and cityscape. Nonetheless: "Finding ways to make the interaction among words, location, and quantitative data more dynamic and intuitive will yield rich insights into complex sociocultural, political, and economic problems, with enormous potential for areas far outside the traditional orbits of humanities research. In short, we should vigorously explore the means by which to advance translation from textual to visual communication, making the most of visual media and learning to create 'fits' between the messages of text and numbers and the capabilities of visual forms to express spatial relationships." Bodenhamer, Harris and Corrigan 2013, 173.

20 The number, or even categorization of digital projects in Egyptology is beyond the scope of this paper and even the volume, as they range from complex digital paleography tools such as AKU (https://aku.uni-mainz.de/) to VR modelling of sites (see other papers in this volume for further references).

tography and other related techniques—the next generation of photographic and related documentation—are illustrated in other papers in this volume and elsewhere.[21] Digital epigraphy has been well-documented at least since the mid-1990s.[22]

Mapping all digital initiatives in Egyptology would, by now, require a small volume, and it can only be drafted here. The texts and digital approaches to text analysis are a prominent concern. Developments of this area of interest were mapped by Rosmorduc.[23] Computational linguistics approaches were considered for Egyptology as early as in the 1960s.[24] Here Egyptology has faced a particular problem of encoding the hieroglyphic script. The Egyptian writing system presents specific issues for the Optical Character Recognition (OCR) or various automated text mining approaches. But presenting texts is becoming less of an issue.[25] Creating digital editions, a powerful tool where collaboration and accessibility are concerned, is helped by TEI and EpiDoc platform.

A general overview of the ongoing Egyptological discussions from the 1970s onwards can be gleaned from the volumes of the series of the workshops "Informatique et Égyptologie."[26] It is of interest to note that "I&E" was contemporary with or even preceding some other digital humanities publications.[27] As a glimpse of their tables of contents may show, making Egyptological practice digital,[28] the digitization of texts, and creation of text corpora, as well as of digital access to museum collections (or to corpora of particular artefacts, coffins, shabtis, sealings, etc.), were at the forefront, with the manual for encoding hieroglyphic texts being developed since the 1980s. The dictionaries and text data banks are by now a stable part of the Egyptology digital landscape and developed in successful viable projects over the years—Thesaurus Linguae Aegyptiae[29] and Ramsés[30] being a case in point.

Other topics were discussed as well, if less prominently. Possibilities of Egyptological publishing in the developing digital environment were presented and

21 Compare Vértesz 2019 for an accessible outline.
22 Der Manuelian 1998.
23 Rosmorduc 2015.
24 See Rosmorduc 2015.
25 Rosmorduc 2015.
26 The Informatique & Égyptologie series now has a number of volumes, the latest published volume being Polis and Winand 2013.
27 For a 1990s overview, contemporary to the quickly developing I&E series, see Greenstein 1994, with references to computer-aided history research reaching to the 1960s.
28 For an introduction see Bergman 1990.
29 http://aaew.bbaw.de/tla/
30 http://ramses.ulg.ac.be/

debated.[31] Artificial Intelligence appeared in the discussion as early as 1986.[32] Digital organization of archaeological data has been addressed since the 1980s as well.[33] Since 1989, specialist presentations on relational databases and data management[34] were discussed by Dag Bergman and Nigel Strudwick.[35] Strudwick and Adams presented an example of a relational database oriented to Egyptian iconography—a model of a viable research infrastructure that was developed in discussions at the I&E. They were also clear that in thinking about information, the digital turn is a new generation, but does not constitute a radical change in work with organized data. "A database is a collection of related data. Not all databases are in computers: the telephone book, dictionary, and encyclopedia are all databases of sorts ... Computerized databases have the advantage of being searched quickly and changed easily. More importantly, they can be organized in several different ways ... creating indexes."[36] At present, this quote may seem superfluous because it apparently states the obvious. Nonetheless, the "manual database system"[37] has been a useful precursor, especially if a manual database had already been well organized.

Equally, it is important to note that past data do not become obsolete with the digital turn—indeed, organized data in bibliographies or encyclopedias can be and are being used as excellent starting points for complex digital infrastructures.[38] Similarly, museum catalogues and registers are incorporated in new digital presentations of museum collections, again growing incrementally.[39]

A step toward the idea of an interconnected information system or infrastructure was proposed by Fathi Saleh,[40] who proposed a universal coding sys-

31 Cf. Strudwick 1987.
32 Marx 1988.
33 Sinclair and Troy 1988.
34 At that point, a large part of discussion on databases in historical research was also oriented at the type of database approach—whether they ought to be method-oriented, or source-oriented. The distinction was spearheaded by Manfred Thaller. "By method-oriented, he means the approach adopted by historians using relational methodologies ... where the historical resource is converted into a set of strict and well-defined conceptual categories. With the source-oriented approach, on the other hand, the historian always enters the original text into the computer, only later deciding on the categories ..." Harvey and Press 1996, 190.
35 Adams and Strudwick 1990; Adams and Strudwick 2008; see also Bergman 2008.
36 Adams and Strudwick 1990, 9–10.
37 Greenstein 1994, 61.
38 See Online Egyptological Bibliography—http://oeb.griffith.ox.ac.uk/.
39 Digitization of the Journal d'entrée of the Egyptian Museum in Cairo—Kamrin 2015.
40 Saleh 1990.

tem for museum databases with Egyptian objects and a plea for a technology-assisted international Egyptological communication was made by Fekri Hassan.[41] Further advancement was proposed by Pawel Wolf[42] who discussed a shared standard of information in a rapidly growing landscape of diverse datasets and databanks. More recently, Vincent Razanajao has underscored the need for consistency in digital datasets and formats used to record and keep them.[43]

However, the tendency to establish data standards and shared points of reference did not come with the digital initiative but again has had a longer tradition. To follow these pre-digital developments, we must return deeper into the Egyptological past. An early Egyptological bibliography was introduced by the Fondation Egyptologique Reine Elisabeth as the so-called "fiches bibliographiques", available on subscription in the 1930s. Its successors became known as the Annual Egyptological Bibliography (after 1947), and the Online Egyptological Bibliography in its current digital incarnation.[44]

The organization of Egyptological knowledge in research infrastructures also has a pre-digital tradition as georeferenced infrastructure—the Topographical Bibliography. In the first decades of the twentieth century, Rosalind Moss developed an idea of a topographical bibliography. Her intellectual effort aimed at a systemization of the Egyptological information. It is remarkable that her project makes Egyptology an early participant of the making of the Information Age,[45] as her infrastructure was aiming at achieving a hub for scholars. It was discipline-specific, but a coeval and a small-size parallel to substantially larger cross-disciplinary projects of Paul Otlet or Wilhelm Ostwaldt.[46] These projects had roots in the social and cultural developments of the late nineteenth century, and did not immediately succeed, but their concepts were poised to suggest a quantitative as well as a qualitative change in research, and circulation of information in general.

Rosalind Moss developed her concept of a topographical bibliography as a categorization of sites, monuments, parts of monuments and finds organized according to sites. The Topographical Bibliography has been developed to include a geo-referencing aspect from its outset, recently fully used in its present incarnation with GIS.[47]

41 Hassan 2007.

42 Wolf 1993.

43 See the Book of Abstracts, the International Congress of Egyptology 2019, V. Razanajao.

44 The transfer begun in 2008 at the I&E Vienna meeting, cf. Preface in Strudwick 2008.

45 Compare Wright 2014 on the topic.

46 On these see Wright 2014.

47 For development of the Topographical Bibliography in Digital Topographical Bibliogra-

The first volume of the Topographical Bibliography was dedicated to the Theban area and appeared in 1927, being followed to date by eight others, moving from geographically situated archaeological locations to unprovenanced finds. The first book was in the making for nearly three decades and required a painstaking collection of bibliographical data, but also an organization according to a geographical-spatial key.

The description of monuments gradually gained more granularity. The 1960 revised edition of the first Theban volume "differs in many respects from its predecessor, and the scope has been enlarged to provide a brief description of all scenes in accessible tombs, many still unpublished, together with tomb-plans, and maps showing their position in the necropolis."[48] The scope of included documentation and records pertaining to individual monuments had also developed: "Besides the references to new publications, it has been thought worthwhile to include certain important series of photographs, notably those taken by Harry Burton for the New York Metropolitan Museum of Art, by the Chicago Oriental Institute, and by Professor Siegfried Schott."[49] Both published and unpublished archive documentation was increasingly included in the concept of this dataset. It was an adventurous mind-set, integrating the

---

phy see http://www.griffith.ox.ac.uk/. Regarding the potential of georeferencing itself, it is widely recognized in archaeology that "GIS programs combine maps and data about maps in ways that bring significant benefits to archaeology. The data about the maps are of two kinds, standard relational data tables (with information about artifacts, flora, fauna, etc.) linked to areas or points on maps, and information derived from map data, such as the steepness of the grade in a given area (from contour lines or point-source elevation data). The crucial benefit of GIS is the connection between bounded portions or individual points on a map and data about them—and the ability to analyze the data according to any of the available criteria. The resulting ability to analyze material remains in concert with the physical environment is extremely powerful" (H. Eiteljorg in Schreibman, Siemens, and Unsworth 2004), see also http://digitalhumanities.org:3030/companion/view?docId=blackwell/9781405103213/9781405103213.xml&chunk.id=ss1-2-2&toc.depth=1&toc.id=ss1-2-2&brand=9781405103213_brand.

Or, in other words "A GIS is a combination of a database and a computer mapping system in which every item of data is georeferenced to give it a real-world location." The potential to use with corpora of finds, or with text corpora, is evident: "This structure offers a number of advantages: it allows the researcher to explore the database by location to ask questions such as What is here? and What is near here?; it allows data to be mapped to summarize the geographies that the database contains; it allows data from different sources to be integrated because all data are underlain by real-world coordinates; and it provides a platform for spatial analysis, a form of statistical analysis in which the locations of the items under study are explicitly included with the analysis (Gregory, Cooper, Hardie and Rayson 2015, 152)."

48 Moss, in Burney et al. 1960, vii.

49 Moss, in Burney et al. 1960, vii.

field and the archive, and not fully implemented until today, as diverse funding, organizational, and institutional settings are not always in position to facilitate a practical and prompt accessibility of emerging fieldwork material, ideally layered with archival data.

These historical, or legacy, records of Egyptology are also an important historiographical resource that feeds into history of Egyptology, its self-fashioning and its public image and reflections. Records preserved in the archives document a visual and written culture of Egyptology[50] influenced by a network of circumstances, interests, and audiences.[51] The digital datasets including legacy records are in a good position to be effective tools enabling access to a combined body of historical and contemporary evidence.[52] A practical example of making unpublished research archives available digitally, building on Moss's legacy, is the Howard Carter archive presented in *Tutankhamun: Anatomy of an Excavation*,[53] a comparable well-organized project is the *Digital Giza*.[54]

The infrastructures, be it the dictionary or the bibliographies, were and are growing organically, in a good practice identified later as crucial for a successful infrastructure: "Infrastructure development and take up is far more successful if it emerges from researchers own practices: if it fills gaps in existing provision, or it is a solution to identified problems and perceived difficulties. Infrastructure will be taken up if it is seen to be integral to the achievement of research goals."[55]

4 A Case for a Secondary Epigraphy Infrastructure—Identifying a Corpus

Is there a case for a secondary epigraphy infrastructure that would be "seen to be integral to the achievement of research goals"? There is, after all, the Topographical Bibliography with site-specific date on rock texts, inscriptions, and select corpora of secondary epigraphy, especially if containing hieroglyphic texts. There is also Trismegistos—"an interdisciplinary portal of papyrologi-

50 Borrowing from the title of a book by John Baines (*Visual and Written culture*, Oxford 2007).

51 Aptly summed up by Lewis 2016, 9.

52 Compare digital collections such as the Griffith Institute Archive http://www.griffith.ox.ac.uk/archive/, or the Digital Giza http://giza.fas.harvard.edu/search/.

53 http://www.griffith.ox.ac.uk/discoveringtut/

54 Cf. above, http://giza.fas.harvard.edu/search/.

55 Anderson 2013, 11.

cal and epigraphical resources."[56] These are valuable resources in operation as of the time of writing, and it is to be hoped that they will remain cornerstones of Egyptological information landscape, linked also to any secondary epigraphy infrastructure that may be developed later. However, they were not intended to cover fully the area of secondary epigraphy in its specific character, namely: strong relation to the site, location, and placement within the monument (requiring georeferencing and visualization), or long-term multilingual character, requiring metadata and digital presentation of texts from multiple languages. At the same time, the need to relate and link information in the Topographical Bibliography or Trismegistos (not to mention other datasets) is also evident, as the following case for SEE will show.

Secondary epigraphy in Egypt has been studied since the early days of Egyptology. The following paragraphs will introduce but a few milestones in its analysis.[57] In the 1890s, Wilhelm Spiegelberg noted that a set of New Kingdom documents relating on Western Thebes indicated that there must have lived a community of specialized artists and artisans. He also noted that there were many inscriptions in the West Theban hills and suggested that these texts and figures could have been related to West Theban communities.[58] He saw graffiti of Western Thebes as a corpus with local historical significance. In the 1920s, Jaroslav Černý followed Spiegelberg's idea and began a long-term, systematic "graffiti hunt" in Western Thebes.[59] Although site-specific graffiti editions were by that time nothing new, the idea that there was a corpus with a particular, localized, historical content that could be connected spatially to other local resources had not yet been implemented elsewhere.

In the 1930s, Battiscombe Gunn observed that the graffiti in the Step pyramid complex at Saqqara were part of a larger corpus that spread across Egypt, and contained repetitive formulae, which changed over time. Gunn thus set a schema for perceiving graffiti as a corpus substantiating social and cultural history evolution. In the Eighteenth Dynasty, something of a historical interest could be seen, contrasting with later times, when prayers and devotional concerns took the front seat.[60] In the 1970s, Ricardo Caminos noted that graffiti

56 Gülden 2008; Verreth 2016.

57 See further Navratilova 2011, also outline of research in Navratilova 2015, and upcoming volume of conference proceedings "Clamour from the Past".

58 Spiegelberg 1895 and 1898.

59 First articulated the idea to Lexa in 1926; the IFAO mission led by B. Bruyère accepted this research plan. On early research plans of Černý see Navratilova 2021.

60 Griffith Institute Archive, University of Oxford Battiscome Gunn Collection, Gunn Mss. XIII.3.

added on the surface of Egyptian monuments "cannot be disregarded." Furthermore, "They often are, or may later prove to be, of highest significance and interest historically, philologically, and in many other ways. They may consist of writing alone, of figures, or a combination of both; they occur in a variety of languages and scripts; and as regards time range, they may be as old as the monument itself and as recent as today."[61] Yet again, graffiti were perceived as an eloquent witness of Egyptian history, locally as well as comprehensively. Caminos was also very sensitive to the context of graffiti: they belonged to their wall, and the wall belonged to them.

These Egyptologists contributed significantly to the study and methodology of study of epigraphic features, and largely took the term "graffiti" for granted. However, the "graffiti" are a complex category, defying a technical description or a strict categorization.[62] In 1999, Richard Parkinson defined text graffiti as "an integral part of Egyptian writing practice and of official culture."[63] In addition, figural graffiti have a wider appeal.[64] They appear everywhere where humans have passed.

The numbers are large: We are looking at hundreds of sites and thousands to tens of thousands of texts and figures, published and unpublished. Essentially every site in Egypt comes with secondary epigraphy attached. The historical examples of graffiti corpora mentioned above referred to texts and figural images, to hieroglyphs as well as hieratic and Demotic texts, and to locations involving temples, tombs, and natural features such as rock cliffs, the latter being part of a landscape substantially shaped and marked by human hand. Eventually, secondary markings appear on statues and stelae.

To name such a diverse material by one name, be it "graffiti," or "secondary epigraphy," may be rather bewildering, especially as neither a precise location nor a technique of execution can be agreed to prevail and characterize graffiti.[65] Yet, the unifying name also has a certain legitimacy. It comprises epigraphical features that are not part of a primary decorative scheme (if such there be), but have been added as derivative, auxiliary, subsequently, even repeatedly as a reaction to a site, and as a factor in change and adaptation of the historical sites.[66] The sites have a life, and edifices have a life cycle, hence by exploring

61 Caminos 1976, 20–22.

62 For Egyptian material see Cruz-Uribe 2008a and b; Navratilova 2010. In more general terms contributions in Taylor and Baird 2011, also Keegan 2014.

63 Parkinson 1999, 92.

64 Staring 2018; Pelt and Staring 2019.

65 Navratilova 2010. Salvador 2020.

66 Ragazzoli 2013; 2017; Frood 2013.

graffiti we are looking for more than just an epigraphy record, but for a biography of a place. But it is the epigraphy record and its data that are our starting points.

Graffiti are written in a range of scripts and a multitude of languages, corresponding to multiple written cultures that have come in and out of being, sometimes overlapping, sometimes consecutively, in the geographic region of the Egyptian Nile valley since pharaonic times. Often by necessity (cf. Caminos 1976), diverse groups of texts and figures on one and the same monument might have been published in separate editions and often without an explicit relation to their location.

Using an example of the temples at Abydos, it is striking that information on graffiti, their readings and spatial analysis of their location is dispersed across a number of publications, which, though they generally acknowledge the presence of graffiti, give no details of number, place, contents, etc. In addition, graffiti in Abydos (as in many Egyptian sites) also have unpublished, legacy records, largely in the Griffith Institute Archive, Oxford (Sayce, Gunn and Černý MSS in the Griffith Institute Archive, University of Oxford), and the Dumbarton Oaks Archive, Washington, DC. The case of Abydos illustrates a more general problem. Egyptian material from one site is frequently split among several disciplines. Hieroglyphic to Demotic graffiti, are studied by Egyptologists and Demotists, Greek and Roman graffiti by Classicists, and so on.[67]

Also, even existing epigraphy databases typically focus on one language, script, or site, and do not integrate evidence from different periods.[68] Whilst this is understandable due to practical demands and disciplinary boundaries, it also inadvertently blurs the historical continuity of the existence of a site. A corpus that would be dedicated specifically to secondary epigraphy would benefit the understanding of this continuity, and the indications offered by secondary epigraphy for the study of social and cultural history as well as localized microhistory of sites or buildings.

5 SEE—A Digital Research Infrastructure

The SEE therefore proposes a different concept compared to the proposal of a visitors' graffiti database[69] as it is multilingual and multicultural and enables

67 for Abydos see Crum, Milne and Murray 1904; Perdrizet and Lefebvre 1919; Gunn in Frankfort 1933; Kornfeld 1978; Rutherford 2003; Bucking 2014; Westerfeld in Choat and Giorda 2017.

68 Compare e.g., Packard Humanities Institute database, https://epigraphy.packhum.org/.

69 Navratilova 2004.

mapping of cultural and religious change, as opposed to an idea of a specific limited text corpus. The previous proposal, albeit it would have enabled a work with a significant corpus (as its embryo phase on spreadsheets had begun to do),[70] would not have provided fully for the cultural biography aspect of study of monuments and landscapes.[71] In the Thaller definition of dataset approaches, the visitors' graffiti database was method-oriented, categorizing a particular set of texts, whereas SEE is intended as a source-oriented infrastructure.

The SEE team believes that there is an urgent need to systemize and synergize the somewhat fragmented ongoing research in secondary epigraphy. Its ontology and methodologies have evolved, but standards for recommended and best practices in recording, archiving, and publication are needed. Also needed are a close attention to individual case studies, and a research practice that is flexible and open to changes in paradigm (as was demonstrated by modern graffiti studies).[72] There is a large body of ongoing graffiti research in Classical studies, using the concepts of materiality of text, and concepts based on or similar to an "anthropology of the text,"[73] embedding the graffiti and graffiti writers in their social world. Regionally based projects also attempt to provide a comprehensive map of textual and visual communication within a geographical region.[74] The secondary epigraphy across Egypt has a comparative perspective to offer, and its impact goes beyond the study of the ancient world. There is in addition a future scope for including of modern Arabic and European epigraphy in Egypt. The tools offered by digital humanities may provide a practical approach.

A pooling of research information on secondary epigraphy among different disciplines that study material from Egypt will make it possible to develop research networking that allows for a timelier circulation of research data, as well as for a consolidation and development of standards. SEE proposes a knowledge-based digital resource with legacy and contemporary records to deliver up-to-date research information, i.e., a research infrastructure. The research infrastructure software and interface solutions will be addressed, integrating the EpiDoc description scheme,[75] and geographical coordinates, with

70 Cf. Navratilova 2006.

71 Articulated by the outcomes of the workshop Walking Dead II, October 2019, Cairo, the Ministry of Antiquities, with a forthcoming publication in 2022.

72 Samutina and Zaporozhets 2015, for instance in interpretive changes of reading graffiti as articulating subversion vs. graffiti articulating establishment values or debating both.

73 So defined by Hilgert 2017.

74 Seidlmayer et al. 2013.

75 EpiDoc is a tool for encoding scholarly and educational editions of ancient documents. It

a URL for each graffiti group or individual graffito and information about its spatial location. The spatial aspect has proved crucial for analysis of secondary epigraphy—the practice is best evaluated in a highly contextualized approach.

As the research resource will contain data that are also open to change with ongoing archaeological and epigraphical work, differentiated levels of description are proposed. The epigraphic features are proposed to be presented with three levels of description, depending on available data and allowing for a modular work on individual sites and monuments.

Level 1 concerns presence and quantity of secondary epigraphy on site, and indication on approximate historical periods, scripts and languages represented in the body of secondary epigraphy evidence. Level 2 concerns individual monuments and adds location and placement of the secondary epigraphy features and their spatial relations if applicable. Level 3 is a description of individual graffiti or graffiti groups and will include a digital edition in EpiDoc. This level may be fully included only for published graffiti for which the copyright and other legal issues have been adequately addressed. In all cases bibliography of graffiti on the site and in any given monument should be included, with links to existing specialized bibliographies. The sites and monuments or individual texts should be also linked to other relevant research infrastructures, including but not limited to Epigraphic Database Heidelberg, EAGLE, Topographical Bibliography, and the Pelagios platform.

6 Technical Summary

SEE is proposing to set up a research infrastructure that is open on both the epistemological and the technological side. The epistemological side contains several historical and philological disciplines—Greek, Egyptian, Coptic, and potentially Arabic material would be made accessible. Technically, the concept envisages API communications with other datasets, as well as an alignment with the Thot project[76] to join the proposed semantic web of Egyptian cultural heritage linking different thesauri.

There are three levels of description in the research infrastructure using XML as the main descriptive language, enabling preservation of data in a non-proprietary form. The ideal technical solution was proposed by the project

uses a subset of the Text Encoding Initiative's standard for the representation of texts in digital form. See https://sourceforge.net/p/epidoc/wiki/Home/ and Bodard 2010.

76 In more detail http://thot.philo.ulg.ac.be/project.html and see Polis and Razanajao 2016.

partner, King's College Digital Lab in London, along the following guidelines:[77] Using the open-source web framework Django, King's Digital Lab would create an online resource that facilitates the creation, storage, and discovery of the three levels of data. Django will interface with the platform EFES (EpiDoc Front End Services), a readily customizable platform for publication and search/indexing of EpiDoc files, based on the Kiln platform developed and maintained by KDL. An online interface will allow team members to enter level 1 & 2 data directly into the resource and KDL will explore the possibility of building an EpiDoc editor in Django to enable direct creation of inscription editions, too. Users will be able to drill down through the levels of data using either faceted browse functionality or map visualizations. Inscription texts will be displayed in diplomatic and edited versions and the underlying XML will also be viewable. The qualities of being user friendly, and open, and linkable, are essential, and in terms of Egyptological context, the aim would be to liaise with the Topographical Bibliography as one of the project's priority goals, but other epigraphy-oriented databases and relevant projects are equally important.

7 Why SEE Matters?

The research infrastructures are still perceived as both a national and an international asset. Anderson listed a Taiwanese, Canadian, and European approach, all coinciding in their emphasis on competitiveness in research and success in economy as motivations for research infrastructures—which is not necessarily boding well for infrastructures in humanities.[78] Thus Anderson: "the European Commission established the European Strategy Forum on Research Infrastructures (ESFRI) to create a roadmap that would guide significant funding for research infrastructures. The Roadmap says that the aim is 'to develop the scientific integration of Europe and to strengthen its international outreach' with the goals of keeping Europe at the forefront of scientific and technological innovation, to help drive economic development. There is also an underlying political aim to support and enhance European unity and to promote a sense of European unity internationally."[79] These are bold aims, and not ones that would always prove helpful to building and developing infras-

77 With many thanks for technical outline to Paul Caton, Pamela Mellen, Ariana Ciula, and Samantha Callaghan, King's College Digital Lab, London.

78 Barker et al. 2012.

79 Anderson 2013, 5.

tructures in the humanities, because they may set short-term, as opposed to long-term, goals, and may impose directives not allowing for an organic growth of the infrastructure.

Consequently, it is fast becoming a trend to build large and small infrastructures with a defined thematic target, to create a tangible, measurable asset.[80] Yet, it is also needed to interconnect these datasets, or to build such infrastructures that are designed purposefully as overcoming disciplinary boundaries. The Thot initiative very helpfully suggested building standards these infrastructures should ideally use in Egyptology.[81]

The core elements of the SEE project—a proof of concept for an Egyptian graffiti research resource, and an identification of recording standards and an analytical and interpretive framework for secondary epigraphy—would answer pressing issues regarding multifaceted Egyptian material in particular, but also research on secondary epigraphy in general. Its target is also to overcome disciplinary boundaries. The vulnerability of secondary epigraphy on site as well as of modern records of it intensifies the need to build an accessible and reliable digital resource, concentrating diverse evidence and references at one port of call.

The project concept has an unusually long chronological scope (from 3000 BC to 400 AD, with potential for extension), and it involves complex textual and visual material. It aims to extend and develop the scope of earlier studies which focused narrowly on graffiti in one particular language by offering a broader, synoptic approach arranged on the basis of geographical context. Its general applicability as a case study goes beyond individual Egyptian locations, fostering interdisciplinary communication relevant for an understanding of history and changes of archaeological sites.

The appeal of the project is to international scholars including Egyptian scholars as key stakeholders and decision makers. The compilation of a significant dataset within the research resource pilot would improve accessibility of material for Classicists, archaeologists, and Egyptologists, and would allow them to explore the geolocation and visualization of the material, offering a

80 Compare Egyptological and related projects, such as Lowe 2016; Dilley 2017; Kalchgruber and Hudáková 2017; Töpfer 2018; Reggiani 2019 to name but a few projects in the last five years. A number of projects was presented in Polis, Winand and Gillen 2013; Hafemann 2003; a Topographical Bibliography-inspired approach but applied on a single site Buongarzone 2003; an outline of earlier text databases focused on religious texts was offered by DuQuesne 2001.

81 Polis and Razanajao 2016 and see following note. An earlier call for standardization Plas 1996.

succinct presentation comprehensible across disciplines. A range of specialized subjects would benefit, from archaeology to anthropology, and to ancient history.

The concept would build on and expand the legacy of existing e-infrastructures and contribute to further integrated use and development of "big and deep data" in ancient history. Existence of a multilingual accessible corpus of secondary epigraphy, showcasing its applicability in historical research, also opens up the possibility of involving communication and media specialists in their analysis. It should also become a helpful resource for future academic activity—from undergraduate studies to advanced research in ancient history, including in context of archaeological fieldwork. Eventually, the project could offer a methodology applicable to other graffiti corpora not only within, but beyond the ancient world, as secondary epigraphy has been produced in all historical periods including the present.

References

Adams, Ernest W., and Nigel Strudwick. 1990. "Relational Database Design: A Tutorial and Case Study for Egyptologists." *Informatique et Égyptologie* 7: 9–24.

Adams, Ernest W., and Nigel Strudwick. 2008. "Relational Database Design: A Tutorial and Case Study for Egyptologists." In *Information Technology and Egyptology in 2008: Proceedings of the Meeting of the Computer Working Group of the International Association of Egyptologists (Informatique et Egyptologie), Vienna, 8–11 July 2008*, edited by Nigel Strudwick, 183–207. Piscataway, NJ: Gorgias.

Anderson, Sheila. 2013. "What Are Research Infrastructures?" *International Journal of Humanities and Arts Computing* 7 (1–2): 4–23. https://doi.org/10.3366/ijhac.2013.0078.

Anderson, Sheila, and Tobias Blanke. 2015. "Infrastructure as Intermediation—from Archives to Research Infrastructures". *Journal of Documentation* 71 (6): 1183–1202. https://doi.org/10.1108/JD-07-2014-0095.

Anonymous 1948. "Nouvelles". *Chronique d'Égypte* 23 (45): 102–105.

Baird, J.A., and Claire Taylor. 2011. *Ancient Graffiti in Context*. New York; London: Routledge.

Barker, Elton, Chris Bissell, Lorna Hardwick, Allan Jones, Mia Ridge, and John Wolffe. 2012. "Colloquium: Digital Technologies: Help or Hindrance for the Humanities?" *Arts and Humanities in Higher Education* 11 (1–2): 185–200. https://doi.org/10.1177/1474022211428311.

Bergman, Dag. 1990. EDOC—*Egyptological documentation on computer*. S.l.: s.n.

Bergman, Dag. 2008. "Using Relational Databases at the Beginning of the 21st Century."

In *Information Technology and Egyptology in 2008: Proceedings of the Meeting of the Computer Working Group of the International Association of Egyptologists (Informatique et Egyptologie), Vienna, 8–11 July 2008*, edited by Nigel Strudwick, 209–219. Piscataway, NJ: Gorgias.

Bodard, Gabriel. 2010. "EpiDoc: Epigraphic Documents in XML for Publication and Interchange." In *Latin On Stone: epigraphic research and electronic archives*, edited by Francisca Feraudi-Gruénais, 101–118. Maryland: Lexington Books.

Bodenhamer, David J., John Corrigan, and Trevor M. Harris. 2015. *Deep Maps and Spatial Narratives [Electronic Resource]*. Spatial Humanities. Bloomington, Indiana: Indiana University Press.

Bodenhamer, David J., Trevor M. Harris, and John Corrigan. 2013. "Deep Mapping and the Spatial Humanities." *International Journal of Humanities and Arts Computing* 7 (1–2): 170–175. https://doi.org/10.3366/ijhac.2013.0087.

Brown, Marina Wilding. 2017. "Agents of Construction: Ancient Egyptian Rock Inscriptions as Tools of Site Formation and Modern Functional Parallels." *Journal of Egyptian History* 10 (2): 153–211. https://doi.org/10.1163/18741665-12340038.

Bucking, Scott. 2014. "Now You See It, Now You Don't: The Dynamics of Archaeological and Epigraphic Landscapes from Coptic Egypt." In: *Archaeologies of Text*, edited by Matthew Rutz and Morag Kersel, 59–79. Oxford: Oxbow Books.

Buongarzone, Roberto. 2003. "Archaeological documentation and Egyptological database." In *The North Saqqara archaeological site: handbook for the environmental risk analysis*, edited by Fabrizio Ago, Edda Bresciani, and Antonio Giammarusti, 122–131. Pisa: Plus, Università di Pisa.

Burney, Ethel W., Bertha Porter, and Rosalind L.B. Moss. 1960. *Topographical Bibliography of Ancient Egyptian Hieroglyphic Texts, Reliefs, and Paintings I: The Theban Necropolis. Part 1: Private Tombs*. 2nd revised. Oxford: Oxford University Press; Griffith Institute.

Caminos, Ricardo A. 1976. "The Recording of Inscriptions and Scenes in Tombs and Temples." In *Ancient Egyptian Epigraphy and Palaeography*, edited by Anonymous, 1–25. New York: Metropolitan Museum of Art.

Charle, Christophe, Jürgen Schriewer, and Peter Wagner. 2004. *Transnational Intellectual Networks: Forms of Academic Knowledge and the Search for Cultural Identities*. Frankfurt: Campus.

Crum, W.E., J. Grafton Milne, and Margaret A. Murray. 1904. *The Osireion at Abydos*. Vol. 9. British School of Archaeology in Egypt and Egyptian Research Account. London: Bernard Quaritch. http://digi.ub.uni-heidelberg.de/diglit/murray1904.

Cruz-Uribe, Eugene. 2008a. *Graffiti (Figural)*. Edited by Willeke Wendrich. https://escholarship.org/uc/item/7v92z43m.

Cruz-Uribe, Eugene. 2008b. *Hibis Temple Project Volume 3: The Graffiti from the Temple Precinct*. San Antonio, TX: Tambotempa; Van Siclen.

Darnell, John Coleman, Deborah Darnell, Renee Friedman, and Stan Hendrickx. 2002. *Theban Desert Road Survey*, Vol. 1, Chicago: Oriental Institute.

Dilley, Paul C. 2017. "The Inscriptions of Late Antique Egypt (ILAE) database: digitizing textual culture." In *From Gnostics to Monastics: studies in Coptic and early Christianity in honor of Bentley Layton*, edited by David Brakke, Stephen J. Davis, and Stephen Emmel, 245–256. Leuven: Peeters.

DuQuesne, Terence. 2001. "Dictionaries and databases of Egyptian religious texts." *Discussions in Egyptology* 50: 15–24.

Foka, Anna, Anna Misharina, Viktor Arvidsson, and Stefan Gelfgren. 2018. "Beyond Humanities qua Digital: Spatial and Material Development for Digital Research Infrastructures in HumlabX." *Digital Scholarship in the Humanities* 33 (2): 264–278. https://doi.org/10.1093/llc/fqx008

Frood, Elizabeth. 2013. "Egyptian Temple Graffiti and the Gods: Appropriation and Ritualization in Karnak and Luxor." In *Heaven on Earth: Temples, Ritual, and Cosmic Symbolism in the Ancient World*, edited by Deena Ragavan, 285–318. Chicago: The Oriental Institute.

Gertzen, Thomas L. 2013. *École de Berlin Und "Goldenes Zeitalter" (1882–1914) Der Ägyptologie Als Wissenschaft: Das Lehrer-Schüler-Verhältnis von Ebers, Erman Und Sethe*. Berlin: de Gruyter.

Grajetzki, Wolfram. 2008. "Review of Darnell et al. Theban Desert Road Survey, volume 1 (Chicago 2002)." *Journal of Egyptian Archaeology* 94: 312–314.

Greenstein, Daniel I. 1994. *A Historian's Guide to Computing*. Oxford Guides to Computing for the Humanities. Oxford: Oxford University Press.

Gülden, Svenja A. 2008. "Trismegistos: An Interdisciplinary Portal of Papyrological and Epigraphical Resources." In *Information Technology and Egyptology in 2008: Proceedings of the Meeting of the Computer Working Group of the International Association of Egyptologists (Informatique et Egyptologie), Vienna, 8–11 July 2008*, edited by Nigel Strudwick, 17–28. Piscataway, NJ: Gorgias.

Hafemann, Ingelore (ed.). 2003. *Wege zu einem digitalen Corpus ägyptischer Texte: Akten der Tagung "Datenbanken im Verbund" (Berlin, 30. September–2. Oktober 1999)*. Thesaurus linguae aegyptiae 2. Berlin: Achet.

Harvey, Charles and Jon Press. 1996. *Databases in Historical Research: Theory, Methods and Applications*. Basingstoke: Macmillan.

Hassan, Fekri. 2007. "Conserving Egyptian Heritage: Seizing the Moment." In *British-Egyptian Relations from Suez to the Present Day*, edited by Ayman Ahmed El-Desouky, and Noel Brehony, 209–233. SOAS Middle East Issues. London: Saqi.

Hilgert, Markus. 2017. *Understanding Material Text Cultures: A Multidisciplinary View*. Materiale Textkulturen; Band 9. Berlin; Boston: Walter de Gruyter, GmbH.

Kalchgruber, Peter and Lubica Hudáková. 2017. "*Crowdsourcing* in Egyptology: images and annotations of Middle Kingdom private tombs." In *Proceedings of the XI Interna-*

tional Congress of Egyptologists, Florence Egyptian Museum, Florence, 23–30 August 2015, edited by Gloria Rosati, and Maria Cristina Guidotti, 303–307. Oxford: Archeopress.

Kamrin, Janice. 2015. "The Egyptian Museum Database, Digitizing, and Registrar Training Projects: Update 2012." *Bulletin of the Egyptological Seminar* 19: 431–440.

Keegan, Peter. 2014. *Graffiti in Antiquity*. London; New York: Routledge.

Kornfeld, Walter. 1981. *Neues Über Die Phönikischen Und Aramäischen Graffiti in Den Tempeln von Abydos*. Wien: Verlag der Österreichischen Akademie der Wissenschaften.

Lewis, Clare. 2016. "Inaugural Lectures in Egyptology: T.E. Peet and His Pupil W.B. Emery." *Bulletin of the History of Archaeology* 26 (1): Art. 9. https://doi.org/10.5334/bha-591.

Lowe, Adam. 2016. "Tomb recording: epigraphy, photography, digital imaging, and 3D surveys." In *The Oxford handbook of the Valley of the Kings*, edited by Richard H. Wilkinson, and Kent R. Weeks, 528–543. Oxford: Oxford University Press.

Marx, Christoph. 1988. "Artificial Intelligence (AI)—Expert Systems (ES): Some Reasons and Proposals for Their Use in Disciplines of Ancient History, Particularly in Egyptology." *Informatique et Égyptologie* 4: 108–116.

Navrátilová, Hana. 2006. "The Phraseology of the Visitors' Graffiti: A Preliminary Report of the Graffiti Data Base." In *Abusir and Saqqara in the Year 2005: Proceedings of the Conference Held in Prague (June 27 July 5, 2005)*, edited by Miroslav Bárta, Filip Coppens, and Jaromír Krejčí, 83–107. Prague: Czech Institute of Egyptology, Faculty of Arts, Charles University in Prague.

Navrátilová, Hana. 2010a. "Gardiner and Graffiti." *Egyptian & Egyptological Documents, Archives, Libraries* 2: 171–186.

Navrátilová, Hana. 2010b. "Graffiti Spaces." In *Egypt in Transition: Social and Religious Development of Egypt in the First Millennium BCE. Proceedings of an International Conference, Prague, September 1–4, 2009*, edited by Ladislav Bareš, Filip Coppens, and Květa Smolárikova, 305–332. Prague: Czech Institute of Egyptology, Charles University in Prague.

Navratilova, Hana. 2015. *Visitors' Graffiti of Dynasties 18 and 19 in Abusir and Northern Saqqara: With a Survey of the Graffiti at Giza, Southern Saqqara, Dahshur and Maidum*. 2nd, revised ed. Wallasey, UK: Abercromby Press.

Parkinson, Richard. 1999. *Cracking Codes: The Rosetta Stone and Decipherment*. London: British Museum Press.

Pelt, W. Paul van, and Nico Staring. 2019. "Interpreting Graffiti in the Saqqara New Kingdom Necropolis as Expressions of Popular Customs and Beliefs." *Rivista Del Museo Egizio* 3. https://doi.org/10.29353/rime.2019.2577

Perdrizet, Paul. 1919. *Les graffites grecs du Memnonion d'Abydos*. Nancy: Berger-Levrault.

Plas, Dirk van der 1996. "Computing and standardization in Egyptology: some proposals." In *Informatica ed Egittologia all'inizio degli anni '90*, edited by Francesco Tiradritti, 165–196. Roma: Bulzoni.

Polis, Stéphane, and Vincent Razanajao. 2016. "Ancient Egyptian Texts in Context: Towards a Conceptual Data Model (the Thot Data Model—TDM)." *Bulletin of the Institute of Classical Studies* 59 (2): 24–41. https://doi.org/10.1111/j.2041-5370.2016.12036.x.

Polis, Stéphane, Jean Winand, and Todd Gillen. 2013. *Texts, Languages & Information Technology in Egyptology: Selected Papers from the Meeting of the Computer Working Group of the International Association of Egyptologists (Informatique & Égyptologie), Liège, 6–8 July 2010*. Vol. 9. Aegyptiaca Leodiensia. Liège: Presses Universitaires de Liège.

Polkowski, Paweł Lech. 2016. *Krajobraz i Sztuka Naskalna: W Palimpseście Egipskiej Oazy Dachla*. Vol. 16. Bibliotheca Fontes Archaeologici Posnanienses. Poznań: Muzeum Archeologiczne.

Quirke, Stephen. 2013. "Exclusion of Egyptians in English-Directed Archaeology 1822–1922 under British Occupation of Egypt." In *Ägyptologen Und Ägyptologien Zwischen Kaiserreich Und Gründung Der Beiden Deutschen Staaten: Reflexionen Zur Geschichte Und Episteme Eines Altertumswissenschaftlichen Fachs Im 150. Jahr Der Zeitschrift Für Ägyptische Sprache Und Altertumskunde*, edited by Susanne Bickel, Hans-Werner Fischer-Elfert, Antonio Loprieno, Sebastian Richter, and Lutz Popko, 379–405. Berlin: Akademie Verlag Berlin.

Ragazzoli, Chloé. 2013. "The Social Creation of a Scribal Place: The Visitors' Inscriptions in the Tomb Attributed to Antefiqer (TT 60) (with Newly Recorded Graffiti)." *Studien Zur Altägyptischen Kultur* 42: 269–323.

Ragazzoli, Chloé. 2017. *La Grotte Des Scribes à Deir El-Bahari: La Tombe MMA 504 et Ses Graffiti*. Vol. 135. Mémoires Publiés Par Les Membres de l'Institut Français d'archéologie Orientale. Le Caire: Institut français d'archéologie orientale.

Ragazzoli, Chloé. 2018. "Graffiti and Secondary Epigraphy in Deir El-Medina: A Progress Report." In *Outside the Box: Selected Papers from the Conference "Deir El-Medina and the Theban Necropolis in Contact" Liège, 27–29 October 2014*, edited by Andreas Dorn and Stéphane Polis, 407–420. Liège: Presses Universitaires de Liège.

Reggiani, Nicola (ed.) 2019. *Greek medical papyri: text, context, hypertext*. Archiv für Papyrusforschung und verwandte Gebiete, Beiheft 40. Berlin; Boston: Walter de Gruyter.

Rosmorduc, Serge. 2015. *Computational Linguistics in Egyptology*. Edited by Julie Stauder-Porchet, Andréas Stauder, and Willeke Wendrich. https://escholarship.org/uc/item/0fk4n4gv

Rutherford, Ian C. 2003. "Pilgrimage in Greco-Roman Egypt: New Perspectives on Graffiti from the Memnonion at Abydos." In *Ancient Perspectives on Egypt*, edited by Roger Matthews and Cornelia Roemer, 171–190. London: UCL Press.

Saleh, Fathi. 1990. "A Proposed Coding System for the Use with Databases in the Egyptology [Sic]." *Informatique et Égyptologie* 7: 99–134.

Salvador, Chiara. 2020. "Graffiti." In The Oxford Handbook of Egyptian Epigraphy and Palaeography, edited by Dimitri Laboury and Vanessa Davies, 434–447. New York: Oxford University Press. http://doi.org/10.1093/oxfordhb/9780190604653.013.30.

Samutina, Natalia and Oksana Zaporozhets 2015. http://www.soclabo.org/index.php/laboratorium/article/view/548/1417

Seidlmayer, Stephan Johannes. 2013. "Rock Inscriptions in the Area of Aswan: From Epigraphy to Landscape Archaeology." In *The First Cataract of the Nile: One Region—Diverse Perspectives*, edited by Dietrich Raue, Stephan J. Seidlmayer, and Philipp Speiser, 205–210. Berlin; Boston: De Gruyter.

Sinclair, Paul J.J., and Lana Troy. 1988. "Organizing Archaeological Data from the Scandinavian Joint Expedition: Preliminary Results of a Multivariate Analysis." *Informatique et Égyptologie* 5: 43–45.

Spiegelberg, Wilhelm. 1895. *Arbeiter Und Arbeiterbewegung Im Pharaonenreich Unter Den Ramessiden (ca. 1400–1100 v. Chr.): Eine Kulturgeschichtliche Skizze*. Strassburg: Trübner.

Spiegelberg, Wilhelm. 1898. *Zwei Beiträge Zur Geschichte Und Topographie Der Thebanischen Necropolis Im Neuen Reich: I. Der Grabtempel Amenophis' I. Zu Drah-Abu'l-Negga; II. Plan Einer Gesamtarbeit Über Die Verwaltung Der Thebanischen Necropolis Im Neuen Reich (Vortrag)*. Strassburg: Schlesier & Schweikhardt. http://digi.ub.uni-heidelberg.de/diglit/spiegelberg1898.

Staring, Nico. 2018. "Products of the Physical Engagement with Sacred Space: The New Kingdom Non-Textual Tomb-Graffiti at Saqqara." In *Decoding Signs of Identity: Egyptian Workmen's Marks in Archaeological, Historical, Comparative and Theoretical Perspective. Proceedings of a Conference in Leiden, 13–15 December 2013*, edited by B.J.J. Haring, D.M. Soliman, and K.V.J. van der Moezel, 79–112. Leiden: Peeters.

Strudwick, Nigel. 1987. *Editorial Foreword. Informatique et Egyptologie* 3 (1987), 1–3.

Strudwick, Nigel. 2008. *Information Technology and Egyptology in 2008: Proceedings of the Meeting of the Computer Working Group of the International Association of Egyptologists (Informatique et Egyptologie), Vienna, 8–11 July 2008*. Vol. 2. Bible in Technology. Piscataway, NJ: Gorgias.

Sullivan, Elaine, and Willeke Wendrich. 2008. "An Offering to Amun-Ra: Building a Virtual Reality Model of Karnak." In *Information Technology and Egyptology in 2008: Proceedings of the Meeting of the Computer Working Group of the International Association of Egyptologists (Informatique et Egyptologie), Vienna, 8–11 July 2008*, edited by Nigel Strudwick, 109–128. Piscataway, NJ: Gorgias.

Töpfer, Susanne. 2018. The Turin Papyrus Online Platform (TPOP): an introduction. *Rivista del Museo Egizio* 2. doi: 10.29353/rime.2018.1916.

Urcia, Alberto, John C. Darnell, Colleen M. Darnell, and Sara E. Zaia. 2018. "From Plastic Sheets to Tablet PCs: A Digital Epigraphic Method for Recording Egyptian Rock Art and Inscriptions." *African Archaeological Review* 35 (2): 169–189. https://doi.org/10.1007/s10437-018-9297-z.

Verreth, Herbert. 2016. "Topography of Egypt online." In *Proceedings of the 27th international congress of papyrology: Warsaw, 29 July–3 August 2013* 3, edited by Tomasz Derda Adam Łajtar, and Jakub Urbanik, 1379–1385. Warsaw: University of Warsaw, Faculty of Law and Administration, Chair of Roman and Antique Law.

Vértesz, Krisztian. 2019. "On the Crossroads between Tradition & Innovation: The Digital Evolution & Expansion of the Chicago House Method." *KMT* 30 (3): 71–79.

Westerfeld, Jennifer. 2017. "Monastic Graffiti in Context." In *Writing and Communication in Early Egyptian Monasticism*. Texts and Studies in Eastern Christianity; Vol. 9., edited by Malcolm Choat, and Mariachiara Giorda, Leiden; Boston: Brill: 187–212.

Wolf, Pawel M. 1993. "Standards Für Den Datenaustausch Zwischen Datenbanken in Der Ägyptologie." *Informatique et Égyptologie* 8: 127–136.

Wright, Alex. 2014. *Cataloging the World: Paul Otlet and the Birth of the Information Age*. New York: Oxford University Press.

CHAPTER 15

SIGSaqqâra: A Digital Project to Understand the Spatial Occupation of Saqqara

Éloïse Noc

Abstract

To understand the organization of the spatial occupation at Saqqara from the Predynastic period to the end of the Old Kingdom, a project called SIGSaqqara was launched in July 2016, supported by LabEx ArcHiMedE of Paul-Valéry Montpellier 3 University in France.[1] To carry out this study, a geographic information system (GIS) linked to a database, including forms for recording semantic data, is used. Data concerning the monuments and also the owners of the tombs are recorded to better understand the necropolis and to try to grasp the factors which determined the locations chosen for construction of its various features. After careful investigation and because we believe the data should be made available, an online publication is planned for the larger public, students and researchers, which we believe will lead to further developments in analysis of this material.

Keywords

Saqqara – Predynastic – Old Kingdom – geographic information system – (GIS) – database

1 Project Genesis

During my PhD entitled *Analyse spatiale à Saqqâra des origines à la fin de l'Ancien Empire. Les exemples des complexes funéraires de Netjerikhet et de Sekhemkhet*,[2] I treated the Saqqara spatial organization and I used the exam-

1 This project was supported by LabEx ARCHIMEDE from "Investissement d'Avenir" program ANR-11-LABX-0032–01.

2 Noc 2015.

ples of two circumscribed spaces: the funerary complexes of Netjerikhet and of Sekhemkhet.

To understand the organization of space and its layout, I recorded the data with tools specifically created for my study: a geographic information system (GIS) linked to a database. In order to analyze these funerary complexes, I recorded all the archaeological remains: structures, accesses, material, decoration and inscriptions both on architecture and on material, and human and faunal remains. Moreover, to find the information sources, a table for bibliographic references was created and another one for the pictures enabling the user to visualize the artifacts.

Thanks to the use of these tools, new elements were revealed about the circulation in the complex, on the choice of the materials of the vases according to the different periods and on the dummy doors of the funerary complex of Netjerikhet, for example.[3]

The results being thus encouraging, I wanted to extend my study to the whole site of Saqqara, because it seemed very interesting to be able to take advantage of the methods implemented as well as the tools created during my PhD.

2 Issue, Aim and Main Purposes

While Saqqara is the subject of many studies, paradoxically there is no global examination of the entire site to understand its spatial organization. Indeed, the main difficulty is that there is no archaeological map, namely a single map of the entire site with all the monuments identified. There are several reasons for that.[4] The *SIGSaqqâra* project aims to know more about the spatial organization of the site of Saqqara from the Predynastic to the end of the Old Kingdom.

To address this problem and to better understand the space and its development during this period, the *SIGSaqqâra* project has three main purposes. The first is scientific: it is necessary to identify all the monuments present during the period studied in order to record them precisely in a database and to georeference them in a GIS, from the publications (monographs, articles, reports, etc.). The second, still scientific, is to analyze the spatial data that means the location

3 The results are in the process of being published.
4 For some of the reasons, see, for example, Noc 2018, 58.

of the monuments, and the semantic or attribute data about the structures and their owners. Finally, the third, more technical, is to plan the online publishing.

The difference with my PhD work is that I focus more on the architecture of the funerary monuments built from the Predynastic Period to the end of the Old Kingdom and their owner and not on archaeological material, decoration and inscriptions discovered *in situ*.

3 Tools and Methods

Archaeology and Egyptology have appropriated IT and are developing digital tools such as databases and GIS. Indeed, these tools can be used to meet several needs in the framework of different studies, including research on networks,[5] on landscapes[6] or work on endangered heritage and sites or monuments with risks,[7] for example.

---

5 About networks, "Desert Networks" Project funded by the European Research Council and hosted by the HiSoMA research center at Lyon (CNRS) works on physical, economic and social networks in the Egyptian Eastern Desert from the mid-second millennium BC until the late third-early fourth century AD, see https://desertnetworks.hypotheses.org/ and https://desertnetworks.huma-num.fr/.

6 For the study of landscapes, in particular Saqqara, the project led by E.A. Sullivan (Associate Professor of History at the University of California, Santa Cruz), *Constructing the sacred. Visibility and Ritual Landscape at the Egyptian Necropolis of Saqqara*, is developing a GIS. It "explores how concepts of sacred space were reinvented as the built and natural landscape changed, creating new meanings as individuals and communities reimagined the form and use of the site over time" (https://constructingthesacred.supdigital.org/cts/introduction), thanks to the possibilities offered by a 3D GIS "to peel away the layers of history at the site, revealing how changes to sight lines, skylines, and vistas at different periods of Saqqara's millennia-long use" (https://constructingthesacred.supdigital.org/cover/index.html), see Sullivan 2020, https://constructingthesacred.supdigital.org/cover/index.html.

7 The project, first called "Enhancement of the Organisation and Capabilities to preserve Cultural Heritage Assets of Egypt", then synthesized into the "Risk Map for North Saqqara Site" started in July 2000, presented on 2 March 2002 in Cairo, and completed in 2003, focused on Saqqara. "The Direzione Generale per la Cooperazione allo Sviluppo (DGCS, General Directorate for Development Cooperation) of the Italian Ministry of Foreign Affairs started a supporting programme for the Egypt Environmental Affairs Agency—Technical Co-operation Office for the Environment in the ambit of the environmental programme of application of the NEAP which, for the part concerning the safeguarding of the cultural heritage, has the SCA (Supreme Council for Antiquities) as its operative counterpart" (Ago, Bresciani, GIammarusti 2003, 76). The aims of this project was to create a "computer model as a support to analysis and for proposals for the minimization of the environmental degradation which is impoverishing Egyptian archaeological sites", to aid the "SCA to improve the archaeological sites' management" and to create "an integrated system of support and development of Egyptian civil

In order to carry out this study, my work uses the digital tools—database and GIS—created during my PhD. Some criteria were considered to choose the software. For different reasons,[8] I chose *Microsoft Access*[9] and *ArcGIS*.[10] One of the reasons for choosing this software is that they are complementary and compatible. Indeed, thanks to the personal geodatabase in *ArcGIS*, the information recorded in one of these tools is directly visible in the other one.

To gather the documentation and thus record the data, I chose to create a database with a database management software. I did so because a database management software makes it possible to create forms which allow better ergonomics for the recording of data but also for their management: editing, deleting or simply for consultation. Each entity is identified by a unique identifier. Furthermore, in a database, it is possible to record entities that could not be represented in the GIS for various reasons such as the fact that the location is lost.

To locate a monument in the space, at its exact position in the site, various ancient maps are used and georeferenced. Since the seventeenth century, Saqqara has been visited by many explorers and the first map of the area was published in *La Description de l'Égypte*, between 1809 and 1822, following Napoleon's campaign. The map of "Memphis et environs" includes a large area but no structure is really recognizable (Figure 15.1).

Since then, a lot of maps and plans have been published, with varying degrees of detail, depending on the different discoveries made by the researchers working on the site, such as Lepsius, for example (Figure 15.2).

---

service for the prevention of environmental risks and the socio-economical development of the country cultural resources" (Ago, Bresciani, GIammarusti 2003, 77). Six hundred monuments were surveyed and the information about them was organized in a *Microsoft Access* database in which environmental, archaeological, tourism and conservation data could be recorded. These monuments were then located on a georeferenced map, in a GIS under *Esri ArcView*. The second phase aimed to assess the vulnerability and dangers that Saqqara and its monuments can withstand. Finally, the third phase was the actual analysis of the different risks experienced by the North Saqqara site through the exploitation and development of the GIS (Ago, Bresciani, GIammarusti 2003, 90–96). More recently, the EAMENA project focuses on the Endangered Archaeology in the Middle East & North Africa, including Roman military sites in the Eastern Desert of Egypt. "EAMENA's primary aim is to rapidly record and evaluate the status of the archaeological landscape of the MENA region in order to create an accessible body of data which can be used by national and international heritage professionals to target those sites most in danger and better plan and implement the preservation and protection of this heritage" (https://eamena.org/background-and-aims).

8 Noc 2019, 175–176.

9 https://www.microsoft.com/en-us/microsoft-365/access

10 https://www.esri.com/en-us/home

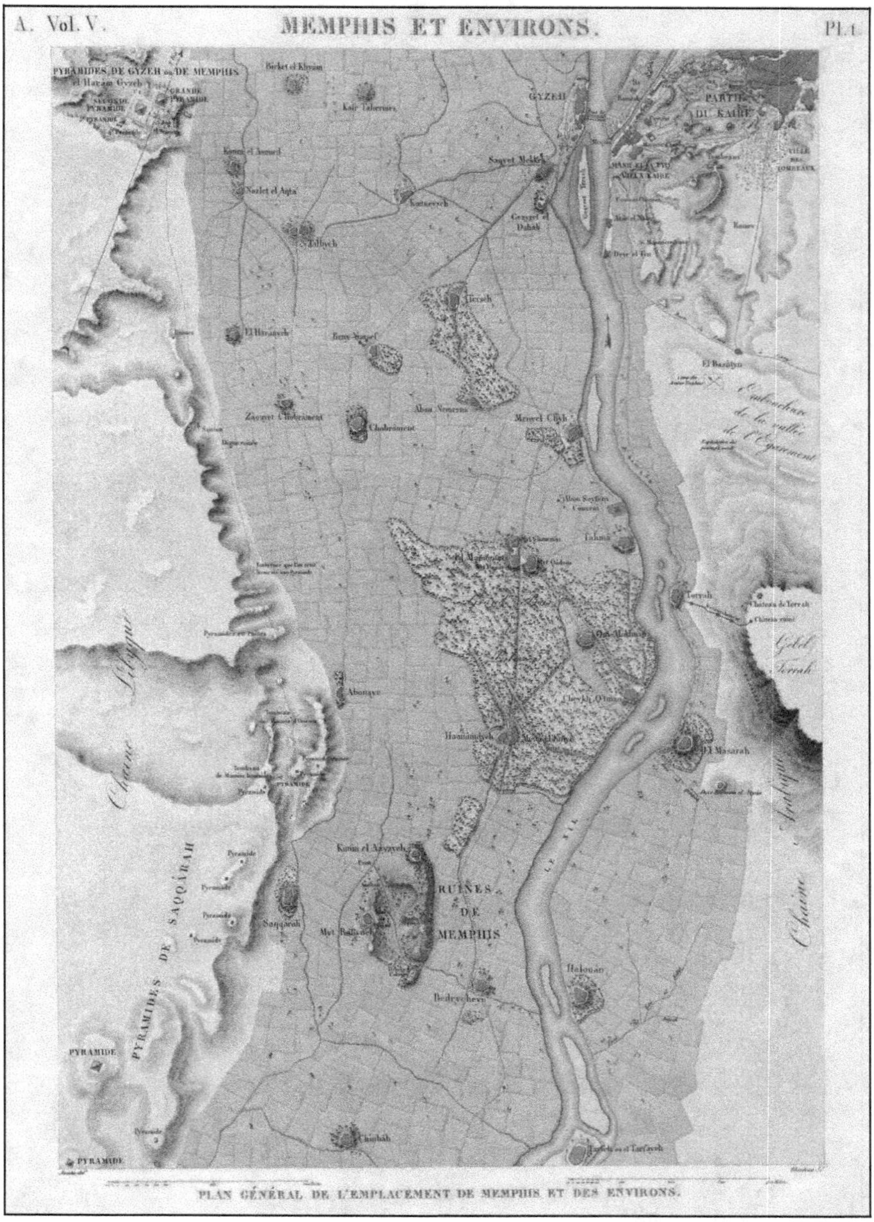

FIGURE 15.1 "Memphis et environs," *Description* 1823, pl. 1

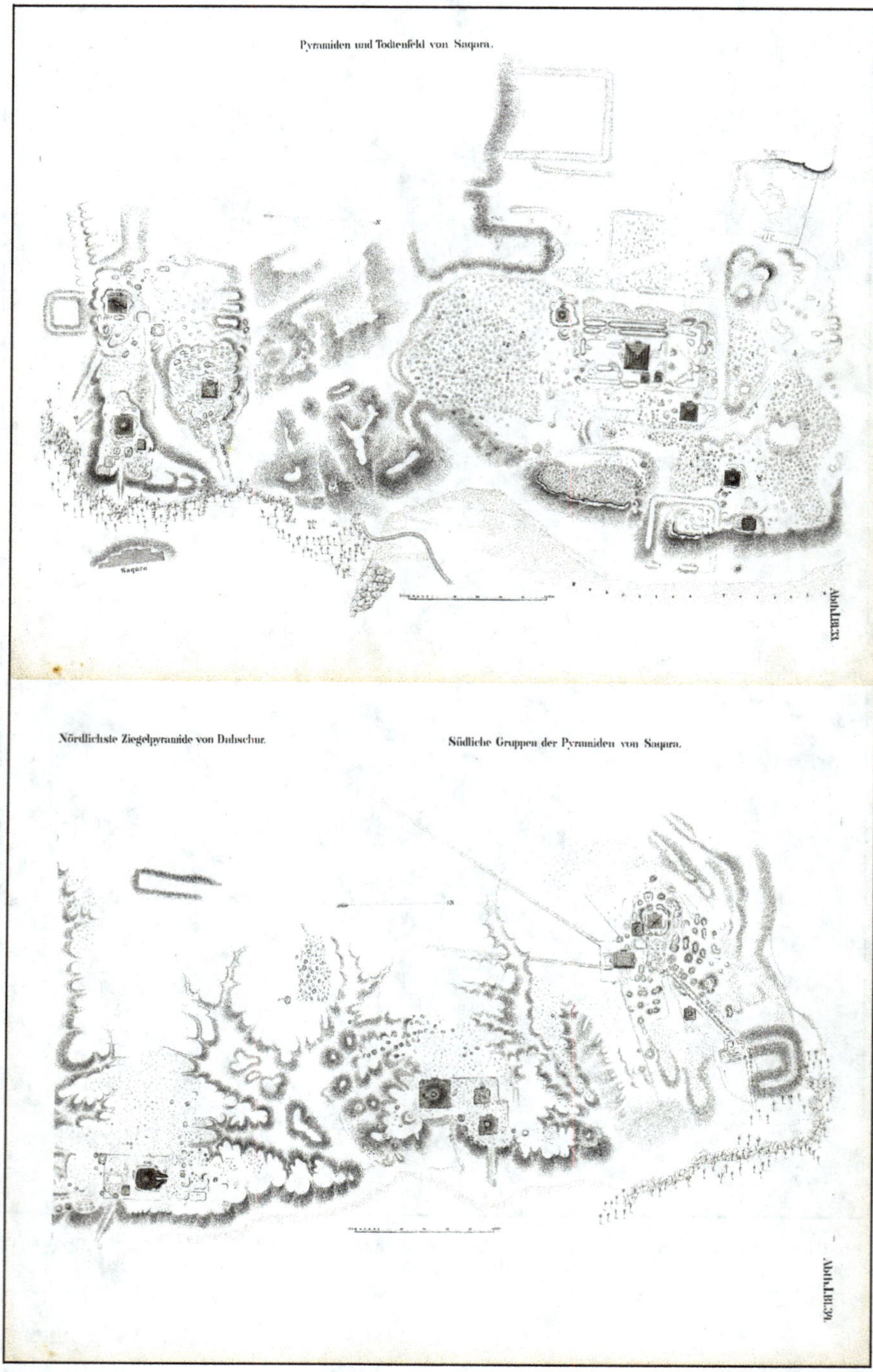

FIGURE 15.2 Lepsius 1848–1859, pls. 33–34

Gradually, North Saqqara was distinguished from South Saqqara. Maps were increasingly accurate with respect to topography, like the map made by W.S. Smith in 1936.[11] It is based on the cadastral plans of the visible monuments produced by the Survey of Egypt in 1932 at a scale of 1/2500.[12]

Several decades later, in 1974, A.J. Spencer mapped all the tombs found at North Saqqara.[13] He associated the monuments with the numbers assigned by the researchers—K.R. Lepsius, A. Mariette, J. De Morgan, J.E. Quibell, C.M. Firth and W.B. Emery—at the time of the discoveries. He also indicated the name of their owner. For burials without a number, he gave their position in relation to other known monuments. A.J. Spencer thus listed the monuments known at the time of publication. This is a first step that allows us today to have a vision of the occupation of North Saqqara. Unfortunately, except for the tomb numbers given by the discoverers and the names of their owners, no other information is available to get a deeper understanding of the spatial occupation of this area. The organization of the necropolis depends on the choices made by individuals who settled there according to various factors. To understand the space studied, it is necessary to examine the interaction of the various elements discovered. As the artifacts are part of an organized system, they are an indicator of occupation. Therefore, all data concerning the context (chronology, environment/landscape, etc.), the social aspect (social status with the titles of the deceased and family relationships, etc.) and also the architecture of the tombs (material used, interior layout, etc.), etc. must be studied.

During the Survey of Egypt in 1977, aerial photogrammetry was produced at a scale of 1/5000 and carried out in 1978 by the SFS/IGN France Consortium for the Ministry of Housing and Reconstruction of the Arab Republic of Egypt. For the Saqqara area, these maps are H22 for North Saqqara and H23 for South Saqqara.[14]

4 Georeferencing and Overlay

With their geographic coordinates, these latter maps are used to georeference plans without geographic coordinates. I use the H22 map in particular because

11 Reisner 1936, carte 2.

12 These cadastral plans are preserved in the Institut français d'archéologie orientale (IFAO).

13 Spencer 1974, tab. 1.

14 These maps are preserved in the IFAO. H22: n° 61396–65429 and H23: n° inv. 56181-56182-65428. They are reproduced in Bárta and Brůna 2006, 29 and 31.

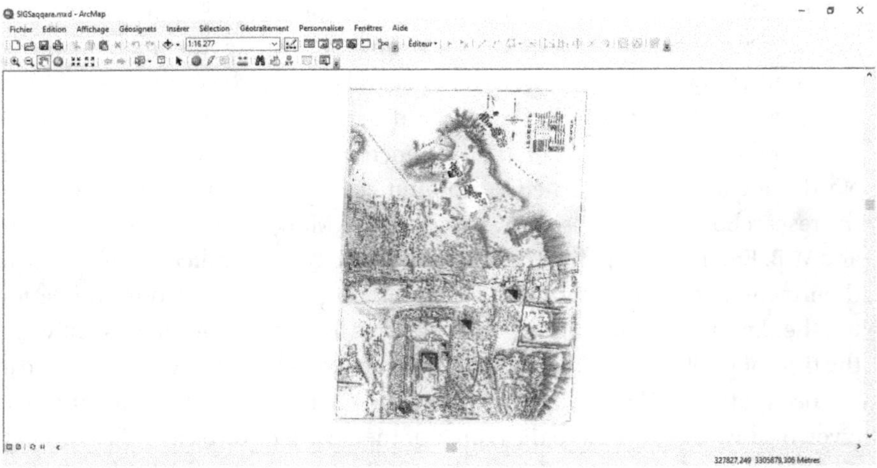

FIGURE 15.3 Overlay in *ArcGIS* of plans and maps in *ArcGIS*

my work is more focused on North Saqqara. De Morgan's map,[15] which does not have any geographic coordinates, has been georeferenced on map H22 like Quibell's two plans,[16] for example (Figure 15.3).

Thanks to the overlay of all these maps, some of them being georeferenced, it is possible to identify the location of a monument with precision, which could then be cross-referenced with the data of surveys made *in situ*.

5 Vector Data

Then, it is thus possible to represent the monuments in the GIS software. To draw an entity, we can choose a point, a line or a polygon according to the scale of the study. For the *SIGSaqqâra* project, I chose to draw each monument as a polygon (Figure 15.4). 494 monuments are georeferenced in the GIS out of 716 recorded in the database.[17]

15 De Morgan 1897, map section 10.
16 Quibell 1923, pl. I and II.
17 Status of data record: March 2020.

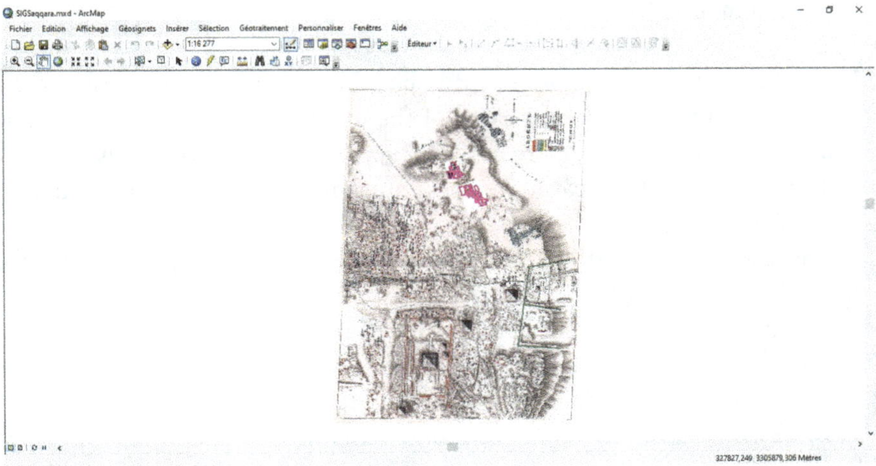

FIGURE 15.4 Vector data: in pink, monuments drawn as polygons on Quibell's plans

6 Semantic Data

In the database, the data are classified by type of entities. These are the tables. In each table, fields are created to record all the data useful for analysis. For this project, I wanted to record the monuments with the numbers attributed by the discoverers, the types of tombs and the materials, among others.

The owners are also considered with their names and titles written on the walls of the monuments. The hypotheses about the filiation are also recorded. Images are also visible with their names, for example. Finally, to find the sources of the information, the bibliographic references are also still recorded in the table "Bibliography" with its own fields. These semantic data are linked to the geometrical data in the GIS, thanks to the unique identifier of each entity.

7 Queries

The point of producing these tools is to be able to analyze the large quantities of data. With these tools, the queries can be carried out on the semantic data as well as on the geographic ones, and therefore on the spatial dimension, which multiplies the possibilities. And it also makes it possible to cross the results of different queries.

8 Semantic Queries

In the framework of this project, it is possible to query the data about the discovery of the monuments: name of the discoverer and date, for example and to cross the results (Figure 15.5, 15.6, 15.77). These kinds of analyses allow us to understand the excavation works and the progression of the different researchers.

It is also possible to query data about the type of the monuments, like the "mastabas," for example (Figure 15.8).

The data on architecture can also been queried, like the "palace-façade motif" (Figure 15.9). Then, the result can be crossed with other data, like the dating of these last monuments (Figure 15.10).

By examining the dating of the monuments with façade-palace motif, we learn that 13 out of these 25 monuments date from the First Dynasty, which is more than half of them.

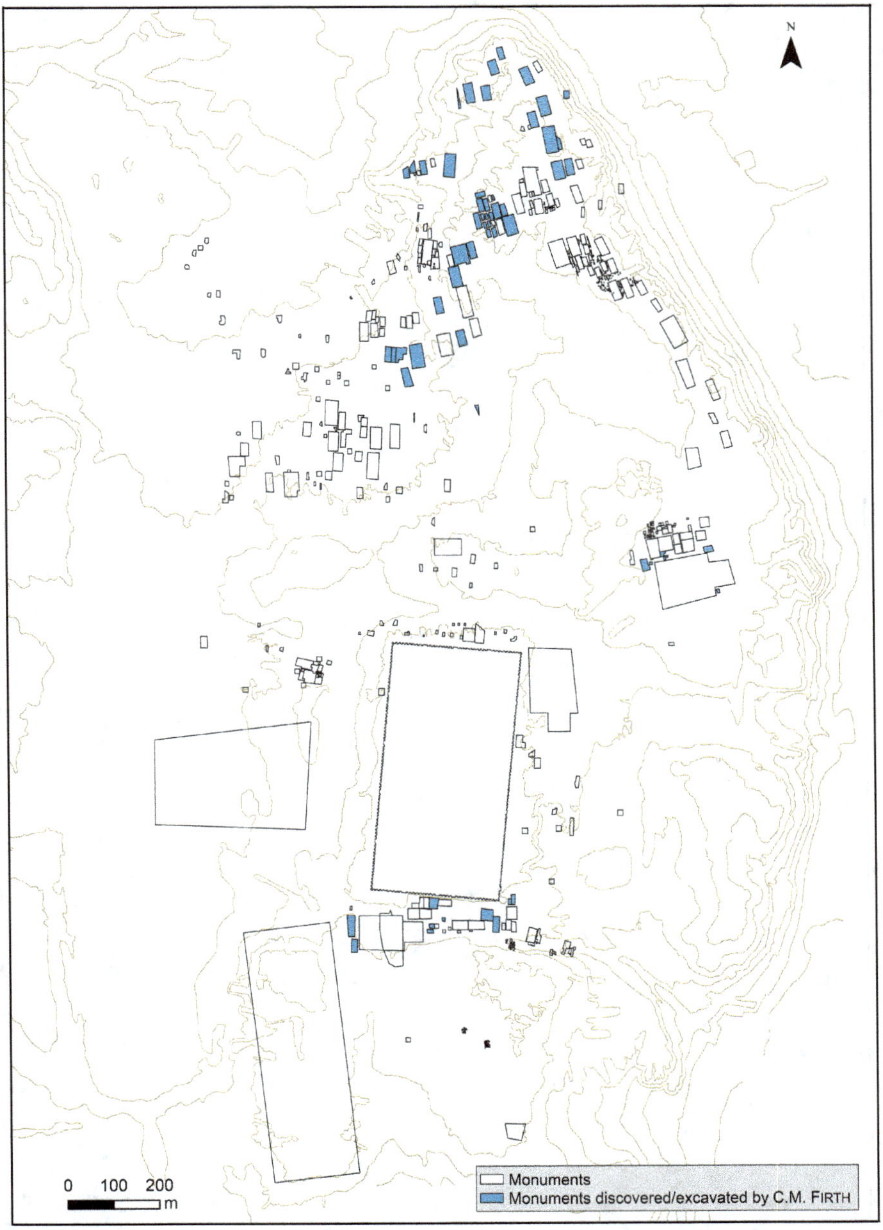

FIGURE 15.5 Monuments discovered by C.M. Firth (Status of data record: March 2020)

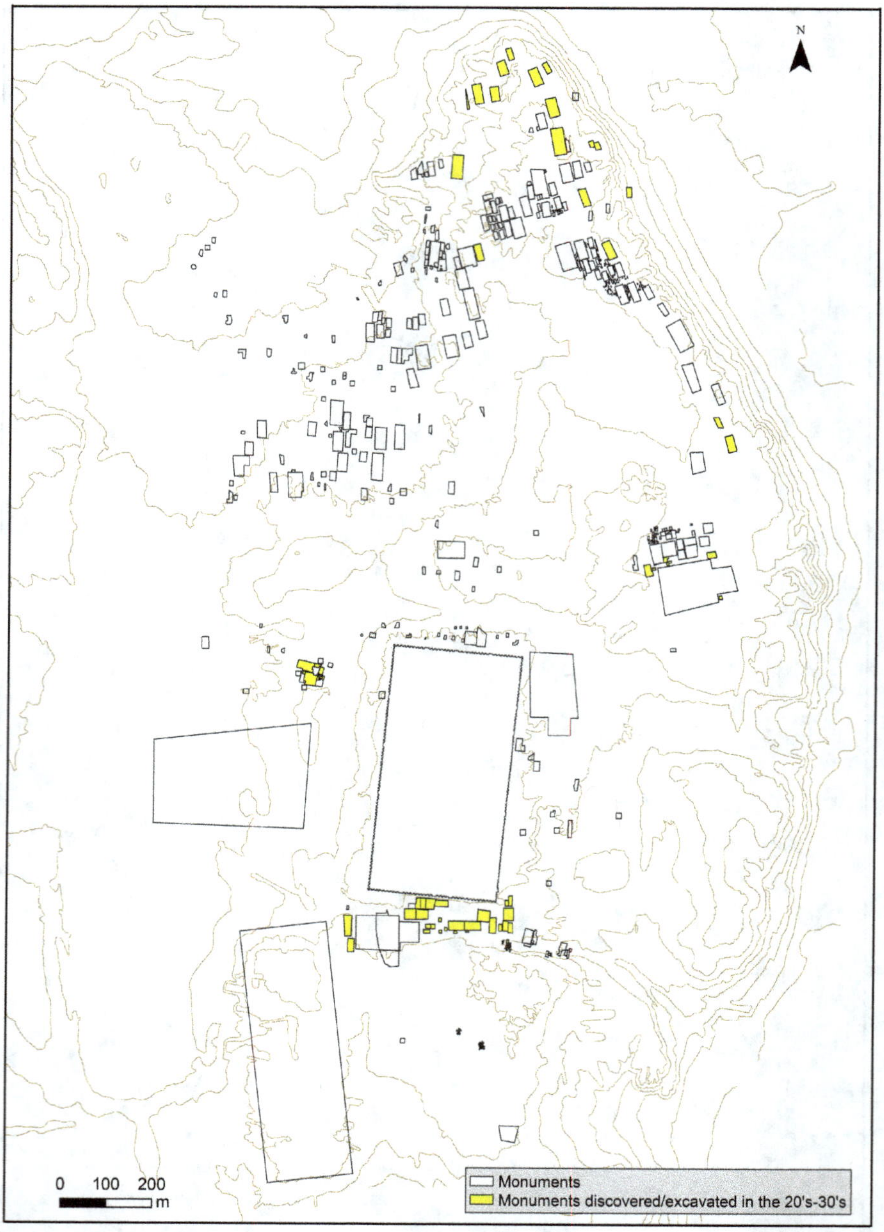

FIGURE 15.6 Monuments discovered in the Twenties and Thirties (Status of data record: March 2020)

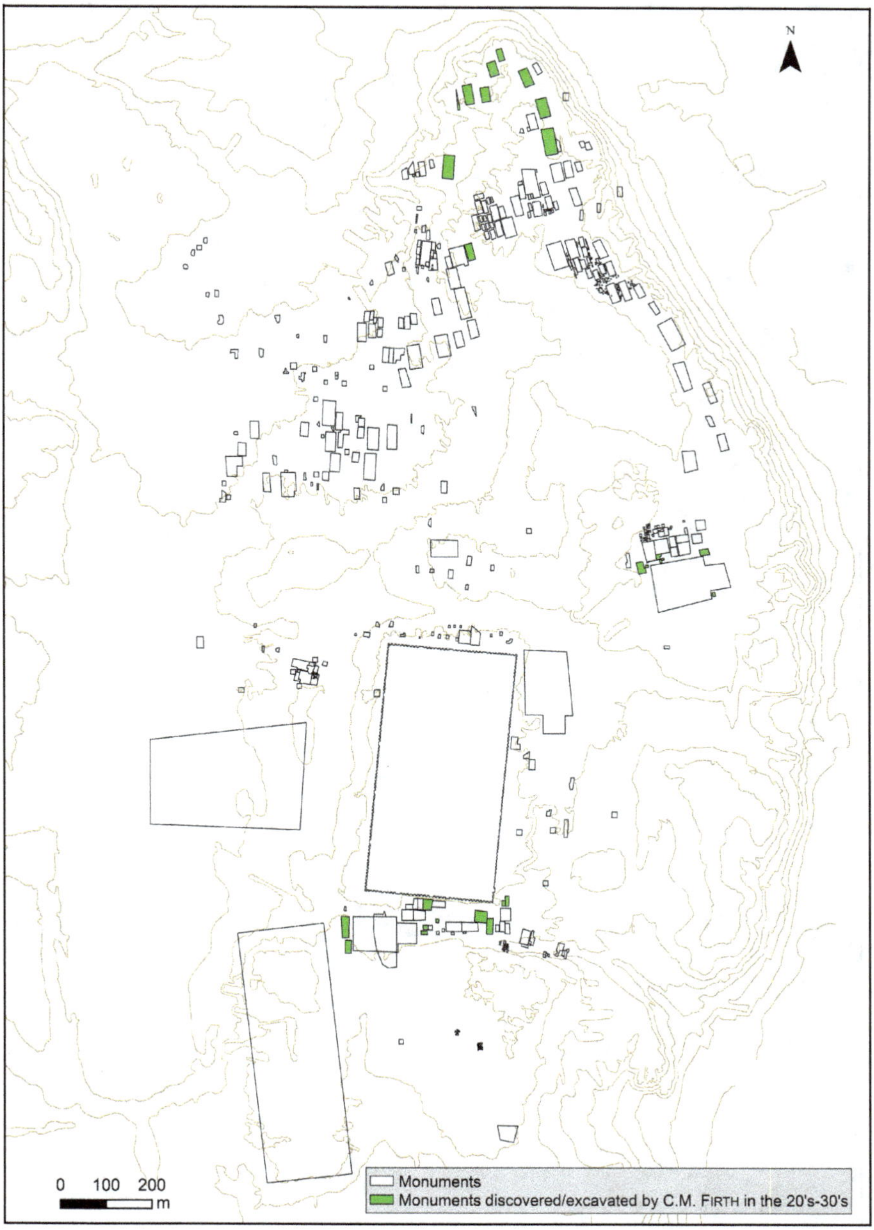

FIGURE 15.7 Monuments discovered by C.M. Firth in the Twenties and Thirties (Status of data record: March 2020)

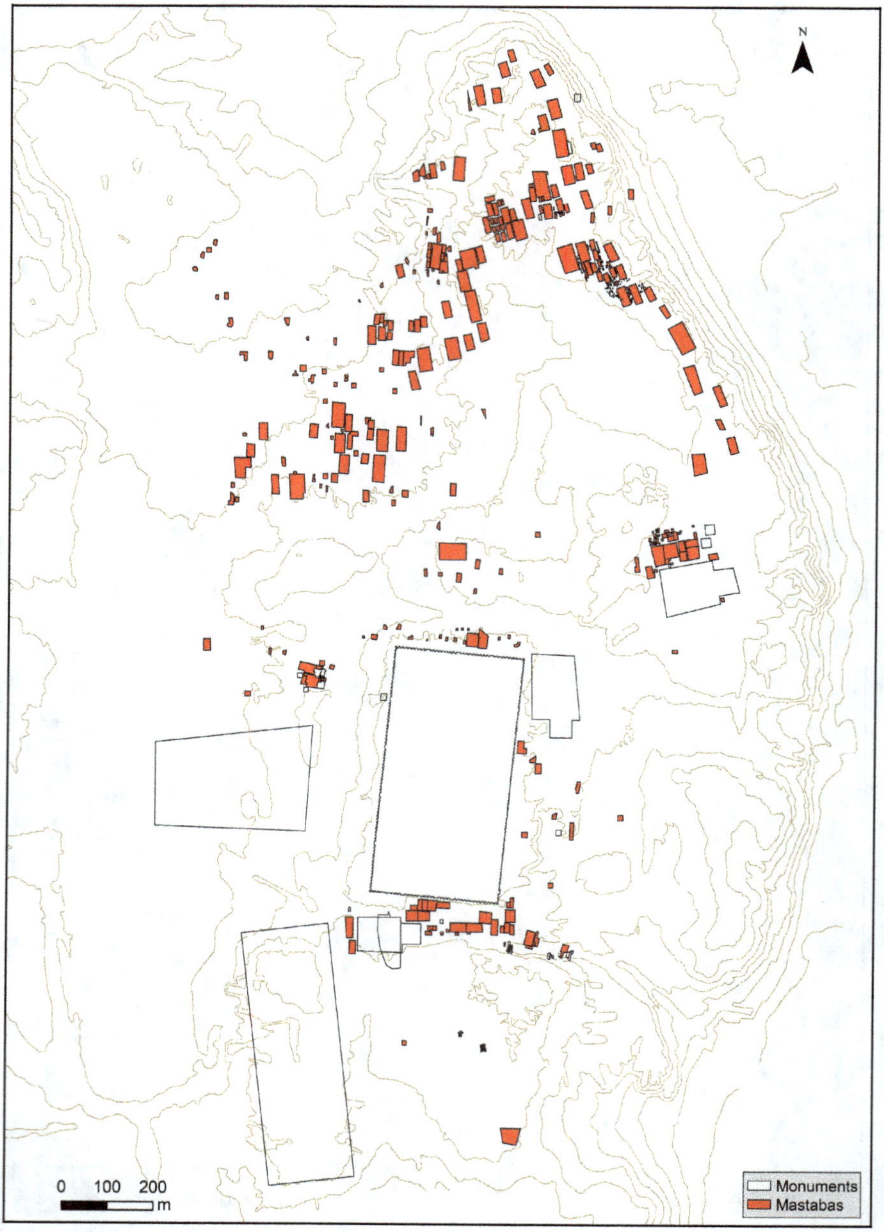

FIGURE 15.8 Mastabas (Status of data record: March 2020)

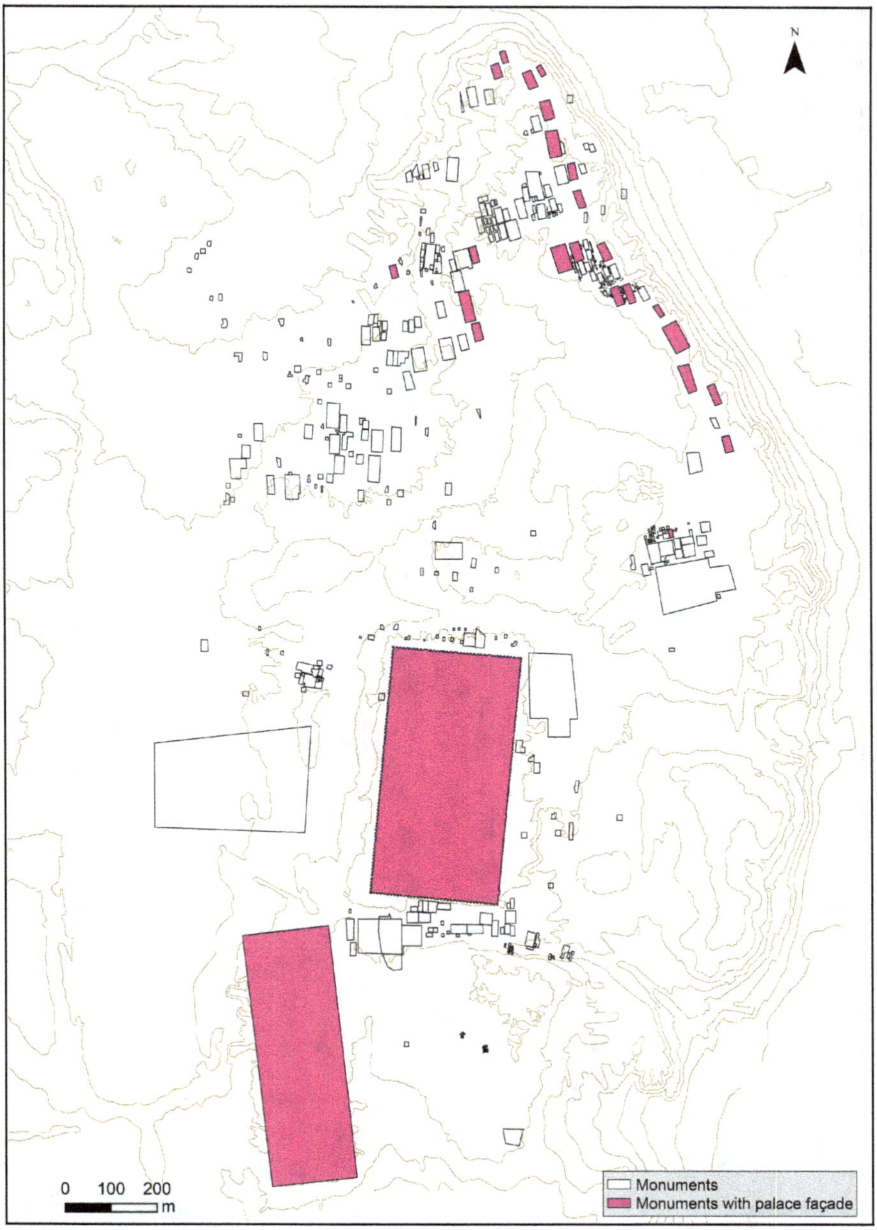

FIGURE 15.9 Monuments with palace-façade (Status of data record: March 2020)

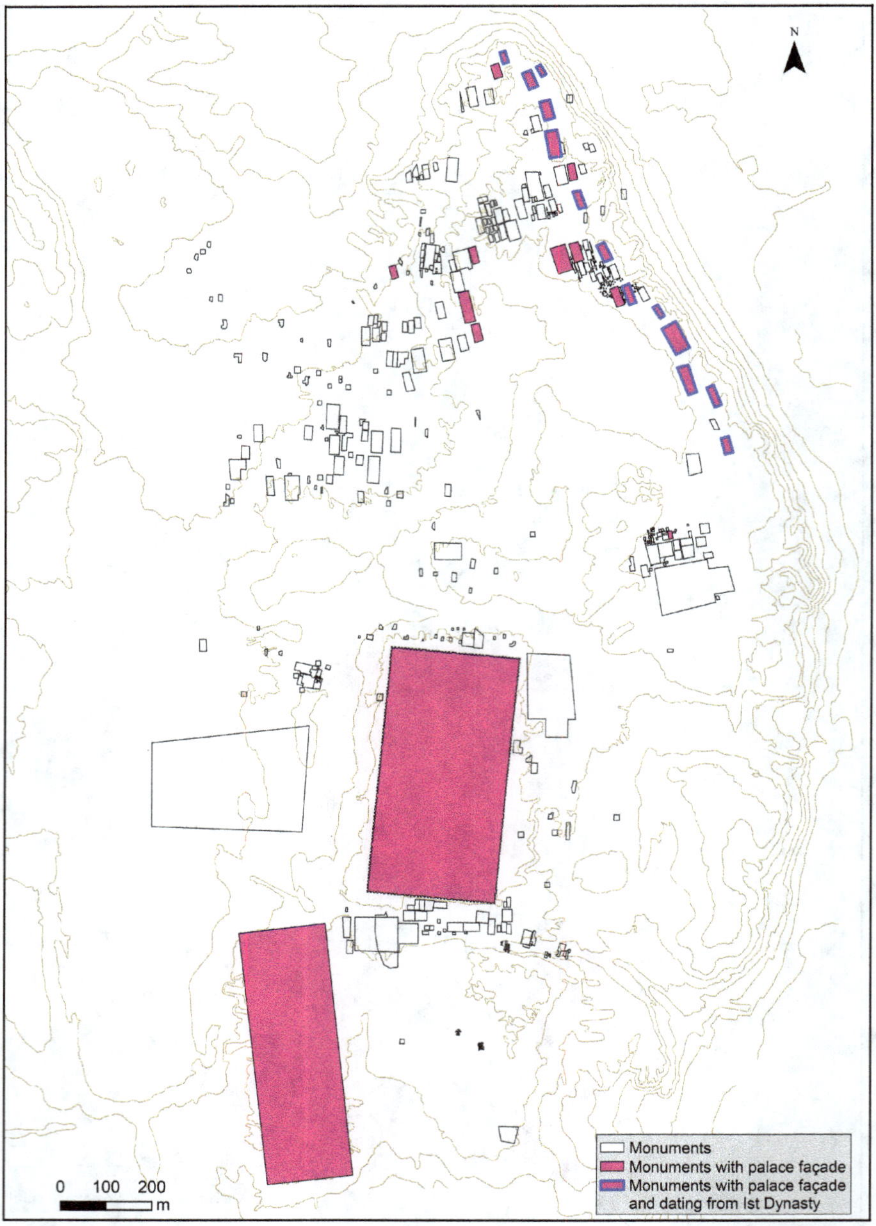

FIGURE 15.10 Monuments with palace-façade and dating from the First Dynasty (Status of data record: March 2020)

9 Digital Terrain Model (DTM)

In order to understand the organization of the monuments, the site topography can be taken into account thanks to a DTM. After a work on contour lines, a tool contained in the GIS software generates the DTM. The colors range from blue for the lowest places to red for the highest places (Figure 15.11). It allows us to better visualize the topography of the site and to know which monuments are built on more or less high ground compared to other monuments.

By crossing the DTM data with the attribute data on the architecture and the dating: the monuments with a palace-façade motif dating from the First Dynasty, the result appears directly in the software. The result is quick and easy to obtain. Moreover, the visual aspect of the result: a map with the location of the selected tombs on a background with different colors according to the terrain levels makes it easy to see and understand the situation. It is possible to appreciate that this motif shaped on these monuments, which is "un symbole de royauté et de pouvoir,"[18] was visible from relatively far away. Then, it allows hypotheses to be put forward, or even new research perspectives to be considered, by cross-referencing further data, for example. Of course, this information is already known but this example allows us to understand the conceptual approach and the process. A query can be made on a large number of data, but the result that appears on a map will also be easily readable and understandable.

10 Geographical Data

The geographical data, which means the location of the tombs in the space, can be examined too. Indeed, it is, for example, possible to query the data to know which monuments are at a distance of less than 75 meters from a royal funerary complex (Figure 15.12).

18 Monnier 2011, 36; See also Wilkinson 1999, 224–229.

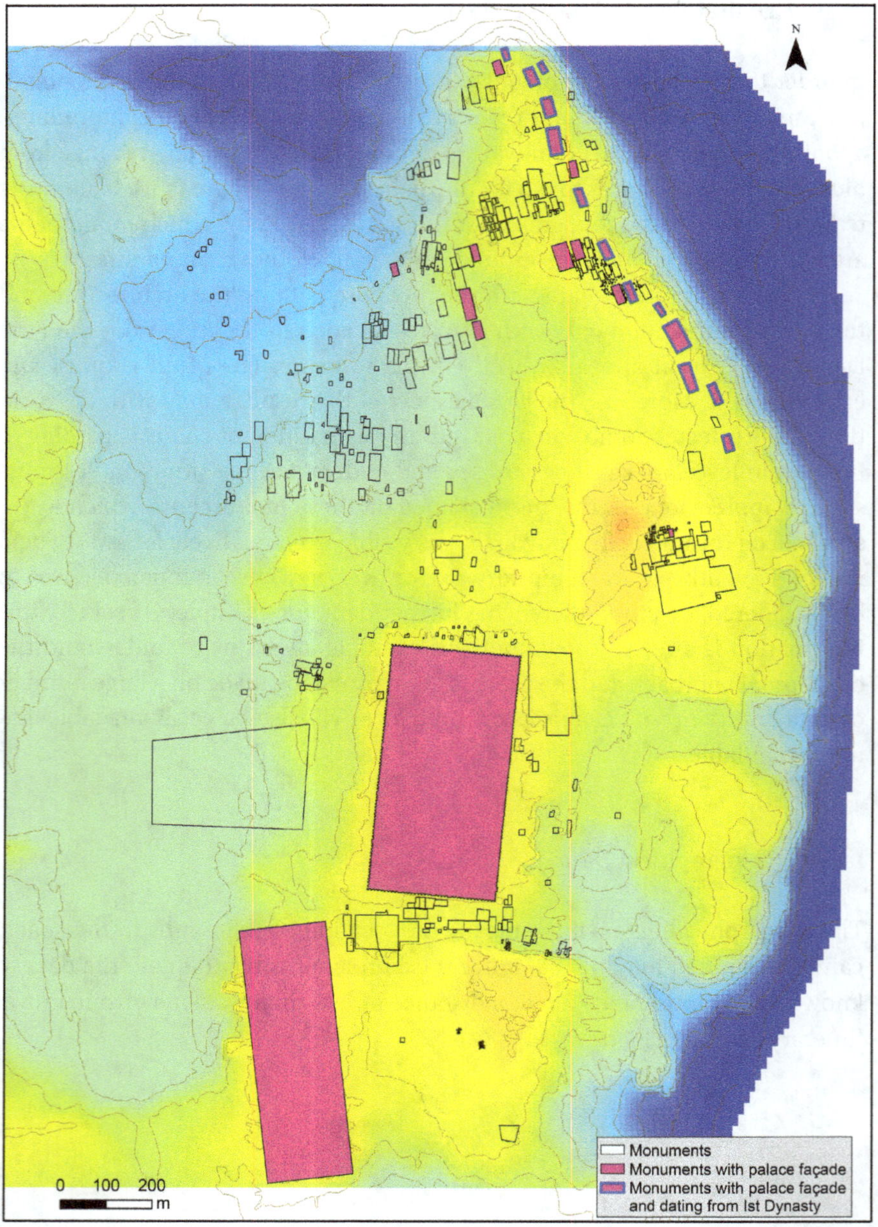

FIGURE 15.11 Monuments with palace-façade and dating from the First Dynasty on the DTM of North Saqqâra (Status of data record: March 2020)

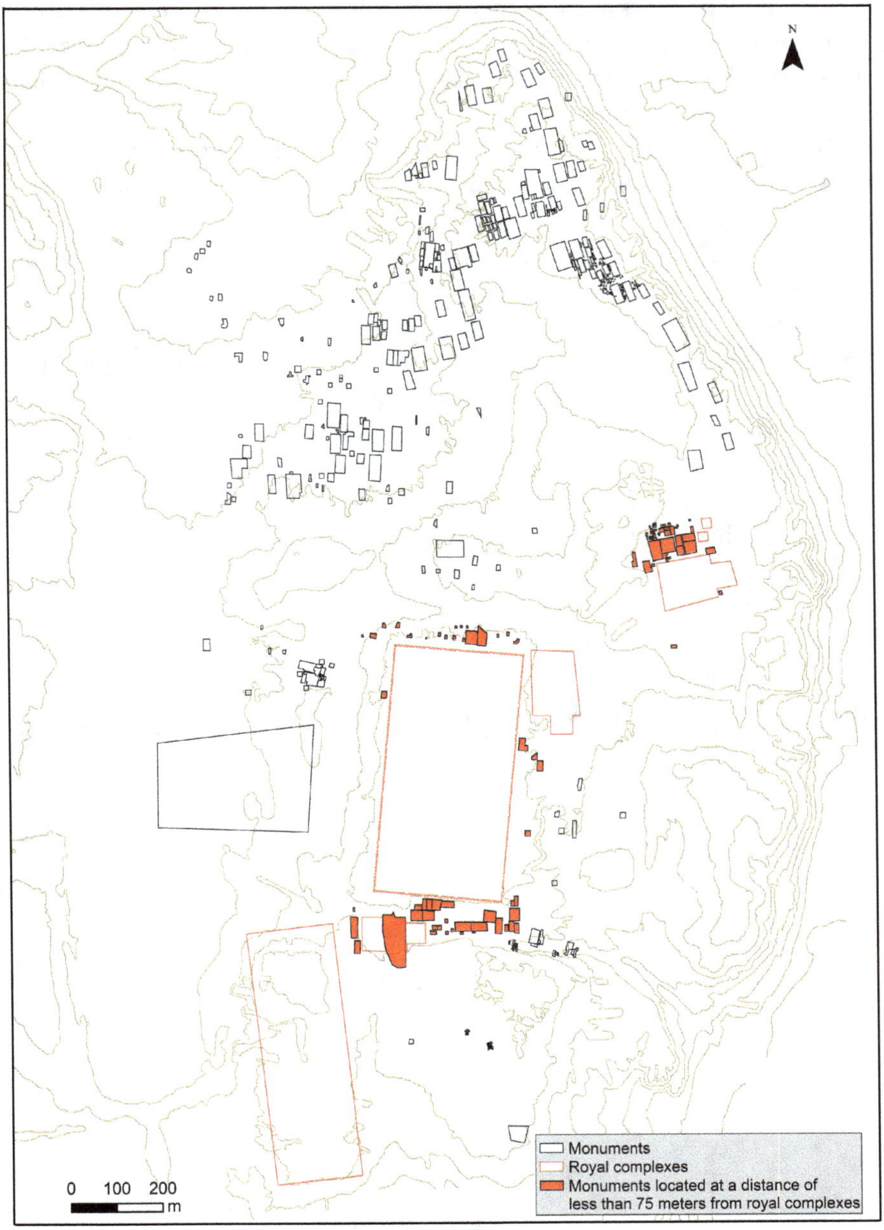

FIGURE 15.12 Monuments located at a distance of less than 75 meters from royal complexes. (Status of data record: March 2020)

The result provides 96 monuments. Then, it is possible to query the status of the owners of these tombs. In the database, 744 individuals are listed. I recorded titles for only 138 of them buried in 127 different monuments, some individuals being buried in the same tomb.

In the GIS, 112 out of these 127 monuments containing deceased persons whose titles are recorded, are represented (Figure 15.13). These monuments host 123 individuals.

Afterwards, the analysis can be refined. If the 96 monuments located at a distance of less than 75 meters from royal complexes (Figure 15.14) are queried on the titles of their owners, it is striking to note that 25 monuments host 27 individuals bearing the title of *smḥr wʿty* (Figure 15.15), which means more than a quarter of the monuments. It is also very important to remember that the titles of all the persons buried in these 96 tombs have not been recorded in the database. Therefore, it is possible that there are more individuals bearing this title of *smḥr wʿty* and therefore more monuments concerned.

In total, there are 33 monuments drawn in the GIS hosting 36 persons bearing the title of *smḥr wʿty* (Figure 15.16).

With recording being in progress, the results are incomplete. But some serious avenues of reflection and work are emerging. The tools provided by the GIS software, make it possible to test some hypotheses and to produce documents that can help to make decisions and help research.

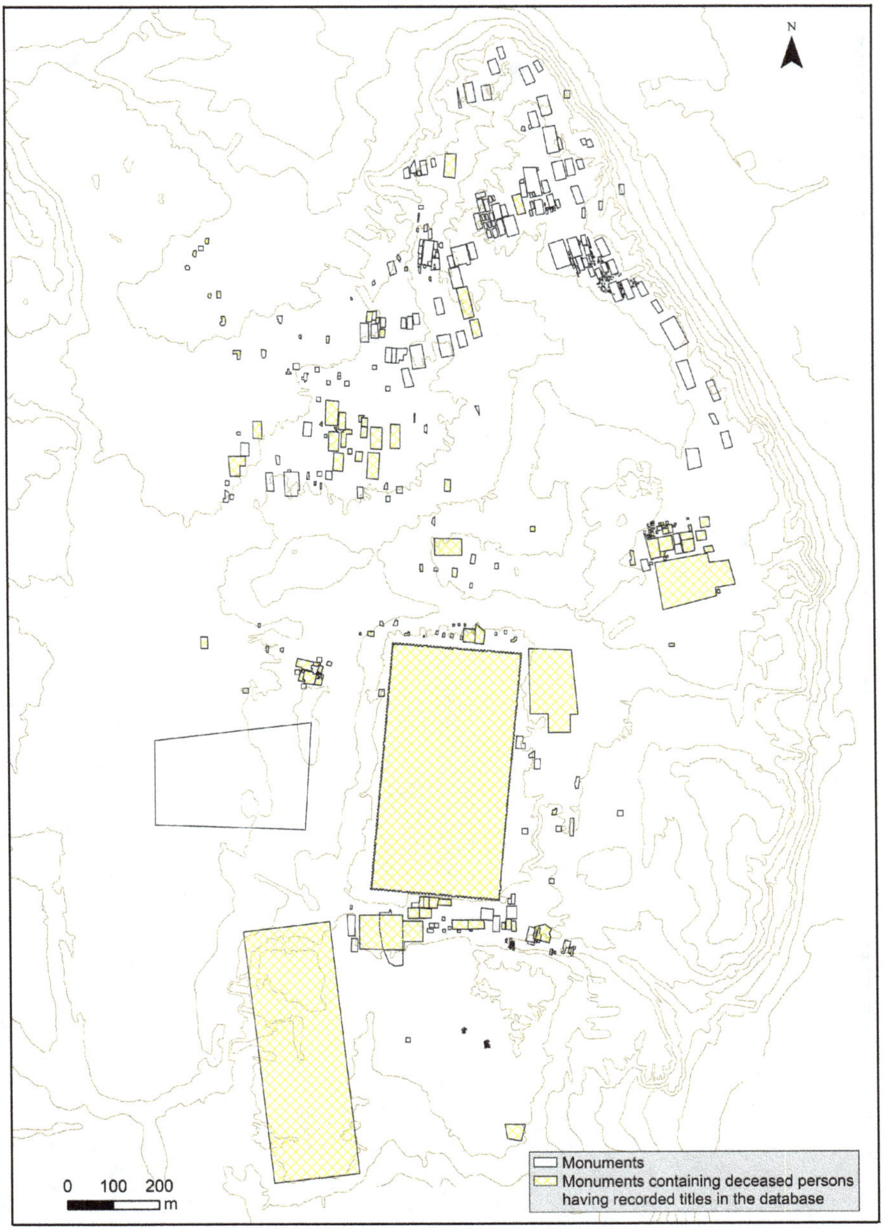

FIGURE 15.13 Monuments hosting individuals having titles recorded in the database (Status of data record: March 2020)

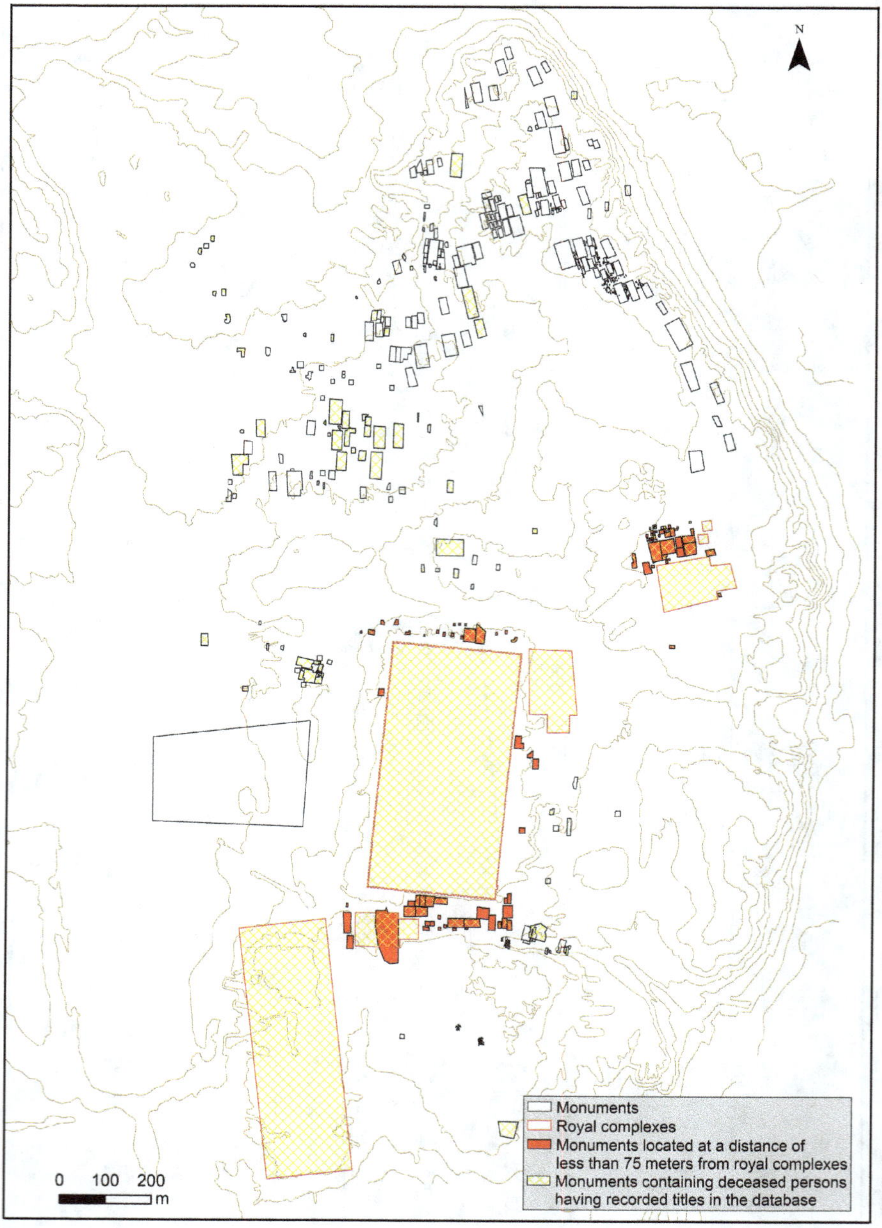

FIGURE 15.14 Monuments hosting individuals having titles recorded in the database and the monuments located at a distance of less than 75 meters from royal complexes (Status of data record: March 2020)

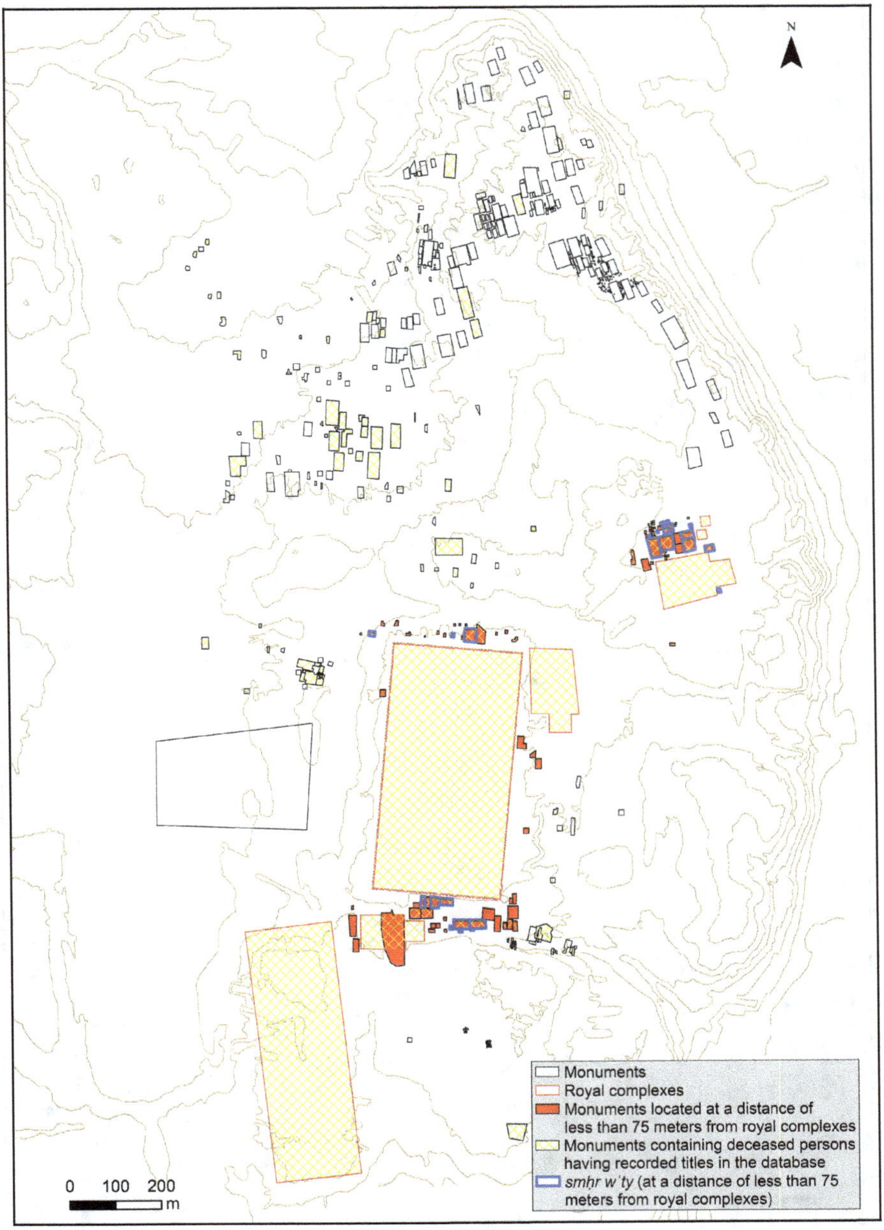

FIGURE 15.15 Monuments hosting individuals bearing the title of *smḥr wʿty* located at a distance of less than 75 meters from royal complexes (blue outlines) (Status of data record: March 2020)

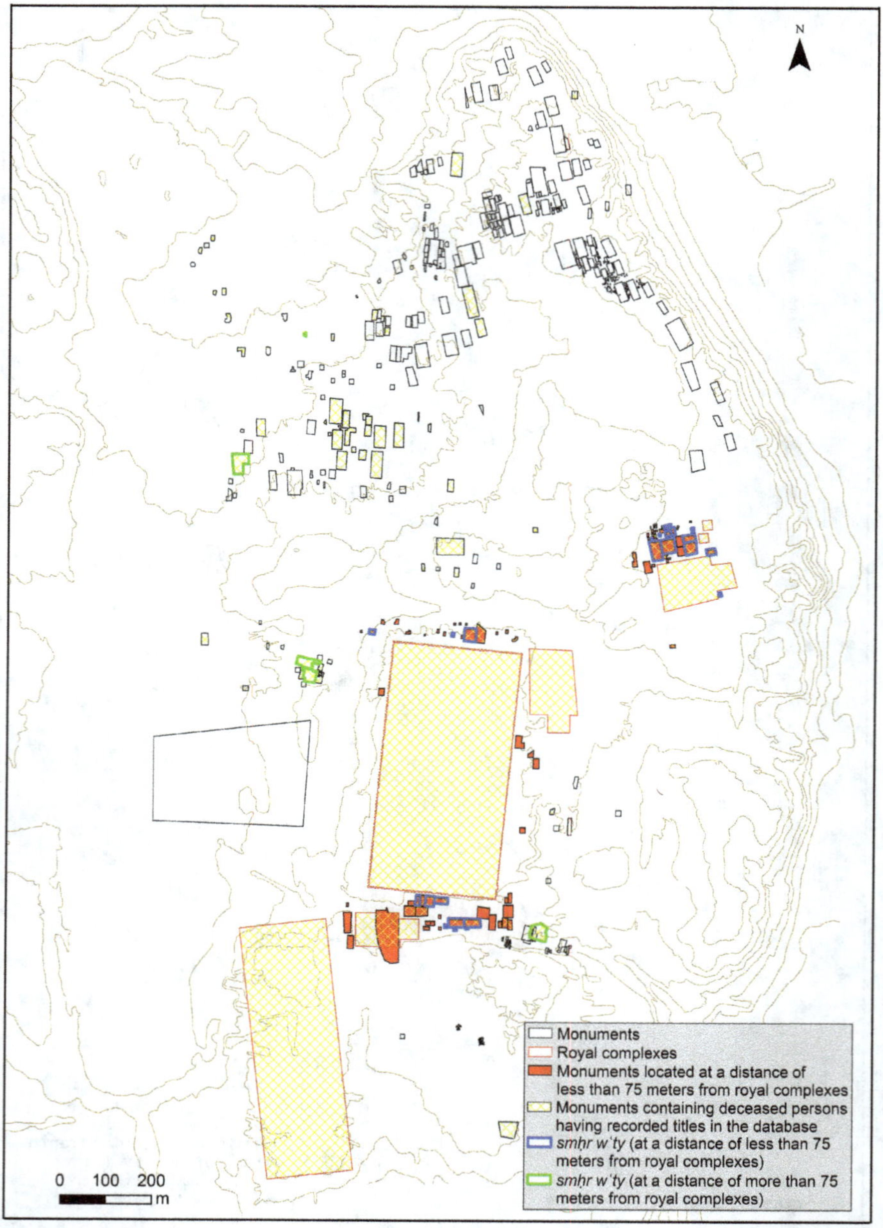

FIGURE 15.16 Monuments hosting individuals bearing the title of *smḥr w'ty* (blue outlines: for monuments at a distance of less than 75 meters from royal complexes; green outlines: for monuments at a distance of more than 75 meters of royal complexes) (Status of data record: March 2020)

11 Analysis

Thanks to the architectural data, the future analyses could reveal some information on techniques, methods, and know-how. By comparing the data, it can then be determined whether the choices were made based on the deceased's preferences, on the social status (thanks to the titles) he had or on compliance with the conventions in use.

Thanks to the data about filiation and genealogy, information about the influence of family ties on the spatial organization of the necropolis can be detected. The interaction between different individuals can be studied. Information may emerge and be updated. Thus, it may be possible to determine whether some architectural conventions (construction techniques, materials, dimensions, etc.) are transmitted from generation to generation in the same family or whether phenomena only related to conventions prevail according to some other criteria (such as the period in which the tomb was built, the geographical location of the tomb or the social status of the deceased, for example). In the longer term, we can imagine doing this analysis on the decorations made in the tombs.

In addition, other data could be aggregated, such as geomorphological data or data on landscape and vegetation, among others. Finally, an atlas could be created.

12 Online Publishing

After the data analyses, the data are planned to be published online to disseminate the data to the scientific community and the general public. Several options are possible in order to share, present, query, store and keep the data to save them.

However, some questions about the accessibility of the information must be studied because the information is sensitive due to the georeferencing of the data. A lot of options were considered in order to choose the most appropriate format and to plan accessibility, given that the data must be FAIR (Findable, Accessible, Interoperable and Reusable).

In order to collect the opinion of the future users and to know what they expect, I elaborated and conducted a survey.[19] The survey was carried out on different publics: researchers, students, Egyptophiles and the general public to know more precisely about their needs and expectations.

19 The survey was carried out between 23/05/2017 and 15/02/2018 among 26 persons. A conference about this survey was given, entitled: "The *SIGSaqqara* Project: around the online

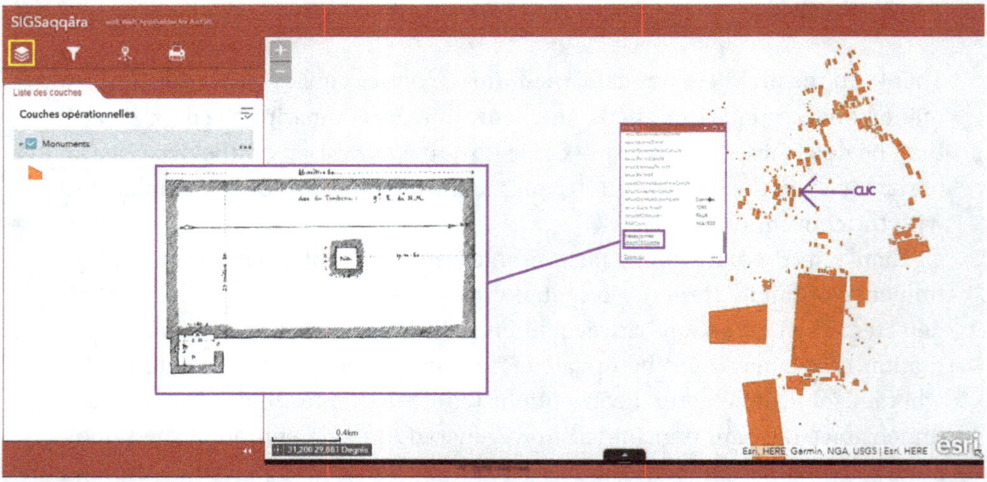

FIGURE 15.17 Web application created with *ArcGIS Online*: layer "Monuments" (Test in August 2018)

After reviewing the responses, I developed a test with *ArcGIS Online* because it can be adapted to the needs of everyone. This tool may be one option among others. But because it offers a lot of options which can meet the needs of the future users, I chose it to show what is possible to do.

In the tool created to test the possibilities, several layers can be displayed (Figure 15.17, button in yellow frame). It is possible to click directly on the monument in order to get its recorded data and then all the detailed information is displayed and pictures, like a plan for example, can be seen (Figure 15.17, in purple).

Filters were also made and are recorded in the application. To have access to them, it is necessary to click on a button (Figure 15.18, yellow frame). There is, for example, a layer with all the monuments containing a false door. Then, it is possible to have a summary of the information by clicking on another button (Figure 15.18, orange frame). The number of entities concerned and the data are displayed (Figure 15.18). Finally, the map can be printed or saved in PDF, for example (Figure 15.18, pink frame).

By clicking on a button, all the data stored in the database can be queried. If we consider possible extension to tourism management, the system brings new opportunities for tourists who could get a map. Indeed, it is easy to have a map with the monuments open for visit, for example, because this information is also recorded in the database.[20]

publishing," BEC 4 *The Fourth British Egyptology Congress*, Manchester, 7–9 September 2018.

20 This is one of the wishes expressed in the survey.

FIGURE 15.18 Web application created with *ArcGIS Online*: filter "monuments containing a false door" and the data about them (Test in August 2018)

These are only a few examples, but anything is possible. The web application will be retrievable online and accessible with computer, tablet, or smartphone. For novices, some maps will be proposed; as for experts, they can make their own queries. To protect the data, a password can be implemented to restrict access. This web application can be integrated to a web site. Two main options are available: a web site dedicated to this project or a platform with other archaeological databases, not specifically in Egyptology. According to the survey investigation, the first option seems to be favored by the potential users.

13 Conclusion

After the data analyses and thanks to the online publishing of the data, everybody will be able to carry out research according to their own interests and thus conduct new investigations. Therefore, these tools could be used to create the georeferenced archaeological map of Saqqara.

But ambitions can be greater. In the long-term, the project could allow the recording of more data, such as the material contained in the tombs or their decoration and inscriptions, for example. The geographical area, but also the chronological period, could be extended, which would allow to have a still wider vision of the logic of the spatial occupation at Saqqara, with South Saqqara and all other periods. Moreover, these tools can be applied to all kinds of sites, which might bring further developments in many archaeological areas under study.

Bibliography

Description 1823. *Description de l'Égypte ou Recueil des observations et des recherches qui ont été faites en Égypte pendant l'expédition de l'armée française Seconde édition dédiée au Roi, publiée par C.L.F. Panckoucke, Chevalier de la légion d'honneur, Antiquités*. Tome Cinquième, Paris. https://www.wdl.org/fr/item/2403/view/1/1/ Accessed 4-07-2019.

Ago, Fabrizio, Edda Bresciani, and Antonio Giammarusti. 2003. *The North Saqqara archaeological site: Handbook for the environmental risk analysis*. Progetti: documenti per l'archeologia eziziana 1, Pisa.

Bárta, Miroslav, and Vladimír Brůna. 2006. *Satellite Atlas of the Pyramids: Abu Ghurab, Abusir, Saqqara, Dahshur*. Plzeň: Dryada.

De Morgan, Jacques. 1897. *Carte de la nécropole memphite: Dahchour, Sakkarah, Abou-Sir*. Cairo: Imprimerie de l'Institut français d'archéologie orientale.

Lepsius, Richard. 1848–1859. *Denkmäler aus Aegypten und Aethiopien nach den Zeichnungen der von seiner Majestät dem Könige von Preussen Friedrich Wilhelm IV nach diesen Ländern gesendeten und in den Jahren 1842–1845. Ausgeführten Wissenschaftlichen Expedition auf Befehl seiner Majestät, Neudruck der Ausgabe 1849*, Tafelwerke Abtheilung I, Berlin: Nicolaische Buchhandlung. http://edoc3.bibliothek.uni-halle.de/lepsius/tafelwa1.html Accessed 4-07-2019.

Mariette, Auguste. 1885. *Les mastabas de l'Ancien Empire / fragment du dernier ouvrage de A. Mariette; publié d'après le manuscrit de l'auteur par G. Maspero*. Paris: F. Vieweg.

Monnier, Franck. 2011. "La signification des enceintes à redans dans les forteresses nubiennes du Moyen Empire." *GöttMisz* 228: 33–49.

Noc, Eloise. 2015. *Analyse spatiale à Saqqâra des origines à la fin de l'Ancien Empire. Les exemples des complexes funéraires de Netjerikhet et de Sekhemkhet*, PhD diss. Paul-Valéry Montpellier 3, University of Montpellier.

Noc, Eloise. 2018. "*SIGSaqqâra*: présentation du projet." *Égypte, Afrique et Orient* 89 (mars-avril-mai): 57–60.

Noc, Eloise. 2019. "*SIGSaqqâra*: Presentation and Applications of the Project." In *Egyptian & Egyptological Documents Archives Libraries* (EDAL) VI, *Old Kingdom Art and Archaeology 7 Proceedings of the International Conference, Università Degli Studi di Milano 3–7 July 2017*, edited by Patrizia Piacentini and Alessio Delli Castelli, 174–179 and plates. Milano.

Quibell, James Edward. 1923. *Excavations at Saqqara (1912–1914), Archaic Mastabas*. Cairo: Service des Antiquités de l'Égypte: Institut français d'archéologie orientale.

Reisner, George Andrew. 1936. *The Development of the Egyptian Tomb down to the Accession of Cheops*. Cambridge: Harvard University Press; Oxford: Oxford University Press, London: Humprey Milford 1936.

Sullivan, Elaine A. 2020. *Constructing the Sacred: Visibility and Ritual Landscape at the*

Egyptian Necropolis of Saqqara. Stanford: Stanford University Press. doi:10.21627/2020CTS. https://constructingthesacred.supdigital.org/cts/introduction Accessed 3-28-2022.

Spencer, Alan Jeffrey. 1974. "Researches on the Topography of North Saqqâra." *Orientalia* 43: 1–11.

Wilkinson, Toby A.H. 1999. *Early Dynastic Egypt*. London-New York: Routledge.

Online Resources

EAMENA Project	https://eamena.org/home
Desert Networks Project	https://desertnetworks.huma-num.fr/
	https://desertnetworks.hypotheses.org/
ArcGIS by ESRI	https://www.esri.com/en-us/home
Microsoft Access	https://www.microsoft.com/en-us/microsoft-365/access

CHAPTER 16

'Where Did THAT Come from?!' The Giza Project's Development of Citation and Referencing Documentation for 3D Archaeological Visualizations

Nicholas Picardo

Abstract

Within a broader mission to comprehensively integrate archival data for the site of Giza, the Giza Project at Harvard University has applied over a century's-worth of archival resources to produce 3D models of the Giza Plateau and many of its major monuments.[1] As technology has expanded the options for documentation and communication of archaeological information to both academic and popular audiences, it has been challenging to adequately preserve fundamental standards of academic transparency and citation for 3D media. Whether intended as visualizations of selected data or as detailed visions of ancient milieus, 3D digital models—along with other media that incorporate them—often include significant elements of reconstruction. Once released, often they become disassociated from the sources and thought processes that informed their creation. By conceiving of 3D archaeological reconstructions as "new data" that integrate with "old data," the Giza Project has developed referencing protocols and documentation that promote necessary, reasonable, and accessible standards of transparency and citation for 3D archaeological reconstructions and visualizations.

1 Some material contained in this contribution has been shared, in part or whole, in two conference papers: Nicholas Picardo, "'Where Did THAT Come From?!' The Giza Project's Development of Citation and Referencing Standards for 3D Archaeological Visualizations" (2017 Annual Meeting of the American Schools of Oriental Research Boston, MA, November 15–18, 2017). Nicholas Picardo, "What Happens Between the Maps and the Models: Developing Referencing Standards for 3D Archaeological Visualizations" (69th Annual Meeting of the American Research Center in Egypt, Tucson, AZ, April 20–22, 2018). Peter Der Manuelian also summarized some material during the *Ancient Egypt—New Technology* Conference keynote address, titled "Who Owns the (digital) Past?" See also Chapter 1.

DOI:10.1163/9789004501294_017

Keywords

3D modeling – 3D media – animation – archaeological visualization – archaeological archives – citation – *Digital Giza* – digital reconstruction – Giza Project at Harvard – information sourcing – intellectual transparency – reference documentation

•••

> *There was no reason for an archaeologist to be uninterested in the creation of theoretical models of their objects of study, for other disciplines (astronomy, medicine, and physics) had been engaged in such studies for a long time. The goal of building this type of model that gathers all the accumulated knowledge of a specific site on a computer is to give a synthetic representation of the object of study that is useful for further research and to communicate it to the public. This double objective required careful consideration and the development of an appropriate methodology.*[2]

∴

1 Introduction

The archaeological site of Giza has been studied via organized excavation for well over a century and as a result has produced massive amounts of archaeological documentation. The Giza Project at Harvard, a collaborative international initiative based at Harvard University in Cambridge, Massachusetts, assembles, curates, and provides free access to records of past archaeological work at Giza. The Project is not an on-the-ground field research program, but rather an archaeological data management, informatics, and visualization project. It currently manages the world's largest digital archive of Giza records and media, the Giza Consolidated Archaeological Reference Database (GizaCARD),[3] and leverages this diverse content to produce powerful online

2 Golvin 2012, 77.

3 The GizaCARD runs on The Museum System collections database software by Gallery Systems (http://www.gallerysystems.com/). Initial set-up and population of the GizaCARD by the Giza Project at Harvard was supported by a Humanities Collections and Reference

and traditional academic research tools along with new teaching technologies.[4] The Giza Project's public-facing point of contact is its *Digital Giza* website (http://giza.fas.harvard.edu), which exists currently as a limited-function (but nonetheless highly functional) proof-of-concept prototype version. The first fully-built iteration is still under construction at the time of this writing.[5]

As part of its overarching mission for the collection, preservation, integration, study, and access for Giza data as comprehensively as possible, the Giza Project has for several years approached the creation of 3D model reconstructions of the Giza Plateau and its archaeological remains with two purposes: (1) as the effective visualization of archival holdings contained within the digital archive, and (2) as highly accessible, interactive interfaces that can serve as effective access points to that core data for many types of users. Thus, through the evolution of the Giza Project 3D modeling has become a pursuit not only *for* visualizing data holdings, but as a very important access point *to* the data itself, especially for non-traditional users of primary documentation. Digital media have become important vehicles for bringing primary documentation to a broad audience, not just through lowering access barriers but also through contextualization, interaction, and direct experience.[6]

2 Brief Problematization: Facts, Figures, and 3D Modeling in Archaeology

Illustration has been ubiquitous in the dissemination of information in archaeology and Egyptology for about as long as the fields have been professionalized. The fast pace of technological evolution in recent decades has expanded options for creation, presentation, distribution, and consumption of visual media dramatically. Three-dimensional modeling and reconstruction are by

---

Resources grant from the National Endowment for the Humanities Division of Preservation and Access (grant #PW-51569-14).

4 A custom application program interface (API) feeds information from the GizaCARD to the *Digital Giza* website. The Digital Giza API is built primarily on Python and JSON coding. This tripartite infrastructural system optimizes querying of the database and frequency of data refreshes for public availability.

5 Presentation of this article's focal content here is, in many ways, a concession to the length of time required for both updates and advancements to the Giza Project's infrastructure. As with any totally new data type, referencing standards—more specifically, the citation documents to be described below—must await updates to all three components of *Digital Giza*'s three-part system for their comprehensive online release.

6 For a brief overview of the Giza Project's approach to, and goals for, 3D model reconstruction (specifically for Giza mastaba tomb G 7530–7540), see Aronin 2018, 48–56.

now established elements of the digital archaeology toolkit. Whether captured through photogrammetry or laser scan, or built up from 2D documentation in modeling software, 3D visualizations and reconstructions comprise the most recent stage—and extension—of visual media for representing information about sites, site components, data, interpretive models, hypotheticals, and conclusions in both academic and popular forums. Increasingly, it has been recognized that the utility of modeling virtual reconstructions rests in more than just in their outputs as visualized ancient moments, monuments, and milieus; rather, the undertaking is itself valuable to the process of inquiry and data interpretation in archaeological research.[7]

Egyptology is not generally expected to be a driving force behind technological applications and advancements. In fact, often it has been slower than some other disciplines at adopting new technologies. Nonetheless, as a field that studies a highly visual material culture that enjoys relatively good preservation, it is an apt beneficiary of the advantages that 3D capturing and modeling technologies proffer. It is a very good candidate for contributing to the disciplinary landscape of "virtual heritage."[8] Three-dimensional visualizations offer an efficiently plastic medium for experimentation, trial-and-error, and comparison of multiple possibilities.[9] These capacities are of great utility for working with the (re)construction of complex subject matter, providing powerful means for testing and displaying ideas.[10] In practical terms, 3D technologies enhance the capacities for graphic models to support intellectual models in ways that push beyond traditional illustration to new modalities of research, learning, and engagement with the past.[11] Even more than the inherent perspectival differences between 2D and 3D media, "virtual" 3D objects and environments add further dimensions to their subject matter. Whereas a two-dimensional figure is perceived relative to the size of a page, to the bounding box of a figure, or to a scale indicator within it, 3D models can afford first-person viewpoints at life-like scale and in all directions. The option of representing the fourth dimension of time is of immense value to archaeological subjects.[12] This capability offers

7 Hermon 2009, 36–42.

8 Donald 2012, 95–103.

9 See, for example, the conclusions of Helena Rua and Pedro Alvito 2011.

10 Aronin 2018, 55; Manuelian 2017, 154–189; Sullivan 2012.

11 Frischer 2009, especially v–ix.

12 This ability is especially significant for exploration and display of archaeological/architectural phasing. See, for example, the Digital Karnak project (http://dlib.etc.ucla.edu/projects/Karnak now only archived at http://wayback.archive-it.org/7877/20160919152116/http://dlib.etc.ucla.edu/projects/Karnak/); Sullivan and Wendrich 2008, 109–128; Sullivan 2016, 80–84. For an example at Giza, also Manuelian 2017, 174–177.

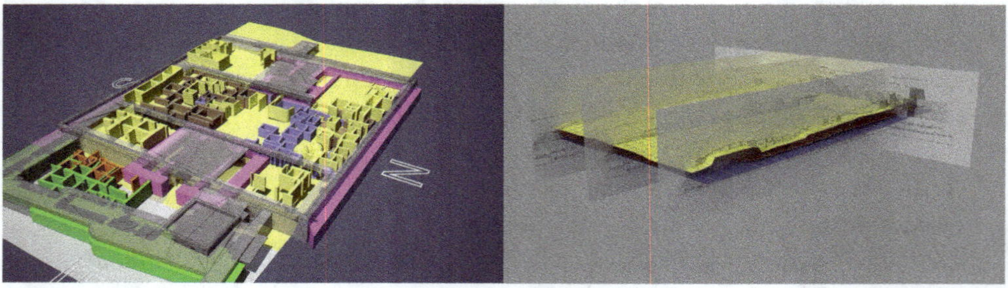

FIGURE 16.1 The Giza 3D Menkaure Valley Temple, work in progress: (left) Superimposed architecture and extrusions of settlement phases and (right) extrapolations and comparisons of ground levels in the temple courtyard. (Original plans of Reisner, *Mycerinus* (1931), Plans VIII–X & Hassan (1960), frontispiece.)

advantages for both creation and presentation of visualizations (Figure 16.1 & Supplemental Media A). Many 3D media formats also allow embedding of traditional media formats (e.g. images, illustrations, videos), further expanding their versatility by permitting juxtaposition of "the virtual," in the form of reconstruction, and "the real," in the form of archaeological records and other traditional data formats.[13]

Visualized reconstructions are thus effective vehicles for productively using archival records from past excavations, as well as, in some cases, enabling research to overcome limitations inherent to past excavations' recording practices.[14] It is with these latter applications of 3D modeling that the Giza Project at Harvard University has almost a decade of experience and experimentation, in particular.[15] As both process and media output, 3D modeling is a vehicle for investigating real spatial relationships among people, objects, buildings, sites, and landscapes that are fundamental to archaeological interpretation. It is fitting that some recent Egyptological research explores first-hand sensory and spatial experiences in—and of—ancient settings, regarding them as integral to understanding the original uses of ancient spaces and landscapes.[16]

---

13 Augmented reality as an extension of 3D media's dialogue between "the real" and "the virtual" is worth mention, but is outside the scope of this article. The Giza Project's explorations in Augmented Reality applications likely will be treated in future publications.

14 Even while acknowledging that not all data is equal, with reliability being variable. See Andreas Georgopoulos 2014, 155–162.

15 Manuelian 2013, 730–734; Manuelian 2017, 124–189.

16 For notable examples, see Sullivan 2012, 1–37; Sullivan 2017, 1227–1255; Sullivan 2016, 71–88.

However, the technology that makes this kind of visualization potentially so useful also can make it a dicey affair. Just as new and improved technologies for digital creation and consumption have expanded the capabilities of illustration and visualization, they have brought significant new potential pitfalls to consider, and consider thoughtfully. As 3D visualizations increasingly become entry points for students and an interested public to engage with archaeology and Egyptology, 3D media easily can contribute to misunderstandings of the subject matter. Three-dimensional models can be photorealistic. They can be immersive, 360-degree Virtual Reality arenas that lend more textured "experiential" aspects to content, readily invoking a palpable sense of "reality." With the current prevalence of electronic modes of transmission, display, and consumption, the significance of this difference from traditional 2D illustration can be underappreciated. Yet, just as with CGI in movies and video games, academic model content is not necessarily subject to physics and other constraining facets of reality. In effect, any idea can "become a reality," especially in the absence of context or annotation to qualify the visual content. Indeed, the lines between high-end academic media and entertainment productions are perhaps blurrier than ever before, even as the gaming industry takes incremental steps into the educational sphere.[17]

The speeds at which 3D media are distributable via electronic transfer only further complicate the situation. Manipulation, repurposing, and re-/de-contextualization occur more rapidly than ever before. Lastly, once academically generated 3D media are released into the world of the internet, they exist in a world that operates outside the structures and value systems of academe. Videos, 3D model files, and other formats are not yet included as a matter of course by journals and other academic publications, those domains wherein academics know the rules because they both created and continue to police them.[18] Supply chains for 3D media on the Internet are less stringently rule-aware, to vastly understate current conditions.

There always will be media produced outside the data-driven world. In fact, there will be more and more as technologies advance. At any point in time Google video or YouTube searches for relevant keywords will yield results that include erroneous, untenable, fantastical, curious, and altogether absurd content. Some highly questionable (at best) examples will be of similar, if not

17 Casey 2018. For an academic perspective on "gamification," see Liarokapis, Petridis, Andrews, and Freitas 2017, 374–376.

18 This statement gradually will become outdated as more publications include provisions for online access to supplemental data and media, as has become a more common practice in fields such as the biological and biomedical sciences.

superior graphic quality and production value to their academic counterparts. Options for clearly marking academically vetted content (including educated hypothesis or speculation) are few, and possibly shrinking in number over time. Once released, 3D media can quickly take on alternate lives of their own, disassociated from the often substantial source materials (e.g. primary documents; empirical data; excavated artifacts) and intellectual processes (e.g., decisions made; theories applied; extrapolation from parallels; informed speculation; artistic necessity or license) that factored into their creation. Effectively, they are born-digital academic works that do not lend themselves easily to the full spectrum of appropriate citation by traditional conventions of academic publications. A priority and long-standing conundrum for academically supported media, then, has been to find straightforward means of conveying intellectual support and explanation.

Whether intended to highlight selected data or to envision more complete ancient settings, 3D models often include reconstruction, sometimes considerable. This fact alone may be lamented or embraced, but there is no denying its necessity for some modeled media to be effective in their desired function(s). It is necessary because the archaeological record is, by definition, incomplete. If, for any number of reasons, the end goal is a model reconstruction that is "complete" in some sense of the term—an object, a building, a site, a geological formation, etc.—empirical data will at some point reach a limit that is well shy of that objective.[19] One will always need to bridge that gap. What transpires between that threshold and an end product accounts for many debates about the value and standards of 3D modeling in archaeology. The major points of contention often boil down to issues of realism versus empirical reality—and ultimately how to separate the two for audiences of 3D media—and the merits of partial or full reconstruction.[20] Both empirical and hypothetical reconstructions of archaeological remains always have been elements of academic research, as have reasonable attempts at approximating their original states.[21] But, by the very nature of academic undertakings, it is imperative to try to capture as much as reasonably possible, while still manageable and desirable, to explain what has been done to fill the gaps between what is documented to a standard of "realness" versus what is not.

19 When the Giza Project started to pursue 3D visualizations based on and related to its archival holdings, much deliberation was given over to defining the desired final products for intended applications at that time. For the first generation of 3D Giza models, the decision was to generate finished versions of all modeled subjects.

20 Wittur 2013; Manuelian 2017, 164–167.

21 Golvin 2012, 77–82.

3 3D Modeling and the Giza Project at Harvard

Project Director Peter Der Manuelian first explored 3D modeling for creating visualized reconstructions of Giza mastaba tombs published in volume 8 of the *Giza Mastabas* series.[22] At that time he was Director of the Giza Archives Project at the Museum of Fine Arts, Boston.[23] In this early instance, then, the end goal of the modeling was to leverage archival excavation records to generate reconstructions for figures in a paper publication. This traditional paper format permitted more or less straightforward citation of underlying archival sources and resources that informed the creation of the mastaba models. A major subsequent initiative called "Giza 3D" dramatically changed the trajectory of Giza modeling. A collaboration between the Giza Archives Project and French modeling software company Dassault Systèmes, Giza 3D would be a fully immersive 3D environment: a model of the Giza Plateau and its major Old Kingdom monuments, with as many as possible modeled in detail to allow for thorough exploration (Figure 16.2).[24]

At that time there was a distinctly experimental aspect to the ambitious aims of Giza 3D. Early in the endeavor a strategy was set in place for all detailed models to be built to "finished" states of completion. Egyptologists working on the Giza 3D were interested in exploring the kinds of questions, decisions, conflicts, and concessions that could—and did—arise along the way. There also was a keen awareness of the stakes involved in addressing these facets of media that was destined for public exposure. As a semi-commercial venture with a small battalion of digital artists logging modeling hours and a very small staff of Egyptologists trying to keep apace, neither schematics nor target outcomes could practically integrate front-facing referencing. The intellectual underpinnings had to be left as understood inherently backed by the level of academic involvement at, first, the Museum of Fine Arts, and, later, at Harvard University. The developmental period of Giza 3D also saw the completion of the Giza Archives Project at the Museum. By then Project Director Peter Der Manuelian had begun his tenure as Phillip J. King Professor of Egyptology at Harvard University.[25] With Manuelian again directing, some veteran members of the Giza Archives Project were brought aboard the newly established Giza Project

22 Manuelian 2009.

23 Manuelian 2002, 319–328; Manuelian 2006, 311–333; Manuelian 2009, 149–159; Manuelian 2005, 68–80.

24 Manuelian 2017, 124–153; Manuelian 2013, 730–734.

25 Peter Der Manuelian has since been named as the Barbara Bell Professor of Egyptology.

FIGURE 16.2 Render of the Giza 3D Plateau model
Note: Several versions of the virtual Giza Plateau model have evolved through the Giza Project's history, with a framework that has incorporated a combination of satellite imagery, digital elevation models, 1978 Egyptian Ministry of Housing and Reconstruction topographic maps, contour data generously provided by Ancient Egypt Research Associates (AERA), and the HU-MFA, and other expedition plans of Giza's cemetery zones. The Giza Project is especially grateful to AERA for fruitful exchanges and collaboration over the years, and especially for the provision of mapping data created by Peggy Sanders, Mark Lehner, Ana Tavares, Farrah Brown, Rebekah Miracle, Camilla Mazzucato, David Goodman, Glen Dash, and others.

at Harvard, providing some continuity to ongoing Giza 3D work with Dassault Systèmes.[26]

The Giza 3D collaboration culminated in the launch of a Dassault-built *Giza 3D* website (see Figure 16.3; https://www.3ds.com/stories/giza-3d/). On a virtual Giza Plateau, site visitors initially had access to 3D reconstructions of Giza's three royal pyramid complexes and a selection of high-profile and/or well-preserved private tombs, all of which could be explored in a real-time virtual environment. The Giza 3D website has since been decommissioned, replaced by a summary page to describe the legacy project. The modeled assets live on to be updated and used in new applications. However, they have required resource-heavy and time-consuming processes of revision and conversion to

26 Egyptological Research Assistant Jeremy Kisala and Research Associates Rachel Aronin and Nicholas Picardo. This charter Giza Project team also included in-house Digital Artists David Hopkins and (later, more briefly) Le Pan, and Technical Artist Rus Gant.

FIGURE 16.3 Homepage of the *Giza 3D* website

a more open standard format (Unity), which is still ongoing.[27] Through these developments "Giza 3D" has been retained as the standard nomenclature for deployment of models on the Giza Project's *Digital Giza* website.[28]

Also during these years a major focus turned to devising a sound strategy for integrating the Project's modeled art assets (both old and new) with GizaCARD holdings, to allow its direct linking to all primary archival sources that contributed to their creation. This turn accompanied a new conceptual treatment of archaeologically informed 3D models. They are not media outputs that rely on data of many forms; they *are data* in and of themselves—new data created

27 The original Giza 3D site and its models used Dassault Systéme's then proprietary VirTools platform. The company's eventual shift away from (and end of support for) VirTools was a major factor in prompting this change of course by the Giza Project.

28 At present writing, five monument models (the pyramid, pyramid temple, and valley temple of King Khafre, the Sphinx, and the Sphinx temple) appear on the prototype *Digital Giza* webpage as proof-of-concept for presentation of 3D models in the site's structure and design.

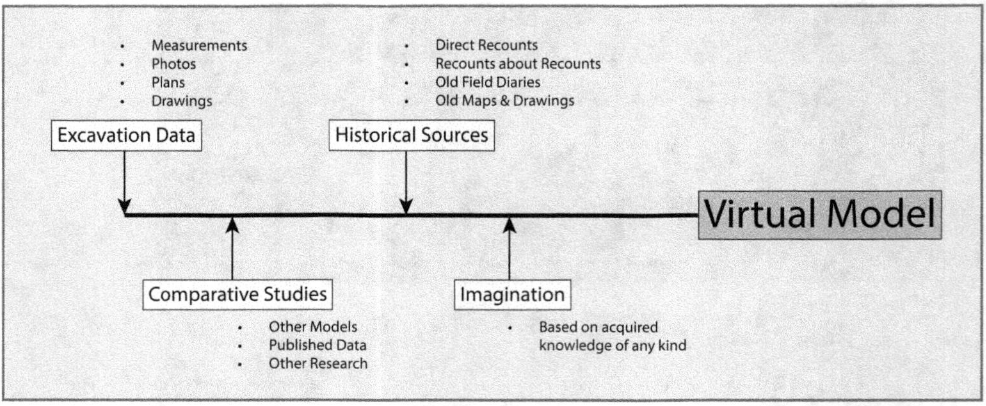

FIGURE 16.4 Schematic diagram of 3D modeling sources and influences (after Hermon 2009, Fig. 4, with adjustments and reformatting)

from old/other data. This frame shift is not a long stretch. Even if understanding a model as the *product* of its resources, it nonetheless becomes a member of the data constellation that produced it (Figure 16.4). An obvious outcome of this viewpoint was new database record types to accommodate 3D models as nodes in the Giza data constellation.[29] Interconnections within the database carry through to the *Digital Giza* website, such that all archived sources relevant to a model's construction are readily apparent in that model's online record view.

In a controlled, self-contained infrastructural system like that of the Giza Project (i.e. database, API, and website) this approach might be more or less sufficient alone. However, it falls short of accounting for much information in model-building that inevitably comes from beyond the scope of the archival records themselves. The range of other "behind the scenes" details can be voluminous.[30] Each 3D model is unique from its inception, in that every model has:
– its own limiting factors related to the primary data and imagery
– its own bibliography of scholarly referencing and interpretation, sometimes internally conflicting
– its own needs for extension, extrapolation, educated speculation, or fabrication for completion

29 Note, however, that this does not mean the model files themselves must be stored within the database itself. The only need is for database records to include relevant file paths for deploying models online.

30 Münster et al. 2017, 313–330.

Regardless of whether end users of models are likely to be non-specialists or scholars, digital media literacy is a major concern. How, then, does one package most, if not all of this underlying information into a format that sufficiently fulfills practical referencing needs of traditional academic publishing, but conducive to 3D media's peculiarities?

4 Low-Tech Solutions for High-Tech Capabilities

The easiest solution to the abovementioned issues might seem clear, at first: annotation within the model itself. This possibility stands up better in theory than in implementation. If a model deserves more than just a handful of notes and citations to cover the necessary reference information, a visual field is likely to get muddled rather quickly, even if, for instance, notations can toggle on and off. Often there is so much detail to cover that, if a model's use is primarily casual or educational, reference annotations will likely drown out didactic material fairly readily. Further, although ideal in concept, in practice the completely thorough, step-by-step chronicling of every single step of model (re)construction is detrimentally cumbersome for both creators and consumers if they are expected to be visible during use of the media itself, or even at all. As a result, development of in-media documentation standards, particularly as they apply to dissemination, has been very slow to advance beyond bear minimum annotation or caption within the media, traditional bibliography, or a slate of "movie credit" attributions and bibliography at the end of a production.[31] With the understanding that Giza 3D media will reach a very broad and diverse user base, it is essential to package referencing and citation information in a broadly accessible format.

The Giza Project's response is consciously low-tech in its approach. As part of larger program of funding from the National Endowment for the Humanities Division of Preservation and Access (2016–2017),[32] the Project developed a referencing system for its models that fulfills the main functions of footnotes/endnotes, bibliography, and explanatory manual. The strategy is comprised of

31 A notable exception for use specifically during academic development of models, as well as in educational settings, is VSIM, which offers "two critical functions for academic use of interactive computer models: the narrative features that allows users to create linear presentations within the virtual space ... and the embedded resources feature that allows users to embed annotations and links to primary and secondary resources within the virtual environment." https://idre.ucla.edu/research/active-research/vsim.

32 National Endowment for the Humanities Division of Preservation and Access Humanities Collections and Reference Resources grant # PW-234775-16.

three referencing documents: a Visual Construction Summary, a Model Sourcing Document, and a Scene Composition Document. Composed as Word files but archived and distributed as PDFs, these three documents themselves also become data objects associated with models in the database, so that (upon future completion of infrastructural updates) they will be readily retrievable as data, along with derivative media applications for which modeled content has been deployed. This linkage ensures that references will be readily available alongside models themselves.[33] When such time comes that 3D model assets themselves are available for immediate download (as opposed to only online viewing), reference documents can be downloaded along with them as accompanying explanatory literature. Templates for these referencing documents have been made available online and can be accessed at https://doi.org/10.6084/m9.figshare.21501591

5 Reference Type #1: The Visual Construction Summary

The first reference type is the Visual Construction Summary (VCS; see Appendix 1A & online Supplemental Media B). It is the least formal of the three documentary devices to be introduced here, the least technical, and most publicly accessible in content. Its purpose is to record a visual "quick reference" sheet for the gross progression that a 3D-graphic model underwent from start to finish in its build—for the Giza Project, usually from base architectural plans to completed model. The objective is not all-encompassing, or even necessarily systematic coverage of the model's development. Rather, it is the digital equivalent of time-lapsed photography of a real-world construction site. Whether arranged in a linear sequence or asynchronously, images and figures are assembled to generally illustrate the major steps of the process, i.e. key benchmarks from laying the first foundations to a finished 3D art asset (e.g. building, landscape, object, avatar, etc.).

The Giza Project has found that, by and large, the simple format of this document offers usefully broad latitude for how much or how little to show based on several factors, including: size and complexity of the modeled subject, special artistic techniques, areas of special difficulty/attention, and even, sometimes

33 Because of resource limitations and the timing of overlapping grant cycles, ongoing upgrades to expand the *Digital Giza* website (and corresponding Digital Giza API) from limited-function prototype to its first full version have not included integration of these reference documents beyond the GizaCARD. At this time of writing, availability of these reference documents via the public interface awaits the next phase of API and website updates.

unfortunately, how (in)attentive digital artists were to photo-documenting their creative and technical processes along the way. Both intent and process are the same in producing these summaries regardless of how large or small the subject of the model is—whether a whole landscape, a building, a person (avatar), an archaeological artifact, or a prop. For most, a basic range of stages that will appear in the document are:

- indications of primary bases of sources documentation
- early-stage extrusions from two-dimensional data (e.g. plans and sections, object illustrations)
- wireframe models of a subject's geometry (complete or partial)
- solid, or "grey" models that fill in the structural geometry as solids
- finished or near-finished modeled subject, often displaying facets of material properties, lighting effects, etc.

Appendix 1B is an example VCS for the Giza 3D Menkaure Valley Temple, for which multiple phases of temple renovation and settlement activity in courtyard spaces were modeled as part of Giza 3D.[34]

6 Reference Type #2: The Model Sourcing Document

The second reference form includes information that collectively fulfills several important needs of archaeological visualization data. Labelled a Model Sourcing Document (MSD; see Appendix 2A & online Supplemental Media C), it functions analogously to the bibliography and footnotes/endnotes in a written academic work. Sixteen categories of information are identified as essential for this document. These categories are based primarily upon the combined recommendations of the Archaeology Data Service and Digital Antiquity, as published in their Guides to Good Practice for "Virtual Reality" projects. More precisely, the Giza Project has adopted and/or adapted several of their recommended elements for general documentation, methods, and techniques.[35] While some categories have been carried over to the Project's documentation scheme as presented, others are adjusted or conflated in the interest of designing a manageable template.

34 Reisner 1931, 34–54, Plans VIII–X; Lehner 1997, 137; Lehner and Hawass 2017, 271–283; Lehner 2015, 227–314. For recent renewed archaeological investigation, see Tavares 2008, 8–11; Lehner 2018, 14–17. Picardo 2012.

35 These guidelines at time of writing: http://guides.archaeologydataservice.ac.uk/g2gp/Vr_5-2; http://guides.archaeologydataservice.ac.uk/g2gp/Vr_5-4. Periodic updates are expected in the future, and the Giza Project intends to stay relatively current with its practices, insofar as staffing and resources can reasonably implement comprehensive updates.

First and foremost, the MSD records the source materials used in the creation of a model—primary data and archival records and images, publications, theoretical interpretations, specialist communication, unusual or non-traditional sources, etc. A section for "Interpretive Specifications & Commentary" accommodates the explanation of decisions and special details that are particular to a given model, especially for instances from which extension from hard data, artistic license, or other non-empirical processes were required—which, as discussed above, can be frequent. This form also requests the listing of accessory files such as images of model surface textures (e.g., materials, ground types and coverings, human and animal "skins," etc.) and sound files (e.g., environmental sounds, activity sound effects, ambient noise, etc.).[36] It also includes some technical metadata for the model as a media object, including the 3D formats in which it exists, the file formats used as underlying or integrated components of the model, and listing of the software tools used to generate the model. It is important to distinguish the MSD as a citation document, as opposed to a technical spec sheet. It is not necessary to chronicle fundamental steps of the modeling and rendering processes, e.g. drawing lines; extruding, editing, and decimating polygons; fine-tuning material properties and lighting settings. Such technical specifications, which are discoverable when a model is opened with the appropriate software, are not primary concerns for documenting the intellectual underpinnings of modeled content.

Appendix 2B is an MSD for the Sphinx Temple, a component of the Khafre Pyramid Complex.[37] This monument model (see Figure 16.5) is a good case example for the importance of developing the documentation introduced in this article. Originally built under the mandate to reconstruct models to completion, the Giza 3D Sphinx Temple presented the unusual circumstances of a monument that itself was unfinished in antiquity.[38] As this example MSD necessarily records, the decision for this version of a Sphinx Temple model was to understand "finished" as equivalent to "as may have been intended."[39]

36 The inclusion of this item anticipates a future when 3D modeling is a substantially more prevalent skill, and sharing of model files as data resources directly from online archives such as *Digital Giza* is more commonplace. The integration of texture images will allow them to be treated as modeling resources in the same manner as traditional archival documents.

37 Hassan 1953, 25–29, 33, 37–39, Pls. XV–XVII, XXXI–XXXII, plan (p. 16). Lehner 1997, 128–130; Lehner and Hawass 2017, 220–226.

38 Lehner 1997, 128–129; Lehner and Hawass 2017, 220–221.

39 The Giza Project is not the first to visualize the Sphinx Temple with red cladding stone that likely was not included in its original construction. For example, see Lehner *1997*, computer reconstruction spanning pages 110 + 115.

FIGURE 16.5 Render of a Giza 3D Sphinx Temple model, hypothetical "as likely intended originally" reconstruction

7 Reference Type #3: The Scene Composition Document

Often archaeological models are combined into vignettes, scenes, animated video productions, and applications. Visualized scenes serve a wide spectrum of exploratory, educational, research, and entertainment (and indeed, "edutainment") purposes. Consequently, their target audiences are equally diverse. However, as with individual models, the importance of providing means to understand the material as combined and presented is nonetheless the same for all viewers/users. The inherent dangers in constructing composite scenes are more significant than with individual models, all the more so because of the nature of archaeology itself. As a field of study, archaeology is fundamentally about spatial and situational relationships among people and things: *who* did *what*; *when* and *how* they did it; *where* they did it; and with *what*? Archaeologists then ultimately address larger interpretive questions of *why* across multiple scales of human activity and culture. Visualizations are powerful tools for communicating and testing these aspects of the past, and 3D media are well equipped for conveying all of them at once. However, by situating people and objects within natural and built environments, realistic relationships among all three are implied. Scenes create archaeological realities, which are still interpretive. Even when not distributed by identifiably reputable creators, they can be misinterpreted as absolute, authoritative realities by many viewers.

To address this situation the Scene Composition Document (SCD; see Appendix 3A & online supplemental media D) adapts the methodology advanced in response to similar considerations in the biological and biomedical sciences.[40] These fields have in recent years turned increasingly to 3D modeling and animated visualization as means of demonstrating biological structures, mechanics, and interactions at the molecular and cellular levels to visualize theoretical models based on laboratory experimentation.[41] The Giza Project's SCD has simplified portions of the published approach to align it better with archaeological datasets. Briefly stated, the purpose of the Scene Composition Document is to thoroughly dissect the content of every scene in an animated media production/application. Doing so first requires identification of all **elements** that comprise a scene. Elements include everything from overall environment to tiny, individual props; from avatar characters to text or arrows on the screen; from generic (or fabricated) items to archival documents embedded in the animation (see Table 16.1).

In produced media of this sort, everything has been put there for a reason, however major or minor it may be. The subsequent sections of the SCD clarify these reasons, and they are dependent upon the scene elements themselves. Wherever applicable, the document records three categories of information for each element:

(1) **Element Properties**: These attributes fully describe an element in the context of the scene, in terms of structure, appearance, activity/interaction, grouping.

(2) **Reference Categories**: This category identifies the nature of the reference sources that have been brought to bear on an element's properties in the scene. A minimalist nomenclature of just four terms—visual, quantitative, qualitative, speculation/artistic license—economically accommodates a broad range of reference types, ranging from strictly empirical to primarily conjectural.

(3) **Reference Uses:** These categories specify the process(es) by which, and extent(s) to which, information in Reference Categories have translated into the element's appearance and use in the scene, whether by direct import, adaptation, interpolation, extrapolation, sampling, and/or reduction (see Table 16.2).

---

40 The citation methodology adapted by the Giza Project is specifically that of Jantzen, Jenkinson, and McGill 2015, 293–297.

41 For relevant summary background and discussion, see "Bio-cinema verité?" 2012, 1127; Iwasa 2015, 84–88; Iwasa 2016, 247–250.

TABLE 16.1 Description and examples of element types in a Scene Composition Document (SCD)

ELEMENT TYPE	*DESCRIPTION*	*EXAMPLES*
ENVIRONMENT	*Settings and surroundings of Scene subject or activity; provides containment, spatial parameters, and context*	*tomb chapel; temple sanctuary; house courtyard; riverbank*
CHARACTERS—primary	*Primary subject(s) of the scene; central to theme or narrative*	*avatar; animal*
CHARACTERS—secondary	*Secondary subject(s) of the scene; peripheral or supportive to the theme or narrative*	*avatar; animal*
OBJECTS	*Non-fixed elements of the Scene, i.e., "props" or scenic elements that are not fixed components of another element such as the environment*	*statue; furniture; truck; boat*
DATA OBJECTS	*Primary data items, included wholly or partially in the scene; may appear statically/dynamically/interactively*	*excavation photograph; object illustration; field diary page; newspaper clipping*
TEXT	*Written, on-screen text*	*caption; speech bubble; label*
ACCESSORIES	*Assistive communication devices (may be textual)*	*arrow; icon; bounding-box; highlighting; floating text instruction*
INFORMATION GRAPHICS	*Visual representations of data synthesis, generated for this Scene (i.e., not primary data objects)*	*table of priests' duty rotation schedule; genealogy tree of a character's family*
AUDIO	*Auditory track(s)*	*narration soundtrack; music; dialogue; animal calls; ambient wind*

TABLE 16.2 Description of reference use types in a Scene Composition Document (SCD)

USE	*DESCRIPTION*
DIRECT IMPORT	*Information is imported directly or copied one-to-one with no modification*
ADAPTATION	*Information is translated into another format (usually visual)*
INTERPOLATION	*Missing information is approximated*
EXTRAPOLATION	*A specific mode of interpolation; known information is extended to encompass aspects of elements to which it likely also applies*
SAMPLING	*Selected pieces of information are used*
REDUCTION	*Some information is removed and/or partially excluded*

For additional definitions and uses of terms, see the SCD template in Appendix 3A.

Appendix 3B is an example SCD for just a single scene in the narrative video "The Wonders of Queen Meresankh's Tomb,"[42] set in the Giza 3D mastaba of Meresankh III (G 7530–7540).[43] This segment is about a minute in duration. The primary delimiting factor in defining it as a discrete scene is spatial coverage. It begins upon entry into a space within Meresankh's tomb chapel and ends upon leaving those spatial confines. Scene elements include a built environment, primary and secondary human avatar characters, modeled objects or props, and pop-out content.

The purpose of the SCD is to document how sources have influenced the composition of the scene, as opposed to every facet of individual model assets that comprise the scene, a need that is satisfied rather by the Model Sourcing Documents described above. Once completed for all scenes in an animated production, a complete set of SCDs function collectively like an annotated storyboard, identifying how information sources have been selectively employed in service of both a narrative (if applicable) and the overall objective of the production. Depending upon the number of scenes in total, a comprehensive sequence of SCDs can result in a very long document.

---

42 http://giza.fas.harvard.edu/videos/87542/full/.

43 Dunham and Simpson 1974. http://giza.fas.harvard.edu/sites/1175/full/ See also Aronin 2018, 48–56.

8 Concluding Perspective

These three reference documents together do not comprise a perfect solution, and expectations are that they will require updates in the future as continually evolving technologies yield new tools and standards.[44] However, in design and format the Visual Construction Summary, Model Sourcing Document, and Scene Composition Document offer a practical balance of thoroughness and transparency, on one hand, and simplicity and ease of distribution, on the other hand. As a solution, not only can this low-tech system be preserved and distributed in the Giza Project's own database–API–website ecosystem (alongside, and integrated with, many primary resources for the models themselves), but it also can be disseminated via external, public venues. If a communication mode is able to transfer a file that contains 3D media, it is a virtual certainty that it is equipped to transfer one, two, or three PDF or Doc files as annotated support. With some obstacles to the transmission of supporting information removed, discussions about data, interpretive choices, artistic license, and notions of reality/realism can be addressed head-on via traditional avenues of scholarly discourse. As a result, far less is left to the realm of assumption, which—whether to the benefit or detriment of 3D model content—often inserts its own kinds of unsubstantiated information.

List of Online Supplemental Media

The following documents and media are available online at https://doi.org/10.6084/m9.figshare.21501591.

A. Animation of successive phases of the Menkaure Valley Temple at Giza and associated settlement occupations (based on and adapted from Reisner, *Mycerinus* (1931) and Harvard University-Museum of Fine Arts, Boston Expedition photography). Animation by David Hopkins and the Giza Project at Harvard University.
B. Visual Construction Summary template (MS Word)
C. Model Sourcing Document template (MS Word)
D. Scene Composition Document template (MS Word)

44 Giza Project activities currently also include modeling via photogrammetry as well as Augmented Reality applications; however, the three reference documents presented here have not been rolled out for these media. But, although the documents were developed to accommodate manual reconstruction of archaeological remains from archival records, they lend themselves readily—with or without modification—for use in 3D model building from photogrammetric data and other processes based on image capture.

Appendix 1

Appendix 1A: VCS Template

VISUAL CONSTRUCTION SUMMARY

GIZA PROJECT at HARVARD

DOCUMENT ID:
GizaCARD record number for this document

DATE:
Date of document compilation or most recent update.

MODEL TITLE:
Name of the model, subject, or project.

MODEL ID:
GizaCARD record number for the model

MODEL TYPE:
3D-graphic model; reconstruction

MAPS/PLANS, WIRE-FRAME, GREY MODEL

Figure 1. [CAPTION]

TEXTURE MAPPING, COLORING, LIGHTING

Figure 2. [CAPTION]

FINISHED MODEL (EXTERIOR)

Figure 3. [CAPTION]

FINISHED MODEL (INTERIOR)

Figure 4. [CAPTION]

APPENDIX 1A

Appendix 1B: Example VCS

VISUAL CONSTRUCTION SUMMARY

GIZA PROJECT at HARVARD

Document ID:
GPH_3D-VCS_Menkaure Valley Temple

Date:
August 9, 2017

Model Title:
Menkaure Valley Temple

Model ID:
GPH_3D_Menkaure Valley Temple

Model Type:
3D-graphic model; reconstruction

MAPS/PLANS, EXTRUSIONS, WIRE-FRAMES, GREY MODELS

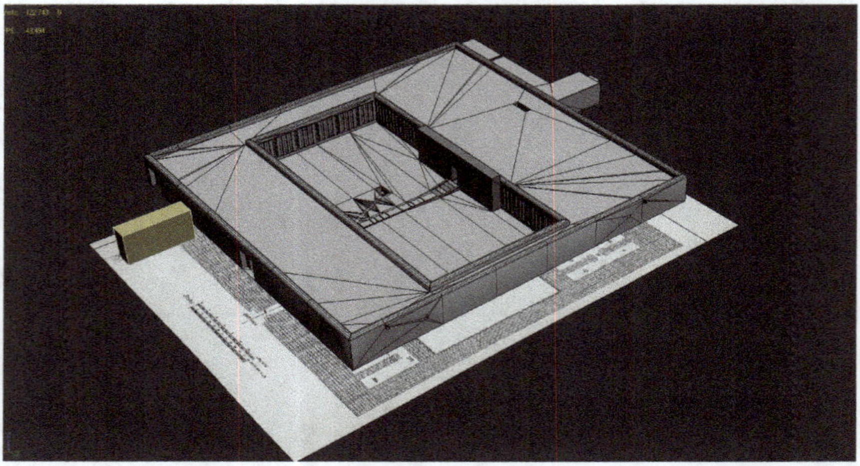

Figure 1. GPH_3D_MVTv1_2011-06-28_wip_002.jpg [phase 1]

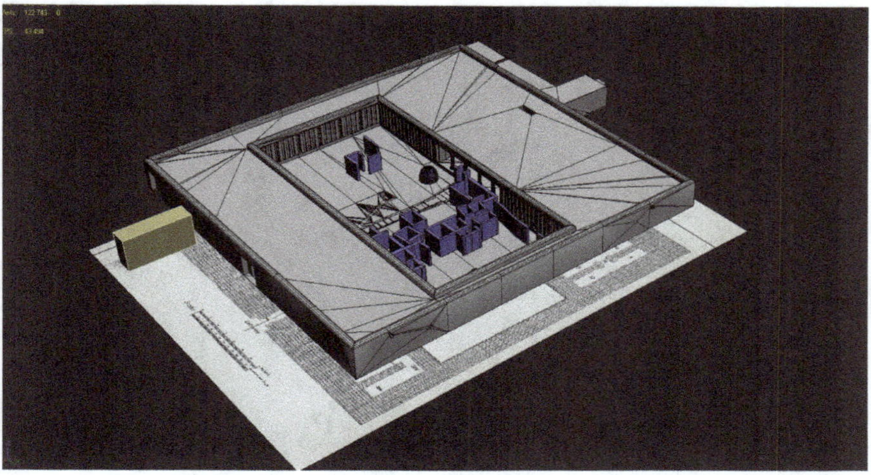

Figure 2. GPH_3D_MVTv2_2011-06-28_wip_001.jpg [phase 2]

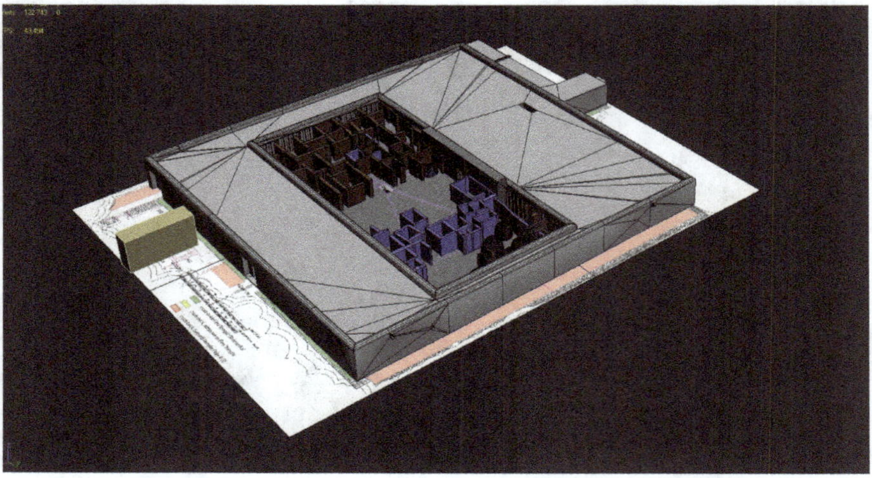

Figure 3. GPH_3D_MVTv2-3_2011-06-28_wip_001.jpg [phase 2-3]

APPENDIX 1B

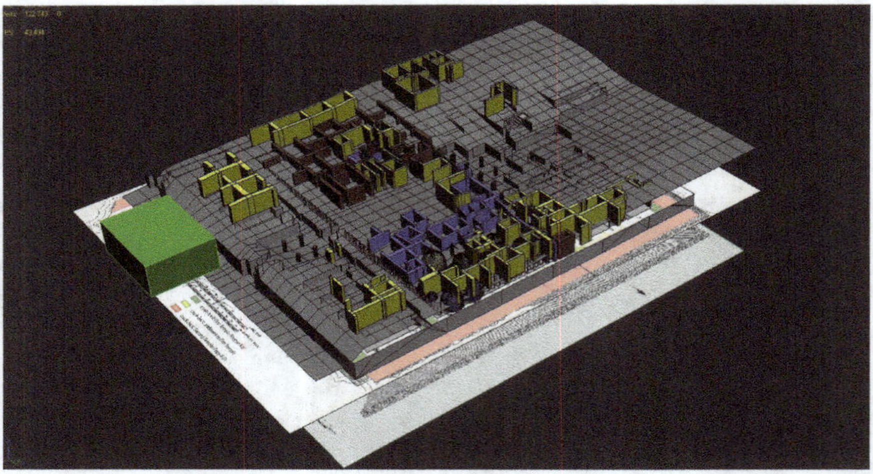

Figure 4. GPH_3D_MVT_2011-06-28_wip_001.jpg [phases 2-3]

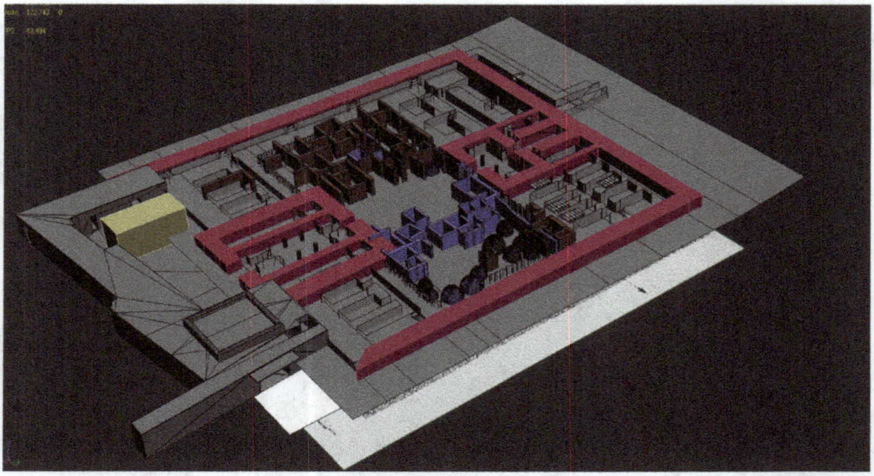

Figure 5. GPH_3D_MVT_2011-06-28_wip_002.jpg

APPENDIX 1B

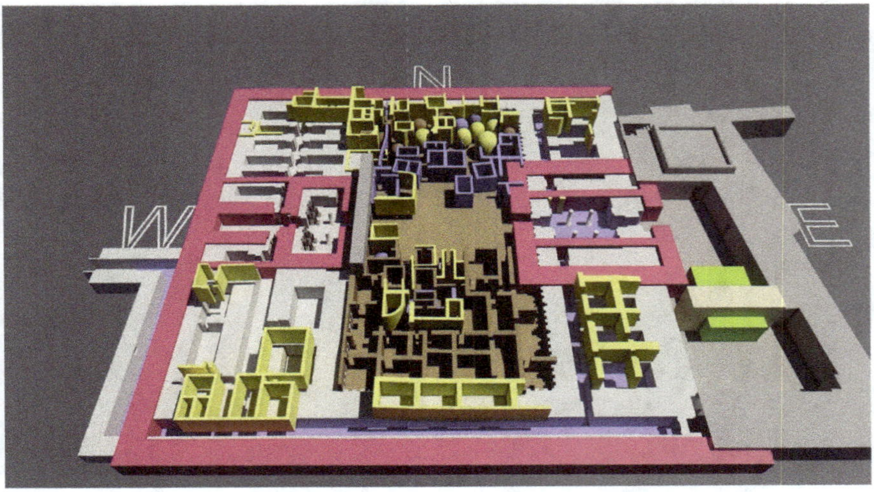

Figure 6. GPH_3D_MVT_2011-06-02_wip_001.jpg [superimposed phasing extrusions]

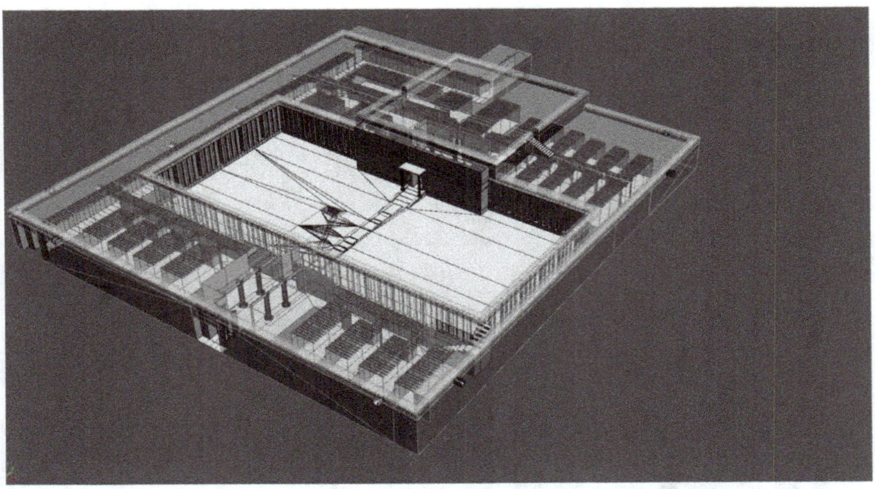

Figure 7. GPH_3D_MVT_2010-10-19_wip_008.jpg [phase 1]

APPENDIX 1B

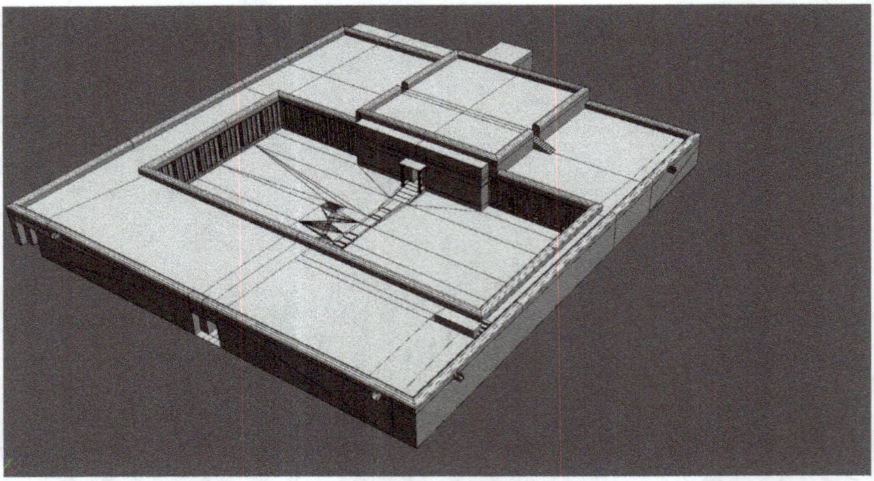

Figure 8. GPH_3D_MVT_2010-10-19_wip_009.jpg [phase 1]

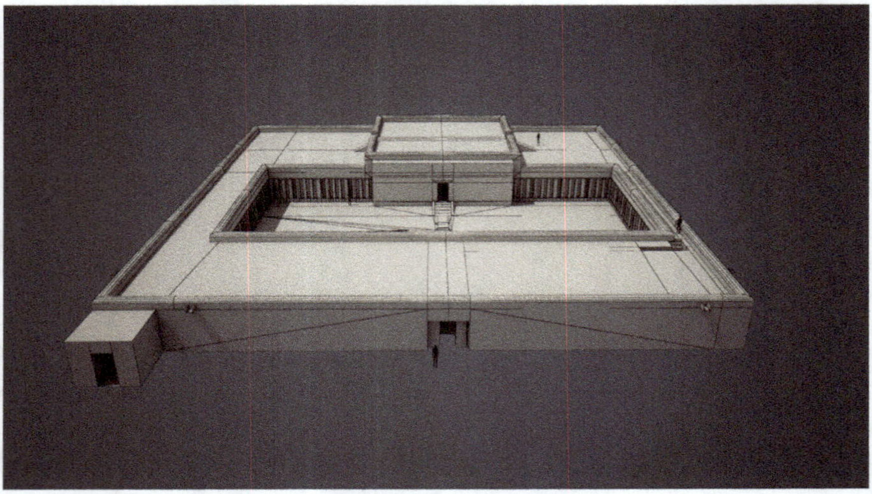

Figure 9. GPH_3D_MVT_2011-11-08_wip_082.jpg [phase 1]

APPENDIX 1B

Figure 10. GPH_3D_MVT_2011-11-08_wip_088.jpg [phase 1, scale]

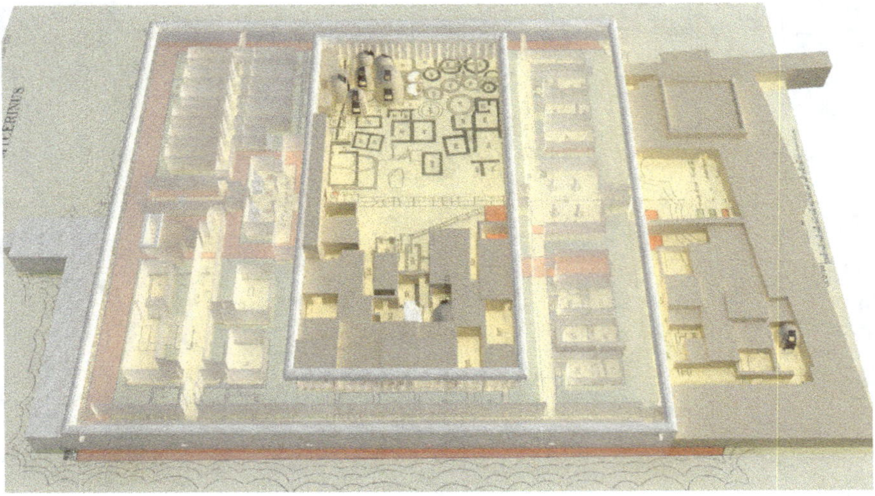

Figure 11. GPH_3D_MVT_2011-11-21_wip_006.png [phases 2-3 work in progress]

APPENDIX 1B

Figure 12. GPH_3D_MVT_2011-11-21_wip_007.jpg [granary detailing]

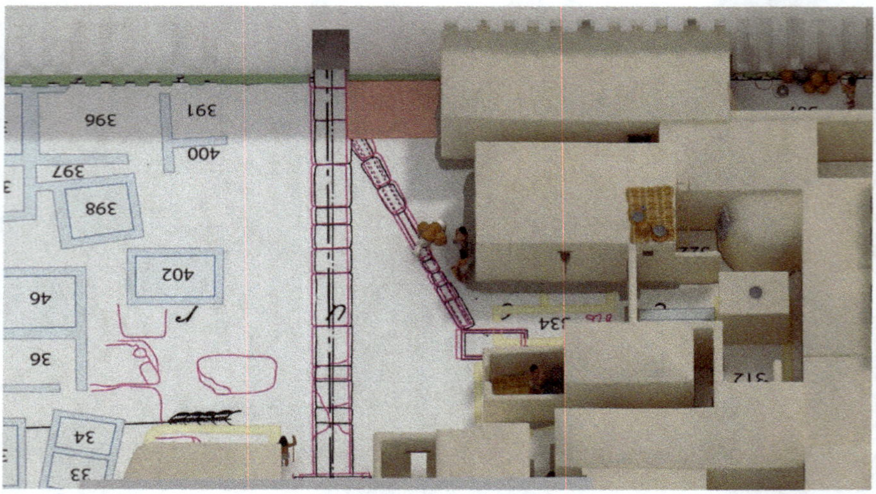

Figure 13. GPH_3D_MVT_2012-03-29_wip_006.jpg [settlement plan alignment]

APPENDIX 1B

Figure 14. GPH_3D_MVT_2012-03-29_wip_007.jpg [settlement plan alignment]

TEXTURE MAPPING, COLORING, LIGHTING

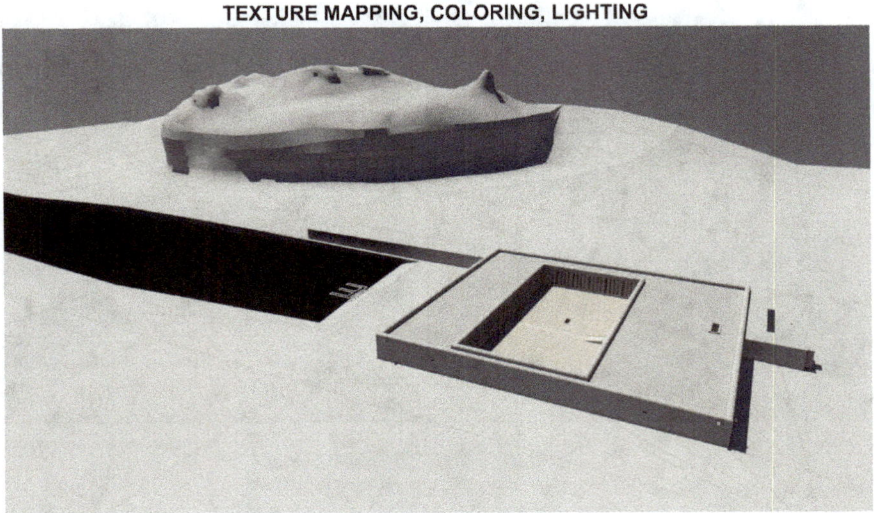

Figure 15. GPH_3D_MVT_2011-05-24_wip_002.jpg

APPENDIX 1B

Figure 16. GPH_3D_MVT_2011-11-08_wip_007.jpg

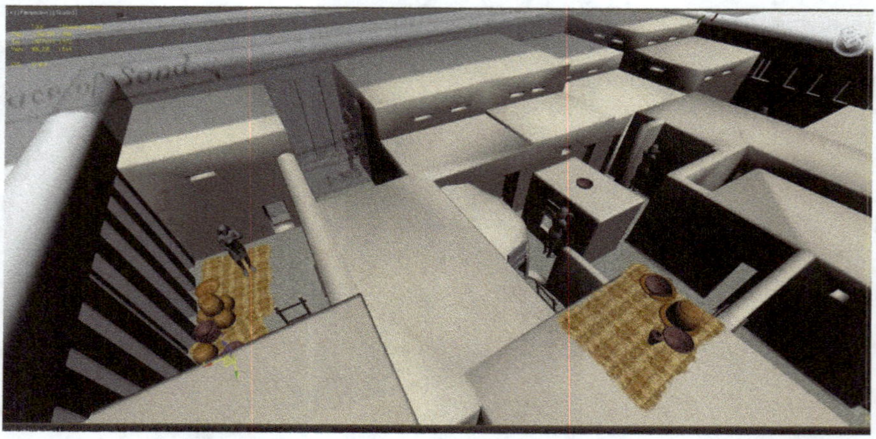

Figure 17. GPH_3D_MVT_2012-03-20_wip_013.jpg [texturing, later phase]

APPENDIX 1B

Figure 18. GPH_3D_MVT_2011-05-05_wip_013.jpg [statue placement and texturing, coloring]

Figure 19. GPH_3D_MVT_2011-05-05_wip_021.jpg [lighting test]

APPENDIX 1B

Figure 20. GPH_3D_MVT_2011-05-05_wip_022.jpg [lighting test]

Figure 21. GPH_3D_MVT_2011-11-08_wip_030.jpg [lighting test]

APPENDIX 1B

Figure 22. GPH_3D_MVT_2011-11-07_wip_008.jpg [coloring]

FINISHED MODEL (EXTERIOR)

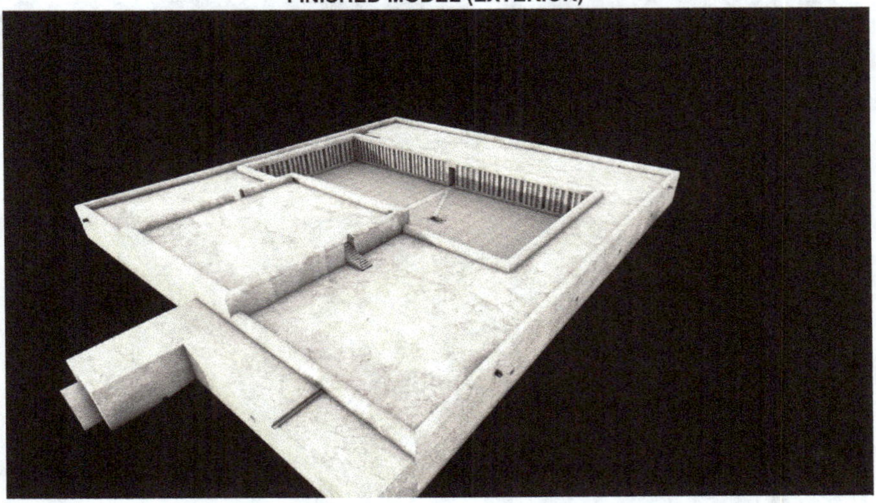

Figure 23. GPH_3D_MVT_2010-03-11_wip_010.jpg [phase 1]

APPENDIX 1B

Figure 24. GPH_3D_MVT_2010-03-11_wip_011.jpg [phase 1]

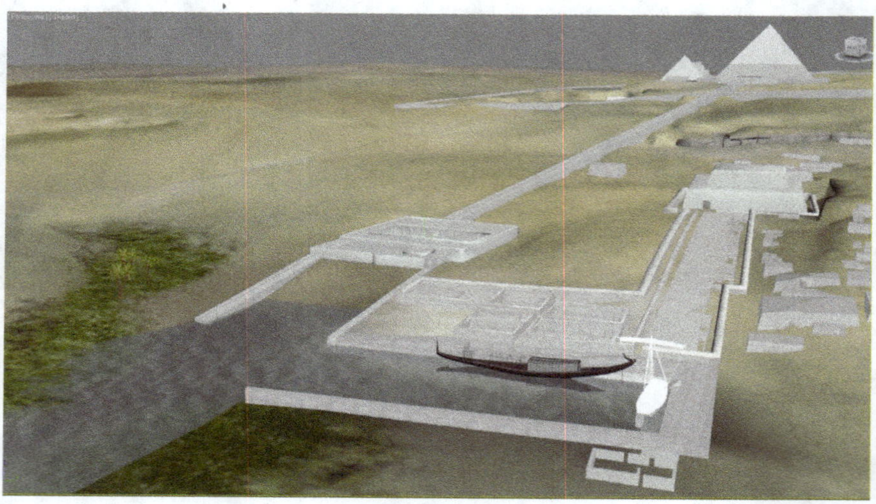

Figure 25. GPH_3D_MVT_2012-05-02_wip_001.jpg [phase 1]

APPENDIX 1B

Figure 26. GPH_3D_MVT_2011-11-07_wip_001.jpg [phase 1]

Figure 27. GPH_3D_MVT_2011-11-07_wip_002.jpg [phase 1]

Figure 28. GPH_3D_MVT_2012-11-29_wip_004.jpg [later phase 2-3]

APPENDIX 1B

Figure 29. GPH_3D_MVTv4_2013-05-08_wip_002.jpg [phase 4]

Figure 30. GPH_3D_MVTv5_2013-05-08_wip_002.jpg [phase 5]

APPENDIX 1B

FINISHED MODEL (INTERIOR)

Figure 31. GPH_3D_MVT_2010-03-11_wip_006.jpg

Figure 32. GPH_3D_MVT_2010-03-11_wip_008.jpg

APPENDIX 1B

Figure 33. GPH_3D_MVT_2011-11-08_wip_125.jpg

APPENDIX 1B

Appendix 2

Appendix 2A: MSD Template

MODEL SOURCING DOCUMENT GIZA PROJECT at HARVARD

DOCUMENT ID:
GizaCARD record number for this document

DATE:
Date of document compilation or most recent update.

MODEL TITLE:
Name of the model, subject, or project.

IDENTIFIER:
GizaCARD record number for the model

MODEL TYPE:
3D-graphic model; reconstruction

IMAGE:
Primary model view,
rendition appropriate to content.

AVAILABLE FORMATS:
File formats in which this model exists (e.g. .unity, .max, .blend, etc.)

FILE FORMATS OF MODEL ASSETS:
Record all file types used as assets in the model (e.g. .jpg, .tiff., .png, .dxt, .max, .obj., .cmo, .nmo., unity, etc.)

SUBJECT:
Keywords or phrases that describe the subject or content of the model. These may include:

- Subject discipline: e.g. archaeology, architecture, geomorphology, etc.
- Subject type: e.g. mastaba tomb, causeway, sculpture, statue, pottery vessel, donkey, furniture, etc.
- Temporal period: e.g. Old Kingdom, ancient Egypt, etc.

APPENDIX 2A

DESCRIPTION / COVERAGE:
A brief description of the model, including its primary purpose, spatial and/or topical extent, and general degree of reconstruction, and any other details appropriate to the subject. Also briefly indicate the model's relationship to real-world entities.

ARTIST(S) / SPECIALIST(S):
List the principal participants in creation of the model; i.e. those who fulfilled primary roles in research, data assembly, interpretation, and build. Specific roles may be indicated.

DATES:

- **Creation:** Inclusive dates of model build.
- **Publication(s):** date(s) of the model's release (open or limited) to public outlet (e.g. online, digital application, museum, etc.). Indicate the outlet to which it was released.
- **Decommissioned:** where applicable, the date after which the model is to be considered outdated or otherwise not for primary use, but rather primarily as a historical/archival holding.

3D DRAWING TOOLS:
Record all known 3D drawing software used in the creation of the model and its assets (e.g. 3ds Max, Blender, SketchUp, Mudbox, etc.), including version.

SOURCES/REFERENCES:

- **Publications:**
- **Images:**
- **Maps/Plans/Drawings:**
- **Field Documents:**
- **Consultants:**
- **Other Sources:**

TEXTURES:
List by file name and format (e.g. .jpg, .tiff, .dxt, etc) all files used for the model's textures.
Where applicable, include also the format in which these files have been incorporated into the model after processing or compression.

AUDIO FILES:
List by file name and format (e.g. .wav, .mp3, etc.) all files used for sound in the model.

INTERPRETIVE SPECIFICATIONS & COMMENTARY:
Include any useful comments and details about the intellectual, interpretive, and/or artistic choices that are particular to this model. Note especially any specifics that directly affected the model's construction process and its final form in noteworthy ways.

APPENDIX 2A

OTHER MODEL CONTENT:
List all individual models (by IDENTIFIER and MODEL NAME) that are used in it. This will apply especially, but not exclusively, to real-time models/environments, scenes, animations, applications, etc.

RIGHTS:
Any known copyrights on the model.

APPENDIX 2A

Appendix 2B: Example MSD

MODEL SOURCING DOCUMENT

DOCUMENT ID:
GPH_3D-MSD_Sphinx Temple

DATE:
6/14/2017

MODEL TITLE:
Sphinx Temple

IDENTIFIER:
GPH_3D_Sphinx Temple

MODEL TYPE:
3D-graphic model; reconstruction

AVAILABLE FORMATS:
.ax, .unity

FILE FORMATS OF MODEL ASSETS:
.jpg

SUBJECT:
Archaeology, ancient Egypt, Giza, Old Kingdom, Dynasty 4, Sphinx, temple, architecture, history

DESCRIPTION / COVERAGE:
This is a model of the Old Kingdom temple associated with the Great Sphinx at Giza, part of the Khafre Pyramid Complex. The Sphinx Temple was never fully completed in antiquity. Most notably, the casing of red granite blocks around the exterior was not constructed. This model reconstruction depicts the temple theoretically as it may have been intended, with red exterior casing applied.

<u>ARTIST(S) / SPECIALIST(S)</u>:
Dassault Systémes sub-contractor digital artists; David Hopkins

<u>DATES</u>:

- **Creation:** 2013-2017
- **Publication(s):** Published online on the Giza 3D website (http://giza3d.3ds.com) in 2013. Published online on the *Digital Giza* website (giza.fas.harvard.edu) prototype release in 2017.
- **Decommissioned:** N/A

<u>3D DRAWING TOOLS</u>:
3ds Max, Unity

<u>SOURCES/REFERENCES</u>:

- **Publications:**

Anthes, Rudolf. "Was Veranlasste Chefren zum Bau des Tempels vor der Sphinx?" In *Zum 70. Geburtstag von Herbert Ricke. Beitraege zur Aegyptischen Bauforschung und Altertumskunde* 12. Wiesbaden: Franz Steiner Verlag, 1971, p. 48.

Lehner, Mark. *The Complete Pyramids: Solving the Ancient Mysteries*. London: Thames and Hudson, 1997, pp.128-130.

Maragioglio, Vito, and Rinaldi, Celeste. *L'Architettura delle Piramidi Menfite, Parte V: Le Piramidi di Zedefra e di Chefren*. Rapallo: Tipografia Canessa, 1966, pl. 14.

Ricke, Herbert. *Der Harmachistempel des Chefren in Giseh. Beiträge zur Ägyptischen Bauforschung und Altertumskunde* 10. Wiesbaden: Franz Steiner Verlag, 1970.

Stadelmann, Rainer. *Die Grossen Pyramiden von Giza*. Graz: Akademische Druck- u. Verlasanstalt, 1990, p. 188.

- **Images:**

ASU_key, ASU_A6_012, ASU_A6_013, ASU_A6_016, ASU_A6_017, ASU_A6_018, ASU_A6_025, ASU_A6_026, ASU_A6_027, ASU_A7_006, ASU_A7_007, ASU_A7_008, ASU_A7_009, ASU_A7_010, ASU_A7_011, ASU_A7_012, ASU_A7_013, ASU_A7_014, ASU_A7_015, ASU_A7_016, ASU_A7_017, ASU_A7_018, ASU_A7_019, ASU_A7_021, ASU_A7_022, ASU_A9_001, ASU_A9_003, ASU_A9_004, ASU_A9_005, ASU_A9_007, ASU_A9_008, ASU_A9_013, ASU_A9_014, ASU_A9_015, ASU_A10_001, ASU_A10_002, ASU_A10_003, ASU_A11_015, ASU_A11_016, ASU_A11_019, ASU_A11_020, ASU_A11_022, ASU_A11_050, ASU_panorama_002, ASU_panorama_002, ASU_panorama_003, ASU_panorama_010, HUMFA_A7814_NS, MFAB_AAW1141, MFAB_AAW1138, MFAB_AAW1139, PDM_06066, PDM_06097, PDM_1999.201.35, PDM_1999.008.24, PDM_2011.01.17_033, PDM_2011.01.17_034, PDM_2011.01.17_035, PDM_2011.01.17_037, PDM_2011.01.17_038, PDM_2011.01.17_039, PDM_2011.01.17_040, PDM_2011.01.17_041, PDM_2011.01.17_042, PDM_2011.01.17_093, PDM_2011.01.17_129, PDM_2011.01.17_147, PDM_2011.01.17_149, PDM_2011.01.17_150, PDM_2011.01.17_151, PDM_2011.01.17_152, PDM_2011.01.17_153, PDM_2011.01.17_154, PDM_2011.01.17_155, PDM_2011.01.17_156, PDM_2011.01.17_169, TUR_LastraGizaA5559, TUR_Giza548
Lacau archive (at Centre Wladimir Golenischeff) photographs C I 111, C I 112, C I 131, C I 138, C I 156, C I 157, C I 158, C I 159, C I 160

- **Maps/Plans/Drawings:**
  HUMFA_EG019924 (maragioglio-rinaldi_v_plate_014)
  Ricke, *Harmachistempel*, plans 1-2
  Stadelmann, *Die Grossen Pyramiden*, fig. 122

- **Other Sources:**
  ASU_A7_video_002, ASU_A7_video_003, ASU_A7_video_004, ASU_A8_video_001, ASU_A8_video_002, ASU_A8_video_003, ASU_A9_video_001, ASU_A9_video_002, ASU_A11_video_001, ASU_A11_video_002, ASU_A11_video_004, ASU_A11_video_006, ASU_A11_video_008, ASU_I2_video_001, ASU_I2_video_002, ASU_I3_video_001
  Airpano website (https://www.airpano.com/360photo/egypt-cairo-pyramids/)
  Assorted photographs identified via Google Image search, used for comparative visual purposes (i.e. surface textures, lighting properties, etc.).

TEXTURES:
GPH_3D_tx_gen_alabaster_001.jpg
GPH_3D_tx_gen_granite_002.jpg
GPH_3D_tx_gen_limestone_rough_001.1.jpg
GPH_3D_tx_gen_wall_mudplaster_white_001.jpg
GPH_3D_tx_SPHT_courtyard_pillars_inscribed_001.jpg
GPH_3D_tx_SPHT_door_001.psd
GPH_3D_tx_SPHT_doorjamb_001.jpg
GPH_3D_tx_SPHT_doorjamb_inscribed_001.jpg

AUDIO FILES:
N/A

INTERPRETIVE SPECIFICATIONS & COMMENTARY:
Although no statues (or fragments thereof) were found in the Sphinx Temple, the presence of statue emplacements in the floor of the courtyard indicate that there likely had been (or at least, were intended to be) statues in the temple. Therefore, we have placed 10 statues (themselves based on statues used in the Khafre Valley Temple) in the Sphinx Temple courtyard, in the emplacement locations following Ricke's reconstruction (*Harmachistempel*, plan 3). Early versions of the Sphinx Temple model also included an inscription on the temple entrances (again see Ricke, ibid., plan 2; and GPH_SPHT_inscription_001), but since no trace of inscription remains in the Sphinx Temple it was ultimately decided to omit these.

OTHER MODEL CONTENT:
N/A

RIGHTS:
Rights for use of this model belong to the Giza Project at Harvard University.

APPENDIX 2B

Appendix 3

Appendix 3A: SCD Template

SCENE COMPOSITION DOCUMENT GIZA PROJECT at HARVARD

DOCUMENT ID:
GizaCARD record number for this document

DATE:
Date of document compilation or most recent update.

PRODUCTION TITLE:
Title of record for the whole media production

PRODUCTION ID:
GizaCARD record number for the media Scene to which this document applies

PRODUCTION TYPE:
animation/movie

SEGMENT NUMBER & TIME:
Segment # of # (0:00:00–0:00:00)

SCENE CITATION THUMBNAILS:
Include one or more images (e.g., screen captures or similar) of this Scene segment. Select views to provide basic reference coverage of all Scene Elements. Size thumbnails as appropriate to content, image resolution, and any other considerations germane to clarity. Caption each as "citation thumbnail" with the segment number and a letter designation to (e.g., Citation thumbnail 1a, Citation thumbnail 1b, etc.).

Place citation thumbnail images in this space.

Citation thumbnail 1a

Number of views will vary based on scene type and content.

Citation thumbnail 1b

Identify Scene Elements of the following types:

ELEMENT TYPE	DESCRIPTION	EXAMPLES
ENVIRONMENT	Settings and surroundings of Scene subject or activity; provides containment, spatial parameters, and context	tomb chapel; temple sanctuary; house courtyard; riverbank
CHARACTERS-primary	Primary subject(s) of the scene; central to theme or narrative	avatar; animal
CHARACTERS-secondary	Secondary subject(s) of the scene; peripheral or supportive to the theme or narrative	avatar; animal
OBJECTS	Non-fixed elements of the Scene, i.e., "props" or scenic elements that are not fixed components of another element such as the environment	statue; furniture; truck; boat
DATA OBJECTS	Primary data items, included wholly or partially in the scene; may appear statically/dynamically/interactively	excavation photograph; object illustration; field diary page; newspaper clipping
TEXT	Written, on-screen text	caption; speech bubble; label
ACCESSORIES	Assistive communication devices (may be textual)	arrow; icon; bounding-box; highlighting; floating text instruction
INFORMATION GRAPHICS	Visual representations of data synthesis, generated for this Scene (i.e., not primary data objects)	table of priests' duty rotation schedule; genealogy tree of a character's family
AUDIO	Auditory track(s)	narration soundtrack; music; dialog; animal calls; ambient wind

In each thumbnail, identify the major Scene elements with (1) "ghosting" (applying a white/light, translucent overlay) and (2) short, abbreviated text labels appropriate to each (e.g. Kh for King Khufu avatar, Ca for cat, Na for navigation arrow, Fd for field diary page, etc.).

APPENDIX 3A

Element Index:
In list form, provide a key for all Scene Elements and their abbreviated text labels used in the Citation Thumbnails above. For example:

Kh: King Khufu
Ca: cat
Fd: field diary page
Na: navigation arrow
Pv: pottery vat

Element Properties:
For each element as appropriate (usually excluding Accessories), list the following properties as applicable.

PROPERTY	DESCRIPTION	EXAMPLES
STRUCTURE	Overall shape/geometry of an element; architectural and spatial data; scaling and absolute/relative proportioning	architecture of a building; shape of avatar; morphology of a pottery jar
APPEARANCE	Visual/surface appearance; coloration, texturing; interaction with light/shadow,	stone; grass; building plaster; skin; fur; carved relief contour/depth; quality/interplay of light and shadow
ACTIVITY / INTERACTION	Movement, activity, interactions, and associations of an element, alone and/or with other elements	actions of avatars baking bread; movements of dancers; three men talking; donkey pulling cart of onions
GROUPING	Presence of multiples/groups of the same or similar elements	five lamps in one room; six priests offering at a shrine; four hundred quarrymen working a rock outcropping; eight farmers in a field

Use the following format or similar:

Element Type: *Element Name*
Structure:
Appearance:
Activity/Interaction:
Population:

Under each property heading, cite relevant source material(s) that informed that Element Property in this Scene. For each reference listed, indicate the ***Reference Category*** *(or Categories) to which it applies for its use in this Scene. Refer to the chart of Reference Categories below as a guide. Also indicate the* ***Reference Use*** *(or Uses) to clarify the manner in which information was applied from the source to this Scene. Refer to the chart of Reference Uses below as a guide.*

Reference Categories:

CATEGORY	DESCRIPTION	EXAMPLES
VISUAL	Based on visual sources	architectural map/plan; technical illustration of artifact; stratigraphy; pottery profile
QUANTITATIVE	Based on numerical data or direct measurement	3D-scan data, recorded max wall height; minimum vessel count figures
QUALITATIVE	Based on non-visual descriptive data	published interpretation of ancient text about daily temple rituals; synthetic study of functions of house rooms; conference lecture about foundation deposit contents
SPECULATION / ARTISTIC LICENSE	No reference; estimated or imagined information	background character(s); reconstructed house ceiling height; arm movements of ceremony officiant; generic Egyptian text written on scroll

APPENDIX 3A

Reference Uses:

USE	DESCRIPTION
DIRECT IMPORT	*Information is imported directly or copied one-to-one with no modification*
ADAPTATION	*Information is translated into another format (usually visual)*
INTERPOLATION	*Missing information is approximated*
EXTRAPOLATION	*A specific mode of interpolation; known information is extended to encompass aspects of elements to which it likely also applies*
SAMPLING	*Selected pieces of information are used*
REDUCTION	*Some information is removed and/or partially excluded*

ADDITIONAL NOTES:
Include in this space any other noteworthy information that relates to the composition of this Scene and the sources of information used in its conceptualization and creation.

APPENDIX 3A

Appendix 3B: Example SCD

SCENE COMPOSITION DOCUMENT

GIZA PROJECT at HARVARD

DOCUMENT ID:
3D-SCD_Wonders_Meresankh_2013

DATE:
July 1, 2015

PRODUCTION TITLE:
The Wonders of Queen Meresankh's Tomb

PRODUCTION ID:
3DP_Wonders_Meresankh_2013

PRODUCTION TYPE:
animation/movie

SEGMENT NUMBER & TIME:
Segment 5 of 8 (0:02:37–0:03:34)

SCENE CITATION THUMBNAILS:

Citation thumbnail 5a

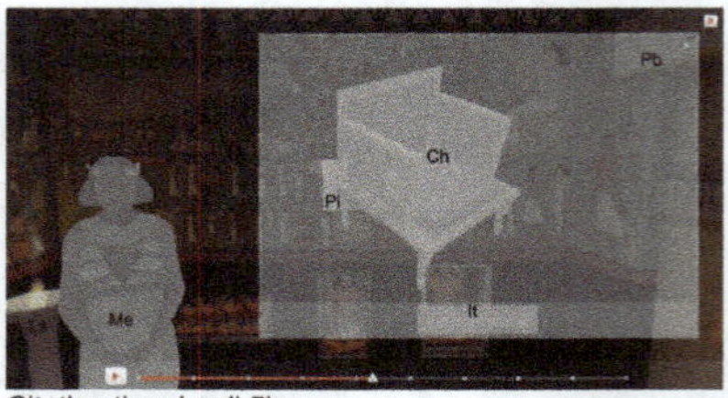

Citation thumbnail 5b

Citation thumbnail 5c

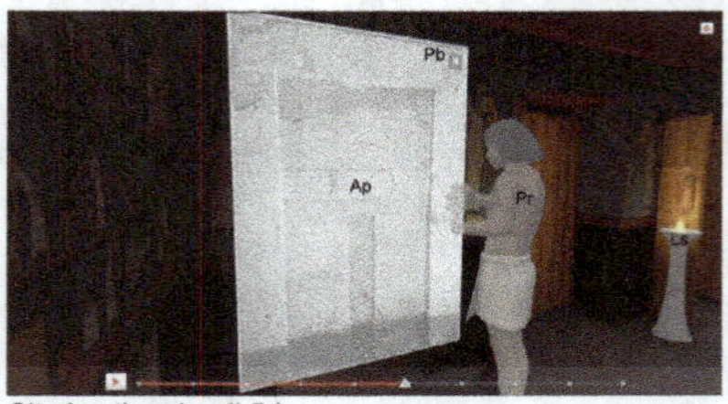

Citation thumbnail 5d

ELEMENT INDEX:

Me: Queen Meresankh III
Pr: censing priest
Ls: lamp stand with bowl and flame
Ch: chair
Pi: pop-out content indicator
Pb: pop-out content containment box
Ap: archival photograph
It: instructional text & graphic for interactive pop-out content

APPENDIX 3B

ELEMENT PROPERTIES:

Environment: Mastaba tomb of Queen Meresankh III (G 7530-7540); Chapel Room A

Structure:

Key references: Dunham, Dows and William Kelly Simpson. *The Mastaba of Queen Mersyankh III (G 7530–7540)*. Giza Mastabas 1. Boston: Museum of Fine Arts, 1974; Model Sourcing Document 3D-MSD_G 7530-7540.

Visual & Quantitative. Adaptation, Artistic License: The architectural structure of this scene's setting is based primarily on measured architectural plans and sections from Dunham & Simpson 1974. Although based on the same source, the engaged statuary in this section of the tomb required more free-hand shaping, with reference to photography published in Dunham & Simpson 1974. Additional details appear as described in reference document 3D-MSD_G 7530-7540.

Speculation/Artistic License. Interpolation: Lighting effects have been applied to suit the scene and its elements.

Appearance:

Key references: Dunham & Simpson 1974; modern field drawings and color photographs as described in Model Sourcing Document 3D-MSD_G 7530-7540.

Visual. Direct Import, Adaptation, Extrapolation, Sampling: Epigraphic line drawings of wall scenes and decorations have been mapped onto modelled wall surfaces, then contours and coloration added. No structural alterations have been made from the full model 3D_G 7530-7540 as described in reference document 3D-MSD_G 7530-7540. Color conventions for the chapel's painted decoration are derived from modern field drawings/paintings and surviving pigment in the tomb, as documented by color photography. Colors were sampled from color photography for improved accuracy, as described in reference document 3D-MSD_G 7530-7540. Areas with no surviving pigment have been restored by extrapolation from these known trends.

Character-primary: Queen Meresankh III (Me)

Structure:

Key reference: Limestone statuette of a woman (possibly Meresankh III?) MFAB_30.1461; Model Sourcing Document 3D_MSD_Meresankh_III.

Visual & Speculation/Artistic License. Adaptation, Sampling, Reduction: Me's facial features are based loosely on those of this excavated statue, with physical stature proportioned from its slight remaining shoulder and upper torso. Me's body reflects considerable speculation and artistic license. Aside from animating the figure, no structural alterations have been made to the avatar as described in reference document 3D_MSD_Meresankh_III.

Appearance:

Key references: limestone statue of woman MFAB_30.1461; beadnet dress MFAB_27.1548.1 + MFAB_27.1548.2; object photographs MFAB_AAW1591 and MFAB_AAW1722; Model Sourcing Document 3D_MSD_Meresankh_III; Dunham, Dows. "An Egyptian Diadem of the Old Kingdom." *Bulletin of the Museum of Fine Arts*, Boston 44, No. 255 (February 1946), pp. 23-29; Dunham & Simpson 1974.

Visual & Artistic License: Adaptation, Sampling, Reduction: Me's hairstyle is based on excavated limestone statue MFAB_30.1461. Her garment choices (linen and beadnet dress) derive from relief depictions of Queen Meresankh III in her tomb chapel (G 7530-7540) as documented in Dunham & Simpson 1974, Model Sourcing Document 3D-MSD_G 7530-7540, and the garment depicted in pair statue excavated from her tomb, MFAB_30.1456, with adjustments for realism as a functional garment. Her beadnet dress is adapted from excavated and reconstructed object MFAB_27.1548.1 + MFAB_27.1548.2. Me's crown/diadem is an artistic reconstruction of excavated object MFAB_37.606a, with adaptations from object photography recorded in the avatar reference documents 3D_MSD_Hetepheres and 3D_MSD_Meresankh_III, with some reference to Dunham 1946.

Activity/Interaction:

Artistic License: all aspects of Me's activities are wholly artistic choices for the scene, including the guided tour of Room A of the chapel, her narration, and her use of the English language.

APPENDIX 3B

Grouping:
Speculation/Artistic License: the simultaneous presence of Queen Meresankh III and a censing priest in her tomb chapel (G 7530-7540) is speculative, composed solely for purposes of the scene.

Character-secondary: censing priest (Pr)
Structure:
Key References: Manuelian, Peter Der. *Mastabas of Nucleus Cemetery G 2100.* Giza Mastabas 8. Boston: Museum of Fine Arts, 2009; excavated object EMC_JE_67601 + EMC_JE_67602; field photograph HUMFA_C13786_N; Model Sourcing Document 3D_MSD_priest_censing.

Visual & Artistic License. Adaptation, Extrapolation: Pr's shape and proportions represent an artistic interpretation of relief depictions of ancient Egyptians in Giza Cemetery G 2100, as they are presented in Manuelian 2009. The form of Pr's censer is an artist's interpretation of excavated bell censer of type EMC_JE_67601 + EMC_JE_67602, for which see field object photograph HUMFA_C13786_NS. Aside from animating the figure, no structural alterations have been made from the avatar as described in reference document 3D_MSD_priest_censing.

Appearance:
Key References: Manuelian 2009; objects EMC_JE_67601 + EMC_JE_67602; field object photograph HUMFA_C13786_NS.

Visual & Artistic License. Adaptation, Sampling: Pr's hair, attire, and coloration are based on artists' assessments of trends observed in relief depictions of ancient Egyptians in Giza Cemetery G 2100, as they are presented in Manuelian 2009. The source of color/texture applied to the censer is unknown/unavailable.

Activity/Interaction:
Speculation/Artistic License. Interpolation: No censing activity is documented in the archaeological record of Chapel A of Tomb G 7530-7540, nor is it known that censing activities occurred in this location near the false door. The movements of the censing priest, although suggestive of an offering ritual, are conjectural also. Only the use of a censer by Pr identifies him as a priest. His attire and appearance do not include any other indications of a priestly role. He is located and animated to provide general context and background activity to the environment.

Grouping:
Speculation/Artistic License: the simultaneous presence of a censing priest and Queen Meresankh III in the chapel of tomb G 7530-7540 is speculative, composed solely for purposes of the scene.

Object: chair (interactive pop-out element) (Ch)
Structure:
Key References: Objects GEM_6365 and (museum reproduction) MFAB_38.957; hundreds of excavation drawings, annotated drawings, and photographs from the Harvard University—Museum of Fine Arts Expedition to Egypt (as recorded in Model Sourcing Document 3D_MSD_EMC_6365).

Visual, Quantitative & Qualitative. Adaptation, Sampling, Extrapolation: Ch is a model of the excavated and restored chair GEM_6365 (from Tomb G 7000 X). The extensive researching and use of hundreds of primary excavation records involved in the modelling the form of Ch (and associated chairs from G 7000 X) is documented in full by Model Sourcing Document 3D_MSD_EMC_6365.

Appearance:
Visual, Qualitative & Artistic License. Adaptation, Interpolation: Choice and distribution of colors and textures for Ch's materials are based upon documentation recorded in Model Sourcing Document 3D_MSD_EMC_6365. However, the actual colors and textures have been sourced from stock "materials" imagery and texture files, as opposed to directly sampled from actual artifactual remains and/or object photographs.

APPENDIX 3B

Activity/Interaction:
As interactive pop-out content, Ch is intended to provide a 360-degree view of an example of a "real" chair like the one depicted in the tomb chapel's painted wall decoration. This use does not necessarily imply that such a chair was originally in tomb G 7530-7540. No excavated remains suggest so.

Object: lamp stand with active flame (Ls)
Structure:
Key References: excavated objects HUMFA_36-2-15 (from Tomb G 2132 C), HUMFA_36-3-43 (from Tomb G 2006), HUMFA_34-12-3 (from Tomb G 1407), EMC_JE_67622; MFAB_21.2616 (from Tomb G 4640); object register pages HUMFA_OR28_p1331, HUMFA_OR28_p1341, HUMFA_OR27_p1233, HUMFA_OR27_p1230; HUMFA_ OR02_p059; object photographs MFAB_AAW1803, MFAB_AAW2011, MFAB_C5450_NS.

Visual. Adaptation, Sampling: This lamp stand is composed of two parts, a pottery base with an alabaster bowl set on it with a burning flame (fueled by oil, not indicated). The form of the base is an artist's rendering that selectively adapts aspects of a range of excavated tall offering stands, including HUMFA_36-2-15, HUMFA_36-3-43, and HUMFA_34-12-3, and EMC_JE_67622, with reference also to technical profile drawings of these objects recorded by excavators in HUMFA_OR28_p1331, HUMFA_OR28_p1341, HUMFA_OR27_p1233, and HUMFA_OR27_p1230, HUMFA_ OR02_p059, respectively, and object photographs MFAB_AAW1803, MFAB_AAW2011. The morphology of the alabaster bowl component is based on bowl MFAB_21.2616, with reference to a technical profile drawing in object register HUMFA_ OR02_p059 and object photographs MFAB_C5450_NS and MFAB_SC27863. The model bowl has been scaled to fit the tall base appropriately.

Appearance:
Visual & Artistic License: Coloration and texturing of materials (red-painted pottery and alabaster) have been sourced from stock "materials" imagery and texture files, as opposed to directly sampled from artifactual remains and/or object photographs. The addition of active flames was accomplished through the use of an application within the rendering engine employed to produce this media production (VirTools).

Activity/Interaction:
Speculation/Artistic License: Tall offerings stands were commonly used to support bowls of many types, while many types of stone bowls are understood as lamps, often having been set on the ground, floor, or other surfaces during use. However, Ls' specific configuration of tall offering stand and this form of alabaster bowl for regular use as lamps is hypothetical.

Grouping:
Speculative/Artistic License: The lamp stand model appears twice in this scene. Distribution and placement are conjectural and have been chosen to support the environment and overall scene.

Data Object: **archival excavation photograph (pop-out content) (Ap)**
Structure:
Key reference: excavation photograph MFAB_A4683_NS

Visual. Reduction & Direct Import: Ap is excavation photograph MFAB_A4683_NS. The photo does not appear in its original form. Rather, for use as an element of this scene, it has been cropped to frame a selected view of the false door.

Appearance:
Key reference: excavation photograph MFAB_A4683_NS

Visual. Direct Import: the uncropped portion of excavation photograph MFAB_A4683_NS that appears is unedited.

APPENDIX 3B

Audio: movie narration/character speech
Artistic License: The audio track for this scene was scripted by the Giza Project at Harvard University solely for purposes of this scene, and within the context of the larger animated production. Recording of the track was contracted by Dassault Systémes, 2013. The audio file of the audio track has not been retained outside of the finished movie production.

ADDITIONAL NOTES:
N/A

APPENDIX 3B

Bibliography

Aronin, Rachel. 2018. "Queen for Eternity: Digital Archaeology and the (after)Life of Meresankh Iii at Giza." *Nile Magazine* 13 (April–May): 48–56.

"Bio-Cinema Verité?". 2012. *Nature Methods* 9 (12): 1127. http://dx.doi.org/10.1038/nmeth.2284.

Casey, Christian. 2018. "Tombs, Temples, and Blood—Assassin's Creed Origins as a Digital and Pedagogical Tool." 69th Annual Meeting of the American Research Center in Egypt, Tucson, AZ, April 22 2018.

Dunham, Dows, and William Kelly Simpson. 1974. *The Mastaba of Queen Mersyankh III, G7530–7540*. Giza Mastabas I. Boston: Deptartment of Egyptian and Ancient Near Eastern Art, Museum of Fine Arts, Boston.

Frischer, Bernard. 2009. "Introduction: From Digital Illustration to Digital Heuristics." In *Beyond Illustration: 2D and 3D Digital Technologies as Tools for Discovery in Archaeology*, edited by Bernard Frischer and Anastasia Dakouri-Hild, v–xxiv. Oxford: Archaeopress.

Georgopoulos, Andreas. 2014. "3D Virtual Reconstruction of Archaeological Monuments." *Mediterranean Archaeology and Archaeometry* 14 (4): 155–154.

Golvin, Jean-Claude. 2012. "Drawing Reconstruction Images of Ancient Sites." In *Picturing the Past: Imaging and Imagining the Ancient Middle East*, edited by Jack Green, Emily Teeter and John A. Larson. Oriental Institute Museum Publications, 77–82. Chicago: Oriental Institute of the University of Chicago.

Hassan, Selim. 1953. *Giza 8: 1936–1937. The Great Sphinx and Its Secrets. Historical Studies in Light of the Recent Excavations*. Cairo: Government Press.

Hermon, Sorin. 2009. "Reasoning in 3D: A Critical Appraisal of the Role of 3D Modelling and Virtual Reconstructions in Archaeology." In *Beyond Illustration: 2D and 3D Digital Technologies as Tools for Discovery in Archaeology*, edited by Bernard Frischer and Anastasia Dakouri-Hild, 36–45. Oxford: Archaeopress.

Iwasa, Janet H. 2015. "Bringing Macromolecular Machinery to Life Using 3D Animation." *Current Opinion in Structural Biology* 31: 84–88. https://doi.org/10.1016/j.sbi.2015.03.015.

Iwasa, Janet H. 2016. "The Scientist as Illustrator." *Trends in Immunology* 37 (4): 247–250. https://doi.org/10.1016/j.it.2016.02.002.

Jantzen, Stuart G., Jodie Jenkinson, and Gael McGill. 2015. "Transparency in Film: Increasing Credibility of Scientific Animation Using Citation." Commentary. *Nature Methods* 12 (4): 293–297. https://doi.org/10.1038/nmeth.3334 http://www.nature.com/nmeth/journal/v12/n4/abs/nmeth.3334.html#supplementary-information.

Lehner, Mark. 1997. *The Complete Pyramids*. London: Thames & Hudson.

Lehner, Mark. 2015. "Shareholders: The Menkaure Valley Temple Occupation in Context." In *Towards a New History for the Egyptian Old Kingdom: Perspectives on the*

Pyramid Age, edited by Peter Der Manuelian and Thomas Schneider, 227–314. Leiden & Boston: Brill.

Lehner, Mark. 2018. "Spring 2019: A Return to the Menkaure Valleytemple." *AERAgram* 19 (2): 14–17.

Lehner, Mark, and Zahi Hawass. 2017. *Giza and the Pyramids: The Definitive History*. Edited by Zahi A. Hawass. Chicago: The University of Chicago Press.

Liarokapis, Fotis, Panagiotis Petridis, Daniel Andrews, and Sara de Freitas. 2017. "Multimodal Serious Games Technologies for Cultural Heritage." In *Mixed Reality and Gamification for Cultural Heritage*, edited by Marinos Ioannides, Nadia Magnenat-Thalmann and George Papagiannakis, 371–392: Springer International Publishing, 2017.

Manuelian, Peter Der. 2002. "An Approach to Archaeological Information Management: The Giza Archives Project." In *Abusir and Saqqara in the Year 2001. Proceedings of the Symposium (Prague, September 25th–27th, 2001)*, edited by Filip Coppens. Archiv Orientální, 319–328. Prague: Oriental Institute, Academy of Sciences of the Czech Republic.

Manuelian, Peter Der. 2017. *Digital Giza: Visualizing the Pyramids*. Metalabprojects. Cambridge, MA; London: Harvard University Press.

Manuelian, Peter Der. 2009. "Eight Years at the Giza Archives Project: Past Experiences and Future Plans for the Giza Digital Archive." *Egyptian and Egyptological Documents, Archives, and Libraries* 1: 149–159.

Manuelian, Peter Der. 2013. "Giza 3D: Digital Archaeology and Scholarly Access to the Giza Pyramids. The Giza Project at Harvard University." In *2013 Digital Heritage International Congress (Digitalheritage)*, 727–734. Piscataway, NJ: Institute of Electrical and Electronics Engineers.

Manuelian, Peter Der. 2006. "The Giza Archives Project." *Egyptian Archaeology* 28: 31–33.

Manuelian, Peter Der. 2009. *Mastabas of Nucleus Cemetery G 2100. Part I: Major Mastabas G 2100–2220*. Giza Mastabas 8. Boston: Department of Art of the Ancient World, Museum of Fine Arts, Boston.

Manuelian, Peter Der. 2005. "Virtual Pyramids—Real Research: The Giza Archive Project Goes Live Online." *KMT* 16 (3): 68–1980.

Münster, Sander, Cindy Kröber, Heide Weller, and Nikolas Prechtel. 2017. "Virtual Reconstruction of Historical Architecture as Media for Knowledge Representation." In *Mixed Reality and Gamification for Cultural Heritage*, edited by Marinos Ioannides, Nadia Magnenat-Thalmann and George Papagiannakis, 313–330: Springer International Publishing.

Picardo, Nicholas. 2012. "Three-Dimensional Graphic Modeling as Archaeological Research Tool: The Menkaure Valley Temple Settlements at Giza." 63rd Annual Meeting of the American Research Center in Egypt, Providence, RI, April 29 2012.

Picardo, Nicholas. 2018. "What Happens between the Maps and the Models: Developing Referencing Standards for 3D Archaeological Visualizations." 69th Annual Meeting of the American Research Center in Egypt Tucson, AZ, April 20–22 2018.

Picardo, Nicholas. 2017. "'Where Did That Come From?!' The Giza Project's Development of Citation and Referencing Standards for 3D Archaeological Visualizations." 2017 Annual Meeting of the American Schools of Oriental Research Boston, MA, November 15–18 2017.

Reisner, George A. 1931. *Mycerinus: The Temples of the Third Pyramid at Giza*. Cambridge, MA: Harvard University Press.

Rua, Helena, and Pedro Alvito. 2011. "Living the Past: 3D Models, Virtual Reality and Game Engines as Tools for Supporting Archaeology and the Reconstruction of Cultural Heritage—the Case-Study of the Roman Villa of Casal De Freiria." *Journal of Archaeological Science* 38 (12): 3296–308. https://doi.org/10.1016/j.jas.2011.07.015.

Sanders, Donald H. 2012. "A Brief History of Virtual Heritage." In *Picturing the Past: Imaging and Imagining the Ancient Middle East*, edited by Jack Green, Emily Teeter and John A. Larson. Oriental Institute Museum Publications, 95–103. Chicago: Oriental Institute of the University of Chicago.

Sullivan, Elaine. 2016, "Potential Pasts: Taking a Humanistic Approach to Computer Visualization of Ancient Landscapes." *Bulletin of the Institute of Classical Studies* 59: 71–88.

Sullivan, Elaine A. 2017. "Seeking a Better View: Using 3D to Investigate Visibility in Historic Landscapes." *Journal of Archaeological Method and Theory* 24: 1227–1255.

Sullivan, Elaine A. 2012. "Visualizing the Size and Movement of the Portable Festival Barks at Karnak Temple." *British Museum Studies in Ancient Egypt and Sudan* 19: 1–37. http://www.britishmuseum.org/research/online_journals/bmsaes/issue_19/sullivan.aspx.

Sullivan, Elaine, and Willeke Wendrich. 2008. "An Offering to Amun-Ra: Building a Virtual Reality Model of Karnak." *Information Technology and Egyptology in 2008*, edited by Nigel Strudwick, 109–128. Piscataway, NJ: Gorgias Press.

Tavares, Ana. 2008. "Two Royal Towns: Old Digs, New Finds." *AERAGRAM* 9 (2): 8–11.

Wittur, Joyce. 2013. *Computer-Generated 3D-Visualisations in Archaeology: Between Added Value and Deception*. BAR International Series 2463. Oxford: Archaeopress.

CHAPTER 17

All Words and No Play: Identifying Paronomasia in New Kingdom Texts with Pattern Matching

Julia Viani Puglisi and Daniel Dakota

Abstract

Word-play has not been studied extensively in Egyptian texts, possibly because the iconographic and consonantal nature of Egyptian orthography complicates the identification and analysis of word-play types. For an Egyptologist, an identifiable example of word-play is paronomasia, or the semantic juxtaposition of similar sounding words. In this regard, research in Egyptian word-play might benefit from pattern matching, a technique commonly used in computational linguistics, to identify consonantal patterns within a transliterated text. Our research aims to understand the mechanics underlying constructions of Egyptian paronomasia in texts with variable occurrences of word-play.

Keywords

Pattern matching – Phonology – Word-Play – Dream Papyrus – Love poetry – Digital Humanities

1 Introduction

The ancient Egyptians believed that word-play contained the magical and productive force that empowers a text. The mechanism that produces this force assumes various forms, generally in phonological, orthographic, and semantic spheres.[1] The most recognizable form is *paronomasia*, or the semantic juxtapo-

1 On ancient Egyptian word-play, see, e.g.: Noegel 2018; Vinson 2018; Winand 2018; Kousoulis 2011; Richter 2016; Barbash 2015; Navrátilová and Landgráfová 2015; Szpakowska 2011; Hays 2010; Landgráfová and Navrátilová 2009; Noegel and Szpakowska 2006; Breyer 2003b; Morenz 2002; Lippert 2001; Loprieno 2000; 2001; Noegel 2000; Redford 2000; Oréal 1998; Guglielmi 1996; 1986; 1984; Fischer-Elfert 1993; Wessetzky 1985; Malaise 1983; Foster 1980; Meltzer 1975; van de Walle 1969.

sition of similar-sounding words.[2] This similitude ranges from the repetition of lexemes (ploce and antanaclasis), roots (alliteration, consonance, and polyptoton), and lexical endings (homoeoteleuton) to the close homophony between lexemes, roots, and endings. Even with an improved understanding of Egyptian phonology,[3] the identification of paronomasia relies on defining phonetic patterns in a text. This task can be accomplished with a computational tool called pattern matching.

Our project investigates the effectiveness of pattern matching for the identification and categorization of paronomasia in Late Egyptian texts.[4] As test corpora, two Ramesside (1295–1070 B.C.E.) texts were selected for their contemporaneity and use of paronomasia: The Dream Manual (Papyrus Chester Beatty III; BM EA 10683)[5] and the "Seven Houses" cycle (Papyrus Chester Beatty I;

2 There might be paronomastic types in ancient Egyptian texts that are "hidden," such as assonance, or unknown to the Western tradition (e.g., vowel harmonies). The following types of paronomasia have parallels in both the Western and ancient Egyptian record: **Ploce**: repetition of words for emphasis (e.g., *I am* that *I am*); **antanaclasis**: repetition of words with different meanings (e.g., Your argument is *sound*, nothing but *sound*); **polyptoton**: repetition of lexical roots with different endings (e.g., *Absolute* power corrupts *absolutely*); **alliteration**: words that repeat the initial consonant (e.g., *busy* as a *bee*); **consonance**: words that share the same consonantal pattern (e.g., *pitter, patter*); **homoeoteleuton**: words that share the same lexical endings (e.g., Humpty Dumpty sat on a *wall*, Humpty Dumpty had a great *fall*); **metathesis**: the transposition of shared letters between words (e.g., *Elvis Lives* in *Evil Levis*).

3 For summaries on major scholarship in Egyptian phonology, see: Allen 2020, 181–194 and Peust 1999, 26. The earliest consensus on Egyptian phonology was Alan Gardiner's *Egyptian Grammar* (1927), which was the product of nearly a century of linguistic inquiry since the decipherment of the language in 1822. The so-called "Berlin School," which saw Egyptian phonology as comparable to Semitic, contributed to the development of this discipline in late 19th century with canonical works like Kurt Sethe's *Das Aegyptische Verbum* (1899–1902). Influenced by German Egyptology, scholars like William F. Albright (1923), Wilhelm Czermak (1931), and Otto Rossler (1971) reconstruct Egyptian phonology as a Semitic equivalent. The use of Coptic to reconstruct the phonetics of Egyptian has been central to the study of Egyptian phonology, most notably Josef Vergote's *Phonetique historique de l'egyptien: les consonnes* (1945) and Carsten Peust's *Egyptian Phonology* (1999). Peust's study is the most comprehensive to-date, covering consonants, vowels, phonotactics, and prosody, and primarily focuses on Late Egyptian and Coptic phonology. More recent studies on Late Egyptian phonology include: Allen 2020; Winand 2018ab; Allen 2013b; Peust 1999; Loprieno 1995; Vernus 1988.

4 While pattern matching can be applied to any ancient Egyptian text, paronomasia tends to occur in non-administrative genres where the rhetorical, magical, or mnemonic force of word-play might be employed, such as in the *belle lettres* (literature), theological texts, and didactic texts. Cf. n. 1.

5 "Dream-book" (Gardiner 1935) and "Dream Manual" (Noegel and Szpakowska 2006) are used interchangeably. The British Museum acquired the Dream Manual, the love-songs, and six other Ramesside papyri from Mr. and Mrs. A. Chester Beatty in 1930 (Hall 1930).

BM EA 10681).[6] Since both texts have high occurrences of paronomasia, we could assess the applicability of pattern matching in literary genres that utilize paronomasia differently. The Dream Manual employs word-play for the interpretation of dreams and the prognosis of the dreamer, whereas love poetry avails itself of the rhetorical features of word-play to capture and entertain the attention of the audience.

The identification of paronomasia from phonetically similar word-groups is a task that involves linguistic interpretation. For example, pattern matching might return two homophonous words that are not word-play (a false positive). Although current computational methods cannot identify the semantic significance of a phonetic pattern in a text, treating the text as quantifiable (i.e., a group of computer-readable characters) can clarify the conditions under which paronomasia occurs. The application of this multidisciplinary approach generated both known (published) and new examples of phonetic patterns from our test corpora, from which we could identify three conditions that determine the presence of paronomasia in a text: phonology, typology, and distance. Our research aims to define these variables and the frequency at which each variable occurs in both the Dream Manual and the "Seven Houses" cycle.

Some constraints to our study include a two-word limit for word-play types (i.e., word-play pairs). Although paronomastic combinations that employ three or more lexemes appear in the Dream Manual,[7] a reduction to two words made our scope more manageable. We additionally restrict the analysis to the identification of word-play with at least two shared consonants. Therefore, true alliteration—the repetition of only one initial consonant—is absent in the returns, but this does not mean that they are missing in the text. In general, the use of true alliteration observes a serial structure without lexical intervals. For instance, an alliterative sequence that repeats the consonant *ḥ* occurs in Column 5 of the Dream Manual:

5.13. ***ḥr ḥms ḥr ḥsp.t*** *rʿ | pw nḏm pw*

Sitting in the garden *in the sun | Good; It means pleasure.*

---

6 Papyrus Chester Beatty I preserves two cycles of love songs on the verso (sections C and G) and one song on the recto: the "Seven Houses" (C 1.1–5.2: "Beginning of the Poems of Great Delight"), the "Come Quickly to the Sister" songs (G 1.1–2.5), and the Nakht-Sobek songs (r. 16.9–17.13: "Beginning of the Beautiful Verses"). Navrátilová and Landgráfová 2015; Landgráfová and Navrátilová 2009; Simpson 2003; Mathieu 1996; Hermann 1959; Schott 1950.

7 e.g., Line 8.12: (*ỉr mꜢꜢ sw s m rsw.t*) *ỉbḥw=f* ***ḫr ẖry=f*** *|| ḏw mt s pw n* ***ẖrỉw=f***. (If a man sees himself in a dream) his teeth **falling out** | Bad; it means death of a man by his **dependents**.

Despite the frequency of sequential alliteration, this typology would have returned an overabundance of data due to the preponderance of repeating initial consonants in the text. Thus, we shall defer the study of word-play groups and true alliteration to subsequent analyses.

2 Pattern Matching and Paronomasia

Pattern matching can be described as taking an input sequence, in this case a string of text, and searching for the presence of a predefined pattern. For identifying paronomasia, the search criterion depends on an initial, manual codification of word-play occurrences known as a schematization. Although schematizations are not computer-readable, they facilitate the categorization of paronomastic types for researchers and their conversion into machine-readable equations known as regular expressions. A regular expression (regex) extracts textual information in a string from a defined search criterion or schematization.[8] The regex indicates whether there is a match by returning a boolean value (i.e., "true" or "false"): if "true," the desired match is returned and shown by the code, and if "false," then nothing is returned. Once patterns are defined, this method quickly searches through large quantities of digital texts and identifies instances of the phenomenon, so that Egyptologists can focus on analyzing returns.

Below is an example of a word-play in Papyrus Chester-Beatty III with its corresponding schematization and regular expression for the paronomasia *wḏ* || *wḏꜣ* in line 6.20. All regexes are written in Python 3.6:

<table>
<tr><th>Line no.</th><th>Word-play</th><th>Schematization</th><th>Regular expression for first word</th><th>Regular expression for second word</th></tr>
<tr><td>6.20</td><td>wḏ || wḏꜣ</td><td>AB || AB?</td><td>(?:^|\s)(?P⟨word⟩(?P⟨twoChar⟩(?P⟨charOne⟩\w)(?P⟨charTwo⟩(?!⟨charOne⟩)\wj?)))(?=(?:(\s|$|...)))</td><td>(?:^|\s|)(?P⟨match⟩('+p.group("twoChar")+"(?!("+p.group("charOne")+")|("+p.group("charTwo")+"))\wj?))(?=(?:(\s|$|...)))</td></tr>
</table>

8 Kübler and Zinsmeister 2015, 207–216.

The simplification of transliterated words into schematizations, or the codification of word-play, facilitates the organization of word-play types and the identification of unknown examples in a text. Schematizations tabulate shared and unshared consonants, grammatical gender and number, and verb forms. When a transliterated character in a word matches with another transliterated character in the line, the shared letters in the paronomastic pair are assigned an identical alphabetic value. Unshared consonants are assigned a "?". For example, the word-play pair *wḏ* || *wḏꜣ* becomes AB || AB?. The first half (AB) searches for a character (A) followed by any other character (B), while the second half (AB?) searches the same character (A) followed by the same character (B) followed by any other character (?) than A and B.

The transliteration of the text and the schematizations of paronomasia distinguish between gender markers and verbal conjugations wherever applicable. A period "." at the end of a lexeme indicates gender and number: e.g., *t*-ending for feminine adjectives and nouns (*mw.t*, "mother"). Words ending in a comma "," designate verb type: e.g., the infinitive *t* for weak verbs (*rdi͗,t* "to place, put, give") or the past tense marker *n* (*sḏm,n=f* "he heard"). These grammatical codes conform to a system that the *TLA* implements for transliterations and dictionary entries.[9]

Returning to the example above, the regular expression for the first word (*wḏ*; AB) begins by stating what can come before the word (^|\s). This expression indicates that the word (*wḏ*) appears either at the beginning of the string (^) or after a space (\s), a string being a sequence of characters (a line of text). Since a sentence does not normally begin with a space, the regex captures a pattern if it is the first word of a string. Next, the regex defines what the word can start with (?P⟨charOne⟩\w). The "\w" states that the first character (A) must be alphanumeric. Then we use a "negative lookahead" (?!⟨charOne⟩)\w) that checks whether the character following the first character (A) is the same in the scheme (e.g., AA). If it is not, it continues the expression. However, if the characters *are identical* (AA), the condition (AB) is *not met* and the regex terminates with no returns. Assuming the second character is different than the first, (\s|$|...) checks what is possible to come after the word; in this case, it is either a space, the end of the string ($), or "..." which symbolizes missing text in the transliteration.

A similar pattern appears for the second word (*wḏꜣ*, AB?). First, the regex defines what can come before the word (?:^|\s|-). Then, the expression looks for the same two characters in the first word (p.group("twoChar")). A negative

9 Seidlmayer and Hafemann 2011, 12.

```
def AB_ABX(text):
    candidates = []
    pattern = re.compile("(?:^|\s)(?P<word>(?P<twoChar>
                         (?P<charOne>\w)(?P<charTwo>
                         (?!<charOne>)\wj?)))(?=(?:(\s|$|…)))")
    for p in pattern.finditer(text):
        patternMatch = re.compile(r'(?:^|\s|-)(?P<match>
                                  ('+p.group("twoChar")+
                                  "(?!("+p.group("charOne")+")
                                  |("+p.group("charTwo")+"))\wj?))
                                  (?=(?:(\s|$|…)))")
        for pm in patternMatch.finditer(text):
            distance = get_distance(p.span("word"), pm.span("match"), text)
            candidates.append(("standard pattern", "AB || AB?",
                               p.group("word"), p.span("word"),
                               pm.group("match"), pm.span("match"),
                               distance, text))
    return candidates
```

FIGURE 17.1 Code block for the schematization AB || AB? (Python 3.6)

lookahead ?!("+p.group("charOne")+")|("+p.group("charTwo"), checks whether the final character is either A (p.group("charOne")) or B (p.group("charTwo")) with capture groups that identify these characters in the first word. If the final character is neither A nor B, then the regex returns the following alphanumeric character (\w). Like the regex of AB, the expression concludes with (\s|$|…).

Figure 17.1 expresses the implementation of the schematization in a code block. The function (AB_ABX) expects "text" as input (a string of characters) and compiles the initial pattern as defined for the first word. The regex subsequently searches the text for matches (*finditer*) and compiles the regex for the second word. Then, the expression iterates over the text looking for patterns based on the schematization. If a match is found, it returns the name of the scheme, the words for the match, locations of the match in the text, distance between the two words, and the text that was searched. Importantly, the *finditer* function allows the expression to find matches in both directions (i.e., AB_ABX and ABX_AB).

3 The Ramesside Dream Manual (Papyrus Chester Beatty III; BM EA 10683)

The Dream Manual (Figure 17.2) is the most comprehensive example of ancient Egyptian dream interpretation prior to the Late Period, preserving a corpus of 226 dream interpretations for two types of men.[10] The text follows a tabular

10 For the cache and its owners: Černý 1978; Pestman 1982. For Egyptian oneiromancy, see

FIGURE 17.2 The Bad Dreams: Columns 7–9 of the Ramesside Dream Manual (Papyrus Chester Beatty III; BM EA 10683). Gardiner 1935, Plate 7A

structure demarcated by the line, "If a man sees himself in a dream," which introduces the protasis-apodosis structure for each dream. Here, the protasis describes the dream; the apodosis defines whether the dream is good or bad and prognosticates the consequences of the dream for the dreamer. The text begins with 139 *nfr* ("good")-dreams (r. 1.12–6.25), continues with an interpretation of 83 *ḏw* ("bad")-dreams (r. 7.1–10.9), followed by the section for the "Followers of Seth."[11] The first ten columns in the manuscript might concern the *šms.w-Ḥr*, the "Followers of Horus," but this section is missing.[12] The final column (r. 11.1–11.24) describes the deviant *ỉmy.w-Stḫ*, "Followers of Seth;" however, it only preserves four fragmentary dreams with no evidence of wordplay.

---

e.g., Quack and Ryholt 2019; Prada 2012a; 2012b; Szpakowska 2011a; 2011b; Noegel 2007; Szpakowska and Baines 2006; Vinogradov 2006; Breyer 2003a; 2003b; Bryan 1991, 144–150; Ray 1987; Zivie 1976, 125–145; Sauneron 1959; Erman 1904.

11 Szpakowska 2003, 74–76. For an alternative interpretation of the text's function: Israelit-Groll 1985.

12 Gardiner 1935, 10. Unlike the Demotic Dream Manual, the text does not refer to the female dreamer. See: Quack and Ryholt 2019; Szpakowska 2007; 2003, 68; Sauneron 1959.

Scott Noegel and Kasia Szpakowska categorize seven different types of wordplay in the first 10 columns: alliteration, antanaclasis, homoeoteleuton, onomatopoeia, polysemy, translexical punning, and visual puns.[13] According to their preliminary study, paronomasia (alliteration/consonance, antanaclasis, and homoeoteleuton) appears 42 times in the Dream Manual.[14] Gardiner (1935) provided restorations to lacunae based on paronomasia, which Noegel and Szpakowska note, but we exclude them in this analysis to remain as faithful as possible to the original text. Here, the application of pattern matching in the Dream Manual increased the count from 42 to 107 *possible* occurrences of paronomasia.

3.1 *Phonology*

The language of the Dream Manual dates to the Ramesside period (Dynasty 19–20; 1295–1069 B.C.E). Gardiner dates the text to the early reign of Ramesses II (c. 1270 B.C.E.), but the Dream Manual also employs a "Middle Egyptian vernacular," possibly to elevate the status of the prose.[15] While the hieratic orthography and spelling date to Dynasty 19 (1295–1185 B.C.E.), the grammar and vocabulary is Middle Egyptian. For example, the manuscript uses *mꜣꜣ*, "to see," instead of the Late Egyptian *ptri* and does not employ the phenomenon of group writing.[16] There are words and expressions in the Dream Manual, however, that only begin to appear in the New Kingdom like *gsgs*, "to overflow (r. 10.5; Erman and Grapow 1926–1963, V 207; Lesko II, 195)," and *d̠nw*, "threshing area (r. 6.7; Erman and Grapow 1926–1963, V 575.6; Lesko II, 269)."[17] Thus, a phonological reconstruction in this study relies on Late Egyptian phonology without ignoring the possibility that the Dream Manual employs earlier Middle Egyptian lexicography as an archaizing tactic.

The aural reconstruction of Egyptian paronomasia is hypothetical. As a morphologically unique branch of the Afroasiatic phylum, vocalic absences and uncertain consonantal values only complicate accurate readings of Egyptian paronomasia. Another factor that inhibits the analysis of paronomasia is the conventionalization of transliteration, which obscures phonological change

13 Noegel and Szpakowska 2006, 193–212; Gardiner 1935.

14 Alliteration/consonance (30 occurrences), homoeoteleuton (5 occurrences), antanaclasis (7 occurrences).

15 Gardiner 1935, 10. On Neo-Middle Egyptian ("*égyptien de tradition*") during the New Kingdom onwards, see: Vernus 2016; 1996; 1982; Jansen-Winkeln 1996; 1995; Peust 1999, 27–28. For linguistic hierarchies and language conventions ("registers"), see: Junge 2005, 20–21; Goldwasser 1990, 200–240.

16 e.g., Shehab El-Din 2006; Depuydt 1988; Israelit-Groll 1970, 59.

17 Szpakowska 2003: 10.

and the phonetic similarities between consonants.[18] Nevertheless, advancements in the field of Egyptian phonology have improved our understanding of the Egyptian language. Appendix, Table 17.1 delineates the phonetic values for the twenty-four consonants of the Late Egyptian phase as proposed by James Allen (2020, 47–48).[19]

Late Egyptian is morphologically distinct from earlier phases of the language for two reasons: the orthographic standardization of vernacular Egyptian from the Amarna Period onwards and the intensifying interactions with the Near East.[20] Phonological changes from the Middle Kingdom mature during this time as well, most notably with omission of consonant *r* in the final position and the phonological reduction of the semi-consonants *w* and *ỉ*.[21] In addition, velar (*g*, *k*, *q*) and dental series (*t*, *ṯ*, *d*, *ḏ*) possibly undergo neutralization and lose distinction between their voiced and voiceless consonants.[22] Other consonants are used interchangeably like the velar fricatives *ḫ* and *ẖ*, or the palatalized fricatives *ẖ* and *š*.[23] These transformations are visible in variant spellings with graphically distinct phonemes (e.g., "Seth" as both *stḫ* and *stẖ*),[24] and later Demotic and Coptic equivalents. Pattern matching can localize paronomastic examples that account for these changes, particularly with homophonous consonants.

A homophonous consonant either shares the same manner or place of articulation with another consonant. According to the data in this study, there are 13 examples in our corpus that use homophony. These examples appear in one of two root structures: AB || AB (antanaclasis) and AB.? || A?B (metathesis). A discussion of these forms is treated in the subsequent section on typology, but it should be noted that the AB || AB type with a homophonous consonant is not true antanaclasis (e.g., line 3.10: *kꜣ* || *qꜣ* in the Dream Manual). Among the homophonous groupings listed in Appendix, Table 17.1, the most frequent returns for shared homophonous consonants occur with the coronal stops (*t*,

18 Also, Peust (1999, 15) warns us that "[e]ven if a phonetic correspondence between genetically related languages is assumed to be certain, the correspondence is of a principally abstract nature and does not allow for a conclusion about the actual pronunciation in one of the compared languages."

19 Other reconstructions include Peust 1999 and Loprieno 1995. Cf. n. 4.

20 e.g., Winand 2017; 1992; Zeidler 1993, 579–590; Schneider 1992. For linguistic exchanges with Semitic languages (loanwords), see, e.g., Schneider et al. 2004; Meeks 1997; Ward 1996; Hoch 1994.

21 Peust 1999, 137–158.

22 Peust 1999, 79–84.

23 Peust 1999, 115–125.

24 Erman and Grapow 1926–1963, IV, 345, 3.

d), nasal (*m*, *n*), liquid (*n*, *r*), velar (*k*, *g*, *q*) and sibilant fricative (*s*, *š*) phonemes. Strikingly, the least frequent is the labial series (*b*, *p*, *f*) with only one occurrence, and none with voiceless laryngeals (*h*, *ḥ*). In short, the incorporation of these equivalences reconciles the graphical differences between consonants, so that pattern matching searches can "hear" homophony.

Similarly, Coptic and Demotic equivalents of lexemes differentiate homophonous pairs from false positives. Although equivalents do not confirm homophony, they do clarify weaknesses in the phonological reconstruction of Late Egyptian. For instance, the *skꜣ* and *sḫ.t* in line 4.21, which Noegel and Szpakowska identify (2006, 201) as a possible alliteration, might not be a homophonous pair because the velar *k* and the velar fricative *ḫ* do not share the same value in Late Egyptian and in Coptic: *skꜣ* becomes ⲥⲕⲁⲓ[25] and *sḫ.t* becomes ⲥⲱϣⲉ.[26] Although the pair might be a true alliteration since they share the initial sibilant, the conflation between fricative and non-fricative velars as homophonous consonants must be treated with caution.[27]

Furthermore, Coptic and Demotic equivalents can reconstruct the syllabic structure, vocalic patterns, and overall prosody of Late Egyptian words.[28] Although these features are not covered in this article, they should be investigated in future analyses of paronomasia precisely because consonantal skeletons cannot be taken at face value.

We chose to make our pattern matches more "flexible" to accommodate the orthographic occurrences of weak verbal and nominal endings. The orthography preserves consonants in the spelling of a word that may have not been pronounced like the feminine *t*-marker or the final *ỉ* of weak verbs. Thus, pattern matching accounts for both the presence and absence of these consonants. For example, a pattern match will search for *snṯrỉ* as both *snṯrỉ* and *snṯr*.

3.2 *Typology*

The typology of paronomasia relies on a combination of shared consonants in a defined sequence. In this study, simple paronomasia consists of phonetic wordplay that conform to a two-consonant minimum and a two-word maximum.

The Dream Manual produces 17 schematizations of simple paronomasia that generate 107 possible occurrences of word-play (Graph 1; Appendix, Tables

25 Crum 1962, 328b; Černý 1976, 149; Demotic *sqꜣ*: Johnson 2001, S (13:1), 468; Erichsen 1954, 467, 4.

26 Crum 1962, 377a; Černý 1976, 170; Demotic *sḫ.t*: Johnson 2001, S (13:1), 362; Erichsen 1954, 450, 4.

27 This example also illustrates the weakness of the final *t*.

28 Cf. n. 3 for references, especially Peust 1999.

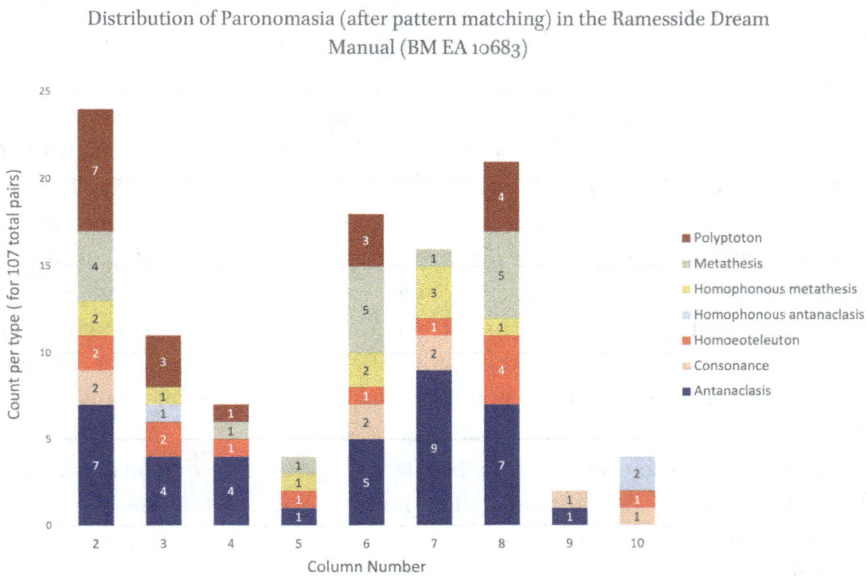

GRAPH 17.1 Distribution of Paronomasia in the Ramesside Dream Manual (BM EA 10683)

17.2–17.3). Five types of paronomasia appear in these returns: consonance, metathesis, homoeoteleuton, polyptoton, and antanaclasis. Variations of root structure for each occurrence of paronomasia suggests a more complex representation of phonemic word-play than previously published for the text. The codification of paronomasia into schematizations not only clarifies the distribution and frequency of typologies in the text, most importantly with unknown cases, but it also helps us understand our own biases when analyzing the phonology and the morphology of paronomasia in Egyptian texts.

The highest returns of paronomasia are metathesis, polyptoton, and antanaclasis. Homoeoteleuton and consonance are fewer with 13 and 8 counts respectively. The apodoses of Column 5 are missing, which possibly contributes to the low occurrence of paronomasia (4 examples). Column 9, however, is well-preserved, but only has two examples of paronomasia and could reflect the constraints imposed by this study. Column 10 also has few returns, but this is due to the length of the column, which only contains nine lines as opposed to the standard 25–28 lines per column in the text.

The most complex form of paronomasia, and the most frequently occurring in this study, is metathesis. Unlike other typologies in the Dream Manual, metathesis jumbles the position of shared consonants in the word-play pair. Frequent combinations for metathesis in the Dream Manual are AB? || A?B (10 counts) and A?B || ?AB (6 counts), but these two types also have high occurrences of false positives. In fact, false positives in our study tend to repeat the liquids *n* and *r* in the word-play pair, particularly with the lexemes *nfr, inr*, and *nṯr*.

Coptic and Demotic equivalents can determine the phonological similarity between words with these consonants. One false positive is *nṯr* || *rmṯ.w* (AB? || A?B) over lines 3.10–3.11. Although it shares the consonants *ṯ* and *r*, it is not homophonous (ⲚⲞⲨⲦⲈ and ⲢⲰⲘⲈ).[29] Lexemes that begin with the consonant *i͗*, a *mater lectionis*, complicate the picture as well.[30] For example, *nfr* || *i͗nr* (A?B || ?AB) in line 6.4, have the Coptic etymologies ⲚⲞⲨϤⲈ[31] for *nfr* and ⲰⲚⲈ[32] for *i͗nr*, which repeat the phonemes [n] and [j] and the labio-velar approximant /w/ (*i͗* > [w], [u]). Future iterations of this study will treat the consonant *r* in the final position of the lexeme as a vowel in Late Egyptian, which would significantly reduce the occurrences of false positives.

Nevertheless, there are metathetic examples that are previously unpublished in this type—perhaps due to their relative lexical distances—that make identification difficult. For example, the word-play *šd.t* || *šꜥd* (AB.? || A?B) appears over lines 8.5–8.6 at a distance of 13 lexemes. The Coptic equivalents for *šd.t*, ϢⲰⲦⲈ,[33] and *šꜥd*, ϢⲰⲰⲦ,[34] indicate a homophonous relationship between these two words in the Dream Manual. Similarly, the pair *wꜣḏ* || *wḏꜣ* (ABC || ACB) appears with a distance of 13 lexemes over lines 6.10–6.11. In Coptic, the relationship is weaker with *wꜣḏ* becoming ⲞⲨⲰⲦ[35] and *wḏꜣ* becoming ⲞⲨϪⲀⲒ.[36] This schematization only occurs once in the Dream Manual.

29 Crum 1962, 294b; Černý 1976, 136; *rmṯ*: Johnson 2001, R (01.1): 37; Erichsen 1954, 247, 5.

30 In ancient Egyptian, a *mater lectionis* ("mother of reading") is a glide ("semi-vowel"; e.g., *w*, *y*, or *ṯ*) in the initial position of a lexeme that indicates the presence of a vowel. Allen 2020, 45, 82; Werning 2016, 29–44.

31 Crum 1962, 240a; Černý 1976, 116; Demotic *nfr*: Johnson 2001, N, 72; Erichsen 1954, 216, 8.

32 Crum 1962, 524a; Černý 1976, 228; Demotic *i͗ny, i͗n*: Johnson 2001, I (11:1): 161, 144; Erichsen 1954, 34, 14.

33 Crum 1962, 595a; Černý 1976, 254; Demotic *šty(t), št.t*: Johnson 2001, Š (10.1): 230; Erichsen 1954, 529, 4.

34 Crum 1962, 590b; Černý 1976, 254; Demotic *šꜥṯ*: Johnson 2001, Š (10:1): 49; Erichsen 1954, 492, 6.

35 Crum 1962, 493a; Černý 1976, 217; Demotic *wt*: Johnson 2001, W (09:1): 189; Erichsen 1954, 104, 5.

36 Crum 1962, 511b; Černý 1976, 224; Demotic *wḏꜣ*: Johnson 2001, W (09:1): 209; Erichsen 1954, 108, 2. The phonetic dissimilarity between *wꜣḏ* || *wḏꜣ* in Coptic might be explained by depalatalization of the *ḏ* > *d* in Late Egyptian (*d* > Ⲧ in Coptic), especially in the final position (e.g., "bird": *ꜣpd* > ˢⲈⲂⲞⲦ; Peust 1999, 124, 156). The retention of the palatal *ḏ* > Ϫ in the medial position might be due to the following laryngeal *ꜣ* (e.g., "guilty": *ꜥḏꜣ* > ⲞϪ), which Osing (1980, 946f) proposes. However, Peust (1999, 124) observes that this does not

Metathesis occurs with homophonous consonants as well. In lines 6.2–6.3, the pair *skꜣ* || *sꜥq* (AB.? || A?B) share an initial position for the sibilant *s*, and the transposition of the velar consonants, *q* ([k]) and *k* ([kʰ], [k]), with the approximants *ꜣ* and *ꜥ*. In line 7.6, the appearance of *sšp.t* in the protasis suggests a metathetic relationship with *ḫpr* as an AB.? || A?B type:

7.6. *ḥr wšꜥ* ***sšp.t***[37] | *<u>dw</u>* ***ḫpr n=f*** *md.wt pw m ḥsy=f*

Snacking on a ***melon*** | *Bad; it means slander* ***will happen*** *and come to meet him.*

Intriguingly, the metathesis does not continue into Demotic and Coptic, but becomes antanaclasis with *sšp.t* in Sahidic as both ϣⲱⲡⲉ/ϣⲱⲃⲉ (Demotic *špy*)[38] and *ḫpr* as ϣⲱⲡⲉ (Demotic *ḫpr*).[39] Since the fricative *ḫ* can shift to a palatal fricative *š* in Late Egyptian, line 7.6 also has a metathetic sequence with the word *wšꜥ* (Coptic ⲟⲩⲱ(ⲱ)ϣⲉ; Demotic *wš*).[40]

The third most common typology in the Dream Manual is polyptoton, with 21 examples. This form is similar to antanaclasis, which shares the same root, but polyptoton differs in the ending of the lexeme.[41] Verbal polyptoton includes the past tense *sḏm,n=f* and the infinitival ending for weak verbs (AB || AB,?). Nominal suffixes (=*f*, =*s*) are considered separate from the construction of the paronomasia, but certainly warrant further study. Nominal endings, which are the most frequently occurring variants in polyptoton, encompass both feminine and plural markers (AB || AB.?; ABC || ABC.?), as well as *nisbe* endings (AB || AB?). Nominal polyptoton often is a correlation between a masculine and feminine pair of the same lexeme. The study returns one example of a *nisbe* in the protasis of line 2.14:

2.14. ***ḥr*** *mꜣꜣ nṯr* ***ḥry*** | *nfr ḏfꜣ.w pw ꜥꜣ*

Seeing *the god* ***above*** | *Good; great are the provisions.*

account for every case, most notably, *wḏꜣ* ("to go away," Erman and Grapow 1926–1963, 1, 403.2–19).

37 For *sšp.t*, see Hudáková 2016.

38 Crum 1962, 580b; Černý 1976, 249. Possible metathetic retention with *sšp.t*, which also occurs as *šwbe*: Johnson 2001, Š (10.1): 67; Erichsen 1954, 503, 2.

39 Crum 1962, 577b; Černý 1976, 249; Johnson 2001, Ḫ (06:1): 50; Erichsen 1954, 355.

40 Crum 1962, 502b; Černý 1976, 221; Johnson 2001, W (09:1): 174; Erichsen 1954, 101, 7.

41 Richter 2016, 26–30; 32–33.

The adverbial construction *ḥr* + infinitive typically begins the protasis of each dream interpretation in the manuscript. In this case, the choice of the *nisbe ḥry* to indicate the location of the god is significant due to its derivation from and resulting phonetic similarity to the preposition *ḥr*. The use of prepositions is not unusual[42] in examples of paronomasia from our corpus, as seen in the alliterative sequence on line 5.13 with *ḥr ḥms ḥr ḥsp.t nʿ | pw nḏm pw*.

Polyptoton also uses the verbal suffix *n* to construct paronomastic relationships. An example published by Noegel and Szpakowska contains a polyptotonic example over lines 3.2–3.3:

> 3.2. *mꜣꜣ[,n=f] g(ꜣ)s | nfr* ***ʿꜣ.w n****(ỉ) ḫ.t=f*
>
> *[He had seen] mourning | Good;* ***the increase of*** *his possessions.*
>
> 3.3. ***ʿꜣ,n*** *šn.w=f | nfr ḫ.t pw ḥḏ ḥr=f n=sn*
>
> *His hair had* ***become long*** *| Good; it means something at which his face will brighten.*

The plural *ʿꜣ.w* and genitival adjective *n(ỉ)*, in which the original *nisbe* ending *ỉ* dropped from pronunciation and writing already during the Old Kingdom, shares a phonetic relationship with the past tense *sḏm,n=f* in line 3.2. It is possible that the scribe or oneiromancer deliberately chose to employ the Middle Egyptian past tense construction in the protasis, which is otherwise uncommon[43] in this text, due to the *nisbe n(ỉ)* in the apodosis of the previous line. Verbal polyptoton takes advantage of the flexibility of lexemes to construct more complex relationships, too. For example, the noun *pr* pairs with the third weak verb *prỉ* in lines 6.5–6.6:[44]

> 6.5. *ḥr ḫꜣʿ ḥbs.w=f r tꜣ | nfr* ***prỉ*** *pw m md.wt*
>
> *Throwing his clothes on the ground | Good; it means* ***coming forth*** *with words.*

---

42 Other examples include lines 5.9: *ḥr … šfd.t || nfr smn s m pr=f;* 8.12: *ỉbḥ=f ḥr ḫry=f || ḏw mt s pw n ḫrỉw=f.*

43 The only other occurrence is on line 2.11: *ʿꜣ.n ḥnn=f || nfr ʿšꜣ n(ỉ) ḫ.t tw=f pw.*

44 A similar relationship occurs in 8.15 (*ḥr sḫ,t (=f) ṯry | ḏw prỉ prr.w pw mt=f*), which Noegel and Szpakowska (2006, 203) discuss, but it is not treated here because the form AB? ABB.C AC fell outside our scope in this preliminary study. The word-play *prr.w pw* (A??.B || AB) occurs in 8.17 as well.

6.6. *ḥr itḥ dp.t* | *nfr mni=f nfr m* ***pr=f***

Towing a boat | *Good; His perfect mooring in his* ***home.***

Although the schematization is not fully developed, since pattern matching does not identify true alliteration (e.g., *pri pw*, above) in this phase, there is a relationship between the word *pr* ("house") and *pri* ("emerging") as polyptoton.[45] However, there might be a more practical reason for the scribe to have spread the paronomastic pair over two lines.

The repetition of lexical roots over multiple lines could be an indication of dream categorization, which antanaclasis exhibits as well. Antanaclasis is the repetition of two lexemes that share the same form but differ in meaning. Pattern matching collects an overabundance of repeating words in the text, so the differentiation between paronomastically significant and non-significant returns is a manual procedure.

We identify two types of antanaclasis that depend on the relative position of constituting words in the word-play pair. Antanaclasis either appears in the same line as true antanaclasis or over two lines as a form of antanaclasis known as ploce. Ploce is the repetition of lexemes for emphasis, which pattern matching cannot distinguish from antanaclasis. We might assume that the frequency of repeating lexemes is inversely proportional to the likelihood of it being antanaclasis. For example, the lexemes *nfr* and *ḏw* repeat throughout the text, but the audience might not interpret this repetition as paronomastically significant because "good" dreams are organized separately from "bad" dreams in the Manual. This distinction is not mutually exclusive since the clustering of antanaclasis and other word-play types (e.g., 7.6 *sšp.t* || *ḫpr*) can be a mnemonic device for the dream interpreter.[46] Lines 2.21–2.22 reflect the subtleties in the function of antanaclasis (or ploce) in the Dream Manual:

2.21. *ḥr* ***wnm*** *iwf n ꜥꜣ* | *nfr sꜥꜣ=f pw*

Eating *the flesh of a donkey* | *Good; it means he becomes great.*

45 Another possible spatial significance between *nfr pri* and *pr=f* with the labial consonants (*f* and *p*) and liquid *r* repeating in both cases.

46 For the *ars memoriae* and the relationship between memory and word-play, see: Bolzoni 1999.

2.22. *ḥr* ***wnm*** *ỉwf n msḥ* | *nfr* ***wnm*** *ḫ.t sr* [*pw*]

Eating *the flesh of a crocodile* | *Good; it means the possessions of an official* ***are consumed.***

The repetition of *wnm* ("to eat") in both protases is not true antanaclasis since the meaning remains the same for both words. The clustering of these two dreams is meaningful for the structural organization of the oneiromantic text—which catalogs both dreams as "good"—and for memorization.

Similarly, the repetition of *wnm* in line 2.22 constructs true antanaclasis in the protasis and apodosis. The subject and use of *wnm* in the apodosis are different than in the protasis. The "eating of possessions" has a more obviously metaphorical tone than the "eating of flesh." Thus, the occurrence of antanaclasis on the same line is significant for the interpretation of a specific dream and often presents a nuanced meaning between constitutive words. On the other hand, the distribution of antanaclasis over two lines demonstrates the organization of dream types in the text and the mnemonic efforts of the scribe, which might explain the frequency of antanaclasis in the Dream Manual.

Consonance and homoeoteleuton are the least common forms of phonetic word-play in the Dream Manual. Unlike true alliteration, which repeats one initial consonant, consonance shares at least two consonants among constitutive words. The repetition occurs in the initial positions of the lexeme as opposed to homoeoteleuton whose consonants repeat at the end of the word. Although consonance resembles polyptoton and antanaclasis, the words in consonance do not share the same lexical root. For example, line 6.20 preserves an occurrence of consonance in the schematization AB?.? || AB?:[47]

2.23. *ḥr* ***nhꜣ.t***[48] *rwḏ,tỉ* || *nfr* ***nhy=f*** *pw m* ...

On a ***sycamore tree*** *that is flourishing* | *Good; it means he* **loses** ...

The repetition of the consonants *n* and *h* in two non-lexically related words realizes the paronomastic relationship between the protasis and apodosis of the line. Possibly, the glottal *ꜣ* and semivowel *y* strengthen the homophony between *nhꜣ.t* and *nhy* since the Coptic equivalents are phonetically identical: ⲛⲟⲩϩⲉ.[49]

---

47 Example published by Gardiner 1935 and Noegel and Szpakowska 2006.

48 Late Egyptian transliteration of sycamore is either nh.t or nꜣA.t. See Lesko 240.

49 Crum 1962, 242b; Černý 1976, 117; Demotic *nhy/nꜣ, nh.w*: Johnson 2001, N: 97; Erichsen 1954, 221, 7. Crum 1962, 241b; Černý 1976, 117; Demotic *nh*: Johnson 2001, N: 96.

Similar to metathesis, consonance runs a risk of false positives which etymological equivalents can clarify. For instance, the preposition *ỉn* ("by") and the noun *ỉnr* ("stone") in lines 6.3–6.4 appear to be a mnemonic pattern, but might in fact be a false positive:

6.3. *ḥr sʿq ỉ3w.t | nfr nwy,t n=f rmṯ **ỉn** nṯr=f*

*Assembling cattle | Good; the gathering of people for him **by** his god.*

6.4. *ḥr b3k **ỉnr** m pr=f | nfr smn s m pr=s*[50]

*Working **stone** in his house | Good; Establishing a man in her house.*

As an example of AB || AB? the proximity of these two lexemes over two dreams suggests a possible example of a typologization or mnemonic device. Again, the *mater lectionis ỉ* complicates the homophony between *ỉn* and *ỉnr*, which become ⲛ[51] and ⲱⲛⲉ,[52] respectively, in Coptic. Nevertheless, pattern matching can help us identify examples in the text that were otherwise overlooked as schematizations in previous publications. Line 10.8 returns an unpublished example of consonance in the schematization AB || AB?:

10.8. *ḥr rdi,t **ḥnq.t** r **ḥnw** | ḏw iṯi,t ḫ.t m pr=f*

*Putting **beer** into a **jar** | Bad; Taking of something from his house.*

The protasis contains a striking arrangement of the consonants *ḥ*, *r*, and *n*, however our constraints only identify consonance between *ḥnq.t* and *ḥnw*. The Coptic equivalents reveal that the consonant *n* remains strong without assimilating into liquid [l] since *ḥnq.t* becomes ϩⲛⲕⲉ[53] and *ḥnw* becomes ϩⲓⲛ.[54] Future developments of pattern matching will consider the metathetic construction

50 There is alliteration and consonance in the apodosis (*smn s m pr=s*). Noegel and Szpakowska publish an identical occurrence in line 5.9 (2006, 201), but not this example.

51 Alternatively, *e-* in Sᶠ, F. Crum 1962, 215b; Demotic *ỉn*: Johnson 2001, I (11:1): 143; Erichsen 1954, 33.

52 Crum 1962, 524a; Černý 1976, 228; Demotic *ỉny, ỉn*: Johnson 2001, I (11:1): 161, 144; Erichsen 1954, 34, 14.

53 Crum 1962, 691a; Černý 1976, 288; Demotic *ḥnq*: Johnson 2001, Ḥ (09:1): 164; Erichsen 1954, 314, 6.

54 Crum 1962, 685a; Černý 1976, 285; Demotic *hnw*: Johnson 2001, H (01.1): 62; Erichsen 1954, 277, 1.

in the protasis. The consonant *ḥ* not only reoccurs in the preposition *ḥr*, but so does the consonant *r*, which repeats in the preposition *r* and the infinitive *rdi,t*.

Homoeoteleuton has a more diverse configuration with its schematizations than consonance. The most popular type of homoeoteleuton in the Dream Manual is paronomasia that plays with the *s*-causative prefix.[55] The following examples include two different occurrences of homoeoteleuton:

2.21. *ḥr wnm iwf n* ***ʿꜣ*** *| nfr* ***sʿꜣ****=f pw*

Eating the flesh of a ***donkey*** *| Good; it means he* ***becomes great.***

8.26 *ḥr rdi,t* ***snṯr*** *ḥr ẖt n* ***nṯr*** *| ḏw bꜣ.w n* ***nṯr*** *r=f*

Placing ***incense*** *on a flame for the* ***god*** *| Bad; the power of the* ***god*** *is against him.*

The roots of the constituent words are identical in *ʿꜣ* || *sʿꜣ* and *snṯr* || *nṯr*, but the paronomastic pair forms with the addition of the *s*-prefix in one of the lexemes. Other forms of homoeoteleuton include the prefixation of semi-vocalic consonants to the lexical root (e.g., 5.21: *wʿḥ* || *iʿḥ*; 10.6: *ʿḥꜣ* || *ḥꜣ*) and weak verbs (7.24–25: *pri* || *iri*).[56] The phonology of glides and *matres lectionis*, particularly with verbal conjugations,[57] warrants a richer integration in future iterations of pattern matching.

3.3 *Distance*

Distance is the lexical interval between two words which pattern matching records for every identified pair (Graph 2). The success of paronomasia relies on the relative distance between constituent words, but it is not necessarily inversely proportional. For instance, the immediacy of an alliterative sequence might have a different effect in the text than the realization of polyptoton over a single stanza:

55 Richter 2016, 26 sees *s*-causative lexemes in a paronomastic pair as polyptoton, whereas we categorize ?AB || AB (e.g., snṯr || nṯr) homoeoteleuton.

56 *pri*: Crum 1962, 267a; Černý 1976, 127; Demotic *pr*: Johnson 2001, P (10:1), 118. Erichsen 1954, 134, 7. *iri*: Crum 1962, 83ab; Černý 1976, 48; Demotic *ir*: Johnson 2001, I (11:1), 174.

57 Allen 2020, 121–144.

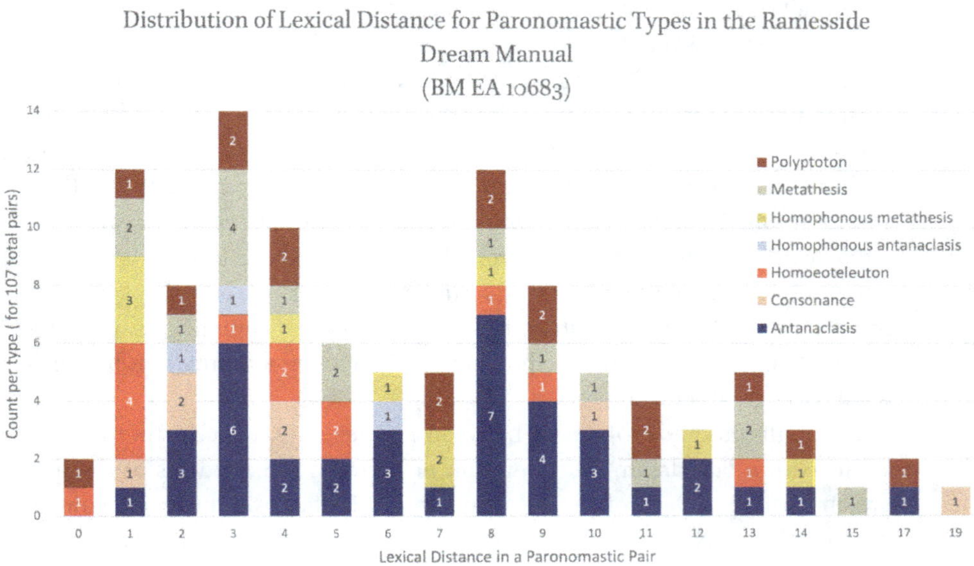

GRAPH 17.2 Distribution of lexical distance for paronomastic types in the Ramesside Dream Manual (BM EA 10683)

4.22. *ḥr rdỉ,t n=f* ***ꜥnḫ.w*** *n ḥw.t nṯr || nfr swꜣḏ n=f* ***ꜥnḫ*** *ỉn nṯr=f*

Giving him ***victuals*** *of the temple | Good; it means* ***life*** *will flourish for him by his god.*

5.13. ***ḥr ḥms ḥr ḥsp.t*** *n šw | nfr nḏm pw.*

Sitting in the garden *in the sun | Good; It means pleasure.*

Strengthened by the constraints of the line, the homogeneity between *ꜥnḫ.w* and *ꜥnḫ* permits a distance of seven lexemes for polyptoton as opposed to the alliterative sequence in line 5.13. In general, returns for polyptoton, antanaclasis, and consonance have a wider range (as discussed below) of lexical distances in the Dream Manual than homoeoteleuton.

Most of the unidentified cases in our study take place over two lines, which Noegel and Szpakowska partially identify.[58] The pattern match searches are limited to a range of two lines, but we also tested distances over one line and over the whole text.

58 Noegel and Szpakowska 2006: lines 2.8–2.9, 3.1–3.2, 3.2–3.3, 5.21–5.22, 5.21–5.22, 9.9–9.10.

Overall, the range of distances vary for each type. We observed that the number of shared consonants in an identical sequence between two words might contribute to wider distances. For example, consonance, which is underrepresented in the text (8 examples), has the highest range of 1–19 lexemes. The occurrence of consonance at a lexical distance of 19 words appears to be an outlier, but a closer look at the pair *ḥmy* || *ḥm.t* in line 7.24 suggests otherwise.[59] Antanaclasis and polyptoton have the second highest ranges (1–17 for antanaclasis and 0–17 for polyptoton) lexemes, but these types are also the most frequent in the text. Notably, antanaclasis does not occur in sequence (no distance). Perhaps, the shared structures between antanaclasis, polyptoton, and consonance improve the chances of making a correlation between words over greater distances.

Paronomasia tends to appear between 1–13 lexemes, which either implies a pair within one dream interpretation or between the apodosis of the first dream and the protasis of the following dream. Metathesis clusters in a range from 1–15 lexemes, but most frequently occurs in a distance of three words with ten total examples.[60] Occurrences of certain typologies (metathesis, consonance, homoeoteleuton) tend to congregate within the protasis and apodosis of one line, with fewer occurrences happening either within the protasis (4 times) or apodosis (3 times).[61]

The Dream Manual operates under the assumption that dreams undergo a diagnosis, which are interpreted for a prognosis. Not only are the dreams catalogued according to type and content, but they also are arranged according to the homophonous and semantic similarities between the dreams themselves and their interpretations, which we have seen with the occurrence of paronomasia over multiple, sequential lines. The scribe takes advantage of the tabular form of the Dream Manual to effectuate the power of paronomasia as a mnemonic or hermeneutic device in the text. The relative proximity of words in paronomasia and the typology influences the function and effect of similar sounding words in a specific genre.

59 *ḥmy*: ϩⲙⲙⲉ (Crum 1962, 677b; Černý 1976, 283); Demotic *ḥmy* (Johnson 2001, Ḥ (09:1), 134; Erichsen 1954, 308, 4.); *ḥm.t*: ϩⲟⲙⲛⲧ (Crum 1962, 678a; Černý 1976, 283); Demotic *ḥm.t*: Johnson 2001, Ḥ (09:1), 127.

60 The number includes ranges between 1, 2, and 3 words for metathesis and homophonous metathesis.

61 Noegel and Szpakowska 2006, 210 for some other examples.

4 Comparison with the "Seven Houses" Cycle (Papyrus Chester Beatty I; BM EA 10681)

The "Seven Houses" cycle is a series of seven poems that play with numerical counts to express longing for a beloved. The text is preserved on the verso of Papyrus Chester Beatty I (BM EA 10681, verso, C1.1–C5.2).[62] In an attempt to discern the flexibility of pattern matches derived from the Dream Manual, we select this text for its preponderance of word-play and structural dissimilarity with the oneiromantic text. Unlike the Dream Manual, the Chester Beatty love poems are arranged in linear, sequenced stanzas demarcated by rubrics, which caption each poem (Figure 17.3). This section focuses on the last stanza, the Seventh Day (C4.6–C5.2):

ḫw.t mḥ sfḫ.wt
sfḫ r sf bw mꜣꜣ=i͗ sn.t
ꜥqꜥq,n=i͗ ḫꜣyt i͗m=i͗
ḫpr,kwi͗ ḥꜥ.w=i͗ wdn
smḫ ḏ.t=i͗ ḏs=i͗
i͗r i͗w n=i͗ nꜣ wr.w swn.w
bw hr i͗b=i͗ pẖr.t=sn
nꜣ ẖry.w-ḥb bn wꜣ.t i͗m=sn
bw wḏꜥ tꜣy=i͗ ḫꜣy.t
pꜣ ḏd n=i͗ mk sw pꜣ nty sꜥnḫ=i͗
rn=s pꜣ nty ṯs=i͗
pꜣ ꜥq pr n nꜣy=s wpw.tyw pꜣ nty sꜥnḫ i͗b=i͗
ꜣḫ n=i͗ sn(.t=i͗) r pẖr.t nb.t
wr sw n=i͗ r tꜣ dmḏ.yt
pꜣy=i͗ wḏꜣ pꜣy=s ꜥq r-bnr
ptr=s kꜣ snb(=i͗)
wn=s i͗r.wt=s rnpy ḥꜥ.t=i͗
md.t=s kꜣ rwd=i͗
i͗w=i͗ ḥp.t=s sḥr=s ḏw.t ḥr=i͗
pr=s m-ꜥ=i͗ ḥr hrw sfḫ.

62 Mathieu 2008, 26–30.

<u>Seventh House</u>:[63]
It has been seven days since yesterday
that I've not seen my lover.
Affliction has consumed me,
my limbs have become heavy,
and my body has forgotten itself.
Even if doctors come to me,
my heart will not be healed by their medicine.
And the priests have no recourse among them;
my ailment cannot be identified.
But the statement, "Look, she is there," would enliven me.
Her name alone is my cure.
(Oh), the comings and goings of messengers are what give me life.
My lover inspirits me more than any medicine.
She is better for me than any therapy.
(C5) My lucky-charm is her stepping outside.
Let her be seen so I can become healthy,
let her open her eyes so my body can become strong,
let her speak that I might flourish.
When I embrace her, she eliminates the wrong in me.
(Alas), it has been seven days since she has left my arm.

Pattern matching produces 91 occurrences of word-play in a stanza of 129 lexical units. The rubricification of "Seventh House" (*ḥw.t mḥ sfḫ.wt*) and the end of the text, which also coincides with the end of the manuscript, specifies the boundaries that delimit the occurrences of word-play in the song. The initial occurrences of word-play have both the closest and furthest lexical distances in the song and revolve around an eponymous play with "seven" (*sfḫ*). The song begins and ends with:

sfḫ r sf bw mꜣꜣ=i͗ sn.t
…
pr=s m-ꜥ=i͗ ḥr hrw sfḫ.

63 I retain the literal translation of *ḥw.t* ("house") here, but *ḥw.t* also translates to "stanza" (Lesko II, 304). Similarly, the word "stanza" in Italian means "room" or "standing place," which reflects the meaning in Egyptian. Note also that in the third-person singular suffix pronoun, I transliterate *s* even though it appears graphically as *st*.

FIGURE 17.3 The "Sixth and Seventh Houses": Love songs from Papyrus Chester Beatty I (BM EA 10681). Gardiner 1931: Plate XXV

> ***Seventh House:*** *It has been* ***seven days*** *since* ***yesterday*** *that I've not seen my lover.*
>
> …
>
> *(Alas), it has been* ***seven*** *days since she has left my arm.*

The consonance between "seven" (*sfḫ*) and "yesterday" (*sf*) establishes the paronomastic correlation between the "Seventh" (*mḥ sfḫ.wt*) House, which observes a polyptonic sequence with the first word of the song, "seven." The play with *sfḫ* only repeats with the completion of the song, "seven," in an example of symploce, a literary device that occurs in every song in the cycle. Symploce is the repetition of a word at the beginning and end of a clause or stanza.

Unlike the oneiromantic text, the composer of the "Seven Houses" cycle uses paronomasia not only to bracket the body of the song, but also as a technique to retain the attention of the audience through the repetition of *sfḫ* over 123 lexemes. The Seventh House takes advantage of distance to define the paronomastic correlation with the title of the song (sequential paronomasia with no distance: *ḥw.t mḥ sfḫ.wt sfḫ r sf*) and to alert the audience to the completion of the song. The body of the song (the interval between the symploce) allows the

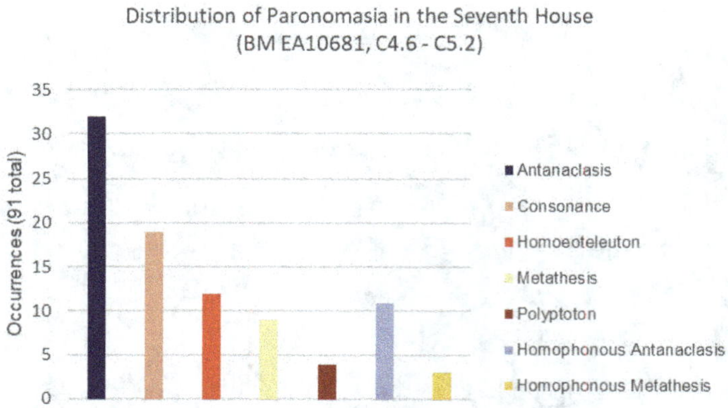

GRAPH 17.3 Distribution of pattern matching returns in the seventh song of the Papyrus Chester Beatty I (BM EA 10681, C4.6–C5.2 (BM EA 10681, C4.6–C5.2)

listener to forget about the initial alliteration with the title until the connection is restored with the final word, *sfḫ*. Symploce incites the audience to reflect on the sentiments of the lover by recalling the paronomasia at the beginning of the song. This reinforces the temporal relationship between the title and "yesterday" (*sf*) creating a structural coherence with earlier songs in the cycle.

Antanaclasis has the widest range of lexical distance and returns more commonly with distances between one to 10 lexemes, although many of these examples may not be paronomastic (Appendix, Table 17.4; Graph 3). For example, most returns in the song are patterns with demonstrative pronouns (*pꜣy, tꜣy, nꜣy*), articles (*pꜣ, tꜣ, nꜣ*), and relative adjectives (*nty*). Their repetition resembles ploce and reinforces the poetic rhythm of the song (e.g., *pꜣy=ỉ wḏꜣ pꜣy=s ꜥq r-bnr*). However, unlike the *nisbe* in lines 3.2–3.3 of the Dream Manual, these classes do not share a paronomastic relationship with verbs and nouns in the song. Other false positives include pairs with words that end with liquid consonants (*r* and *n*).[64]

The body of the poem contains numerous examples of word-play types that replicate occurrences in the oneiromantic text, but we are aware that there are overlooked and more complex examples in the "Seven Houses" cycle that would benefit from pattern matching.[65] Aside from the preponderance of ploce and antanaclasis in the text, pattern matching returns notable examples of metathesis, consonance, and homoeoteleuton. For example, the pair *sḥr* and *ḥr* (?AB || AB) in the following line:

64 False positives: *pr* || *rn*; *hr* || *rn*; and *wr* || *rn*.

65 Mathieu 2008, 30.

sḥr=s ḏw.t ḥr=ỉ

She ***eliminates*** ***(sḥr)*** *the wrong* ***in*** ***(ḥr)*** *me.*

Proximity between constitutive words seems to be a denotive feature of paronomasia in this text. Perhaps, the homophony between two words enables more complex combinations over the song. For instance, the antanaclasis *ꜥq* || *ꜥq* and the metathesis *pr* || *ptr* in the following clauses reveals an intriguing homophonous pattern:

ꜥq pr *n nꜣy=s wpw.tyw pꜣ nty sꜥnḫ ỉb=ỉ* ***ꜣḫ*** *n=ỉ sn(.t=ỉ) r* ***pẖr.t*** *nb.t wr sw n=ỉ r tꜣ dmḏ.yt pꜣy=ỉ wḏꜣ pꜣy=st* ***ꜥq*** *r-bnỉ* ***ptr****=s kꜣ snb(=ỉ).*

(Oh), the ***comings*** ***(ꜥq)*** *and* ***goings*** ***(pr)*** *of messengers are what give me life. My lover inspirits (****ꜣḫ****) me more than any* ***medicine*** ***(pẖr.t)****. She is better for me than any therapy. (C5) My lucky-charm is her* ***coming*** ***(ꜥq)*** *inside. Let her be* ***seen*** ***(ptr)*** *so I can become healthy.*

The pattern plays with the proximity between two paronomastic groups: *ꜥq* || *ꜣḫ* || *ꜥq* and *pr* || *pẖr.t* || *ptr*. First, the verbal sequence *ꜥq* and *pr* establishes the juxtaposition between these two groups. The subsequent clause begins with *ꜣḫ*[66] and reiterates the association with a metathetic pair between *pr* and *pẖr.t* (Coptic ⲡⲉⲓⲣⲉ[67] and ⲡⲁϩⲣⲉ,[68] respectively). Then, the second *ꜥq* realizes the antanaclasis, which also references the geminated *ꜥqꜥq* at the beginning of the song. Similarly, *ptr* completes the sequence as a homophonous group with *pr* and *pẖr*, but it also begins the final refrain of the song, a series of imperative and *kꜣ* + Prospective constructions.

This passage also contains a playful articulation of the *ꜥnḫ, wḏꜣ, snb* expression, of which *ꜥnḫ* and *snb* undergo further paronomastic constructions.[69]

66 There might be a homophonous antanaclasis between *ꜣḫ* and *ꜥq* (ⲁⲉⲓⲕ: Černý 1962, 3; Crum 1976, 3a) here, albeit tenuous. Although the velar stops and velar fricatives are not homophonous, Peust (1999, 118) addresses rare occasions where *ḫ* > ⲕ. The closest etymological equivalent to *ꜣḫ* is *ꜣḫy* ("plants, vegetation," Erman and Grapow 1926–1963,. I 18, 8) > ᴮⲁϧⲓ/ⲁⲭⲓ (Černý 1962, 17; Crum 1976, 25a), which accounts for the depalatization of *ḫ* in late Egyptian (ᴮⲁϧⲓ) as well as a rare alternate where *ḫ* > χ, an aspirated equivalent to ⲕ.

67 Crum 1962, 267a; Černý 1976, 127. Demotic *pr*: Johnson 2001, P (10:1), 118. Erichsen 1954, 134, 7.

68 Crum 1962, 282b; Černý 1976, 131; Demotic *pẖr.t*: Johnson 2001, P (10:1), 157; Erichsen 1954, 139, 5.

69 Mathieu notes this expression in 2008, 44, no. 100.

> *ꜥq pr n nꜣy=st wpw.tyw pꜣ nty s***ꜥnḫ***-ib=i ꜣḫ n=i sn(.t=i) r pẖr.t nb.t wr sw n=i r tꜣ dmd.yt pꜣy=i* ***wḏꜣ*** *pꜣy=s ꜥq r-bni ptr=s kꜣ* ***snb****(=i).*
>
> *(Oh), the comings and goings of messengers are what* ***give me life (sꜥnḫ).*** *My lover inspirits me more than any medicine. She is better for me than any therapy. (C5) My* ***lucky-charm*** *(****wḏꜣ****) is her stepping outside. Let her be seen so I can become* ***healthy (snb).***

The pair *sꜥnḫ* || *ꜣḫ* appears to be a homophonous metathesis:[70]

> *ꜥq pr n nꜣy=s wpw.tyw pꜣ nty s***ꜥnḫ***-ib=i* ***ꜣḫ*** *n=i* ***sn(.t****=i) r pẖr.t* ***nb.t*** *wr sw n=i r tꜣ dmd.yt pꜣy=i wḏꜣ pꜣy=s ꜥq r-bnr ptr=s kꜣ* ***snb****(=i).*
>
> *(Oh), the comings and goings of messengers are what* ***give me life (sꜥnḫ).*** *My* ***lover (sn.t) inspirits (ꜣḫ)*** *me more than* ***any*** *(****nb.t****) medicine. She is better for me than any therapy. (C5) My lucky-charm is her stepping outside. Let her be seen so I can become* ***healthy (snb).***

Pattern matching returns the pair *nb.t* || *snb* as homoeoteleuton (AB.? || ?AB), but does not recognize the pair *sn.t* || *nb.t*. Although the schematization for *sn.t* || *nb.t* does not exist in the study, pattern matching would not have been able to identify this pair because the scribe does not include the feminine *t*-ending of *sn(.t)* in the manuscript. The transliteration is devoid of modern edits, which are incorporated in future iterations of the study. However, the occurrence of *snb* strengthens the relationship between *sn.t* and *nb.t*, each of which pattern matching returns (*sn* || *snb*; *nb.t* || *snb*). This is a case where pattern matching not only recovers paronomasia in texts, but also highlights missing relationships that overlap homophonously with returned pairs.

5 Future Directions and Conclusions

Pattern matching is a powerful tool, but it requires the input of an Egyptologist to define and recognize phonetic patterns in a text. However, the methodology presented here exposes the biases and oversights of the specialist so that approaches to Egyptian paronomasia may be improved. Pattern matching

70 ⲱⲛϩ (Crum 1962, 525a; Černý 1976, 228), Demotic *ꜥnḫ:* Johnson 2001, C (03.1), 80; [A2]ⲓϩ (Crum 1962, 89a; Černý 1976, 50), Demotic *iḥy* (Erichsen 1954, 42, 3).

returns both new and published examples of paronomasia in Papyrus Chester Beatty III; it allows for the visualization of linguistic data and presents information on false positives, the use and percentage of paronomastic types, and the lexical distances between word-play constituents in both the Dream Manual and the Seventh House. Schematizations from the Dream Manual occur in the latter text as well, which suggests that word-play types appear across different genres. Although new occurrences of paronomasia confirm the applicability of pattern matching in Egyptian texts, our methodology can be improved with the implementation of phonotactics—the rules governing the position and values of phonemes in a word—as well as the expansion of the typological repertoire and the application of pattern matching to new texts.

The computational identification of paronomasia highlights how the Egyptians constructed and positioned similar sounding words to increase the magical significance of the text, entertain an audience, or mnemonically assist the scribe. Nevertheless, schematizations are a simplification of the mechanism; thus, they do not account for the complexities of sounds in speech. The use of phonotactics in this research will broaden the range of paronomastic influence beyond the simple, lexical group.[71] This includes the influence of grammatical forms in the execution of paronomasia (e.g., lines 3.2–3.3: *ꜥꜣ n(i͗)* || *ꜥꜣ,n* from the Dream Manual).[72] Additionally, a concordance of Demotic and Coptic does not automatically wean false positives from phonetic word-plays; there is always a need for manual interpretation at this phase. A better implementation of Egyptian phonology, even by means of etymological equivalents, reinforces what constitutes a paronomastic pair.[73] Another measurement of accuracy might include Egyptian renderings of foreign words, especially of Northwest Semitic origin, which is central to understanding Late Egyptian phonology.

It is unclear why in both texts certain typologies are more popular than others. We observe that specific typologies (e.g., antanaclasis and ploce) appear at greater distances: as the distance between constituent words becomes longer, word-play pairs share more consonants. For the Dream Manual, the lexical categorization and distribution of specific dreams accounts for the use of one typology over another. More complex variations of paronomasia play with

71 Allen 2020, 85–94.

72 Such as, but not limited to, the phonological analysis of final weak consonants, verb forms (infinitives; *n*-suffix of *sḏm,n=f*; statives), Late Egyptian conjugation bases (e.g., negative aorist *bw* in the Seventh House), pronominal suffixes (*=f*, *=s(t)*, *=sn*), pronouns, and nominal endings (plural and feminine).

73 For instance, Late Egyptian onomastica are an excellent resource for alternative lexicographic spellings. See, e.g., Quaegebeur 2019; Nims 1950; Gardiner 1947.

typologies and distance to invoke an effect in the text, such as the mnemonic symploce in the Seventh House. Nevertheless, the picture becomes more complicated once we include schematizations that distribute shared consonants over three or more words. Among the 42 examples published by Noegel and Szpakowska, complex examples exhibit a combination of basic paronomasia in multipart schematizations:

> 4.3 ***ḥr mꜣꜣ myw ꜥꜣ || nfr šmw pw ꜥꜣ r ḫpr n=f***
>
> **Seeing a large cat** || Good, it means **a large harvest will happen** to him.

Each bolded lexeme in the line shares at least one consonant with another word and nine words share at least two. The codification of this line into pattern matching is possible once the building blocks of complex paronomasia are understood. These building blocks consist of the simplest examples of paronomasia, which are the focus of this paper. Although not representative of all paronomastic structures, the experiment has taught us that the mechanics of paronomasia are convoluted and multifaceted, and that literary genres discernibly affect the function and distribution of paronomastic types in a text.

The most dynamic avenue for future exploration is the application of pattern matching to texts with unpublished or unknown occurrences of word-play. A potential corpus is the repertoire of funerary spells, which opens investigation to older corpora and earlier phases of the Egyptian language. Lacunae and scribal omissions could also benefit from pattern matching, particularly when modern reconstructions include paronomasia (e.g., Gardiner 1935). Our study shows that pattern matching is useful to assess biases of paronomastic structures in other texts, but it also reveals noteworthy discoveries. In fact, we tested our pattern matches on an Eighteenth Dynasty manuscript, Papyrus Turin 8438, specifically Book of the Dead Spell 175.[74] The experiment returns surprising data, of which an example is shown here:

> *ỉrỉ.n=sn* ***ḥḏỉ.t*** *ỉmn.t m ỉrỉ.t n=k nb.t ỉmỉ ỉr=k wỉ ꜥꜣ* ***Ḏḥw.tỉ*** *ḫrw-fỉ ỉtm*
>
> *"They have created hidden* ***destruction*** *of all that you have done. Cause that you make me great things,* ***Thoth,****" says Atum.*

74 The papyrus dates to the reign of Amenhotep III; its provenience is TT8 of the architect Kha and his wife Meret (Schiaparelli 1927). The language for religious texts during the 18th dynasty is late-Middle Egyptian, as it is in the Dream Manual.

Pattern matching returns a metathesis between *ḥḏi.t* (destruction) and *Ḏḥw.ti* (Thoth), an exciting correlation between the statements of Atum to Thoth. It is possible, then, to use regular expressions to identify paronomasia over a variety of genres with uncertain (or unknown) counts of word-play. It is only by improving our methodology and experimenting with diverse data sets that pattern matching can capture more accurate returns. This process should be flexible and playful, as Egyptian word-play typically is, but determining whether these returns are significant, a colon, or magical, depends on the expertise and attention of a human audience.

6 Appendix, Tables

TABLE 17.1 Late Egyptian phonology (based on Allen 2020, 47–48; 83–84)

Category	Grapheme	Value
Coronal approximant, glottal, or *mater lectionis*	*ꜣ*	[j],—,[ʔ]
	ỉ	[ʔ],—
Laryngealized approximant	*ꜥ*	[ʕ]
Palatal approximant	*y*	[j]
Labial approximant or unstressed vowel	*w*	[w], [u]
Labial phonemes (unaspirated stop *b*, aspirated stop *p*, affricate *f*)	*b*	[b], [β]
	p	[p], [p^{h}]
	f	[ɸ], [f]
Nasal phonemes (labial *m* and coronal *n*)	*m*	[m]
Coronal liquids (nasal *n* and tap *r*)	*n*	[n], [l]
	r	[ɾ],[j],[l]
Voiceless laryngeal (glottal fricative *h* and pharyngeal fricative *ḥ*)	*h*	[h]
	ḥ	[ħ]
Velar fricative phonemes (fricative *ḫ* and palatalized fricative *ẖ*)	*ḫ*	[x], [x^{j}]
	ẖ	[x], [x^{j}]
Sibilants (fricative coronal *s* and fricative palatal *š*)	*s*	[s]
	š	[ʃ]
Velar phonemes (unaspirated stop *q*, aspirated stop *k*, and palatalized stop *g*)	*q*	[k]
	k	[k^{h}], [k]
	g	[k^{j}]
Coronal stops (aspirated stop *t* and unaspirated stop *d*)	*t, tw, tỉ*	[t^{h}], [t]
	d	[t]
Palatal phonemes (aspirated stop *ṯ* and unaspirated stop *ḏ*)	*ṯ*	[t^{hj}]
	ḏ	[t^{h}]

TABLE 17.2 Word-play in Papyrus Chester Beatty III

<table>
<tr><th>Typology, schematization, and paronomasia</th><th>Count of pattern</th></tr>
<tr><td>Antanaclasis</td><td>39</td></tr>
<tr><td>AB || AB</td><td>17</td></tr>
<tr><td>dr || dr</td><td>1</td></tr>
<tr><td>gm || gm</td><td>1</td></tr>
<tr><td>ḥd̠ || ḥd̠</td><td>1</td></tr>
<tr><td>ỉn || ỉn</td><td>3</td></tr>
<tr><td>ỉt || ỉt</td><td>1</td></tr>
<tr><td>nk || nk</td><td>3</td></tr>
<tr><td>pr || pr</td><td>3</td></tr>
<tr><td>tp || tp</td><td>1</td></tr>
<tr><td>tꜣ || tꜣ</td><td>1</td></tr>
<tr><td>ꜥš || ꜥš</td><td>1</td></tr>
<tr><td>ꜥꜣ || ꜥꜣ</td><td>1</td></tr>
<tr><td></td><td></td></tr>
<tr><td>ABC || ABC</td><td>20</td></tr>
<tr><td>ḥnw || ḥnw</td><td>1</td></tr>
<tr><td>ḫpr || ḫpr</td><td>4</td></tr>
<tr><td>ḥwỉ || ḥwỉ</td><td>1</td></tr>
<tr><td>ỉwf || ỉwf</td><td>1</td></tr>
<tr><td>nḫt || nḫt</td><td>1</td></tr>
<tr><td>nṯr || nṯr</td><td>4</td></tr>
<tr><td>psḥ || psḥ</td><td>1</td></tr>
<tr><td>sd̠m || sd̠m</td><td>1</td></tr>
<tr><td>smꜣ || smꜣ</td><td>1</td></tr>
<tr><td>wnm || wnm</td><td>5</td></tr>
<tr><td></td><td></td></tr>
<tr><td>ABCD || ABCD</td><td>1</td></tr>
<tr><td>d̠ꜣỉs || d̠ꜣỉs</td><td>1</td></tr>
<tr><td></td><td></td></tr>
<tr><td>Consonance</td><td>5</td></tr>
<tr><td>AB? || AB.?</td><td>1</td></tr>
<tr><td>ḥmy || ḥm.t</td><td>1</td></tr>
<tr><td></td><td></td></tr>
<tr><td>AB || AB?</td><td>1</td></tr>
<tr><td>ỉn || ỉnr</td><td>1</td></tr>
</table>

TABLE 17.2 Word-play in Papyrus Chester Beatty III (*cont.*)

Typology, schematization, and paronomasia	Count of pattern
AB?.? || AB?	3
ḥnq.t || *ḥnw*	2
nhꜣ.t || *nhy*	1
Homoeoteleuton	13
?AB || ?AB	3
pri͗ || *i͗ri͗*	2
wꜥḥ || *i͗ꜥḥ*	1
?AB || AB	1
ꜥḥꜣ || *ḥꜣ*	1
?ABC || ABC	3
snṯri͗ || *nṯr*	3
A.B || ?A.B	3
ẖ.t || *sẖ.t*	3
AB || ?AB	2
ꜥꜣ || *sꜥꜣ*	2
AB.? || ?AB	1
ḏꜣ.t || *wḏꜣ*	1
Homophonous antanaclasis	3
AB || AB	3
qꜣ || *kꜣ*	1
tw || *ḏw*	2
Homophonous metathesis	10
AB? || A?B	10
bi͗n || *fꜣi͗*	1
ḫpr || *šsp.t*	3
nṯr || *rmṯ*	1
nṯr || *rmṯ.w*	1

TABLE 17.2 Word-play in Papyrus Chester Beatty III (*cont.*)

<table>
<tr><th>Typology, schematization, and paronomasia</th><th>Count of pattern</th></tr>
<tr><td>sḏm || sšd</td><td>2</td></tr>
<tr><td>skꜣ || sꜥq</td><td>1</td></tr>
<tr><td>snf || sḏm</td><td>1</td></tr>
<tr><td>Metathesis</td><td>17</td></tr>
<tr><td>?AB || AB?.?</td><td>1</td></tr>
<tr><td>ḏbꜣ || bꜣk.w</td><td>1</td></tr>
<tr><td>?AB || AB.?</td><td>1</td></tr>
<tr><td>wḏꜣ || ḏꜣ.t</td><td>1</td></tr>
<tr><td>A?B || ?AB</td><td>6</td></tr>
<tr><td>myw || šmw</td><td>1</td></tr>
<tr><td>nḏm || wnm</td><td>1</td></tr>
<tr><td>nfr || ỉnr</td><td>3</td></tr>
<tr><td>nṯr || ỉnr</td><td>1</td></tr>
<tr><td>AB || A?B</td><td>3</td></tr>
<tr><td>ḫr || ḫpr</td><td>1</td></tr>
<tr><td>pw || pẖw</td><td>1</td></tr>
<tr><td>ꜥꜣ || ꜥšꜣ</td><td>1</td></tr>
<tr><td>AB,? || A?B</td><td>1</td></tr>
<tr><td>ꜥꜣ,n || ꜥšꜣ</td><td>1</td></tr>
<tr><td>AB? || A?B</td><td>1</td></tr>
<tr><td>ḏꜣy || ḏfꜣ.w</td><td>1</td></tr>
<tr><td>AB.? || A?B</td><td>3</td></tr>
<tr><td>bn.t || bỉn</td><td>1</td></tr>
<tr><td>šd.t || šꜥd</td><td>1</td></tr>
<tr><td>wꜥ.t || wšꜥ</td><td>1</td></tr>
<tr><td>ABC || ACB</td><td>1</td></tr>
<tr><td>wꜣḏ || wḏꜣ</td><td>1</td></tr>
</table>

TABLE 17.2 Word-play in Papyrus Chester Beatty III (*cont.*)

Typology, schematization, and paronomasia	Count of pattern
Polyptoton	21
AB || AB,?	3
ḫb || *ḫb,t*	1
sd || *sd,t*	1
ꜥꜣ || *ꜥꜣ,n*	1
AB || AB?	9
ḫr || *ḫr.w*	1
ḥr || *ḥry*	3
pr || *pri͗*	3
wḏ || *wḏꜣ*	1
AB || AB.?	4
ḥḏ || *ḥḏ.t*	2
ḥm || *ḥm.t*	2
AB? || AB,?	1
ꜥꜣw || *ꜥꜣ,n*	1
ABC || ABC.?	5
nfr || *nfr.t*	1
nṯr || *nṯr.t*	3
ꜥnḫ || *ꜥnḫ.w*	1
Grand Total	107

TABLE 17.3 Potential, newly identified paronomasia from the Ramesside Dream Manual (BM EA 10683)

Typology, schematization, and paronomasia	Line no.	Count
Antanaclasis (+ homophonous)		36
AB || AB		18
ꜥꜣ || *ꜥꜣ*	4.3	1
dr || *dr*	8.13–8.14	1

TABLE 17.3 Potential, newly identified paronomasia from the Ramesside Dream Manual (BM EA 10683) (*cont.*)

Typology, schematization, and paronomasia	Line no.	Count
gm || *gm*	4.19–4.20	1
in || *in*	7.18; 7.18–7.19	3
it || *it*	6.24	1
nk || *nk*	3.7; 3.7–3.8	3
pr || *pr*	5.15–5.16; 6.4; 6.6–6.7	3
qꜣ || *kꜣ*	3.9–3.10	1
tꜣ || *tꜣ*	8.19–8.20	1
tp || *tp*	6.23–6.24	1
tw || *ḏw*	10.4; 10.4–10.5	2
ABC || ABC		**18**
ḥnw || *ḥnw*	10.8–10.9	1
ḥwi || *ḥwi*	8.19–8.20	1
sḏm || *sḏm*	2.24–2.25	1
iwf || *iwf*	2.21–2.22	1
nṯr || *nṯr*	4.22; 8.26	3
psḥ || *psḥ*	7.18–7.19	1
smꜣ || *smꜣ*	4.8	1
wnm || *wnm*	2.20–2.22; 7.16	5
ḫpr || *ḫpr*	6.16–6.17; 7.5–7.7; 7.19–7.20	4
Consonance		**4**
AB || AB?		1
in || *inr*	6.3–6.4	1
AB? || AB.?		1
ḥmy || *ḥm.t*	7.24–7.25	1
AB?.? || AB?		2
ḥnq.t || *ḥnw*	10.8; 10.8–10.9	2
Homoeteleuton		**2**
?AB || ?AB		2
pri || *iri*	7.24–7.25; 8.6	2

TABLE 17.3 Potential, newly identified paronomasia from the Ramesside Dream Manual (BM EA 10683) (*cont.*)

Typology, schematization, and paronomasia	**Line no.**	**Count**
Metathesis (+ **homophonous)**		24
ABC ǁ ACB		1
wꜣḏ ǁ *wḏꜣ*	6.10–6.11	1
A?B ǁ ?AB		5
nḏm ǁ *wnm*	5.11–5.12	1
nfr ǁ *ỉnr*	6.4–6.5	3
nṯr ǁ *ỉnr*	8.18	1
AB ǁ A?B		2
pw ǁ *pẖw*	6.13–6.14	1
ḫr ǁ *ḫpr*	8.11–8.12	1
AB.? ǁ A?B		10
bỉn ǁ *fꜣỉ*	8.3–8.4	1
ḏꜣy ǁ *ḏfꜣ.*w	2.13–2.14	1
nṯr ǁ *rmṯ*	6.3	1
nṯr ǁ *rmṯ.w*	3.10–3.11	1
šd.t ǁ *šꜥd*	8.5–8.6	1
skꜣ ǁ *sꜥq*	6.2–6.3	1
snf ǁ *sḏm*	5.2–5.3	1
ḫpr ǁ *ššp.t*	7.5–7.6	3
AB.?ǁ A?B		1
wꜥ.t ǁ *wšꜥ*	7.9–7.10	1
Polyptoton		6
AB ǁ AB?		3
pr ǁ *prỉ*	6.4–6.6;	3
ABC ǁ ABC.?		5
ꜥnḫ ǁ *ꜥnḫ.w*	4.22	1
nfr ǁ *nfr.t*	2.7–2.8	1
nṯr ǁ *nṯr.t*	2.25–2.26	3
Total		72

TABLE 17.4 Pattern matching returns from the Seventh House (BM EA 10681, verso, C1.1–C5.2)

Typology, schematization, and paronomasia	Count of pattern
Antanaclasis	**32**
AB || AB	**25**
ꜥq || ꜥq	4
bw || bw	3
ḥr || ḥr	1
ꞽb || ꞽb	1
ꞽm || ꞽm	1
ꞽw || ꞽw	1
kꜣ || kꜣ	1
nꜣ || nꜣ	1
pꜣ || pꜣ	10
pr || pr	1
sw || sw	1
ABC || ABC	**6**
nty || nty	3
pꜣy || pꜣy	2
sfẖ || sfẖ	1
ABCD || ABCD	**1**
sꜥnḫ || sꜥnḫ	1
Consonance	**19**
AB || AB?	**18**
bn || bnj	1
hr || hrw	1
nꜣ || nꜣy	2
pꜣ || pꜣy	10
sf || sfẖ	2
sn || snb	1
tꜣ || tꜣy	1
AB? || AB.?	**1**
snb || sn.t	1

TABLE 17.4 Pattern matching returns from the Seventh House (BM EA 10681, verso, C1.1–C5.2) (*cont.*)

Typology, schematization, and paronomasia	**Count of pattern**
Homoeoteleuton	**12**
?AB || ?AB	**9**
nꜣy || *pꜣy*	1
nꜣy || *tꜣy*	1
pꜣy || *nꜣy*	2
pꜣy || *tꜣy*	2
tꜣy || *nꜣy*	1
tꜣy || *pꜣy*	2
?AB || AB.?	**1**
snb || *nb.t*	1
AB || ?AB	**2**
ḥr || *sḥr*	2
Homophonous Antanaclasis	**11**
AB || AB	**11**
ḏs || *ṯs*	1
ḥr || *ḥr*	2
hr || *rn*	3
ỉr || *rn*	1
pr || *rn*	2
ṯs || *ḏs*	1
wr || *rn*	1
Homophonous Metathesis	**3**
AB.? || A?B	**3**
sfḫ || *snb*	2
snb || *sḥr*	1
Metathesis	**9**
A?B || ?AB	**3**
tꜣy || *nty*	3

TABLE 17.4 Pattern matching returns from the Seventh House (BM EA 10681, verso, C1.1–C5.2) (*cont.*)

Typology, schematization, and paronomasia	Count of pattern
AB || A?B	**3**
pr || *ptr*	2
wn || *wdn*	1
AB.? || ?AB	**1**
nb.t || *snb*	1
AB(.t) || A?B	**2**
ḥw.t || *ḥʿw*	1
wꜣ.t || *wḏꜣ*	1
Polyptoton	**4**
AB || AB.?	**4**
ḥʿ || *ḥʿ.t*	1
sn || *sn.t*	2
wr || *wr.w*	1
Total	**90**

Abbreviations

CDD *Chicago Demotic Dictionary* or *The Demotic dictionary of the Oriental Institute of the University of Chicago.*

Wb *Wörterbuch der aegyptischen Sprache 1–7*

References

Albright, William F. 1923. "The Principles of Egyptian Phonological Development." *Recueil de Travaux Relatifs à La Philologie et à l'archéologie Égyptiennes et Assyriennes* 40: 64–70.

Allen, James P. 2005. *Ancient Egyptian Pyramid Texts*. Atlanta: Society of Biblical Literature.

Allen, James P. 2013a. *A New Concordance of the Pyramid Texts: PT 247–421*. Vol. 3. Providence: Brown University.

Allen, James P. 2013b. *The Ancient Egyptian Language: An Historical Study*. Cambridge; New York: Cambridge University Press.

Allen, James P. 2020. *Ancient Egyptian Phonology*. Cambridge: Cambridge University Press.

Altenmüller, Hartwig. 1967. "Ein Opfertext der 5. Dynastie." *Mitteilungen Des Deutschen Archäologischen Instituts, Abteilung Kairo* 22: 9–18.

Assmann, Jan. 2001. *The Search for God in Ancient Egypt*. Translated by David Lorton. Ithaca, NY: Cornell University Press.

Austin, John L. 1975. *How to Do Things with Words*. 2nd ed. Edited by James O. Urmson and Marina Sbisà. *William James Lectures* 1955. Cambridge: Harvard University Press.

Bamman, David, and Gregory Crane. 2009. "Computational Linguistics and Classical Lexicography." *Digital Humanities Quarterly* 003 (1).

Barbash, Yekaterina. 2015. "Wordplay's Place in Mortuary Liturgies: Scribal Devices in Papyrus W551." In *Liturgical Texts for Osiris and the Deceased in Late Period and Greco-Roman Egypt / Liturgische Texte Für Osiris Und Verstorbene Im Spätzeitlichen Ägypten: Proceedings of the Colloquiums at New York (ISAW), 6 May 2011, and Freudenstadt, 18–21 July 2012*, edited by Burkhard Backes and Jacco Dieleman, 203–215. Studien zur spätägyptischen Religion 14. Wiesbaden: Harrassowitz Verlag.

Bolzoni, Lina. 1999. "The Play of Memory between Words and Images." In *Memory & Oblivion: Proceedings of the XXIXth International Congress of the History of Art Held in Amsterdam, 1–7 September 1996*, edited by Wessel Reinink and Jeroen Stumpel, 11–18. Dordrecht: Springer Netherlands.

Breyer, Francis. 2003a. *Tanutamani: Die Traumstele und ihr Umfeld*. Ägypten und Altes Testament 57. Wiesbaden: Otto Harrassowitz.

Breyer, Francis. 2003b. "Wortspiele in zwei hymnischen Abschnitten der Traumstele." *Lingua Aegyptia* 11: 13–22.

Brunner, Hellmut. 1988. "Name, Namen, Namenlosigkeit Gottes Im Alten Ägypten." In *Das hörende Herz: Kleine Schriften zur Religions- und Geistesgeschichte Ägyptens*, edited by Hellmut Brunner and Wolfgang Röllig, 130–146. Freiburg Schweiz; Göttingen: Universitätsverlag; Vandenhoeck & Ruprecht.

Bryan, Betsy M. 1991. *The Reign of Thutmose IV*. Baltimore: Johns Hopkins University Press.

Busa, Roberto. 1974. *Index Thomisticus: Sancti Thomae Aquinatis operum ominium indices et concordantiae in quibus verborum omnium et singulorum formae et lemmata cum suis frequentiis et contextibus variis modis referuntur quaeque auspice Paulo VI Summo Pontifice consociata plurium opera atque electronico IBM automato usus digessit Robertus Busa SI*. Stuttgart: Frommann-Holzboog.

Buurman, Jan. 1985. Manuel de codage des textes hiéroglyphiques en vue de leur saisie sur ordinateur. *Informatique et Égyptologie* 2. Paris: CNRS.

Buurman, Jan, Nicholas Grimal, Michael Hainsworth, Jochen Hallof, and Dirk van der Plas. 1988. *Manual for the Encoding of Hieroglyphic Texts for Computer-Input. Informatique et Égyptologie.* 3rd, consolidated ed. Vol. 2. Informatique et Égyptologie. Paris: Institut de France.

Černý, Jaroslav. 1976. *Coptic Etymological Dictionary*. Cambridge: Cambridge University Press.

Černý, Jaroslav. 1978. *Papyrus hiératiques de Deir el-Médineh I: Nos I–XVII.* Edited by Georges Posener. Documents de fouilles de l'Institut français d'archéologie orientale 8. Cairo: Institut français d'archéologie orientale.

Chiarcos, Christian, Émilie Pagé-Perron, Ilya Khait, Niko Schenk, and Lucas Reckling. 2018. "Towards a Linked Open Data Edition of Sumerian Corpora." In *Proceedings of the Eleventh International Conference on Language Resources and Evaluation* (*LREC 2018*). Miyazaki, Japan: European Language Resources Association (ELRA). https://www.aclweb.org/anthology/L18-1387. Accessed 1-24-2022.

Crum, Walter E. 1962. *A Coptic Dictionary*. Oxford; New York: Clarendon Press.

Depuydt, Leo 1988. "Die 'Verben des Sehens': semantische Grundzüge am Beispiel des Ägyptischen." *Orientalia* 57 (1): 1–13.

Erichsen, W. 1954. *Demotisches Glossar*. Kopenhagen: Ejnar Munksgaard.

Erman, Adolf. 1904. "Die Sphinxstele." *Sitzungsberichte Der Preußischen Akademie Der Wissenschaften Zu Berlin* 1904: 428–444.

Erman, A., and H. Grapow (eds.) 1926–1963. *Wörterbuch der aegyptischen Sprache 1–7*. Leipzig: Akademie Verlag.

Faulkner, Raymond O. 1933. *The Papyrus Bremner-Rhind,* (*British Museum no. 10188*). Brussels: Fondation égyptologique Reine Élisabeth.

Faulkner, Raymond O. 1937. "The Bremner-Rhind Papyrus: III: D. The Book of Overthrowing 'Apep." *The Journal of Egyptian Archaeology* 23 (2): 166–185.

Fischer-Elfert, Hans-Werner. 1993. "Vermischtes II." *Göttinger Miszellen* 135: 31–37.

Foster, John L. 1980. "Sinuhe: The Ancient Egyptian Genre of Narrative Verse." *Journal of Near Eastern Studies* 39 (2): 89–117.

Gardiner, Alan H. 1916. *Notes on the Story of Sinuhe*. Paris: Librairie Honoré Champion.

Gardiner, Alan H. 1931. *The Library of A. Chester Beatty: Description of a Hieratic Papyrus with a Mythological Story, Love-Songs and Other Miscellaneous Texts, by Alan H. Gardiner. The Chester Beatty Papyri, No. 1 with 31 Plates in Monochrome and 30 in Line by Emery Walker, Ltd.* London: Oxford University Press, Emery Walker.

Gardiner, Alan H. 1935. *Hieratic Papyri in the British Museum: Third Series: Chester Beatty Gift Vol. I. Text; Vol. II. Plates.* 2 vols. London: British Museum, printed by order of the Trustees.

Gardiner, Alan H. 1947. *Ancient Egyptian Onomastica*. 3 vols. Oxford: Oxford University Press.

Guglielmi, Waltraud. 1984. "Zu einigen literarischen Funktionen des Wortspiels." In *Studien Zu Sprache Und Religion Ägyptens: Zu Ehren von Wolfhart Westendorf, Überreicht von Seinen Freunden und Schülern*, edited by Friedrich Junge, 491–506. Göttingen: Herbert & Co.

Guglielmi, Waltraud. 1986. "Wortspiel." In *Lexikon Der Ägyptologie* VI, edited by Wolfgang Helck, 1287–1291. Wiesbaden: Otto Harrassowitz.

Guglielmi, Waltraud. 1996. "Der Gebrauch Rhetorischer Stilmittel in Der Ägyptischen Literatur." In *Ancient Egyptian Literature: History and Forms*, edited by Antonio Loprieno, 465–497. Probleme der Ägyptologie 10. Leiden; New York: Brill.

Hall, Henry R. 1930. "The Chester-Beatty Egyptian Papyri." *The British Museum Quarterly* 5 (2): 46–47.

Hays, Christopher B. 2010. "The Covenant with Mut: A New Interpretation of Isaiah 28: 1–22." *Vetus Testamentum* 60 (2): 212–240.

Hellum, Jennifer. 2015. "In Your Name of Sarcophagus: The 'Name Formula' in the Pyramid Texts." *Journal of the American Research Center in Egypt* 51: 235–242.

Hermann, Alfred. 1959. *Altägyptische Liebesdichtung*. Wiesbaden: Otto Harrassowitz.

Hoch, James E. 1994. *Semitic Words in Egyptian Texts of the New Kingdom and Third Intermediate Period*. Princeton, NJ: Princeton University Press.

Hudáková, Lubica. 2016. "Gardening at Deir el-Bersha in the Middle Kingdom: A Unique Representation of Cucurbitaceae Cultivation." *Ägypten und Levante* 26: 313–327.

Israelit-Groll, Sara. 1985. "A Ramesside Grammar Book of a Technical Language of Dream Interpretation [Papyrus Chester Beatty III]." In *Pharaonic Egypt, the Bible and Christianity*, edited by Sara Israelit-Groll, 71–118. Jerusalem: Magnes Press.

Israelit-Groll, Sara. 1970. *The negative verbal system of Late Egyptian*. London; New York: Oxford University Press.

Jansen-Winkeln, Karl. 1995. Diglossie und Zweisprachigkeit im Alten Ägypten. *Wiener Zeitschrift zur Kunde des Morgenlandes* 85: 85–115.

Jansen-Winkeln, Karl. 1996. *Spätmittelägyptische Grammatik der Texte der 3. Zwischenzeit*. Ägypten und Altes Testament 34. Wiesbaden: Harrassowitz.

Johnson, Janet H. 2001. *The Demotic dictionary of the Oriental Institute of the University of Chicago*. Chicago: Oriental Institute, University of Chicago. https://oi.uchicago.edu/research/publications/demotic-dictionary-oriental-institute-university-chicago. Accessed 3-23-2022.

Kilani, Marwan. 2019. *Vocalisation in Group Writing: A New Proposal*. Lingua Aegyptia. Studia Monographica 20. Hamburg: Widmaier Verlag.

Kousoulis, Panagiotis. 2011. "'Stop, O Poison, That I May Find Your Name According to Your Aspect': A Preliminary Study on the Ambivalent Notion of Poison and the Demonization of the Scorpion's Sting in Ancient Egypt and Abroad." *Journal of Ancient Egyptian Interconnections* 3: 14–26.

Kuebler, Sandra, and Heike Zinsmeister. 2015. *Corpus Linguistics and Linguistically Annotated Corpora*. 1st ed. London: Bloomsbury Academic.

Landgráfová, Renata, and Hana Navrátilová. 2009. *Sex and the Golden Goddess: Ancient Egyptian Love Songs in Context. I.* Prague: Czech Institute of Egyptology.

Lippert, Sandra L. 2001. "Komplexe Wortspiele in der demotischen Chronik und im Mythus vom Sonnenauge." *Enchoria* 27: 88–100.

Loprieno, Antonio. 1995. *Ancient Egyptian: A Linguistic Introduction.* Cambridge; New York: Cambridge University Press.

Loprieno, Antonio. 2000. "Puns and Word Play in Ancient Egyptian." In *Puns and Pundits: Word Play in the Hebrew Bible and Ancient Near Eastern Literature*, edited by Scott B. Noegel, 3–20. Bethesda: CDL Press.

Loprieno, Antonio. 2001. *La pensée et l'écriture: Pour une analyse sémiotique de la culture Égyptienne. Quatre séminaires à l'École Pratique Des Hautes Études, section des sciences religieuses.* Edited by Christiane Zivie-Coche. Paris: Cybele.

Malaise, Michel. 1983. "Calembours et mythes dans l'Égypte ancienne." In *Le Mythe: Son langage et son message: Actes du colloque de Liège et Louvain-la-Neuve 1981*, edited by Henri Limet and Julien Ries, 97–112. Louvain-la-Neuve: Centre d'histoire des religions.

Mathieu, Bernard. 1996. *La poésie amoureuse de l'Egypte ancienne: recherches sur un genre littéraire au Nouvel Empire.* Bibliothèque d'étude 115. Cairo: Institut français d'archéologie orientale.

Meeks, Dimitri. 1997. "Les emprunts égyptiens aux langues sémitiques durant le Nouvel Empire et la Troisième Période Intermédiaire: les aléas du comparatisme." *Bibliotheca Orientalis* 54 (1–2), 32–61.

Meltzer, Edmund S. 1975. "A Possible Word-Play in Khaimuas I?" *Zeitschrift für Ägyptische Sprache und Altertumskunde* 102: 78.

Mitchell, Tom M. 1997. "Concept Learning and the General-to-Specific Ordering." In *Machine Learning.* New York: McGraw-Hill.

Morenz, Ludwig D. 2002. "Zu Formen, Gründen Und Wegen Der Textüberlieferung in Der Altägyptischen Kultur: Teil 1." *Zeitschrift für Ägyptische Sprache und Altertumskunde* 129: 130–141.

Navrátilová, Hana, and Renata Landgráfová. 2015. *Sex and the Golden Goddess II: World of the Love Songs.* Czech Institute of Egyptology.

Nederhof, Mark Jan. 2009. "Automatic Alignment of Hieroglyphs and Transliteration." In *Information Technology and Egyptology.* Gorgias Press. https://research-repository.st-andrews.ac.uk/handle/10023/1667. Accessed 1-24-2022.

Nederhof, Mark Jan. 2013. "The Manuel de Codage Encoding of Hieroglyphs Impedes Development of Corpora," In *Texts, languages & information technology in Egyptology*, edited by Stéphane Polis and Jean Winand, 103–110. Liège: Presses Universitaires de Liège.

Nims, Charles. 1950. "Egyptian Catalogues of Things." *Journal of Near Eastern Studies* 9: 253.

Noegel, Scott B. 2000. *Puns and Pundits: Word Play in the Hebrew Bible and Ancient Near Eastern Literature*. Bethesda, MD: CDL Press.

Noegel, Scott B. 2007. "Enigmatic Dreams in Egypt." In *Nocturnal Ciphers: The Allusive Language of Dreams in the Ancient Near East*, edited by Scott B. Noegel. American Oriental Series. 89. New Haven: American Oriental Society, 89–106.

Noegel, Scott B. 2018. "Appellative Paronomasia and Polysemy in the Tale of Sinuhe." *Lingua Aegyptia* 26: 233.

Noegel, Scott B., and Kasia Szpakowska. 2006. "'Word Play' in the Ramesside Dream Manual." *Studien Zur Altägyptischen Kultur* 35: 193–212.

Oréal, Elsa. 1998. "'Noir parfait': Un jeu de mots de l'Égyptien au Grec." *Revue Des Études Grecques* 111 (2): 551–565.

Pagé-Perron, Émilie, Maria Sukhareva, Ilya Khait, and Christian Chiarcos. 2017. "Machine Translation and Automated Analysis of the Sumerian Language." In *Proceedings of the Joint SIGHUM Workshop on Computational Linguistics for Cultural Heritage, Social Sciences, Humanities and Literature*, 10–16. Vancouver, Canada: Association for Computational Linguistics. https://doi.org/10.18653/v1/W17-2202. Accessed 1-24-2022.

Pestman, Pieter W. 1982. "Who Were the Owners, in the 'Community of Workmen,' of the Chester Beatty Papyri." In *Gleanings from Deir El-Medîna*, edited by Rob J. Demarée and Jac J. Janssen, 155–172. Egyptologische Uitgaven 1. Leiden, Nederland: Nederlands Instituut voor het Nabije Oosten.

Peust, Carsten. 1999. *Egyptian Phonology: An Introduction to the Phonology of a Dead Language*. Monographien zur Ägyptischen Sprache 2. Göttingen: Peust & Gutschmidt.

Polis, Stéphane, Jean Winand, International Association of Egyptologists, and International Association of Egyptologists, eds. 2013. *Texts, Languages & Information Technology in Egyptology: Selected Papers from the Meeting of the Computer Working Group of the International Association of Egyptologists (Informatique & Égyptolgie), Liège, 6–8 July 2010*. Aegyptiaca Leodiensia 9. Liège: Presses Universitaires de Liège.

Prada, Luigi. 2012a. "Classifying Dreams, Classifying the World: Ancient Egyptian Oneiromancy and Demotic Dream Books." In *Current Research in Egyptology 2011: Proceedings of the Twelfth Annual Symposium Which Took Place at Durham University, United Kingdom, March 2011*, edited by Maria Correas-Amador, Nathalie Andrews, Veronica Tamorri, James Taylor, and Heba Abd El Gawad, 167–177. Oxford; Oakville: Oxbow Books.

Prada, Luigi. 2012b. "Papyrus Berlin P. 8769: A New Look at the Text and the Reconstruction of a Lost Demotic Dream Book." In *Forschung in der Papyrussammlung: Eine Festgabe für das Neue Museum*, edited by Verena M. Lepper, 309–328. Berlin: Akademie Verlag.

Quaegebeur, Jan. 2019. *Egyptian language in Greek sources:* scripta onomastica *of Jan Quaegebeur*. Edited by Willy Clarysse and Ana I. Blasco Torres. Orientalia Lovaniensia Analecta 280. Leuven: Peeters.

Quack, Joachim F., and Kim Ryholt. 2019. *The Carlsberg Papyri 11: Demotic Literary Texts from Tebtunis and Beyond.* 2 vols. CNI Publications 36. Copenhagen: Museum Tusculanum Press.

Ray, John D. 1987. "Phrases Used in Dream-Texts." In *Aspects of Demotic Lexicography: Acts of the Second International Conference for Demotic Studies, Leiden, 19–21 September 1984*, edited by Sven P. Vleeming, 85–93. Leuven: Peeters.

Redford, Donald B. 2000. "Scribe and Speaker." In *Writings and Speech in Israelite and Ancient Near Eastern Prophecy*, edited by Ehud Ben Zvi and Michael H. Floyd, 145–218. Atlanta: Society of Biblical Literature.

Richter, Barbara A. 2016. *The Theology of Hathor of Dendera: Aural and Visual Scribal Techniques in the Per-Wer Sanctuary*. Wilbour Studies in Egyptology and Assyriology 4. Atlanta, GA: Lockwood Press.

Ritner, Robert K. 1993. "The Mechanics of Ancient Egyptian Magical Practice." *Studies in Ancient Oriental Civilization*, No. 54. Chicago: Oriental Institute of the University of Chicago.

Rosmorduc, Serge. 2015. "Computational Linguistics in Egyptology." *UCLA Encyclopedia of Egyptology*. https://escholarship.org/uc/item/0fk4n4gv. Accessed 1-24-2022.

Rudin, Cynthia. 2018. "Stop Explaining Black Box Machine Learning Models for High Stakes Decisions and Use Interpretable Models Instead." http://arxiv.org/abs/1811.10154. Accessed on 1-24-2022.

Sauneron, Serge. 1959. "Les songes et leur interprétation dans l'Égypte ancienne." In *Les Songes et leur interprétation: Egypte ancienne, Babylone, Hittites, Canaan, Israël, Islam, peuples altaïques, Persans, Kurdes, Inde, Cambodge, Chine, Japon*, edited by Anne Marie Esnoul and Serge Sauneron, 18–61. Sources orientales 2. Paris: Éditions du Seuil.

Schiaparelli, Ernesto. 1927. *Relazione sui lavori della missione archeologica Italiana in Egitto (Anni 1903–1920), Volume Secondo: La tomba intatta dell'architetto Cha nella necropoli di Tebe*. Torino: Giovanni Chiantore.

Schneider, Thomas. 1992. *Asiatische Personennamen in ägyptischen Quellen des Neuen Reiches*. Orbis Biblicus et Orientalis CXIV. Freiburg-Göttingen: Vandenhoek & Ruprecht.

Schneider, Thomas, Francis Breyer, Oskar Kaelin, and Carsten Knigge (eds). 2004. *Das Ägyptische und die Sprachen Vorderasiens, Nordafrikas und der Ägäis: Akten des Basler Kolloquiums zum ägyptisch-nichtsemitischen Sprachkontakt*, Basel 9–11. July 2003. Münster: Ugarit-Verlag.

Schott, Siegfried. 1959. *Liebeslieder der Pharaonenzeit*. Lebendige Antike. Zürich: Artemis.

Seidlmayer, Stephan J. and Ingelore Hafemann. 2011. *Handbuch zur Benutzung des Thesaurus Linguae Aegyptiae* (TLA). *Auf der Grundlage der Hilfetexte des Thesaurus Linguae Aegyptiae* (TLA). BBAW, Berlin.

Simpson, William K., and Robert K. Ritner. 2003. *The Literature of Ancient Egypt: An Anthology of Stories, Instructions, Stelae, Autobiographies, and Poetry*. 3rd ed. New Haven; London: Yale University Press.

Shehab El-Din, Tahia. 2006. "Some metaphorical meanings of the Egyptian verbs of seeing and knowing." *Annales du Service des Antiquités de l'Égypte* 80, 311–321.

Szpakowska, Kasia. 2003. *Behind Closed Eyes: Dreams and Nightmares in Ancient Egypt*. Swansea, Wales: Classical Press of Wales; Distributed in the US by David Brown Book Co.

Szpakowska, Kasia, ed. 2006. *Through a Glass Darkly: Magic, Dreams and Prophecy in Ancient Egypt*. Swansea, Wales; Oakville, CT: Classical Press of Wales; Distributed in the US by David Brown Book Co.

Szpakowska, Kasia, ed. 2007. "Flesh for Fantasy: Reflections of Women in Two Ancient Egyptian Dream Manuals." *Egyptian Stories: A British Egyptological Tribute to Alan B. Lloyd on the Occasion of His Retirement*, =393–404.

Szpakowska, Kasia. 2011a. "Demons in the Dark: Nightmares and Other Nocturnal Enemies of Ancient Egypt." In *Ancient Egyptian Demonology: Studies on the Boundaries between the Demonic and the Divine in Egyptian Magic*, edited by Panagiotis Kousoulis, 63–76. Leuven: Peeters; Departement Oosterse Studies.

Szpakowska, Kasia. 2011b. "Dream Interpretation in the Ramesside Age." In *Ramesside Studies in Honour of K.A. Kitchen*, edited by Mark Collier and Steven Snape, 509–517. Bolton: Rutherford.

Van den Berg, Hans. 1997. "'Manuel de Codage': A Standard System for the Computer-Encoding of Egyptian Transliteration and Hieroglyphic Texts." http://www.catchpenny.org/codage/. Accessed on 1-24-2022.

Van de Walle, Baudouin. 1969. *L'humour dans la littérature et dans l'art de l'ancienne Égypte*. Scholae Adriani de Buck Memoriae Dicatae 4. Leiden: Nederlands Instituut voor het Nabije Oosten.

Vernus, Pascal. 1982. "Diachronie et synchronie dans la langue égyptienne." In *L'égyptologie en 1979: Axes prioritaires de recherches, Colloques internationaux du Centre National de la Recherche Scientifique* 595, Vol. I, 17–18. Paris: Éditions du Centre National de la Recherche Scientifique.

Vernus, Pascal. 1988. "L'Égypto-copte." *Les langues danse le monde ancien et modern III: Les langues chamito-sémitiques: Textes réunis par D. Cohen*. Edited by Jean Perrot, 161–206. Paris: Éditions du Centre Nationalde la Recherche Scientifique.

Vernus, Pascal. 1996. "Langue littéraire et diglossie." In *Ancient Egyptian literature: History and forms, Probleme der Ägyptologie* 10, edited by Antonio Loprieno, 555–564. Leiden: Brill.

Vernus, Pascal. 2016. "Traditional Egyptian I (Dynamics)." UCLA *Encyclopedia of Egyptology*. https://escholarship.org/uc/item/0bg342rh. Accessed 1-24-2022.

Vinogradov, Alexey K. 2006. "The Puzzles of the Dream Stele." *Beiträge Zur Sudanforschung* 9: 119–154.

Vinson, Steve. 2018. *The Craft of a Good Scribe: History, Narrative and Meaning in the First Tale of Setne Khaemwas*. Harvard Egyptological Studies 3. Leiden; Boston: Brill.

Ward, William. 1996. "A new look at Semitic personal names and loanwords in Egyptian." *Chronique d'Égypte* 71: 17–47.

Werning, Daniel A. 2016. "Hypotheses on Glides and Matres Lectionis in Earlier Egyptian Orthographies." In *Coping with Obscurity: The Brown Workshop on Earlier Egyptian Grammar*, edited by James P. Allen, Mark A. Collier, and Andréas Stauder, 29–44. Atlanta: Lockwood Press.

Wessetzky, Vilmos. 1985. "Table d'offrandes à inscription démotique au Musée Des Beaux-Arts." *Bulletin Du Musée Hongrois Des Beaux-Arts* 65: 11–15.

Winand, Jean. 1992. *Études de Néo-Égyptien, 1: La Morphologie Verbale* 2. Aegyptiaca Leodiensia. Liège: Centre Informatique de Philosophie et Lettres.

Winand, Jean. "Identifying Semitic Loanwords in Late Egyptian." *Greek Influence on Egyptian-Coptic: Contact-Induced Change in an Ancient African Language*, 481–511.

Winand, Jean. "Identifying Semitic Loanwords in Late Egyptian." *Greek Influence on Egyptian-Coptic: Contact-Induced Change in an Ancient African Language*, 2018a. "Dialectal, Sociolectal and Idiolectal Variations in the Late Egyptian Texts from Deir El-Medineh and the Theban Area." In *Outside the Box: Selected Papers from the Conference "Deir El-Medina and the Theban Necropolis in Contact" Liège, 27–29 October 2014*, edited by Andreas Dorn and Stéphane Polis, 493–524. Liège: Presses Universitaires de Liège.

Winand, Jean. "Identifying Semitic Loanwords in Late Egyptian." *Greek Influence on Egyptian-Coptic: Contact-Induced Change in an Ancient African Language*, 2018b. "Late Egyptian." UCLA *Encyclopedia of Egyptology*. https://escholarship.org/uc/item/3fr419rk. Accessed 1-23-2022.

Winter, Thomas. 1999. "Roberto Busa, S.J., and the Invention of the Machine-Generated Concordance." *The Classical Bulletin* 75 (1): 3–21.

Zeidler, Jürgen. 1994. "A new approach to the Late Egyptian 'syllabic orthography.'" *Sesto Congresso Internazionale di Egittologia. Atti II*, edited by Jean Leclant, 579–590. Turin: International Association of Egyptologists.

Zivie, Christiane M. 1976. *Giza au deuxième millénaire*. Bibliothèque d'étude 70. Cairo: Institut français d'archéologie orientale.

CHAPTER 18

Gaining New Perspectives on the Hypostyle Hall at Karnak through the Use of an Unmanned Aerial Vehicle (UAV) and Other Emerging Techniques

Jean Revez, Peter J. Brand, Emmanuel Laroze and Owen Murray

Abstract

Since 2011, the Université du Québec à Montréal (UQAM)—University of Memphis joint epigraphic mission at Karnak has been studying the decoration carved on the 134 columns that originally stood inside the Hypostyle Hall of the temple of Amun-Ra at Karnak.[1] Using photogrammetry, we have produced unrolled orthophotos that can show individual columns in their entirety. During the course of our season at Karnak in November–December 2017, we were able to operate a flying camera or unmanned aerial vehicle (a DJI Phantom 4 Pro UAV, commonly known as a 'drone') around the abaci, the top of the adjacent walls, and the open-air areas of the temple where loose blocks originally attached to the Hall are stored. By taking pictures from different angles and building 3D models of these hitherto lesser investigated parts of the Hall, we were able to improve our recording methods and to gain new insights into the architecture and inscriptions carved on the monument.

---

1 The project discussed in this chapter is a joint University of Quebec in Montreal—University of Memphis epigraphic mission that has been ongoing since 2011. For current and past financial support, we thank the Social Sciences and Humanities Research Council of Canada, the National Endowment for the Humanities, The American Research Center in Egypt, The University of Memphis, The University of Quebec in Montreal, the Tandy Institute at Southwestern Baptist Theological Seminary, and the Centre national de la recherche scientifique (CNRS). For their invaluable collaboration and assistance during our field seasons in Karnak, we are very grateful to the Centre franco-égyptien d'étude des temples de Karnak (CFEETK. MAE-CNRS USR 3172), the Egyptian Ministry of Antiquities, and the Chicago House in Luxor for having allowed us to use their flying camera (UAV) in December 2017.

Except for Figure 18.4 and the archival photo shown in Figure 18.12 (picture on the left), all the pictures and illustrations included in the present article were produced specifically for the joint University of Quebec in Montreal—University of Memphis epigraphic mission. Owen Murray, our photographer, took most of the pictures on site.

Note that the in-text citations are kept to a minimum and the reference list at the end of the article comprises entries that are only closely related to our project.

DOI:10.1163/9789004501294_019

Keywords

Karnak – Hypostyle Hall – digital epigraphy – drone photography – Ramesside Period – Ramesses II – 3D model – anastylosis

1 Introduction

Since 2011, the Université du Québec à Montréal (UQAM)—University of Memphis joint epigraphic mission at Karnak has been recording and studying the complex decoration on the columns of the Great Hypostyle Hall in the Temple of Amun-Re at Karnak. With generous funding from the Social Sciences and Humanities Research Council of Canada (SSHRC), the US National Endowment for the Humanities (NEH) and other agencies, we have employed emerging technologies to gain new insights into the elaborate decorative program and complex epigraphic history of the columns[2]

Through digital photogrammetry, we have produced unrolled orthomosaic photographs (*déroulés*) showing the entire surface of each column, enabling researchers to appraise their complex array of decoration in one image and to understand the spatial layout of the scenes and stereotyped texts that adorn them. These orthomosaic *déroulés* improve our comprehension of the interrelationship between the main ritual scenes and the various layers of stereotyped inscriptions on the column shafts, capitals, and abaci.

The great height of the columns, architraves, and clerestory roof of the Great Hypostyle Hall presents us with the daunting challenge of how to record and study them. The sheer elevation of the 12 great open-bud papyriform columns of the central axis is an insurmountable obstacle, while their wide campaniform capitals largely obstruct the view of their abaci from ground level.

During our November–December 2017 field season at Karnak, we overcame this difficulty by using an unmanned aerial vehicle (UAV), or "drone," to fly above and around the upper reaches of the columns, abaci, architraves, and clerestory roof, along with blockyards in the Karnak precinct where hundreds of loose blocks originally from the Hall are now stored. By photographing these more remote and less well studied parts of the Hall from various angles and building 3D models of them, we have gained a powerful new method for recording these inaccessible parts of the building and gained new insights into its architecture and inscriptions.

2 Revez 2020; Egels et al. 2020.

This chapter offers a case study illustrating the advantages of using emerging technologies, especially UAVs, which provided us with an unprecedented view of the Hypostyle Hall, to record and analyze inscribed monumental structures.

2 General Overview of the Great Hypostyle Hall at Karnak and Main Project Objectives

Situated in the heart of the temple of Amun-Re at Karnak—the most important religious monument of the New Kingdom located inside Thebes, the ancient 4th nome of Upper Egypt—the Hypostyle Hall is the largest and best preserved pharaonic architectural feature of its kind in Egypt. Measuring 103 m wide and 52 m long, the Hypostyle Hall originally contained 134 sandstone columns (Figure 18.1). A double row of 12 giant columns with campaniform capitals and exceeding 20 m in height stand along the main East-West axis. The average diameter of their shafts is 3.5 m; their circumference, replete with inscriptions, measures 10 m. To the north and south of the central aisle are two wings, each with 61 closed-bud papyrus columns divided into seven rows. Their average diameter is 2 m, with a circumference, also filled with carvings, of approximately 6.5 m.

The columns inside the Hypostyle Hall are unique, both in their scale and numbers, and also in the complexity of their decorative program. It is the first building in ancient Egyptian architectural history to have columns inscribed from top to bottom with inscriptions. This unprecedentedly complex decorative program,[3] with multiple layers of reliefs, many of them palimpsests, is the work of no fewer than five kings who left their mark on the columns: Sety I, Ramesses II, Ramesses IV, Ramesses VI and, (more modestly) Herihor. The inscriptions each pharaoh left vary significantly both in terms of their position on the columns and the overall quantity of decoration. As we have discussed elsewhere, the complex arrangement and chronology of these reliefs, many of them being palimpsests, follows the notion of "prime space."[4] According to this logic, kings gave priority to sections of the columns most visible from the processional axes, and these were the first to be decorated. Understanding the successive phases of decoration on the columns is challenging. Ramesses II alone engraved them in four different stages, leaving palimpsests where he erased earlier decoration of Sety I and his own initial reliefs. Some columns have been

3 Brand 2000, 192–218.

4 Revez and Brand 2015.

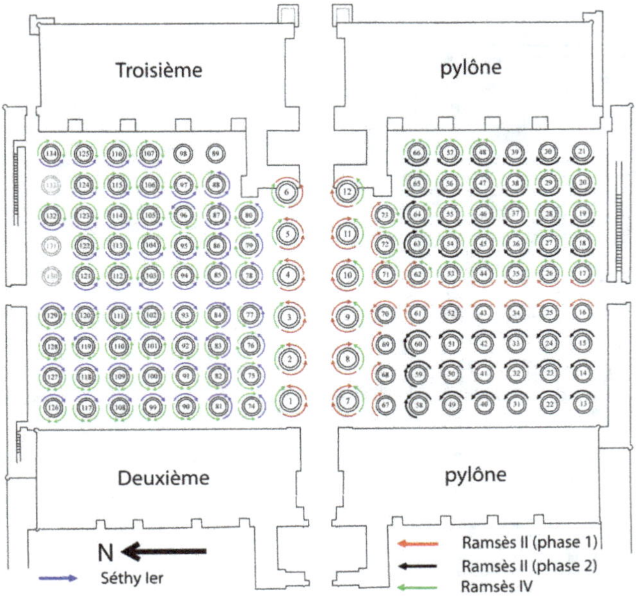

FIGURE 18.1 Plan of the Hypostyle Hall at Karnak (E. Feleg)

inscribed and re-inscribed in six or seven phases during the Nineteenth and Twentieth Dynasties.

There are two categories of column decoration. In the first are approximately 350 ritual scenes depicting Sety I, Ramesses II, or Ramesses IV offering to various gods and glossed with hieroglyphic captions. Most columns have three offering scenes around their circumferences, offering the richest variety of iconographic detail: the costume and regalia of the king and gods, and the offerings (Figure 18.2 area ②; Figure 18.3 area ②).

More abundant is the stereotyped decoration of these kings and of Ramesses VI and Herihor, covering every other available space on the columns. This diverse assortment includes floral and *rekhyet*-bird motifs engraved at the bottom and/or top of each column (the work of Sety I in the north wing and Ramesses II in the south aisle); friezes of royal cartouches at different levels, at the basis of the columns and on the upper shafts and capitals (by Sety I, Ramesses II, IV, and VI); and multiple sets of bandeaux texts containing royal titles (Ramesses II and IV, Herihor) (Figure 18.2 area ①; Figure 18.3 area ①).

Of the 134 original columns, 129 remain standing in varying degrees of preservation. Less remains of the roof. The ceiling slabs have vanished, except for a few, still in place next to the vestibule of the Third Pylon (Figure 18.4 area ①). The upper shafts and abaci of a cluster of columns are missing inside the southwest quadrant of the Hall, swept away when the bulk of the neighboring

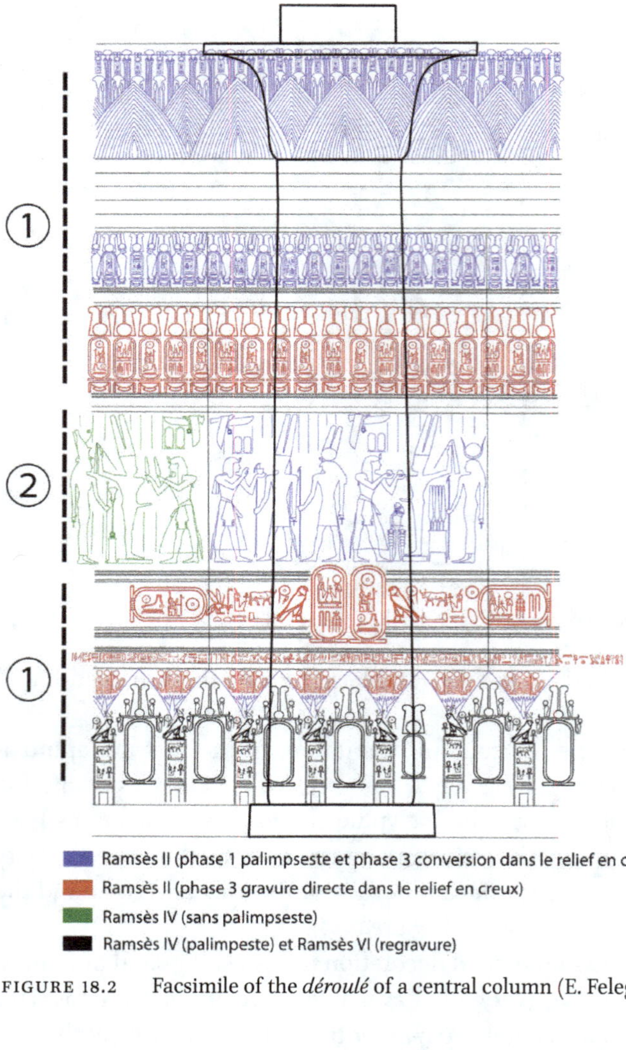

Ramsès II (phase 1 palimpseste et phase 3 conversion dans le relief en creux)
Ramsès II (phase 3 gravure directe dans le relief en creux)
Ramsès IV (sans palimpseste)
Ramsès IV (palimpeste) et Ramsès VI (regravure)

FIGURE 18.2 Facsimile of the *déroulé* of a central column (E. Feleg)

southern tower of the Second Pylon collapsed sometime before the mid-19th century (Figure 18.4 area ②). In the northern aisle, eleven columns with their abaci and architraves collapsed in 1899 in a devasting domino effect. Legrain replaced most of the original architraves in the north wing with metal and concrete pastiches when he rebuilt the northern wing in the early 20th century (Figure 18.4 area ③). Only the bases of columns 89 and 98 remain *in situ* (Figure 18.4 area ④), and the lower half of column 74 in the north now stand (Figure 18.4 area ⑤). Beside the north wall, columns 130, 131, and 133 are missing entirely (Figure 18.4 area ⑥). Drum sections belonging to some of these latter columns are stored just outside of the Hall, to the north (Figure 18.4 area ⑦).

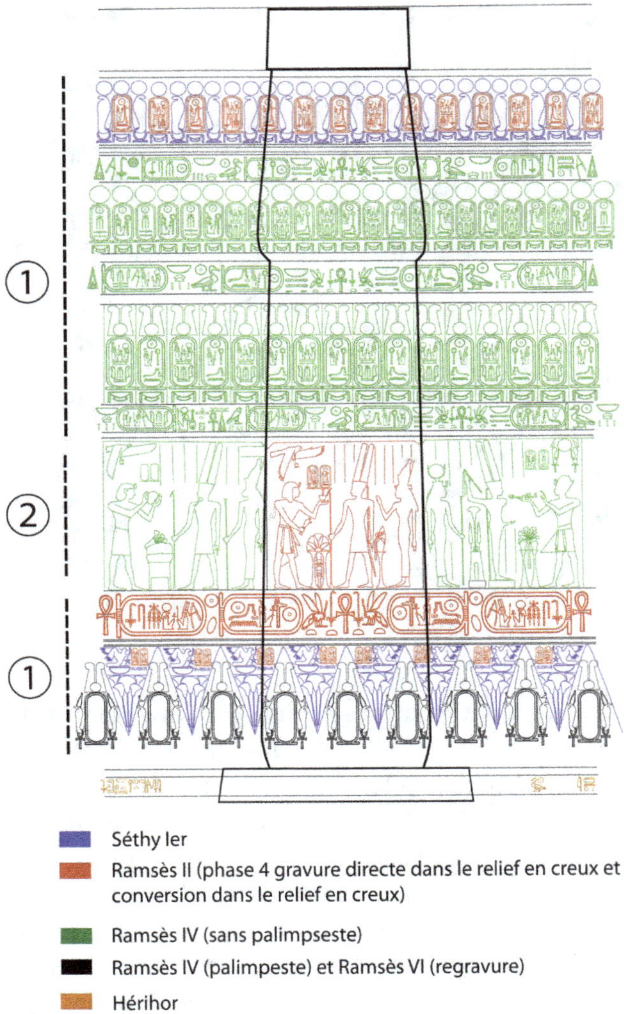

FIGURE 18.3 Facsimile of the *déroulé* of a lateral column (E. Feleg)

Although of paramount importance to our understanding of Ancient Egyptian religion and royal ideology, the columns of the Hypostyle Hall had never been the object of systematic study until we began our study in 2011.[5] Nelson and Caminos produced an unpublished archive of hand copies of the column

5 Revez and Brand 2012, 11–13 for a short historiography on the Hall; see also now Biston-Moulin 2016.

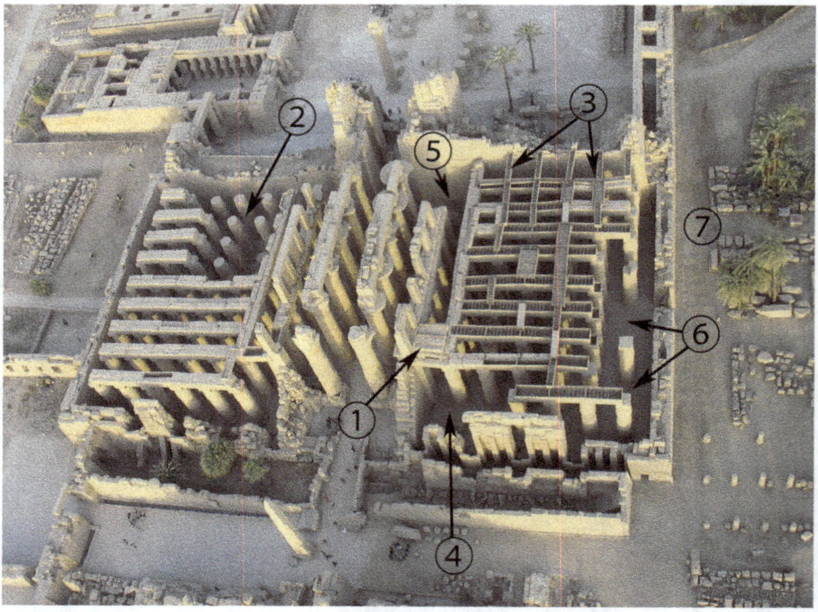

FIGURE 18.4 Aerial view of the Hypostyle Hall
©C. ZANDONA

scenes and abaci inscriptions. L.-A. Christophe published an inventory of the deities and ritual acts depicted in the scenes.[6] Otherwise, previous studies and documentation efforts centered on the wall scenes and architraves.[7]

The gap in the scientific investigation of the columns can be ascribed to their huge size and geometry, which long stymied traditional methods of epigraphic documentation. The apparently redundant stereotyped inscriptions, much of it engraved well out of reach, and the impossibility of recording hundreds of large scenes wrapping around the cylindrical surfaces of the columns using traditional photography all deterred scholars from recording them until the emergence of 21st century digital technologies.

3 Project Aims, Methodology, and Results

The UQAM—University of Memphis joint epigraphic mission at Karnak has endeavored to record and publish the column scenes systematically, with a focus on transcription, transliteration, translation, and commentary of the

6 Christophe 1955.

7 Epigraphic Survey 1986; Rondot 1997; Brand, Feleg, and Murnane† 2018.

texts. Other areas of research include the architecture of the columns and roof, the paleography of hieroglyphic signs, the study of quarrymen's marks left on the stone blocks from the Hall, the study of palimpsest inscriptions, of iconoclasm, and chromatology (color studies) of pigments artists used on the columns.[8]

Between 2011 to 2019, we completed eight field seasons to achieve these goals, ranging from one to three months per season. In our first seasons, we collated the stereotyped column decoration and ritual scenes with the aid of unpublished hand copies and notes assembled by H.H. Nelson, R. Caminos and W.J. Murnane. Although tedious, this process was highly rewarding, leading to several vital insights into how the Egyptians conceptualized the vast decorative program of the columns. Most significant was the notion of "prime space," i.e., the prioritization of the most visible spaces in a three-dimensional ritual environment.[9] We also gained nuanced understanding of iconographic and paleographic norms and exceptions in the arrangement of decoration executed on a vast scale on the 134 columns and of the interrelationship between quality of workmanship and location of decorative elements. Recent seasons were also devoted to sorting out errors made in the placement of inscribed blocks during George Legrain's directorship, when he had much of the northern wing of the Hall reconstructed at the turn of the 20th century. In 2019, we developed a new method for rebuilding dismantled columns from unrolled images of individual blocks no longer *in situ*.

4 First Generation *déroulés* of Entire Columns: Objectives, Methodology, and Results

Our future publications, online and in print, will include photographic *déroulés* of all the columns. In 2008, before the UQAM/University of Memphis project began, Emmanuel Laroze, who was then Director of the Franco-Egyptian Centre for the Study of the Temples at Karnak, used laser scanning and photogrammetry to create a 3D model of the Hypostyle Hall in collaboration with a private French firm called ATM-3D and the National School of Geographical Sciences near Paris.[10] This undertaking included a full three-dimensional lasergrammetric reconstruction of one of the great columns and the first *déroulé* orthomosaic image of its entire surface. This technique unwrapped a composite image of

8 Brand et al. 2013.

9 Revez and Brand 2015.

10 Chandelier et al. 2009; Laroze and Chazaly 2009.

FIGURE 18.5 First generation orthophotographic *déroulé* of a central column (ATM-3D/Y. Egels/E. Laroze)

the full circumference of a column from a cylindrical 3D digital model, allowing the entire decoration of a column to be seen at a glance (Figure 18.5 and Figure 18.6).

Using photogrammetry, the French team developed a method in which georeferenced digital images of a column, using a local coordinate system, are stitched together and assembled in 3D, then unrolled (not unlike the method for projecting a three-dimensional image of the earth's globe into a two-dimen-

FIGURE 18.6 Reprocessed orthophotographic *déroulé* of a lateral column (Y. Egels/E. Laroze/O. Murray)

sional world map). Prior to carrying out this difficult operation, the team acquired a data set by taking some 4,000 pictures of all the standing columns. They devised an ingenious method for taking pictures of each column by mounting four digital cameras at set heights along an 8 m high pole that they moved to eight positions around each column, taking multiple, overlapping shots, section by section.

This approach produced astonishing results: *déroulés* presenting the entire decoration of the column in one composite image, allowing us to gain a much better grasp of the sequence of scenes and stereotyped decoration on various levels of the column shafts. The orthophotos of unwrapped columns served as templates for producing scaled diagrams of a column's decoration, with the freedom to distinguish between the different chronological phases of decoration (Figure 18.2 and Figure 18.3).

Admittedly, there were some drawbacks to the results this pioneering method produced. In 2008, the French team used small commercial digital cameras that produced low resolution images. Moreover the project, whose aim was to document all standing 129 columns in the Hall, was completed in just one month. As time was of the essence, photographs were taken in shade, in order to avoid contrast between shadow and sunlit areas of the columns, which combined with the low resolution of the camera, made it difficult to read damaged hieroglyph signs and palimpsest inscriptions in the resulting *déroulés*. Furthermore, it was necessary to collect more data in later years for the post-processing treatment of the images to be truly effective. Since ATM-3D was no longer involved in the project, E. Laroze and Y. Egels took these images and measurements between 2013 and 2017. Through their meticulous radiographical treatment of the first generation of *déroulés*, the final images are much more legible, but the limitations of the original images in terms of lighting and resolution remain.

5 Second Generation *déroulés* of the Main Scenes and 3D Reconstruction of Columns

To overcome the limitations of the first generation *déroulés*, we decided to produce a new set of rectified orthophotos of more than 300 ritual scenes at the mid-level of the standing columns in the Hall, these having the greater historical and religious significance and variety than the stereotyped inscriptions. These second generation *déroulés* consist of much higher resolution orthophotos, enabling the user to zoom in on all the relevant details of the scenes in unprecedented detail. In the 2014 and 2015 field seasons, photographer Owen Murray shot a sequence of up to 150 images of each of the smaller 117 standing closed-bud papyrus column scenes. This method of documentation relied on a set of moveable scaffolding from which a series of 7–8 photographs along the vertical, Z axis of the column were taken from 16 positions around the column (X,Y axes). In the 2016 field season, O. Murray finished photographing the remaining 12 large column scenes applying a similar methodology, this time using a hydraulic scissor lift from which a series of 13–14 photographs along the vertical, Z axis of the column were taken from 21 positions around the column

FIGURE 18.7 Use of softbox for picture taking (Y. Egels)

(X, Y axes). The result was a sequence of up to 350 images of each of the scenes on the 12 great open-bud papyriform columns.

In this second generation of *déroulés*, O. Murray used controlled lighting to bring out the fine details and palimpsest traces of the reliefs, creating a "soft box" to diffuse natural light and a studio strobe to evenly light the column scenes. This facilitated optimal raking light photographs of the reliefs (Figure 18.7). Combining the new photos with the 3D georeferenced data compiled since 2008 to generate a virtual model of the Hall, we could now produce a new set of orthomosaic *déroulés* of the column scenes. With the invaluable help of Agisoft's photogrammetry software Metashape (previously called Photoscan), we were able to streamline the complex and tricky process for fabricating the *déroulés* by aligning the individual pictures (Figure 18.8), producing a dense cloud, a mesh, and finally a high-resolution texture of the model. After processing the images in Photoshop, the final composite *déroulés* are spectacular (Figure 18.9).[11]

Since a 3D model must be built in order to generate the flattened *déroulés*, Metashape also allows a user to rotate the columns in a virtual environment and view them from different angles (Figure 18.10). In this case, since the scenes are located mid-level on the columns, one can view them virtually in a manner

11 Revez 2020, 394–397.

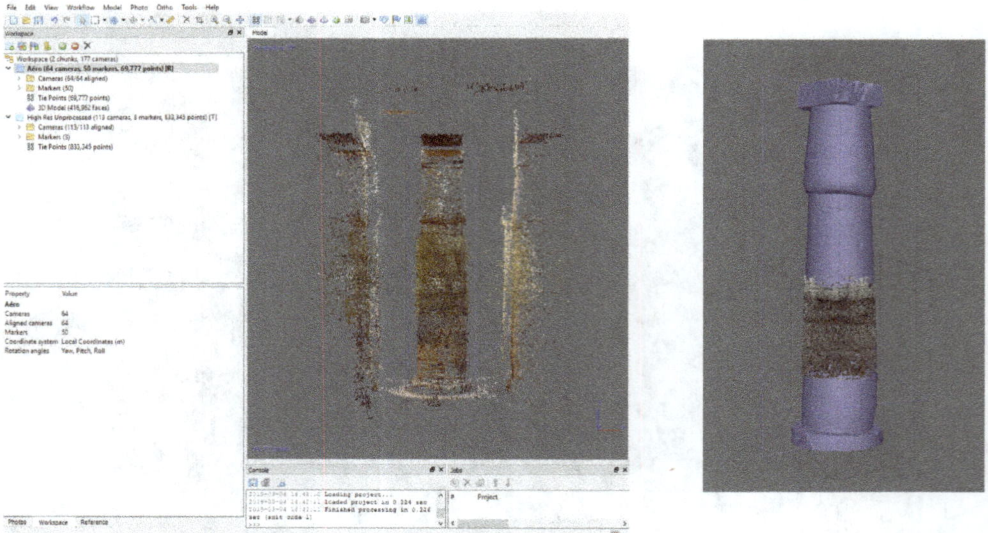

FIGURE 18.8 Construction of 3D model of column (P. Brand)

FIGURE 18.9 *Déroulé* of a complete column scene (Y. Egels/E. Laroze/O. Murray)

not physically possible at the temple. Worthy of mention is that the production of the *déroulés* and the 3D projections were all done by graduate students of the history department at UQAM, under the technical supervision of E. Laroze and O. Murray. This fact shows that the expertise necessary to master these sophisticated tools can be transferred to people who have no prior experience in the field. To our great pleasure, three life-size second generations *déroulés* of columns from the Hypostyle Hall were exhibited in Canada in 2019, at the Montreal Fine Arts Museum in Montreal (MFAM) (Figure 18.11) and at Gallery@501 in Edmonton.

FIGURE 18.10 Three-dimensional view of a column (O. Murray/E. Laroze/Y. Egels)

FIGURE 18.11 Life-size *déroulé* of a column scene from our mission at the MFAM (J. Revez)

6 Assembling Loose Column Drum Fragments with the Help of Orthophotography and 3D-Reconstruction

Applying this successful method of orthophotographic *déroulés* developed to produce the scenes carved around the entire circumference of standing columns has also opened up new and exciting avenues of research for column drum fragments scattered throughout the blockyards around the Karnak Temple precinct. Anastylosis, as an art and science, can gain tremendously from the use of photogrammetry to unroll individual loose column blocks lying far apart from one another, and in turn, enables us to understand the sequence and relationship of these fragments to one another.

In 2019, we successfully tested the method on all the loose blocks identified as being part of column 131, a column that is no longer standing *in situ* inside the Hall. The results have been extremely promising. The first step was to identify and record the 20 column drum sections belonging to column 131, then to draw sketches and determine the measurements for each of them. These half-drums measured about 2.5 m in diameter and 0.93 m in height.

Identifying and reconstructing the scattered fragments that make up columns formerly standing inside the Hypostyle Hall, such as column 131, is a complex operation based on several iconographic, epigraphic, and architectural criteria.[12] From an iconographic and epigraphic viewpoint, determining the *type of carving* (raised or sunk relief) in which the artists cut the figures and inscriptions is crucial, since Sety I used *raised relief* exclusively in his decorative program. Ramesses II (except for the first year of his reign) and Ramesses IV always carved their inscriptions in *sunk relief*. The *orientation of the figures* inside a scene is also generally consistent. Sety I's scenes always face the processional ways in the northern section of the Hall, while Ramesses II's did the same in the south wing. Ramesses IV's decoration faced away from the axis. Thus, the orientation of the reliefs is also crucial to establishing their original position. The scenes carved along the main axes show the gods, who were perceived as residing within the temple, looking out towards the different gates and greeting the king, who faces towards the interior of the temple.

Another crucial epigraphic criterion is the presence of *traces of palimpsest* inscriptions, which are typical of reliefs Ramesses II recarved, over both Sety I's in the north wing, and his own early raised relief in the south. These are only found on columns adjacent to the main east-west processional way or along the secondary north-south axis in the southern half of the Hall. *Traces of icon-*

12 Egels et al. 2020, 47–49.

oclasm can be seen in various parts of the Hall, but the specific sections of the body of the king or the gods that were hacked, and the type of damage left by the iconoclast's tools, can be ascribed to particular sectors inside the Hall. Last but not least, the *conventional sequence of the successive registers of stereotyped decoration* made up of friezes of cartouches and bandeaux texts with their royal and divine titles and epithets, makes it fairly easy to assign the position of a block on any given column. The same can be said of original *stereotyped plant motifs* (the stylized triangular leaf patterns and floral design at the bottom of the columns and the 'binding motif' on the upper portion of the shafts representing the idea of the cluster design of the papyrus bundles), used in carving the closed-bud papyriform columns of the Hall.[13] Their location in space can be assigned quite clearly along the vertical axis of each column. Ramesses IV never added stereotyped decoration on most columns in the southwest quadrant of the Hall, in which case the vegetation motifs are all that remains for us to determine the respective position of each drum. Finally, since a single scene and its accompanying legend extended over several connecting drums, *observing which elements of the overall decoration are carved along the edges of each individual block* makes it possible to determine how separate fragments join.

On a more architectural plane, the *position of the mortises* on the upper surface of the blocks, where *wooden dovetails* were inserted to join two half-drums together, is a key element in the anastylosis of pairs of column drums. Since each column is made up of *courses of half-drums set up alternately lengthways and sideways from one course to the next*, it is also possible to determine how each drum is positioned in connection to the one set up just above and below it.

The reassembly process described above was greatly helped by reference to 19th century photographic archives that show the state of the columns inside the Hall just prior to its tragic collapse in 1899. Since the Hypostyle Hall at Karnak is perhaps second only to the Great Pyramids in Giza as the most photographed monument in Egypt, we can rely on numerous archival photos to assist us in the anastylosis of the columns. In the case of column 131 for instance, Figure 18.12 not only shows us that the upper section of the column was no longer *in situ* at the end of the 19th century. It helped us tremendously in identifying and locating four half drums that were physically set aside on concrete platforms (commonly known as *mastabas*) just outside of the Hall. A drawing by R. Lepsius from the mid-19th century led us to confirm the assemblage of lower sections of the column that are hidden from view in the photo. Finally, one can also turn to the invaluable study of the columns carried out in the

13 Brand 2018.

FIGURE 18.12 Northeastern section of Hall in 1895 and in 2017
©CFEETK/O. MURRAY

1950s by L.-A. Christophe who provided useful insights for reconstructing missing columns. His hypotheses have proved to be highly accurate.

Moving the heavy blocks so as to reassemble them is impractical. Such an operation would be cumbersome and costly (to say nothing of the fragile condition of many of these ancient blocks that have lain in contact with ground water for a century and are often in dire need of conservation). This is not a viable approach for collating their decoration. Virtual reconstruction of *déroulé* images of the decorated surfaces of these column drums offers an ideal, non-invasive method for reconstructing them via photogrammetry.[14]

To achieve this, two approaches were tested. The first was to make individual flat *déroulés* of each block, as we did in our approach to develop complete scenes carved on columns that are still *in situ*. The second method involved producing a three-dimensional reconstruction of each fragment. This method is more complex. In either case, the 2D- or 3D-model allows for digital manipulation of each decorated block from the drums, which could be moved around a

14 Egels et al. 2020.

FIGURE 18.13 2D reconstruction of column 131 with individual *déroulés* of loose blocks (Y. Egels/E. Laroze/O. Murray)

standard column template in order to test fit matching blocks (Figure 18.13 and Figure 18.14). By building a gigantic two- or three-dimensional jigsaw puzzle, block joins can be identified by trial and error.

We used *Cumulus*, an open-source software program for 3D reconstruction of archaeological sites developed by Y. Egels, a member of our team, for producing orthomosaic *déroulés* of the standing columns inside the Hall. Transporting blocks virtually was done through human-machine interfaces (HMI), by attempting to interact with the software in such a way as to enact real life situations as much as possible. Thus, the process of maneuvering and rotating blocks is similar to what crane operators would do if physically moving the blocks themselves.

FIGURE 18.14 3D reconstruction of column 131 made up of loose blocks (Y. Egels/E. Laroze/O. Murray)

This new methodological approach, based on emerging technologies, offers several advantages. The greatest one is the speed and ease with which the reconstruction of the column can be completed. Once the recording in the field was done, we could instantly and without great effort visualize whether sets of blocks matched and could be joined together, or on the contrary, if they did not fit together. The theoretical series of criteria for identifying matching blocks for anastylosis that we itemized in the previous section becomes much easier, since they can be observed first-hand in a virtual workspace. As with the *déroulés* of the *in situ* columns, we have produced a composite image of the three adjoining scenes on column 131. Assembling the small flattened *déroulés* of the column fragments in various combinations made it possible to determine whether the elements of a scene overlapping several blocks fit accurately.[15]

15 Egels et al. 2020, 48–50.

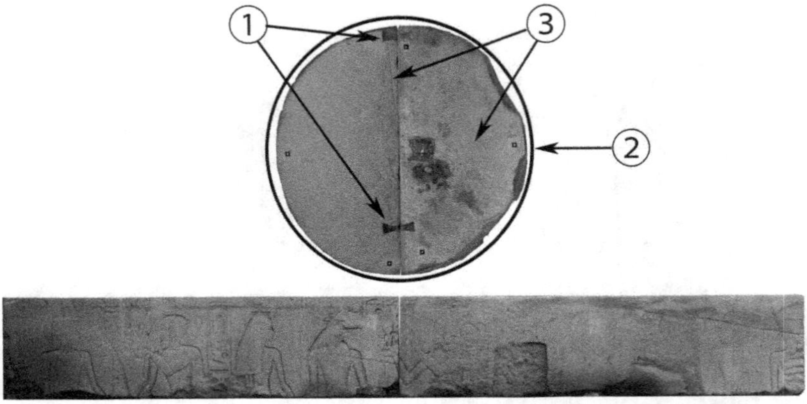

FIGURE 18.15 2D reconstruction of two matching half-drums belonging to column 131 (Y. Egels/E. Laroze/O. Murray)

In addition, the ability of this system to create two- and three-dimensional images of the column drums is profitable in other ways. Matching the two halves of the mortises with the wooden dovetails (Figure 18.15 area ①) so that the circumferences of both half-drums align perfectly (Figure 18.15 area ②) allows us to confirm that the two blocks match. Another criterion for a perfect fit becomes obvious: the two intersecting lines incised on top of the two blocks (these lines guide the stone workers when positioning the row of half-drums located just above) cross each other at a right angle (Figure 18.15 area ③). Other means that contribute to the virtual reassembly of blocks, which would be practically impossible to accomplish by manipulating the blocks themselves, include the fact that the column shafts are slightly conical rather than perfectly cylindrical Thus the diameter of a layer of blocks varies according to its vertical position on the column. Such a feature can be much readily visible with the model.

7 Using an Unmanned Aerial Vehicle (UAV) to Document the Hypostyle Hall

Early in the 2017 field season, experiments with a camera-on-pole technique to photograph the large column abaci from atop the central axis architraves of the Hall proved viable, though requiring much time, as well as attention to safety while working at such heights (Figure 18.16). The permission to use a drone (UAV) to acquire imagery greatly changed the scope of work possible for the season (it is believed this is the first time a UAV was flown at a major temple

FIGURE 18.16 Use of a camera-on-pole technique on top of an architrave to take pictures of abaci (E. Laroze)

complex in Luxor, and used for archaeological research and cultural heritage documentation purposes). It was quickly determined that the best course of action would be to produce scaled photogrammetric models of the architraves and abaci atop the large columns, as well as the accompanying clerestories on either side, as rectified orthomosaic imagery from any desired perspective could be extracted from these models. The ability to scale the models was due to CFTEEK survey efforts during the initial 2008 *déroulés* campaign; these same measurement and control points were used with newly acquired imagery from the UAV.

We used a a DJI Phantom 4 (2016) UAV mounted with a 12mb digital camera, kindly provided to our project by the Epigraphic Survey (Chicago House) (Figure 18.17). Though initial firmware update issues were cause for great consternation, DJI technical support and forum boards provided solutions quickly and resolved the issue within 48 hours.

Due to the large volume of tourists and the desire to acquire imagery in the best possible lighting conditions, flights in the Hypostyle Hall were conducted from approximately 6am to 8am each morning from December 6 to 13, 2017. A staging area in the first court between the first and second pylons clear of obstacles was used as a base for a launching and landing zone, and as a

FIGURE 18.17 The DJI Phantom 4 (2016) UAV mounted with a 12mb digital camera (E. Laroze)

return-to-home location in case the data link between the UAV and the Remote Control device (RC) failed during flight. We determined that a flight radius of 500 m and a maximum altitude settings of 40 m, with a preferred 25 m ceiling, were optimal and were input in the UAV DJI Go 4 control application before launch, along with setting the camera for automatic acquisition of JPEG images at two seconds intervals during flight. Owen Murray piloted the UAV by Visual Line of Sight (VLOS) at all times, with Emmanuel Laroze co-piloting and giving instructions as to directional movements of the UAV and adjustments to the camera position as needed. Although wind patterns within the temple were of concern, this work method proved quite effective, with work in the aforementioned areas completed within the morning flight window each day.

8 Recording the Hitherto Inaccessible Upper Portions of the Hypostyle Hall

From the onset of our project, one of the important goals was to record all of the cartouches carved on all four sides of the abaci resting on top of the columns. The analysis of the spelling of the large-size royal names and the type

FIGURE 18.18 Scaffolding and Scissor Lift used by the project (E. Laroze)

of engraving used (raised or sunk relief, with or without traces of palimpsest) encapsulates the history of the decoration of the Hall under the reigns of Sety I and Ramesses II. After having shot points with a Laser Theodolite on predefined reference points located on the abaci faces to produce rectified images, we were able to improve on our method by pairing the georeferenced data from the initial 2008 CFEETK survey with the images taken from the drone.

Flying the drone around and above the abaci located on top of the central colonnade, as well as over other sections of the Hall, enabled us to take pictures of areas hitherto inaccessible by other means, whether with the help of a scissor lift or of scaffolding (Figure 18.18), two *modi operandi* we had turned to successfully to reach the upper sections of the smaller lateral columns which are only 13 meters high. The sheer height—over 20 meters—was one problem;

FIGURE 18.19 Orthorectified abacus on top of large column (O. Murray/ E. Laroze/Y. Egels)

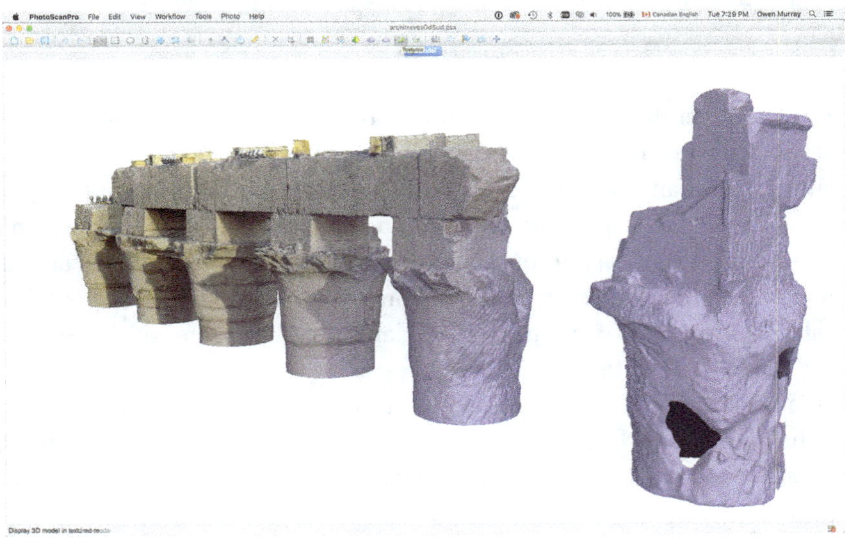

FIGURE 18.20 3D model of upper section of central colonnade (O. Murray/ E. Laroze/Y. Egels)

another was the fact that the abaci on top of the large east-west central columns were resting on large open-bud capitals that hid the view of the abaci from ground level. The UAV allowed us to observe and record many heretofore unfamiliar epigraphic and architectural details of the abaci (Figure 18.19), capitals, and clerestory roof (Figure 18.20).

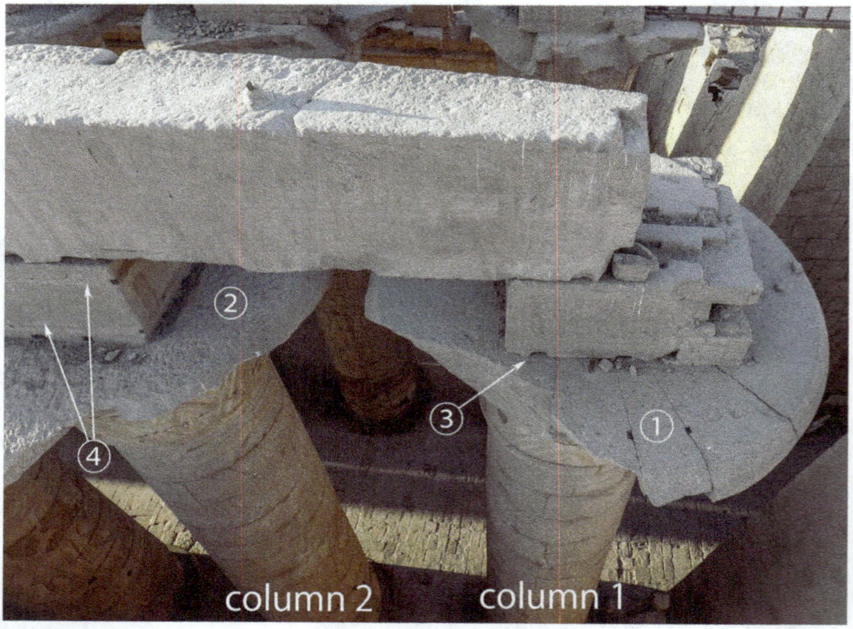

FIGURE 18.21 Upper section of large columns no. 1 & 2 (O. Murray/E. Laroze)

Pictures taken from the flying camera corroborated and reinforced some observations previously made about the ancient repairs of the westernmost column capitals and abaci under Greco-Roman rulers. V. Rondot and J.-C. Golvin[16] observed that some parts of the campaniform capitals of columns 1 and 2 lacked any trace of decoration, in contrast to other sections of these capitals that were inscribed with cartouches of Ramesses II. They also pointed out that the abacus sitting atop column 2 was left blank (Figure 18.21 area ④) where one would have expected to see the cartouches of Sety I, as is the case with all the other abaci from the same row of columns running to the east. They concluded that the undecorated fragments were later additions from Greco-Roman times, made after the partial collapse of an architrave resting atop the abacus, destroying it and part of the capital when it fell.

With the assistance of our drone, we were not only able to confirm the accuracy of these remarks, but we could also clearly identify similar restoration blocks that were inserted inside the upper section of the capital of column 1 as well, a detail that is not visible in photos included in the article by Rondot and Golvin due to the angle from which the photograph was taken.[17] Images

16 Rondot and Golvin 1989.

17 Rondot and Golvin 1989, pl. 31.

FIGURE 18.22 Elevated platform under abacus and pryholes (O. Murray/E. Laroze)

from the drone show that ancient restorers neatly cut a small vertical hole into both sides of a replacement block on column 1 (Figure 18.21 area ①). These two perforations were probably inserted to allow a rope to run through them to help support the new block as it was inserted in between two adjacent blocks.

Images taken from the drone also provided insights into other architectural peculiarities. The center of the upper surface of the capital of column 2 was slightly cut out in order to allow the restored Greco-Roman abacus to fit into the gap. Notice on the picture that the stonecutters did not deem it necessary to smooth the upper surface of the capital, which is only roughly hewn (Figure 18.21 area ②). Usually, the opposite occurs: the abacus is set on top of a small flattened outgrowth that protrudes upwards from the upper surface of the capital. Once set into its place, both the sides of the abacus and the edges of the excess are cut back, smoothed, and aligned. A closer look at the images taken from the drone shows that all four corners of the elevated platform, sticking out from the upper surface of the capital in order to support the abacus, were hollowed out (Figure 18.22 area ①). The small canals that were thus carved out probably crisscrossed through the platform diagonally and were used to control the accuracy of the cutting, assuring in this case that the platform was of even height throughout, and that its top side was levelled off correctly in order to receive the lower surface of the new abacus. It is worth noting that the upper surface of the capital measures more than 30 square meters. Cutting flat and horizontal bedding joints on such a large surface was no small feat. Such a construction technique is well attested in architecture generally speaking, but evidence for this method in ancient Egypt is relatively scant. One can also detect in Figure 18.22 area ② pry-holes that were carved at the base of blocks in order to secure the position of a wooden lever that was used to set the blocks into the right position. A small patch stone was later put into place to plug in the pry-hole at the base of the architrave.

FIGURE 18.23 Longitudinal section of central row of large columns (O. Murray/E. Laroze/Y. Egels)

On a different plane, scaled orthorectified longitudinal sections of the northern row of the clerestory made it possible to record the measurements of all the different architectural features that make up the upper section of the colonnade. Anomalies or deviations from the norm in some of these measurements proved to be quite significant in meaning. Thus, one can make out that the abacus resting on top of column 1 was not set precisely on the vertical axis of the column. It is, in fact, off center from the axis by 25 cm to the west, as Figure 18.23 area ① clearly shows. One could at first conclude from this observation that the abacus, which is an original one dating back to Sety's reign, could have been accidentally moved during the later restoration campaign of the Classical Period. Such hypothesis turns out to be very unlikely, since the abacus on top of column 7 that lays symmetrically directly across column 1, to the south, presents a similar pattern, with a 25 cm departure from the norm. However, the capital on top of column 7 did not undergo any repairs during Greco-Roman times, as opposed to the capital of column 1. In fact, the lines, that were incised on the upper surface of the capital in order to guide the workers when laying the abacus in its right position, are still clearly visible on the picture and show that the abacus in its current state rests in its original place, since the sides of the abacus run along those lines squarely (Figure 18.21 area ③).

Thus, the reason for this 25 cm discrepancy must be found elsewhere. It probably has to do with the relative position of both abaci on top of the two columns (no. 1 and no. 7) that were erected right next to the gateway of the Second Pylon at the western end of the Hypostyle Hall. Because of the outward slope (or batter) of the eastern face of the pylon, there is a greater distance

FIGURE 18.24 Upper section of large colonnade with remains of roof-slabs (O. Murray/ E. Laroze /Y. Egels)

between the top of the column (which stands about 20 m above ground level, lest we forget) and the top section of the gate structure, than between the base of the column and the base of the gate. By placing the abacus 25 cm closer to the inner wall of the second pylon, the ancient architects of the Hypostyle Hall most likely wanted to reduce the size of the architrave that spans the distance between the top of the column and the gate of the pylon. This distance is indeed shortened from 6.31 m to 6.06 m.

During past seasons at Karnak, two types of New Kingdom quarry marks (also called masons' marks) were newly identified in no fewer than 17 spots throughout the Hall. The first type of mark is a circle connected to a small cross (similar to a *nfr* hieroglyphic sign and not unlike the modern female sign); the other is an 'hourglass'-shaped symbol. These marks were carved presumably on rough-hewn faces of blocks in order to identify the team responsible for their extraction in the quarry and transportation to the temple. Up to now, these marks were found on both the northern and southern walls of the Hall, in the masonry of the east wall, and of more direct concern to us, on the drums and abaci of the small columns only. The conspicuous absence of these marks on the large central columns led us to believe at first that their erection could perhaps be of a different date than the smaller ones. However, pictures taken from the drone during the 2017 field season made it possible to detect *nfr*-shaped

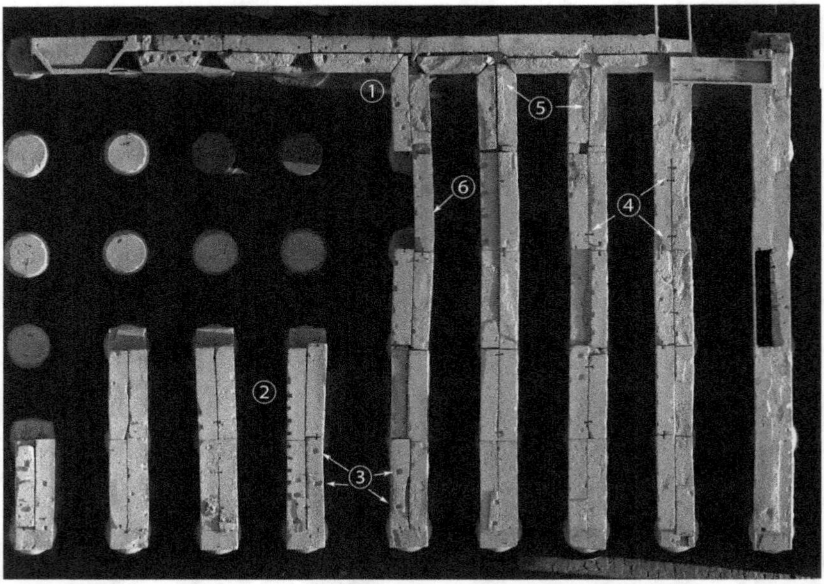

FIGURE 18.25 Overhead view of south-west quadrant of the Hypostyle Hall (O. Murray/E. Laroze)

quarrymen's marks on four different blocks supporting the north and south clerestories of the Hall and another one, on the eastern face of the abacus sitting on the large column no. 7.

Other normally inaccessible areas of the temple were also surveyed with the help of the drone. In the highest part of the Hall, the UAV flew over the only remains of the roof slabs that covered the building. From the clear traces left by the action of cutting tools on the edge of the remaining slabs to the north of the Hall (Figure 18.24), one can gather that the roof slabs connecting the north clerestory to the large northern colonnade did not cave in by accident, but were intentionally broken off before collapsing, probably in order to reuse the fallen material for reconstruction purposes elsewhere.

The top view of the southern part of the Hypostyle Hall offers a wealth of technical architectural features. One can see the system of clamping of six blocks on top of a column (Figure 18.25 area ①). Notches facing each other and intended to support wooden beams are also visible (Figure 18.25 area ②), as well as pry-holes used to move roof-slabs (Figure 18.25 area ③). Sockets for wooden cramps (or dovetails) are visible; their emplacement is random (Figure 18.25 area ④). From the extra thickness of stone and rough surface discernible on top of the architraves, one can deduce that the roof-slabs did not cover the entire surface of the architrave but rested solely on the smoothed

FIGURE 18.26 Palimpsest inscription on North clerestory (O. Murray/E. Laroze)

parts (Figure 18.25 area ⑤). Incised lines at the edge of some architraves border the area of stone that was due to be cut back (Figure 18.25 area ⑥).

The most inaccessible zone of the Hypostyle Hall in terms of elevation is the clerestory, where the highest inscriptions were carved some 22 m above ground level. Reliefs and texts carved on the vertical piers and architraves encasing the huge window grilles of the clerestory roof are of great interest but are largely unpublished. On the northern clerestory for instance, just underneath the roof slabs, Ramesses II engraved a line containing his royal titles over those of his father, Sety I (Figure 18.26). Though the hieroglyphic signs are large in size and can thus be observed with binoculars from below, the fine details of the palimpsest inscription can best be examined fully with close-up shots taken from the drone. In the picture shown, traces of the original raised relief text of Sety I (Figure 18.26 area ②) can be quite easily deciphered under the recarved sunk relief inscription of Ramesses II (Figure 18.26 area ①). Noteworthy of mention here is the use of the same 𓅓 *m*-owl sign (Gardiner sign-list G17) in both texts (for the word 𓄙𓅓 *wḥm* in the suppressed raised relief text of Sety I and the verb 𓅓𓎡 *mk* in the later sunk relief of Ramesses II) (Figure 18.26 area ③). In this case, the sculptor reworking the inscription on Ramesses II's orders had only to convert the carving of the glyph from raised to sunk relief to the other, without the need to erase the whole sign, as it is the case with the other glyphs in the inscription that could not be "recycled" in this manner.

Finally, flying the UAV over the Hall gives one the ability to look at the state of the building from the perspective of its conservation. For instance, it is easier to determine from up close whether fragments of roof blocks that had fallen onto the clerestory level need to be removed or if cracks and fissures within the architraves require urgent restoration or consolidation.

9 Recording and Building 3D Models of the Loose Blocks Stored Outside of the Hall

In 1899, the fragile state of the column foundations led to the collapse of 11 of them in the northern half of the Hall.[18] Georges Legrain, the then *maître d'œuvre* at Karnak, proceeded to rebuild most of them, along with other columns that had fallen previously. This operation, carried out over many years, was a truly amazing feat, considering the size of the columns and the extent of their damage. Some columns inside the Hall that were either truncated or in poor shape at the time of the catastrophe, were not repaired, while others, such as columns 130 and 131, were dismantled, but never reconstructed. These latter two columns stood immediately next to the northern gate, in the first row inside the Hall lining the eastern section of the northern wall, where there is now a large empty space. For decades, the loose drums belonging to these hitherto disassembled columns, and to other columns belonging to the northern half of the Hall, were stashed away just north of the monument, before being partly removed further east, next to the Osiris-Heqa-Djet chapel. More blocks, belonging to the upper sections of columns originally in the southern half of the Hall—themselves fallen casualties of blocks toppled down from the upper portions of the Second Pylon when it caved in—now rest on *mastaba*s lined up in the large open air storage area between the Khonsu temple and the Hypostyle Hall.

All in all, no fewer than eight large areas spread across the Karnak Temple precinct contain hundreds of pieces of various sizes belonging to drums and abaci that are no longer *in situ* inside the Hall. Consequently, a major (and exciting) challenge of the project—and one that has been the focus of our efforts for the past three seasons—is the anastylosis of the dismantled, broken down or decayed sections of the monument. The virtual, and hopefully in the future the physical, reassembly of all the fallen portions of the columns is one of the project's goals. For this purpose, several related objectives had to be met: first,

18 Azim and Réveillac 2004, vol. I, 130–178; vol. II, 39–83.

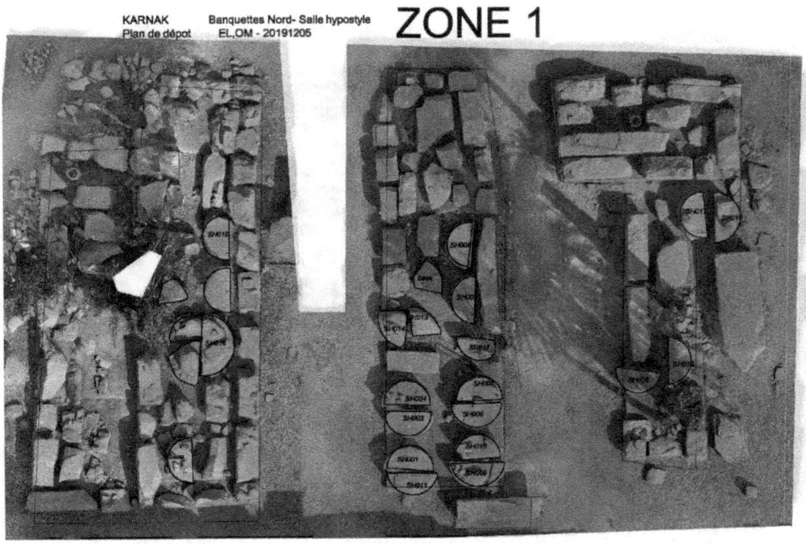

FIGURE 18.27 Sketch-map of zone 1 comprising loose blocks belonging to the Hall (O. Murray/E. Laroze)

identifying, recording, and tagging all the loose drums and abaci that were part of the Hall; second, determining their original position along the vertical axis that typically make up the eleven layers of a column; and third, determining the original location of these column fragments within the Hall.

Using a general plan of the site made by the Centre Franco-Égyptien d'Étude des Temples de Karnak (CFEETK), we first identified the areas in the Karnak Temple precinct where the loose blocks were located. We then flew the drone over those areas in order to generate detailed high-resolution sketch maps of each zone, onto which we plotted and labelled each and every loose column fragment from the Hall (Figure 18.27). This undertaking proved extremely useful when, chart in hand, we needed to locate individual blocks in the field among thousands of sandstone masonry blocks. Perfectly square-shaped abaci and semi-circular drums have characteristic contours, so the pictures taken from above allowed us to identify fragments that had hitherto escaped our attention.

Another interesting result of our experience working with the drone was our ability to use the image data to build 3D models of the *mastaba*s and their blocks in each zone where the loose fragments lay. O. Murray produced such a model for zone 2, located in the northeast quadrant of the temple, consisting of three *mastaba*s where the greatest concentration of loose fragments and column drum sections can be found (Figure 18.28). Off site, we could use these models to navigate virtually over the *mastabas*, moving around the blocks as

FIGURE 18.28 3D model of *mastabas* in zone 2 comprising loose blocks (O. Murray)

one might in person in order to examine their inscriptions, or whatever other details we deemed of interest, more closely. For missions like ours that can only stay in Egypt for relatively short field seasons each year, the ability to navigate and explore the site virtually and experiment with the 3D model to test fit pieces together while working in our office back home, saves valuable time in the field in future seasons.

The high-resolution maps of each zone were generated by flying multiple passes of the drone over the areas in question in a linear fashion at two heights (approximately 8 m and 15 m) with the camera facing down, but moving parallel to the ground (Figure 18.29). In addition to these topographic views, the 3D model of zone 2 used multiple passes of the drone over the *mastabas* with the camera at an oblique angle to the ground (approx. 45°), as well as a perimeter pass around each *mastaba*, with the camera facing inwards towards the fragments, from a height of 3–5 m. The camera was then rotated to a 90° angle so as to face the block fragments, and another perimeter pass was flown around each *mastaba* at the lowest height possible from the ground; 1–1.5 m, and as close to fragments on the *mastabas* as possible without triggering the close proximity warning feature of the drone: 2.5 m.

In order to augment the quality of the model, in addition to the photographs captured using the drone camera, a hand-held perimeter pass around each *mastaba* was also completed using a Nikon D800 with a Nikkor 28 mm lens. These additional facing photos were captured at a distance of 1–1.5 m from the fragments and allowed a greater degree of precision in shooting pieces with

FIGURE 18.29 O. Murray piloting the drone over the *mastabas* in zone 2 (E. Laroze)

more complex and detailed features. All of the photographs were post processed using Adobe Lightroom and exported as high-resolution jpegs to Agisoft Metashape (then Photoscan) and aligned, retaining the drone metadata for accurate position and scale information. A high-quality dense cloud of approximately 66.8 million points was generated from the aligned photos, which in turn was run on a high setting to produce a model with approximately 13.3 million polygons. This was textured at 16,492 pixels and the resulting model allowed researchers and students the ability to virtually visit zone 2 from their offices back home using the same software (Agisoft Metashape).

3D Models *mastabas* with their column block fragments and virtual anastylosis of columns like no. 131 can also prove interesting and invaluable ways of conveying information to both professionals and laymen. On site temple QR code signage that allows anyone with a mobile phone to visit these models, hosted on the relevant research institutions websites, provides both guides and tourists alike with a more in-depth understanding of the temple structure, and a glimpse into what in many cases is no longer physically standing.

10 Conclusion

The use of an unmanned aerial vehicle (UAV) and other emerging techniques has been invaluable to the Université du Québec à Montréal (UQAM)—University of Memphis joint epigraphic mission at Karnak. Producing series of orthophotographic *déroulés* of entire columns, of their most meaningful sections (the middle registers containing the ritual scenes) and of individual loose blocks has made it possible for us to overcome successfully the obstacles associated with producing 2D and 3D representations of the non-planar surfaces of columns. As for the UAV, it has given us access to the upper portions of the Hall hitherto out-of-reach by conventional means. Emerging techniques, such as drones, photogrammetry and photo rectification, represent a great scientific asset to generate high-quality images of the columns of the Great Hypostyle Hall at Karnak. But even more importantly, these sophisticated technologies have led to phenomenal progress in terms of procedure and methodology, especially in the field of anastylosis, where emerging techniques serve as an inestimable testing ground to validate hypotheses of reconstruction.

Bibliography

Azim, Michel, and Gérard Réveillac. 2004. *Karnak dans l'objectif de Georges Legrain: Catalogue raisonné des archives photographiques du premier directeur des travaux de Karnak de 1895 à 1917*. 2 vols. Paris: CNRS editions.

Biston-Moulin, Sébastien. 2016. *Inventaire des monuments, objets, scènes et inscriptions des temples de Karnak*. Montpellier.

Brand, Peter J. 2000. *The Monuments of Seti I: Epigraphic, Historical and Art Historical Analysis*. PdÄ 16. Leiden: Brill.

Brand, Peter J. 2018. "Patterns of innovation in the monumental art of Ramesside Thebes: the example of the Great Hypostyle Hall of Karnak." In *The Ramesside period in Egypt: studies into cultural and historical processes of the 19th and 20th dynasties*, edited by Sabine Kubisch and Ute Rummel, 45–61. Berlin; Boston: Walter de Gruyter.

Brand, Peter J., Emmanuel Laroze, and Jean Revez. 2011–2012. "Le projet américano-canadien des colonnes de la salle hypostyle de Karnak. Objectifs globaux et méthodologie." *Journal of the Society of the Study of Egyptian Antiquities* 38: 17–34.

Brand, Peter J., Jean Revez, Janusz Karkowski, Emmanuel Laroze, and Cedric Gobeil. 2013. "Karnak Hypostyle Hall Project. Report on the 2011 field season for the University of Memphis & the Université du Québec à Montréal." *Les Cahiers de Karnak* 14: 193–229.

Brand, Peter J., Rosa Erika Feleg, and William J. Murnane†. 2018. *The Great Hypostyle Hall in the temple of Amun at Karnak*. 2 vols. Oriental Institute Publications 142. Chicago: Oriental Institute of the University of Chicago.

Chandelier, Laure, Bertrand Chazaly, Yves Egels, Emmanuel Laroze, and Daniel Schelstraete. 2009. "Numérisation 3D et déroulé photographique des 134 colonnes de la Grande Salle Hypostyle de Karnak." *Revue XYZ, Revue de l'association française de topographie* 120: 33–39.

Christophe, Louis-André. 1955. *Temple d'Amon à Karnak: Les divinités des colonnes de la grande salle hypostyle et leurs épithètes*. Bibliothèque d'Étude 21. Cairo: Institut français d'archéologie orientale.

Egels, Yves, Emmanuel Laroze, Jean Revez, Owen Murray, and Peter Brand. 2020. "Une nouvelle colonne dans la salle hypostyle de Karnak. Récit d'une méthode de restitution." *Revue XYZ. Revue de l'association française de topographie* 164: 45–50.

Epigraphic Survey, The. 1986. *Reliefs and Inscriptions at Karnak, Vol. 4: The Battle Reliefs of King Sety I*. Oriental Institute Publications 107. Chicago: Oriental Institute of the University of Chicago.

Laroze, Emmanuel, and Bertrand Chazaly. 2009. "Relevés des colonnes de la grande salle hypostyle de Karnak." *Comptes rendus de l'Académie des Inscriptions et Belles-lettres* 153: 669–685.

Revez, Jean. 2020. "3D scanning, photogrammetry, and photo rectification of columns in the Karnak hypostyle hall." In *The Oxford Handbook of Egyptian epigraphy and palaeography*, edited by Vanessa Davies and Dimitri Laboury, 387–404. New York: Oxford University Press.

Revez, Jean, and Peter J. Brand. 2012. "Le programme décoratif des colonnes de la grande salle hypostyle de Karnak: Bilan de la mission canado-américaine de 2011." *Bulletin de la Société française d'Égyptologie* 183:10–38.

Revez, Jean, and Peter J. Brand. 2015. "The Notion of Prime Space in the Layout of the Column Decoration in the Great Hypostyle Hall at Karnak." *Les Cahiers de Karnak* 15: 253–310.

Rondot, Vincent. 1997. *La Grande Salle Hypostyle de Karnak: Les architraves*. 2 vols. Paris: Éditions Recherche sur les civilisations.

Rondot, Vincent, and Jean-Claude Golvin. 1989. "Restaurations antiques à l'entrée de la salle hypostyle ramesside du temple d'Amon-Rê à Karnak." *Mitteilungen des Deutschen Archäologischen Instituts, Abteilung Kairo* 45: 249–259.

CHAPTER 19

Representing Ancient Egyptian Inscriptions of the Old Kingdom Digitally: Dynamic Visualizations of Poetic Form and Inscriptional Layout

Julie Stauder-Porchet

Abstract

Texts, and texts as inscriptions in particular, are complex objects, posing challenges to digital representation. What may be defined as the linguistic text (the text as an ordered composition of words) is not only a matter of lexicon and grammar. Relevant dimensions also include intertextual relations with other texts (notably in terms of formulae and formulaic schemes), as well as the poetic form, or structure, of the text, which can be complemented by features of inscriptional layout. More broadly, what may be called the inscriptional text (the text as a physically dimensional and architecturally localized artefact inscribed in stone) often carries further significations. Based on two practical examples of Old Kingdom inscriptions (Harkhuf, Hezi), the paper will discuss which dimensions should be present in a digital corpus of inscriptions as complex, multi-dimensional objects.

Keywords

Text – materiality – Old Kingdom – Hezi – Harkhuf – visualization – corpus

1 Introduction

Ancient Egyptian inscriptions are complex objects. They consist in words in a particular arrangement, a "verbal text." But they are also physically dimensional and localized artifacts inscribed in stone, making an "inscriptional text." A schematic outline of relevant dimensions is the following:

- "The verbal text" (the text as an arrangement of words):
  - the words, strung together by grammar
  - phraseology and more specific intertextual relations with other inscriptions

  - structure, poetic form
- "The inscriptional text" (the text as a physically dimensional and localized artifact):
  - positioning and visibility in built or natural space
  - materiality (material, epigraphy, etc.)
  - relation to images (when given)
  - inscriptional layout

Print publications often focus on the verbal text, and particularly on the first and second dimensions above. All the other dimensions can be described verbally as well, and print publications remain the natural medium for analysis and interpretation. Digital textual corpuses present the wording of the verbal text, which can be enriched by translations, grammatical annotations, and philological notes. They permit navigating between texts within a corpus or, ideally, across linked corpuses, and thereby provide an implicit representation of phraseology and intertext, through lexical tagging or, additionally, through a controlled lexicon of motifs. Discussed by others at the conference from which the present volume derives, 3D technology permits visualizations of inscriptions as localized material artifacts in context. The present paper addresses two other dimensions for which a digital presentation offers added value: poetic structure and inscriptional layout.

The first is a dimension of the "verbal text" as defined above, the second a dimension of the "inscriptional text." For both these dimensions, I illustrate the value of a digital presentation with two examples from the Old Kingdom.[1] One is Hezi's short, but complex, autobiography (Saqqara, reign of Teti), inscribed in five columns on the right thickness of the doorway leading into the official's funerary chapel. The other is the extensively inscribed façade of Harkhuf's funerary chapel at Qubbet el-Hawa (early reign of Pepi II), a monument that inaugurated the hieroglyphic inscription of that place on a massive scale. My discussion is entirely prospective, reflecting on how the digital tools that I am evoking would be of interest.

2 Representing the Poetic Structure of Inscriptions

In a simple representation, the text of Hezi's event autobiography can be displayed in a way that reflects its segmentation in clauses, thereby easing the reading of what the words stand for (Figure 19.1). This representation of the

1 I draw on the more detailed analyses in Stauder-Porchet 2015 (Hezi) and Stauder-Porchet 2020a–b (Harkhuf). A general discussion of the dimension of inscriptional layout in the Old Kingdom is offered in Stauder-Porchet 2021.

(1)ḏd.f
jnk zꜣb zš n rk jzzj
jnk zꜣb sḥḏ zš n rk wnjs
jn ttj nb(.j) (w)d w(j) m zꜣb ʿḏ-mr (2)(w)d w(j) m ḫrj-tp nsw
rḏ.n ḥm.f jr.t(j) n(.j) ø
n rḫ ḥm.f rn(.j) m jṯt zš n ʿw.f
ny wnt ḥꜣ nb sḫꜣ n.f w(j) ḏd.n.f sꜣ(r)
(3)jr.n(.j) zš ḫr ḥm.f m-ḥꜣt zšw
jr.n(.j) sr ḫr ḥm.f m-ḥꜣt srw
(4)wn rḏ ḥm.f hꜣ(.j) r wjꜣ ʿꜣ stp-zꜣ jwt(.j) r wꜣwt jr.t(j) ꜣwt mr m ḫrj-tp nsw
sk w(j) m zꜣb ʿḏ-mr
ny wnt jry.t(j) mrtt n mrt(j)(.j) nb
wn ḥm.f nḏ.f ḫt m-ʿ(.j) m rꜣ-ʿ (5)mm srw
sk w(j) m zꜣb zḥḏ zš
n rḫ ḥm.f rn(.j) ṯny r bꜣk nb

FIGURE 19.1A Hezi, a flat representation

(1)ḏd.f

jnk zꜣb zš n rk jzzj   *jnk zꜣb sḥḏ zš n rk wnjs*

jn ttj nb(.j) (w)d w(j) m zꜣb ʿḏ-mr   *(2)(w)d w(j) m ḫrj-tp nsw*
rḏ.n ḥm.f jr.t(j) n(.j) ø
n rḫ ḥm.f rn(.j) m jṯt zš n ʿw.f
ny wnt ḥꜣ nb sḫꜣ n.f w(j) ḏd.n.f sꜣ(r)

(3)jr.n(.j) zš ḫr ḥm.f m-ḥꜣt zšw
jr.n(.j) sr ḫr ḥm.f m-ḥꜣt srw

(4)wn rḏ ḥm.f hꜣ(.j) r wjꜣ ʿꜣ stp-zꜣ jwt(.j) r wꜣwt jr.t(j) ꜣwt
mr m ḫrj-tp nsw
sk w(j) m zꜣb ʿḏ-mr
ny wnt jry.t(j) mrtt n mrt(j)(.j) nb
wn ḥm.f nḏ.f ḫt m-ʿ(.j)
m rꜣ-ʿ (5)mm srw
sk w(j) m zꜣb zḥḏ zš
n rḫ ḥm.f rn(.j) ṯny r bꜣk nb

FIGURE 19.1B A representation with structure

verbal text, often favored in text editions and in digital corpuses similarly, is close to being flat, as the only level of structure made visible here is clausal structure. Going further, a general representation of the poetic structure of the text could be suggested through empty lines and indents (Figure 19.2). Based on this, a dynamic navigation could permit a visualization of the various levels of structure involved.

A basic level in Hezi's inscription is that all things come in pairs (Figure 19.2a): twice, "I was a (*jnk*) … during the time of (*n rk* …)"; then, "It was Teti, my lord (*jn ttj nb*(*.j*)), who … me (*w*(*j*)) …" and "His Person had (*rḏ.n ḥm=f*) … for me (*n*(*.j*)) …"; then twice again, "I have acted as (*jr.n*(*.j*)) … with His Person (*ḫr ḥm=f*) at the head of (*m-ḥꜣt*) …"; and, finally, "His Person used to have me (*wn rḏ ḥm=f*) … as if (*mr m*) … even though I was only a (*sk w*(*j*) *m*) …" and "His Person used to consult (*wn ḥm=f nḏ=f*) … in the manner (done) among (*m rꜣ-ʿ mm*) … even though I was only a (*sk w*(*j*) *m*) …" This structure in pairs is not just formal, but expressive of the tight relationship between Hezi and the king: in the first and third pairs, the speaker, Hezi, acts for kings, twice; in the second and fourth pairs, reciprocally, the king acts for the speaker.

Beyond this structure in alternating pairs, Hezi's autobiography presents palindromic elements. The clauses "… for His Person knew my name (*n rḫ ḥm=f rn*(*.j*)) …" and "… without there being any (*ny wnt … nb* …) …" occur in the first part of the text, then again in the final part of the text, in reverse order (Figure 19.2b). By a similar token, a sequence of three titles is found in the first part of the text, then again in the final part of text, in reverse order (Figure 19.2c). Together, these two palindromes define a form that concentrically focuses the central element, where Hezi states, in lapidary simplicity: "I have acted as a scribe with His Person at the front of scribes (*jr.n*(*.j*) *zš ḫr ḥm=f m-ḥꜣt zšw*); I have acted as an official with His Person at the front of officials (*jr.n*(*.j*) *sr ḫr ḥm=f m-ḥꜣt srw*)" (Figure 19. 2d).

In a print publication, a figure could be constructed to provide a synthetic representation of these elements in addition to their discussion in the verbal commentary (Figure 19.3). Such a representation, however, is bound to remain static and much condensed. A digital environment would permit the unpacking of the various elements of structure just discussed, allowing the viewer to navigate between these. A similar dynamic visualization could easily be applied to a translation of the text and to photographs (not given here), allowing the user to switch back and forth between all of these.

Another example may illustrate the same proposal. Inscribed in three columns on the right of the doorway leading into Harkhuf's funerary chapel, a titulary of the official is expanded into a short poetic form. A flat representation, be it in transcription or in translation, would provide a continuous,

(1) ḏd.f

jnk zꜣb zš n rk jzzj *jnk zꜣb sḥḏ zš n rk wnjs* Hezi > king

jn ttj nb(.j) (w)d w(j) m zꜣb ʿḏ-mr *(2)(w)d w(j) m ẖrj-tp nsw* king > Hezi
rḏ.n ḥm.f jr.t(j) n(.j) ø
n rẖ ḥm.f rn(.j) m jṯt zš n ʿw.f
ny wnt ḥꜣ nb sẖꜣ n.f w(j) ḏd.n.f sꜣ(r)

(3)jr.n(.j) zš ẖr ḥm.f m-ḥꜣt zšw Hezi > king
jr.n(.j) sr ẖr ḥm.f m-ḥꜣt srw

(4)wn rḏ ḥm.f hꜣ(.j) r wjꜣ ʿꜣ stp-zꜣ jwt(.j) r wꜣwt jr.t(j) ꜣwt king > Hezi
mr m ẖrj-tp nsw
sk w(j) m zꜣb ʿḏ-mr
ny wnt jry.t(j) mrtt n mrt(j)(.j) nb
wn ḥm.f nḏ.f ẖt m-ʿ(.j)
m rꜣ-ʿ (5)mm srw
sk w(j) m zꜣb zḥḏ zš
n rẖ ḥm.f rn(.j) ṯny r bꜣk nb

FIGURE 19.2A Hezi, dynamic representation of various levels of form - 1

(1) ḏd.f

jnk zꜣb zš n rk jzzj *jnk zꜣb sḥḏ zš n rk wnjs*

jn ttj nb(.j) (w)d w(j) m zꜣb ʿḏ-mr *(2)(w)d w(j) m ẖrj-tp nsw*
rḏ.n ḥm.f jr.t(j) n(.j) ø
n rẖ ḥm.f rn(.j) m jṯt zš n ʿw.f
ny wnt ḥꜣ nb sẖꜣ n.f w(j) ḏd.n.f sꜣ(r)

(3)jr.n(.j) zš ẖr ḥm.f m-ḥꜣt zšw
jr.n(.j) sr ẖr ḥm.f m-ḥꜣt srw

(4)wn rḏ ḥm.f hꜣ(.j) r wjꜣ ʿꜣ stp-zꜣ jwt(.j) r wꜣwt jr.t(j) ꜣwt
mr m ẖrj-tp nsw
sk w(j) m zꜣb ʿḏ-mr
ny wnt jry.t(j) mrtt n mrt(j)(.j) nb
wn ḥm.f nḏ.f ẖt m-ʿ(.j)
m rꜣ-ʿ (5)mm srw
sk w(j) m zꜣb zḥḏ zš
n rẖ ḥm.f rn(.j) ṯny r bꜣk nb

FIGURE 19.2B Hezi, dynamic representation of various levels of form - 2

(1)*ḏd.f*

jnk z3b zš n rk jzzj *jnk z3b sḥḏ zš n rk wnjs*

jn ttj nb(.j) (w)d w(j) m z3b ʿḏ-mr (2)*(w)d w(j) m ḫrj-tp nsw*
rḏ.n ḥm.f jr.t(j) n(.j) ø
n rḫ ḥm.f rn(.j) m jṯt zš n ʿw.f
ny wnt ḫ3 nb sḫ3 n.f w(j) ḏd.n.f s3(r)

(3)*jr.n(.j) zš ḫr ḥm.f m-ḥ3t zšw*
jr.n(.j) sr ḫr ḥm.f m-ḥ3t srw

(4)*wn rḏ ḥm.f h3(.j) r wj3 ʿ3 stp-z3 jwt(.j) r w3wt jr.t(j) 3wt*
mr m ḫrj-tp nsw
sk w(j) m z3b ʿḏ-mr
ny wnt jry.t(j) mrtt n mrt(j)(.j) nb
wn ḥm.f nḏ.f ḫt m-ʿ(.j)
m r3-ʿ (5)*mm srw*
sk w(j) m z3b zḥḏ zš
n rḫ ḥm.f rn(.j) ṯny r b3k nb

FIGURE 19.2C Hezi, dynamic representation of various levels of form - 3

(1)*ḏd.f*

jnk z3b zš n rk jzzj *jnk z3b sḥḏ zš n rk wnjs*

jn ttj nb(.j) (w)d w(j) m z3b ʿḏ-mr (2)*(w)d w(j) m ḫrj-tp nsw*
rḏ.n ḥm.f jr.t(j) n(.j) ø
n rḫ ḥm.f rn(.j) m jṯt zš n ʿw.f
ny wnt ḫ3 nb sḫ3 n.f w(j) ḏd.n.f s3(r)

(3)*jr.n(.j) zš ḫr ḥm.f m-ḥ3t zšw*
jr.n(.j) sr ḫr ḥm.f m-ḥ3t srw

(4)*wn rḏ ḥm.f h3(.j) r wj3 ʿ3 stp-z3 jwt(.j) r w3wt jr.t(j) 3wt*
mr m ḫrj-tp nsw
sk w(j) m z3b ʿḏ-mr
ny wnt jry.t(j) mrtt n mrt(j)(.j) nb
wn ḥm.f nḏ.f ḫt m-ʿ(.j)
m r3-ʿ (5)*mm srw*
sk w(j) m z3b zḥḏ zš
n rḫ ḥm.f rn(.j) ṯny r b3k nb

FIGURE 19.2D Hezi, dynamic representation of various levels of form - 4

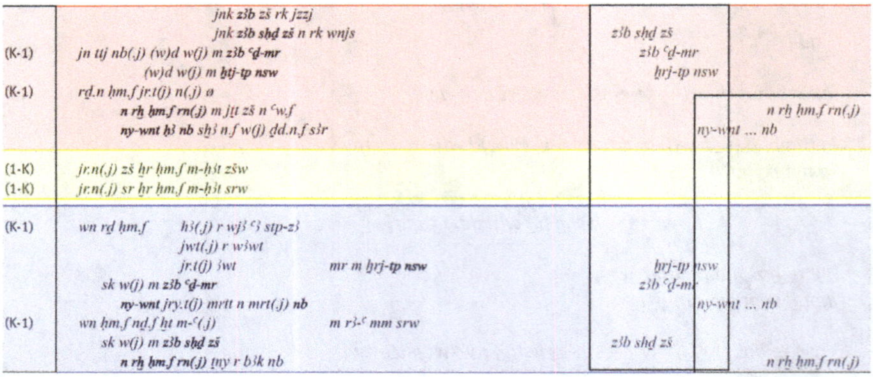

FIGURE 19.3 Hezi, attempt at a synthetic visualization of structure in a print publication

one-dimensional, string of titles and epithets. A representation with slightly more structure would group the titles and epithets based on their nature and contents. Going further, a representation could show visually how the titulary falls in two parts, each fourfold, and how the second part is an expansion, or elaboration, of the first (Figure 19.4a). The result is a short poetic form in which a number of key terms are repeated three times (Figure 19.4b; note that "produce" and "bring" are based on the same root, *jnj*, in Egyptian), giving the expanded titulary a celebratory dimension while defining an immediate biographical identity of Harkhuf.

The event autobiography that follows the titulary is generally read for its historical contents, as a continuous narrative text, and represented accordingly as a linear sequence of words. Going beyond this flat representation, empty lines and indents can be used to provide a hierarchical representation of recurrent structural elements, for example the recurrent tripartite generative scheme [king's command—Harkhuf's action—king's praise or statement of exceptionality] on which each of the three journeys by Harkhuf is based, as well as extensive parallelism throughout the text. Going further yet, a digital environment would make it possible to visualize how each of the three modules (corresponding to Harkhuf's three journeys) is an expansion of the preceding one (Figure 19.5a–b). In the first journey, the king sends Harkhuf out with his father (a) to open a road to Iam (b). Acting in response, Harkhuf accomplishes the trip in seven months (c) and brings back rare produce (d); he is praised by the king (e).

In the second module, corresponding to Harkhuf's second journey, the king's initiating command is a variation of that in the first module: Harkhuf is now sent out alone (a'). Harkhuf's action in response, the event of "accomplishing a trip" (c) is elaborated into its constituent parts, "going out" and "descending"

(1)The foremost, sole companion, lector priest, the one who is in the chamber, guardian of Nekhen, chief of Nekheb,

the seal-bearer of the King of Lower Egypt, sole companion, lector priest, overseer of foreign auxiliaries,
he who is privy to the secrets of all matters of the Head of the South,
the favorite of his lord,
Harkhuf.

(2)[The seal-bearer of the king of Lower Egypt], sole [companion], lector priest, overseer of foreign auxiliaries,
who brings the produce of all foreign lands for his lord,
who brings imported royal luxuria,

overseer of all foreign lands of the Head of the South,
who places the dread of (3)[Horus in] the foreign lands,

who does what [his] lord praises,

the seal-bearer of the king of Lower Egypt, sole companion, lector priest, overseer of foreign auxiliaries,
the imakhu before Ptah-Sokar,
Harkhuf.

FIGURE 19.4A Harkhuf, titulary, dynamic representation of one level of form

(1)The foremost, sole companion, lector priest, the one who is in the chamber, guardian of Nekhen, chief of Nekheb,

the seal-bearer of the King of Lower Egypt, sole companion, lector priest, **overseer of foreign auxiliaries**,
he who is privy to the secrets of all matters of the Head of the South,
the favorite of **his lord**,
Harkhuf.

(2)[The seal-bearer of the king of Lower Egypt], sole [companion], lector priest, **overseer of foreign auxiliaries**,
who **brings** the **produce** of all **foreign lands** for **his lord**,
who **brings** imported royal luxuria,

overseer of all **foreign lands** of the Head of the South,
who places the dread of (3)[Horus in] the **foreign lands**,

who does what **[his] lord** praises,

the seal-bearer of the king of Lower Egypt, sole companion, lector priest, **overseer of foreign auxiliaries**,
the imakhu before Ptah-Sokar,
Harkhuf.

FIGURE 19.4B Harkhuf, titulary, dynamic representation of another level of form

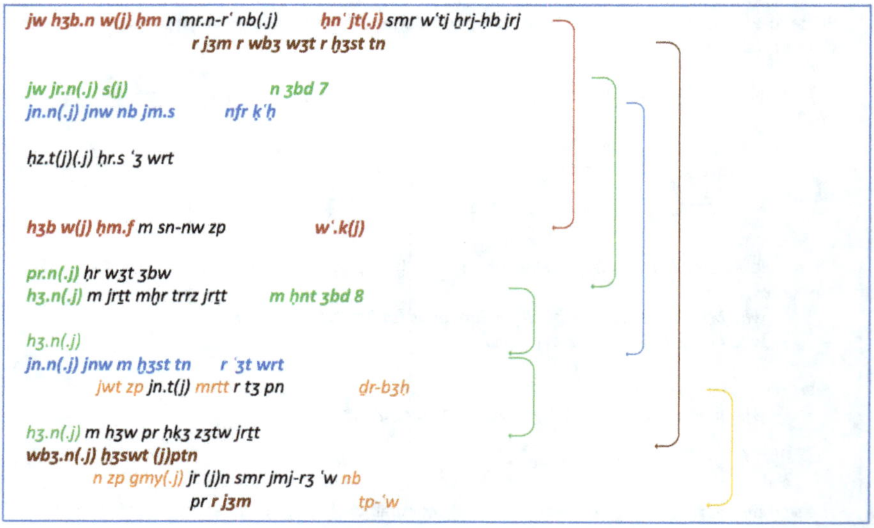

FIGURE 19.5A Harkhuf, beginning of event autobiography, representation of one level of form, in transliteration

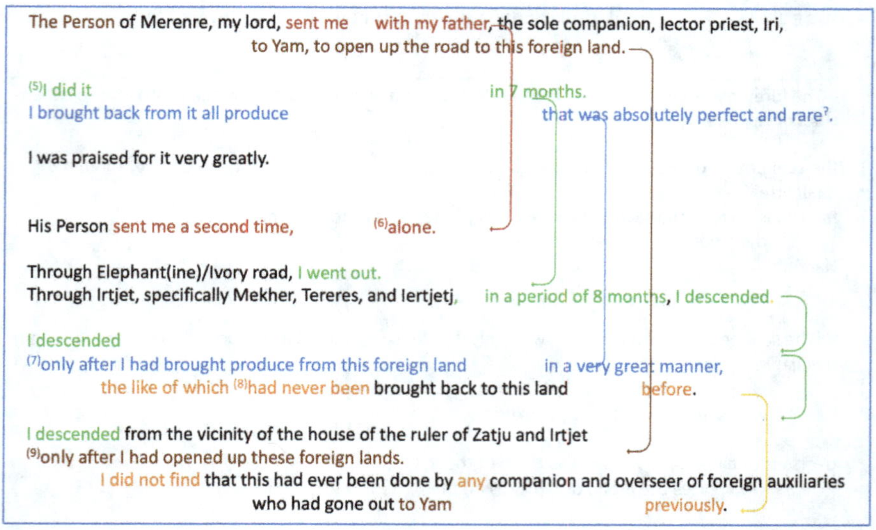

FIGURE 19.5B Harkhuf, beginning of event autobiography, representation of one level of form, in translation

(i.e., returning) (c'). The event of "descending" is itself triplicated, and serves as an anchor for elaborations of the actions of bringing back produce (d') and opening a new road to Iam (b'). Through this triplication and elaboration, the successful return to Egypt is celebrated. Such dimensions, and many more that are present in the text, can be described verbally in all manner of analytical details. To be communicated in a more intuitively appealing manner, however, they call, complementarily, for a dynamic visual representation.

3 Representing Inscriptions as Visual Compositions

Inscriptions are often laid out in careful ways. Through layout, salient articulations or expressions can be foregrounded, the poetic structure of a text can be mirrored visually, and core meanings can be projected over the inscribed surface. In some cases, the complexity of the layout suggests that the texts were probably composed directly with a view on how they would exist visually on the inscribed surface. More generally, inscriptions appear to be visual compositions as much as compositions of words.

As noted above, the poetic form of Hezi's autobiography is tripartite and centrally focused. On the monument, the inscription is laid out in such a way that parts I and III correspond to columns 1–2 and 4–5, respectively, while the central middle part II corresponds exactly to the middle column 3 (Figure 19.6). Layout thus mirrors one level in the poetic form of the inscription.

In part III, Hezi says that he has performed duties for the king and been granted benefactions well beyond his rank. This is expressed verbally through contrastive comparisons: "… *as if* (*mr m*) a royal chamberlain, even though I was (*m*) a territorial administrator" (col. 4); … *in the manner* (*done*) ⁽⁵⁾*among* (*m rꜣ-ꜥ mm*) officials, even though I was (*m*) a judiciary inspector of scribes …" (col. 4–5). The rhetoric articulation is based on the prepositional expressions. These are emphasized visually through occupying the whole breadth of their columns and/or being surrounded by empty space (see Figure 19.5, with close-up for … *mr m* … *m* … in col. 4 and … *mm* … on top of col. 5).

Turning to my second example, Harkhuf's façade, the architrave above the doorway has four offering formulae (ll. 1–3, 8). Each begins with the word *nsw* "king" and ends with Harkhuf's name, itself associated with the seated figure of the official. In the ideal autobiography in between the offering formulae (ll. 4–7), all four lines begin with references to Harkhuf (Figure 19.7). Both horizontally and vertically, the king and Harkhuf are thus set in relation to one another on the architrave that forms the visually most exposed part of the inscribed monument.

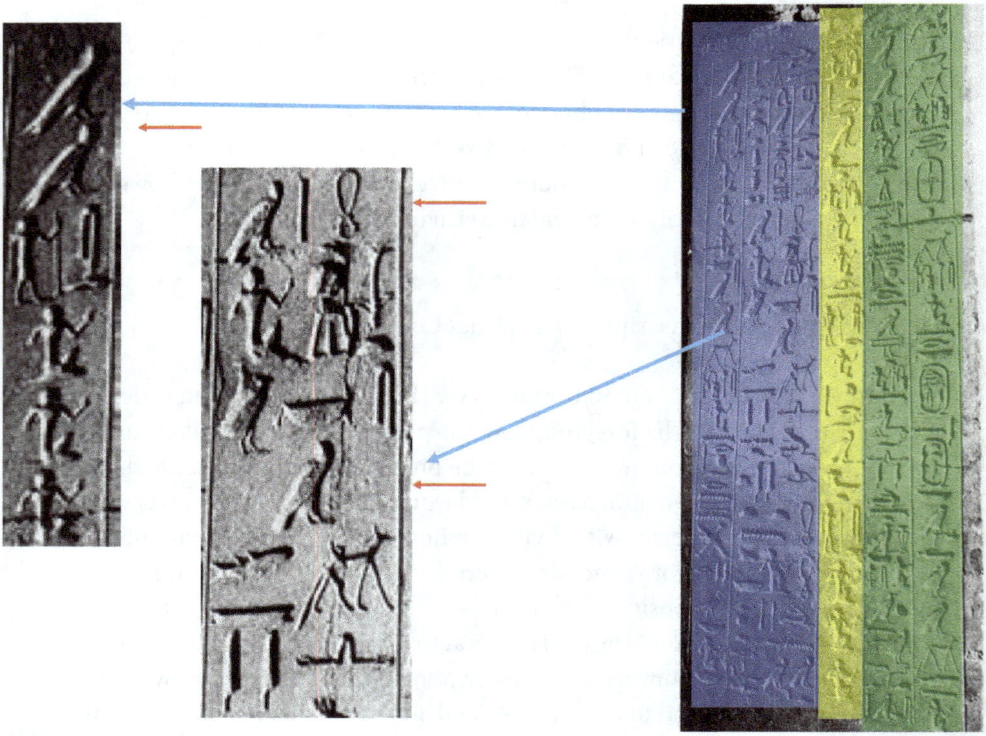

FIGURE 19.6 Hezi, inscriptional layout

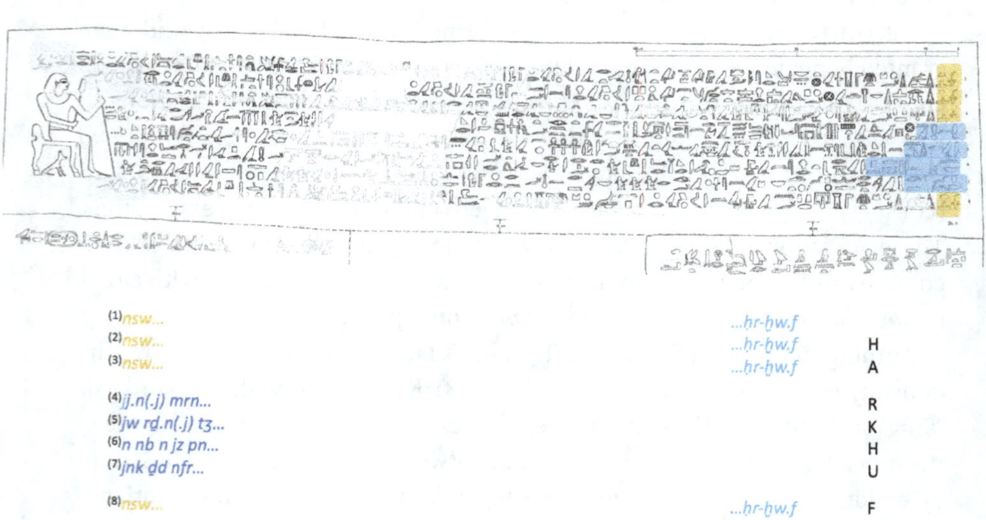

FIGURE 19.7 Harkhuf, architrave, representation of one level of inscriptional layout

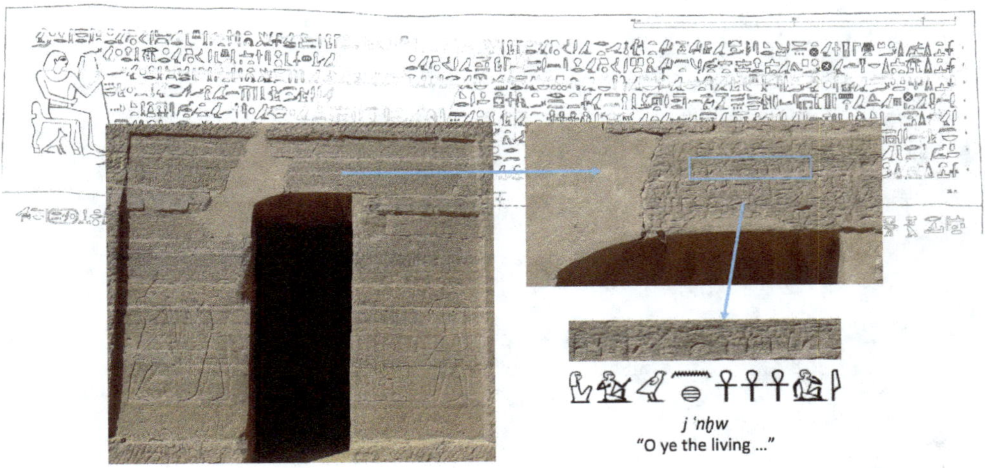

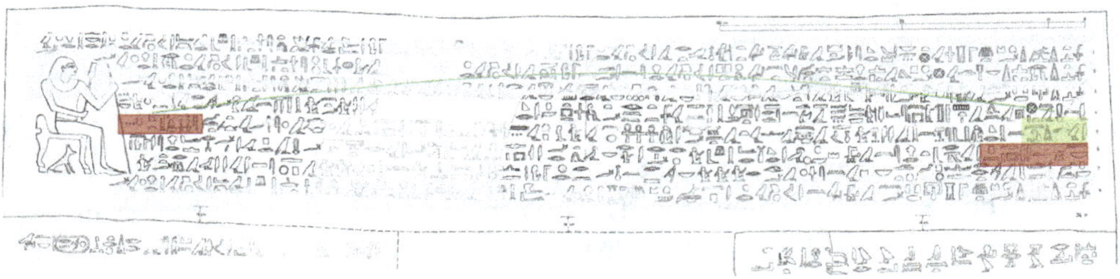

[(4)]*jj.n(.j) mrn...*
[(5)]*jw rḏ.n(.j) tꜣ...* — **1000 tꜣ 1000 ḥnḳt**[(5, end)]
[(6)]***n nb n jz pn***
[(7)]... — *...m zp sšwy zꜣ m ḫrt ỉt=f*[(7, end)]

[(4)]I have come here today...
[(5)]I have given bread... — **'A thousand bread and a thousand bear**[(5, end)]
[(6)]**to the lord of this tomb!'** ...
[(7)]... — ...in an occasion of depriving a son of his father's share.[(7, end)]

FIGURE 19.8A–B Harkhuf, architrave, representation of further levels of inscriptional layout

In the ideal autobiography, the vocative address to "the living" (l. 5) is inscribed just above the doorway through which the visitor is to enter the tomb (Figure 19.8a). The plural is written in the longest way possible, lending further graphic emphasis to the address. Another reference to the visitor is also inscribed above that doorway (l. 6). The beginning of the invocation offering that the visitor is to pronounce, "A thousand bread ...," at the end of line 5, is set in correspondence to Harkhuf's similarly worded statement "I have given bread ...," at the beginning to the same line (Figure 19.8b). The invocation offering is split over two lines in a way that underscores the action itself, an action of

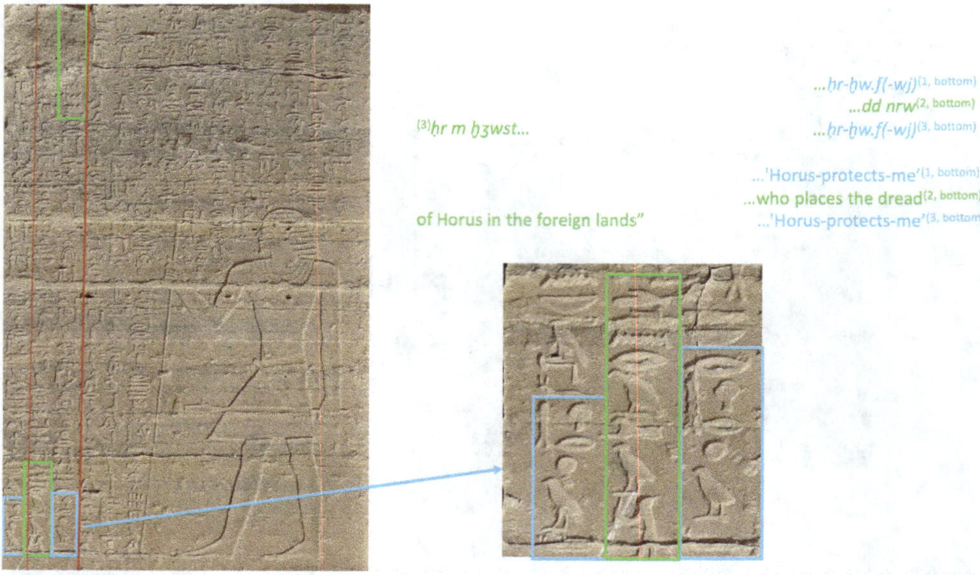

FIGURE 19.9 Harkhuf, titulary

giving: "A thousand bread and a thousand beer (6)*for* the owner of this tomb!" These words, to be pronounced by the ritually acting visitor, sit precisely in the middle (end of l. 5 and beginning of l. 6) of the ideal autobiography (ll. 4–7). The layout thereby underscores the ritual dimensions of the monument.

In the developed titulary on the right side of the entrance (see above), one epithet is split over two columns, R(ight).2–3: "… who places the dread of (R.3)[Horus in] the foreign lands …" (Figure 19.9). The epithet occupies the exact middle position of the second part of the titulary, in columns R.2–3. Moreover, the one "who places …," Harkhuf, is fittingly on bottom of R.2, while the king is on top of R.3. Going further yet, the king, on top of R.3, is set in relation to the name of Harkhuf, at the bottom of the same column, as well as of R.1. Through layout, a relation of reciprocity is expressed between Harkhuf's action of projecting the king's dread into foreign countries and Harkhuf's name, a sentential name meaning "Horus (the king) protects me."

The event autobiography (R.4–14, as well as lines 1–9 on the left side of the façade) is introduced by "He says: …," set on top of R.4. It is linked to the preceding titulary (R.1–3) by horizontal adjacencies across columns R.3 and R.4 (Figure 19.10a). These associate Harkhuf and his father, and Harkhuf and the king. Another link between titulary and event autobiography involves a relation over a distance. At the end of the right side of the façade, the last column is self-contained and reads: "(R.14, top)I satisfied him (= the ruler of Iam) in such

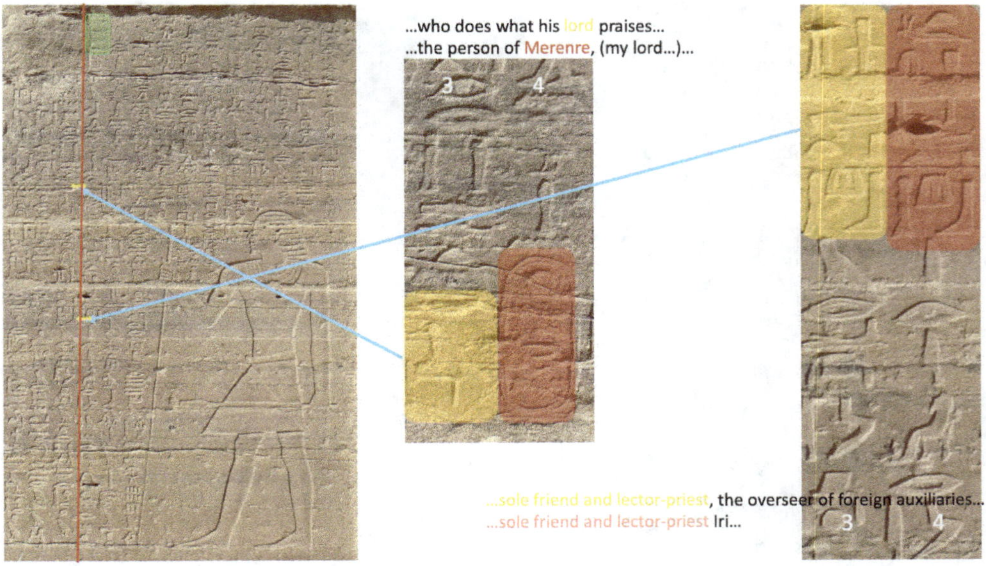

FIGURE 19.10A–B Harkhuf, right side of façade, representations of multiple relations between titulary and event autobiography

ways that he would keep praising all the gods for the Sovereign.(R.14, bottom)" (Figure 19.10b). The statement echoes the epithet just discussed, "... who places the dread of (R.3, top)[Horus in] the foreign lands ..." Both are emphasized through layout. In echoing one another, they frame the right side of the façade as a whole.

The first part of the event autobiography, on the right side of the façade (R.4–14), is laid out in such a way that the top of most columns consists in expressions of Harkhuf's agency in acting for the king (Figure 19.11): "(R.5)I have accomplished it ... (R.6)me acting alone ... (R.7)after I have brought back produce ... (R.9)after I have opened these foreign lands ... (R.11)I have gone out through Ta-wer ... (R.14)I have satisfied ..." In direct visual contact with Harkhuf's staff of authority, the lower part of column R.6 consists of locales travelled by Harkhuf and indications of duration (Figure 19.12a): "... through Irtjet, specifically Mekher, Tereres, and Iertjetj, in a period of 8 months. I descended ..." The iconic determinatives and the numerical notations make this a visual message accessible also to non-fully literate viewers. The shorter columns R.8–13 all end with further toponyms, looked at by Harkhuf or surrounding his head (Figure 19.12b). The last column (R.14) ends with an uncommon designation of the king (*jtw*). For an expedition leader, "who places the dread of Horus in the foreign lands" (R. 2–3), and "pacifies" foreign rulers for the "Sovereign" (R.14), this makes a fitting visual biography, surrounding his standing figure.

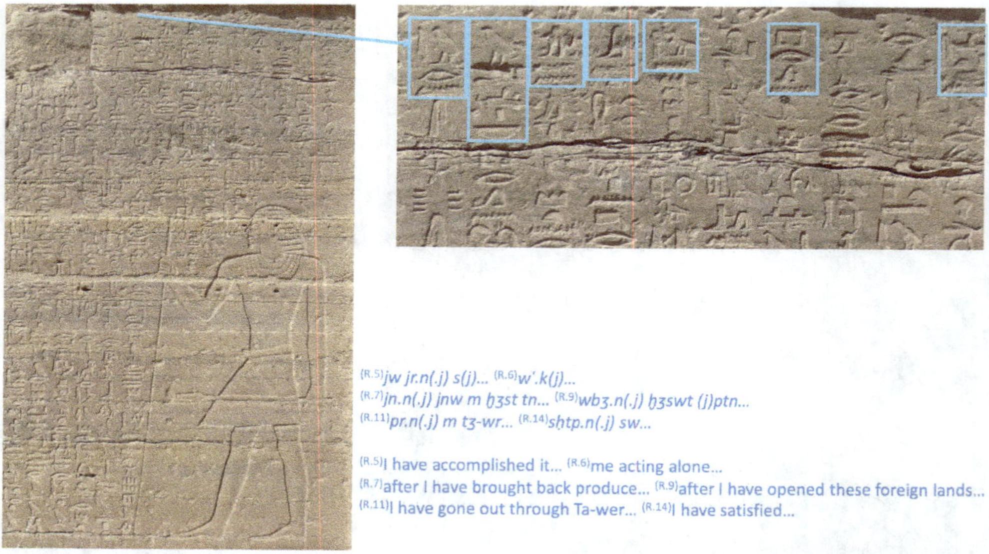

FIGURE 19.11 Harkhuf, autobiography, right side: representation of one level of inscriptional layout

The royal letter inscribed on the outer right of the façade adopts the expected format of the royal decree. After two horizontal lines (1–2), the letter proper consists of twenty-two columns (3–24) with a concluding note on the implementation of the royal order (25–26). In this letter, the king Neferkare praises Harkhuf for having brought back a dwarf from Iam and compares this action to that of another expedition leader, Werdjededba, in the time of another king, Izezi, a century or so earlier. The names of the two kings occur twice each, as does that of Werdjededba. Five of six occurrences are either on top or on the bottom of the columns, alternating for each name (Figure 19.13a, with a representation of the names of Neferkare, in red, and Izezi, in yellow). The visual result is a chiastic arrangement, binding the present to the past.

A number of horizontal adjacencies underscore what the letter is about by associating elements with one another across columns (Figure 19.13b): the "dwarf," the names of the two kings, the dwarf's condition ("alive, whole, and in good health"), and its destination ("the god's dances"). In his letter, the king speaks of how future generations will hear of the benefactions that he pledges to bestow on Harkhuf (col. 13–14: Figure 19.13a). The key term, "as they (= the people to come) hear …" (*sḏm.sn* …) is inscribed on top of column 13. Moreover, these columns 13–14, concerned with the reception to come of the royal letter (and, by extension, of the monument as a whole), occupy the exact center of the letter proper after the first ten columns (3–12) and before the next ten (15–24).

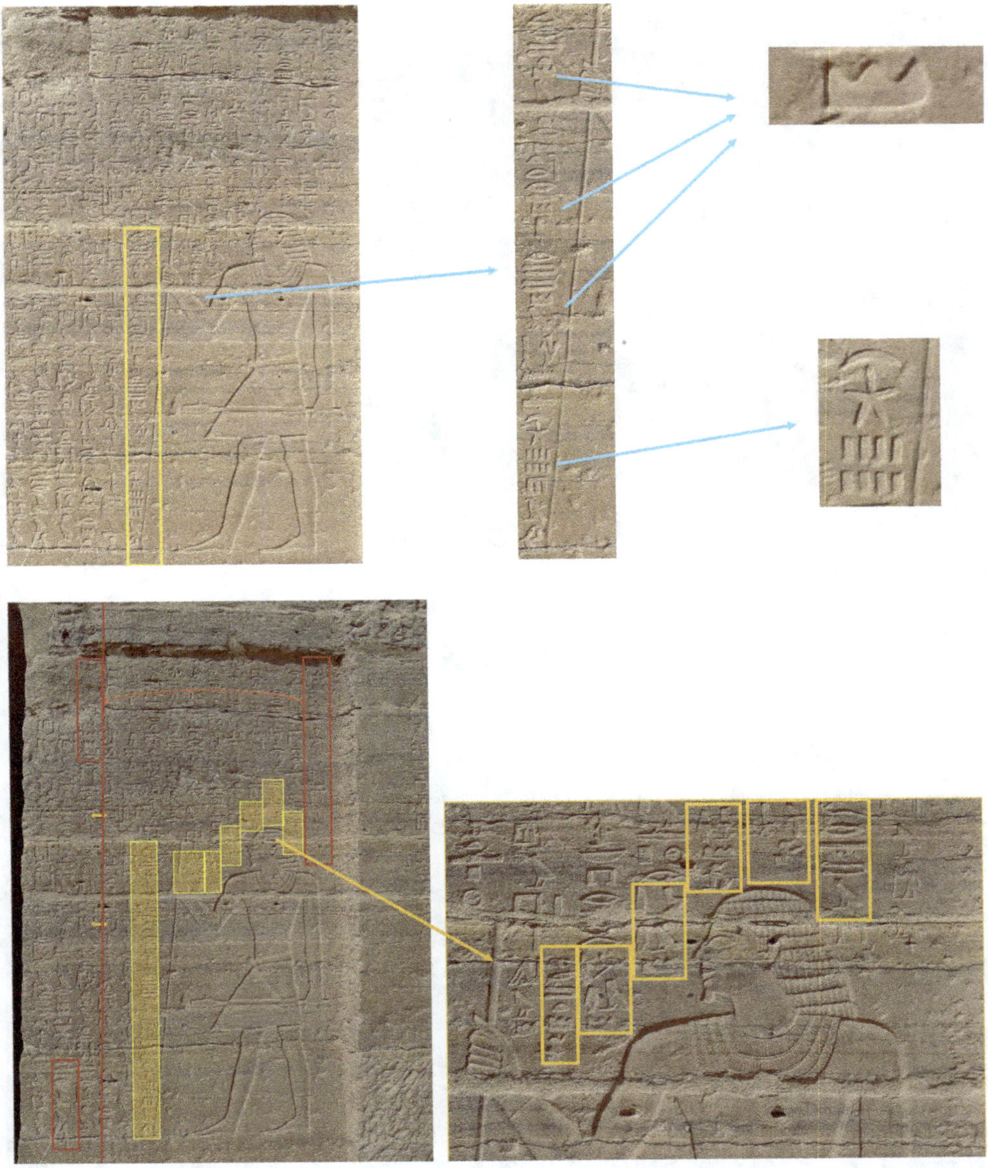

FIGURE 19.12A–B Harkhuf, autobiography, right side: representations of further levels of inscriptional layout

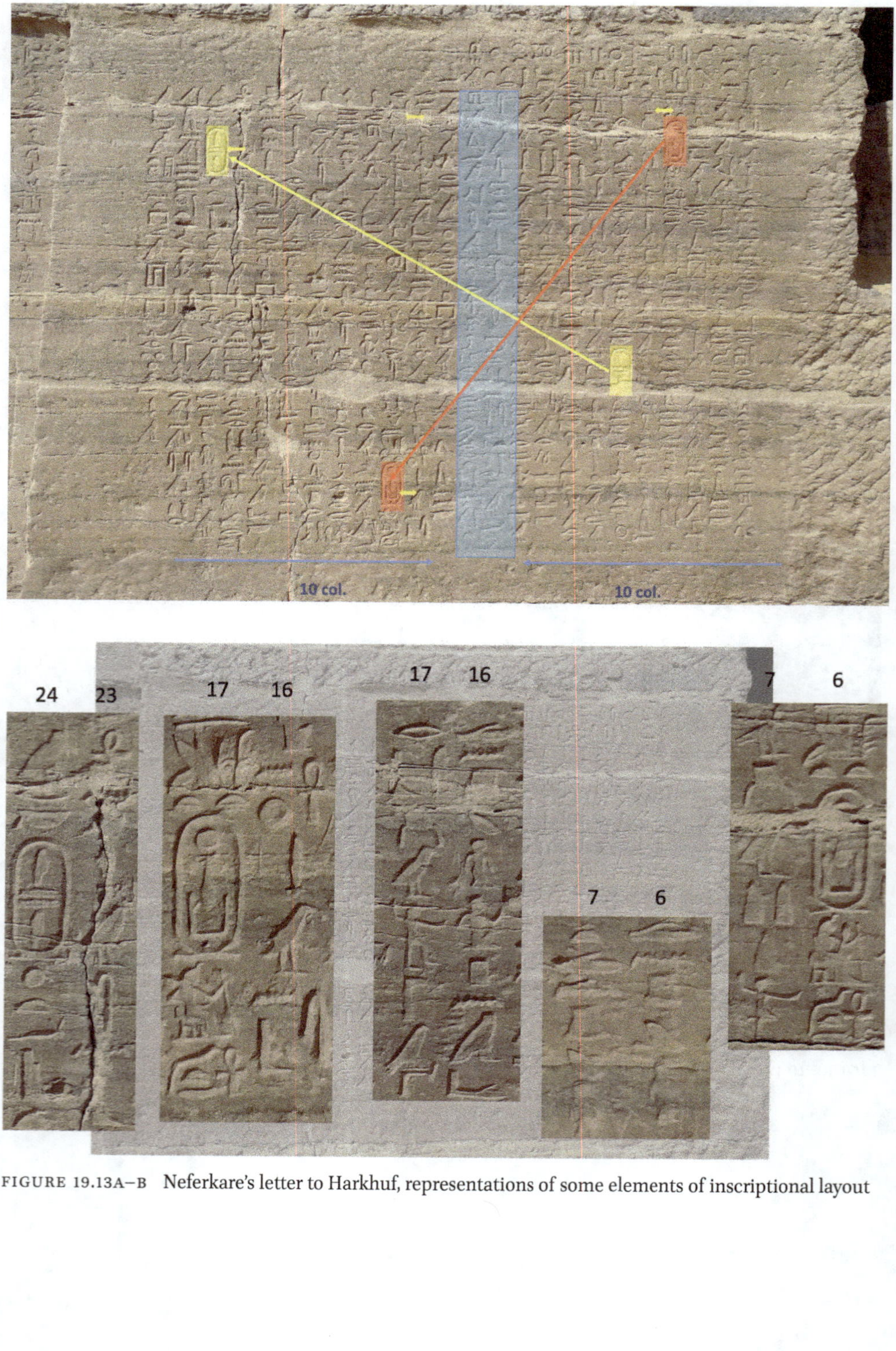

FIGURE 19.13A–B Neferkare's letter to Harkhuf, representations of some elements of inscriptional layout

FIGURE 19.14 Harkhuf's façade, representation of overall layout

The façade as a whole is laid out in a manner that is meaningful geographically (Figure 19.14). On the right (northern) side of the façade, motion "to Iam" (to the south) is emphasized, while the left (southern) side is about motion "to the Residence" (to the north). The royal letter replicates that configuration, with motion "to Iam" emphasized in its right (northern) part and motion "to the Residence" is its left (southern) part. Viewed as a whole, this royal letter is inscribed on the outer right (north) of the façade, pointing to its origin, in the far-northern Memphite Residence. The royal letter is set in correspondence to a previous letter, not inscribed but mentioned: that sent by Harkhuf to Neferkare's predecessor, Merenre. The mention of this previous letter by Harkhuf is inscribed on top of the left (southern) side of the façade (Left, lines 1–2), pointing to its own place of origin, in the far south, the place of maximal distance from both Elephantine and the Residence (Left 1–2). The layout thereby links Harkhuf, "who brings the produce of all foreign lands for his lord," and the king, who command, praises, and writes from the north. The real and ideological geography of the text is thus projected over the bi-dimensional surface of the inscribed monument.

4 Conclusion

The features of poetic form and inscriptional layout presented above are not just formal but meaningful: for instance, of the deeply rhetorical nature of Hezi's inscription; or of how Harkhuf's façade has a strong ritual dimension, as well as projecting royal agency in the far southern locale of Qubbet el-Hawa. The value of a digital presentation lies in making these dimensions more directly accessible and to more people. For inscriptional layout, referring to how a text exists in the bi-dimensional visual space of the inscription, the value of visualization is immediate. The poetic form concerns the "verbal text" but it calls no less for visualization when complex enough, as the examples above illustrate. As these examples also illustrate, multiple levels are involved in both the poetic structure and the inscriptional layout, and the two dimensions are interrelated in meaningful ways. A digital presentation would permit navigation between visualizations of multiple interrelated levels of both poetic structure and inscriptional layout, all of these in transliteration, translation, and, ideally, on photographs of the inscription itself.

Both the poetic form and the inscriptional layout can be described and analyzed verbally. This description and analysis is itself an interpretative procedure, and therefore one that must be justified verbally. Print publications with lengthy verbal commentary therefore are and will remain the natural forum to study the inscriptions. But the results of such research must also be communicated in more immediate ways and disseminated to a broader public in the field and outside it. Beyond verbal commentary (of which the text of the present article is a condensate), a digital visualization (such as suggested by the figures of the present article) is desired for objects as complex as inscriptions. Through visualization and navigation, it can present poetic structure and inscriptional layout in a more intuitively synthetic and interactively experienceable way. Moreover, dynamic visual presentations can stimulate the imagination, and thereby raise further questions on form and meaning of both individual inscriptions and groups of inscriptions.

Acknowledgments

Research for the present paper was carried out in the project "Monumental Discourse in 3d Millennium Egypt" (Université de Genève & Fonds national Suisse, SNSF Professorship PP00P1 163767). I thank Andréas Stauder for reading through my English.

References

Stauder-Porchet, J. 2015. "Hezi's Autobiographical Inscription: Philological Study and Interpretation." *Zeitschrift für Ägyptische Sprache und Altertumskunde* 142.2: 191–204.

Stauder-Porchet, J. 2020a–b. "Harkhuf's Inscriptions. I. Genres, Compositional Schemes, Rhetoric." *Zeitschrift Für Ägyptische Sprache Und Altertumskunde* 147.1: 57–91; "Harkhuf's Inscriptions II. The Inscribed Façade." *Zeitschrift für Ägyptische Sprache und Altertumskunde* 147.2: 197–222.

Stauder-Porchet, J. 2021. "Inscriptional Layout in Continuous Texts of the Old Kingdom." *Bulletin de l'Institut français d'archéologie orientale* 121: 441–474.

CHAPTER 20

Puzzling Tombs: Virtual Reconstruction of the Middle Kingdom Elite Necropolis at Dayr al-Barsha (Middle Egypt)

Toon Sykora, Roberto de Lima, Marleen De Meyer, Maarten Vergauwen and Harco Willems

Abstract

With the interdisciplinary project "Puzzling Tombs,"[1] a team of Egyptologists and engineers of KU Leuven aims to document and reconstruct the architecture and iconography of the Middle Kingdom elite cemetery at Dayr al-Barsha in a virtual environment. Such a tool allows the manipulation of the thousands of tomb fragments, and virtually solve the puzzle they represent. Structured light technology is used to accurately record the fragments, while the remaining standing architecture is documented by terrestrial laser scanning. The resulting 3D meshes are processed on a game engine which allows for coping with the dense polygonal structures that compose the digital models. In combination with this virtual reconstruction, digital epigraphy following the Chicago House method is carried out in the well-known tomb of the Twelfth Dynasty governor Djehutihotep. This detailed documentation will be integrated as an additional layer in the virtual environment.

Keywords

Digital Epigraphy – Chicago House method – laser scanning – digital modeling – virtual environment – Middle Kingdom – Djehutihotep – Dayr al-Barsha – structured-light scanning – game-engine

---

1 The research presented here features within the project 'Puzzling Tombs: Virtual reconstruction of the Middle Kingdom elite necropolis at Dayr al-Barsha (Middle Egypt)' (nr. 3H170337), funded by the KU Leuven Special Research Fund.

1 Introduction

At the outset of the Middle Kingdom, Ahanakht I, governor of the Hare Nome, selected a panoramic plateau on the desert hill north of the Wadi Nakhla to serve as his final resting place. His successors followed his example by building their own decorated funerary chapels with underground burial apartments nearby (Figure 20.1). These monumental governors' tombs were surrounded by the more modest graves of their chief officials, who sometimes had decorated funerary chapels of their own,[2] or were included into the complex of their lord.[3] Together, they make up the Middle Kingdom elite necropolis of Dayr al-Barsha,[4] with an uninterrupted sequence of mid-Eleventh to late Twelfth Dynasty burials. The last dateable tomb on the plateau is the funerary chapel of Djehutihotep, which was constructed and decorated during the reign of Senwosret III.[5] Although it is the largest and most intricately decorated monument at the site, the entire area seems to have been abandoned as a burial site soon after its completion.[6]

Not long after the cemetery fell into disuse, the first threats to its preservation started to emerge. The high quality of limestone, which had initially attracted the Middle Kingdom governors to the site, became a renewed subject of interest during later phases of pharaonic history. Several gallery quarries were excavated in the rock cliffs surrounding the Wadi Nakhla, with an early New Kingdom quarry[7] opening up only a few meters to the east of the tomb of Nehri II (Figure 20.1). An even more extensive quarry complex was dug out directly behind the decorated funerary chapels of the Middle Kingdom governors, and smaller scale limestone extractions appear throughout the area. In

2 For examples, see Willems et al. 2007.

3 One such group of subsidiary burials consists of five funerary shafts in front of the forecourt of the tomb of Djehutihotep (Willems 1988, 75–77; Pommerening and Willems 2021).

4 Zone 2 in the KU Leuven numbering system. Zone 1, which consists of two Middle Kingdom burial shafts on the hill summit, can also be associated with the cemetery. For an overview of the archaeological zones in Dayr al-Barsha, see Willems et al. 2004, 238–240.

5 His *cartouche* and *serekh* are represented on the facade of the outer chapel. While the royal names of Amenemhat II and Senwosret II also occur in this chapel, they are mentioned in reference to earlier stages in the governor's life (Newberry [1894], 6 and Pl. V).

6 The suggestion that Djehutihotep would have been followed in office by a Djehutinakht (VIII; Brovarski 1981, 25) is difficult to prove, as the coffin which would be the only evidence for the existence of such a governor can easily be assigned to another owner, perhaps Djehutihotep himself (Willems 2021, 477–479).

7 A stela at the entrance mentions Thutmose III as its benefactor. For a discussion of this stela and its content, see Luft 2010, 333–374.

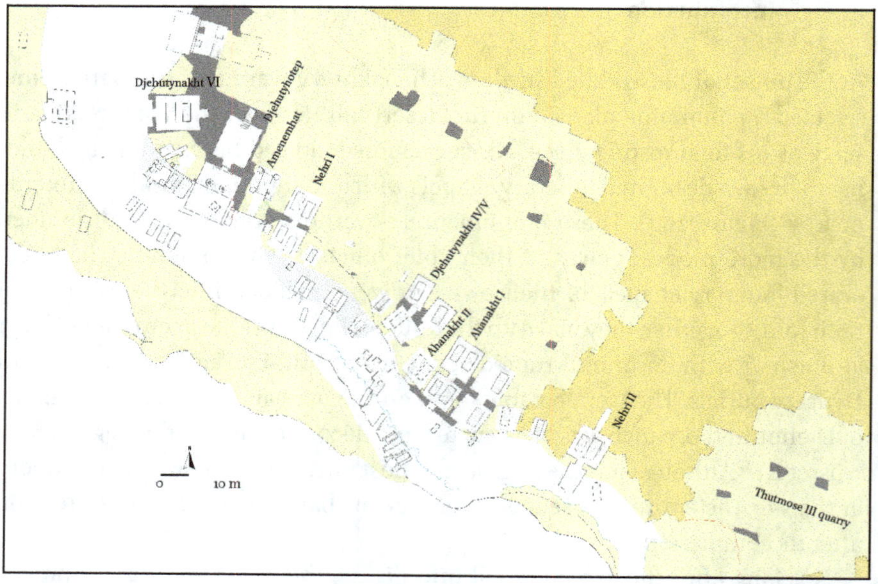

FIGURE 20.1 Plan of the Middle Kingdom nomarchal cemetery of Dayr al-Barsha (zone 2). Still visible portions of the original rock surface are indicated in dark grey and dark yellow; quarried, collapsed and inaccessible parts of the Middle Kingdom site are indicated in light grey and light yellow.

several cases, quarrymen cut straight through tomb floors or through decorated walls and ceilings.[8] Apart from the immediate destructive effect this inflicted on the rock-cut chapels, it greatly destabilized the overlying geological layer, causing it to crack and partially slide forward during a subsequent collapse. As a result, many of the superstructures of the cemetery partially collapsed, leaving only the tomb of Djehutihotep largely unharmed, with the exception of its east wall and outer chamber ceiling.

The fragmented monumental necropolis became a region of interest again during the Coptic era. Christian hermits and associates of the local monastery repurposed the still-accessible quarries and tomb chapels as dwelling places and places of worship.[9] The occupants made several adjustments to the architecture, cutting out separation walls, adding holes for doorposts and other such amenities. They also painted a large number of red Coptic crosses on the site, in many cases overlying original pharaonic decoration (Figure 20.2). This was most visibly done in the tomb of Djehutihotep, where the painting of hundreds

8 Evidence of this can clearly be seen in the tombs of Nehri I & II and Ahanakht I & II.

9 Clédat 1902, 66–67; van Loon 2016, 19–20.

FIGURE 20.2 Two Coptic crosses overlying the image of an official of Djehutihotep and his hieroglyphic legend (inner chamber, north wall)
MARLEEN DE MEYER

of Coptic crosses and the defacement of several human figures seems to have transformed the funerary chapel into a Coptic church.[10]

The "rediscovery"[11] of the pharaonic cemetery by early 19th century European travelers brought along a third wave of threats to its survival. The high quality of the preserved decoration in paint and relief made the tombs, especially the one of Djehutihotep, an obvious target for antiquities dealers. Already during the first decades after its rediscovery, clear traces of such looting activities became apparent, with "holes having been picked in the walls, and a considerable part of the rest nearly obliterated by the rain getting in".[12] The

10 Michael Johann Wansleben mentions this designation in his seventeenth century travel account of the site (Wansleben 1678, 238).

11 Although it is traditionally claimed (Newberry [1894], 3) that the first European travelers to discover the site were Charles Irby and James Mangles (Irby and Mangles 1823, 164–165), the site was already described a century before by Claude Sicard (Sicard 1717, 217–223), and even earlier by aforementioned Wansleben in 1672 (Wansleben 1678, 396–401). See also Berman et al. 2009; De Meyer and Willems 2016–2017, 37–44. These earliest descriptions of the tomb have remained largely unnoticed, however, and it was not until the nineteenth century rediscovery that the tombs of Dayr al-Barsha gained renown among western travellers.

12 Joseph Bonomi, in a letter dated July 28, 1833 (courtesy of the British Library; ADD 29859 ff. 30). He remarks that the tomb was in much better shape when he first saw it. He then follows up this comment with "by dint of scraping and sponging we have succeeded in getting the principal part of the subject of greatest interest," which cannot have improved the preservation of the already damaged decoration.

next documented major destructive event happened in late 1889 or early 1890, when large portions of the decoration of the funerary chapel were cut out to be sold on the antiquities market. It was at this moment that the famous colossus scene, which until that point in time had remained practically intact, was severely damaged.[13]

All these events have left the monumental site in a dilapidated state, with many tomb chapels and shafts inaccessible. Because of this, early archaeologists on the site resorted to extreme measures to gain access to them—on occasion even using dynamite—only further aggravating the problem.[14] Mountains of collapsed tomb debris were moved over and over again, scattering the remnants of single tombs across the plateau and its adjacent rock slopes. The most beautiful decorated wall fragments were selected and shipped to museums in Egypt and the western world.[15]

Although the rock-cut tombs have clearly been reshaped by these events (Figure 20.3), much of the information they contained can still be accessed. The many thousands of decorated wall fragments and *in situ* remains of architecture complement each other and allow us to perceive the outlines of what is now irretrievably lost. With this in mind, we launched a project in 2017 which aims to virtually document and reconstruct the remnants of pharaonic architecture in the area. The limited accessibility of the plateau, lack of stability of the remaining tombs, and danger of inflicting more damage to the fragile remains, rule out an on-site anastylosis. While the elite cemetery of Dayr al-Barsha is now a protected site, its fragmented remains are still under treat of vandalism and natural degradation, making it essential to document the current situation as completely as possible, so as to prevent any further loss of information. This documentation process is threefold: first, the standing architecture at the site is recorded in 3D; second the individual decorated wall fragments are scanned in 3D to allow for a virtual search of matches; and third, digital epigraphic drawings are made of all decorated wall surfaces.

13 Davies 1999, 29; De Meyer and Willems 2016–2017, 39 n. 31.

14 Berman et al. 2009, 96; De Meyer 2015, 107.

15 These can now be found in the British Museum in London (https://research.britishmuseum.org/research/collection_online/ [accessed on 28/04/2020] BM EA1147, 1150–1152, 7521, 71517–70, 90619, 90628, 90631, 90633 and 90662), the Egyptian Museum in Cairo (Smith 1951, 322: TR 10.4.22.2; 18.4.47.3–4 and 19.4.22.12), the Museum of Fine Arts in Boston (https://collections.mfa.org/collections [accessed on 28/04/2020] MFA 47.1659–1660 and 1972.984) and the Museo Egizio in Florence (De Meyer and Willems 2016–2017, fig. 11: N° 7596–7597).

FIGURE 20.3 The current state of the Middle Kingdom elite cemetery in Dayr al-Barsha (zone 2), as seen from the south hill. Note the debris heaps from previous excavations which line the edges of the plateau
MARLEEN DE MEYER

2 Towards a Virtual 3D Reconstruction of the Governors' Tombs

The 3D documentation of heritage is by now an established tool for researchers to enhance their study. However, documentation and registration are only the first steps in a process that should provide the researcher with new possibilities to interpret and work with the data in ways that were impossible without such type of documentation. Therefore, the 3D models of both the standing architecture as well as of loose wall fragments are integrated into a newly developed interface with a game engine that allows the reassembly of the thousands of fragments in a virtual three-dimensional space. This virtual space is further enhanced by adding layers of digitally produced line drawings of the decoration, both in terms of what is still preserved today, as well as by adding archival records[16] that provide information on parts of the monuments that no longer exist.[17]

16 These archival records generally consist of old sketches and drawings, as archival photographs of the tombs at Dayr al-Barsha are hardly preserved. Several squeezes taken by the Lepsius expedition in 1843 (preserved at the Berlin-Brandenburgische Akademie der Wissenschaften) and by Prisse d'Avennes in 1859–1860 (preserved at the Bibliothèque Nationale de France in Paris) will also be integrated into the digital model.

17 See for instance Bassier et al. 2018.

2.1 *Terrestrial Laser Scanning*

The first stage of the process consists of the 3D digitization of the site. Working in a remote and difficult to access area such as the Middle Kingdom necropolis on top of a hill poses an unusual set of challenges in this regard. Portability was an essential condition in the choice of hardware since it also needed to perform in narrow underground burial shafts and chambers. Another challenge to deal with is the occlusion caused by the many collapsed tomb fragments. Therefore, terrestrial laser scanner (TLS) data, structure from motion (SfM) and total station (TS) control points are combined to comprehensively record all the accessible architectural elements of the site in 3D.[18] A Leica BLK-360 was chosen for the terrestrial laser scanning, because its small form factor combined with its accuracy give it the flexibility and portability that are essential elements in the field. This device allows for capturing a high-resolution colorized point cloud within a range from 0.5 up to 60 m with a maximum point accuracy of 6 mm. The recorded scans were accurately registered on the software Leica Cyclone.

Given the numerous occluded areas, over 50 scans were necessary to capture all the architectural elements of the tombs above ground. In order to mitigate registration errors, the point clouds were aligned with respect to the TS control points. Although the resulting point cloud (Figure 20.4) is geometrically accurate, its texturing is subpar due to different light conditions preventing the capturing of homogeneous colors. To enhance the texture, a photogrammetry-based approach was conducted to blend the 3D points with photorealistic colors. The process of registering TLS data with photogrammetric models was performed using the software package Reality Capture. To complete the post-processing stage, high-resolution mesh models are created from the enhanced point cloud, an example of which is depicted in Figure 20.5, being a polygonal model of the outer chamber of the tomb of Djehutihotep that is made of 997,797 faces and 499,656 vertices.

2.2 *Object Scanning*

Not only the remaining standing architecture of the tombs is recorded, so are the thousands of broken decorated wall fragments that come from them. In this case, the recording poses a different set of challenges, pertaining mainly to the fragility, varying sizes and erratic shapes of the blocks. Hence, structured light technology is employed to digitize them. The EinScan Pro+ scanner was chosen

---

18 De Lima and Vergauwen 2018, 293–298.

FIGURE 20.4 Screen shot of the point cloud of Zone 2, showing both the above ground landscape as well as the subterranean shafts and burial chambers

FIGURE 20.5 Screen shot of the mesh model of Djehutihotep's outer chamber obtained by combining TLS with photogrammetry

for recording because of its versatility and specifications. It offers the opportunity to employ two different scanning modes: hand-held and fixed scan using a turning table. This is useful to digitize both small and medium-sized fragments with irregular surfaces. Moreover, it is capable of generating dense polygon meshes since its recording rate is up to 550,000 points per second and is equipped with a separate camera to capture the texture in sufficient detail. The resolution of the digitized models ranges from 300K to 1.3M triangles and point density from 0.2 to 3 mm. This high level of detail is sufficient to apply sophisticated 3D processing algorithms. However, it entails a cumbersome rendering process to visually explore the fragments' properties. Therefore, the models are loaded into a game engine for smooth visualization and interaction.

2.3 *Towards a New Integrated Interface*

To handle all recorded data, the scanned fragments are instantiated into a virtual interface. This is developed on the game engine Unity[19] to cope with the high mesh density of the 3D models. On the one hand, the virtual environment allows the user to explore high level properties of the models by adjusting rendering features such as shades, light conditions, textures and so on. On the other hand, the user is able to intuitively alter alignment of the fragments in 3D space. To facilitate this task, the decorated surface of the fragments is aligned parallel to an imaginary XY plane, restricting translation motion in the z-axis and rotation in x and y-axes.[20]

To automatically implement these steps, we assume that the decoration footprint is located on the dominant planar surface of the fragment. The robust parametrized algorithm RANSAC[21] is deployed to find the dominant and best fitting plane in the fragment's point cloud. The fitted points are projected into an XY plane. Finally, the translation vector and rotation matrix that relate the original model's position to the planar projection are computed. This transformation is applied to the fragments in the global space of the virtual interface so that the user has an isometric view of the decorated surface. This way of displaying the fragments notably reduces the complexity of inferring matching cues. Additionally, the engine extracts the contour region of the decorated planar surface in order for the user to intuitively align counterpart matching regions. Once the user proposes a rough alignment, the Iterative Closest Point[22] algorithm is used to reduce the distance between the matched fragments, thus refining the previous alignment. As a result, the developed virtual environment not only serves as a platform to visualize the recorded data but also empowers experts with tools to reassemble broken fragments.

To test the developed virtual platform, the engine was employed to reassemble the handpicked fragments depicted in Figure 20.6. In total, the input data is encompassed of 1,736,656 vertices and 3,477,954 faces. On average, the engine renders the models at 30 frames per second, ensuring a seamless user-computer

19 Nicoll and Keogh 2019, 1–21.

20 De Lima et al. 2019.

21 Chum et al. 2003.

22 Rusinkiewicz and Levoy 2001.

FIGURE 20.6 Example of a puzzle of decorated wall fragments from the tomb of Nehri I, solved through the developed virtual interface on Unity

interaction. As for the alignment tools, the plane extraction method successfully extracted both the main decorated surface and its contour for all fragments. The proposed tools allow experts to easily identify matching cues based on relief decoration traces as well as on the contour's shape. In spite of the fact that the alignment accuracy and reassembly effectiveness still rely on user expertise, the registration tools proved to be a useful asset to the virtual environment.

3 Beyond Newberry: Digital Epigraphy in the Tomb of Djehutihotep

In addition to automated digital recording of the monuments, a second type of documentation brings in the knowledge and expert eye of the Egyptologist to register and interpret their decoration by using digital epigraphy. Here the focus lies on the remains of elements in two dimensions.[23]

Of the originally at least ten tomb chapels with decoration in zone 2,[24] the funerary chapel of Djehutihotep is the only larger one which retains most of its painting and relief.[25] For this reason it was selected to serve as a pilot study for the documentation of Middle Kingdom tomb decoration. Moreover, the scenes in paint and relief contain a high level of detail (e.g. Figure 20.7), making this tomb a suitable candidate to experiment with various techniques. Its deco-

23 No historical monument contains exact plane surfaces. Even if the decorated surface approaches a plane, decoration is often carved in relief. Instead, we use the term here to include surfaces which can easily be converted into a two-dimensional plane without significant loss of information.

24 Most of these decorated tombs are listed in Griffith and Newberry [1895].

25 For a recent overview of the tomb's decorational scheme, see: Sykora 2015.

FIGURE 20.7 Detail of one of the marsh scenes in the tomb of Djehutihotep (inner chapel, north wall)
MARLEEN DE MEYER

rative program not only stands out because of its execution, but also because of its subject matter, containing unique scenes such as the well-known scene of the dragging of the colossus.

3.1 *Early Epigraphy at the Tomb and Its Limitations*

The tomb was already visited by several early Egyptologists who copied parts of its decoration.[26] A first attempt to completely document the tomb was undertaken by a mission of the Egypt Exploration Fund, led by Percy E. Newberry in 1891/1892. In less than six weeks' time, they not only finished their facsimile of the tomb of Djehutihotep, but also completed their renderings of the other known decorated tombs on the plateau. Their activities resulted in two publications: *El Bersheh* I[27] & II,[28] which are still the most complete and accurate copies of the tombs in Dayr al-Barsha, and remain the basic source for scholars.

Despite the great efforts exercised by the EEF team members, these facsimile drawings cannot be considered to be a complete record. As has been shown already, Newberry's version omits complete scenes[29] and often ignores portions of the decoration that were difficult to access. A clear example of this can be seen in the drawing of the upper part of the south wall of the inner chamber,

26 Especially notable are the descriptions and copies of Nestor l'Hôte (1840, 46–52, "Deyr Naçarah") and Lepsius (1849, Bl. 134–135).

27 Newberry [1894].

28 Griffith and Newberry [1895].

29 De Meyer and Willems 2016–2017.

where only a small part of the—broken but largely preserved—title sequence is published (Figure 20.8). The fragments above the break were never recorded. Repetitive ornamental patterns, like the ceiling decoration and *ḫkr*-frieze, only appear as abbreviated sketches, or not at all. Furthermore, the mission focused exclusively on the original Middle Kingdom decoration, leaving out any record of later additions such as ancient pictorial graffiti, painted Coptic crosses and nineteenth century travelers' inscriptions. And finally, damage patterns were not recorded, which can also be informative to highlight deliberately erased images. This is for instance the case for the eldest son of Djehutihotep, who has been erased from all scenes in which he was originally depicted.[30] In Figure 20.8a (from Newberry's Plate X), Newberry still partially suggests the presence of this figure with a single line, while on the photograph below the entire outline of Djehutihotep's eldest son is clearly visible. In other instances where traces of the erased son are still preserved on the wall, Newberry merely leaves a void in the drawing.

When studied in more detail, the EEF line drawings are also lacking in accuracy and precision. Although especially the drawings made by Howard Carter display a high degree of detail,[31] specific elements such as clothing, hair styles, feather patterns or interior details of hieroglyphs often do not match with their original painted counterparts. In other cases, no interior details at all are indicated, and figures and hieroglyphs are shown as monochrome black characters.[32] Moreover, the line drawings lack any information regarding the color scheme of the decoration. This is especially unfortunate in a tomb like that of Djehutihotep, where a broad color palette was used in masterful ways. Newberry already acknowledged this problem, which is why Carter was sent back a second season to produce watercolors of selected scenes.[33] While the twenty-three watercolors from the tombs of Djehutihotep and Djehutinakht VI form stunning renderings of the scenes,[34] they only highlight a small fraction of the complete decoration, and remain largely unpublished.

---

30 The erased son was originally present in all scenes which also depict his brothers (Newberry [1894], Pls. VIII–XI, XIII, XX and XIX).

31 For the differences in methodology between Newberry and Carter, see Reeves and Taylor 1992, 27–30; James 2001, 25–27; Cortebeeck and Willems 2015, 66–70.

32 This is the case for drawings made by Newberry; see Newberry 1894, Pls. V–VI, XIV, XVI, XXIV, XXVII–XXXI; Griffith and Newberry 1895, Pls. V, VIII–IX, XI, XIV–XVII.

33 Newberry [1894], VII.

34 Now stored in the Griffith Institute in Oxford (GI w&d 67; 72; 148–167; 214); see http://www.griffith.ox.ac.uk/archive/GI-watercolours/Deir-el-Bersha/GI_wd_Deir_el_Bersha_Djehutihotep_1.html (accessed on 14-05-2020); Cortebeeck and Willems 2015.

FIGURE 20.8A–B Detail from Newberry's [1894] Plate x (above, Figure 20.8a) as compared to a photograph of the same scene (below, photograph by Marleen De Meyer, Figure 20.8b), showing an example of an undocumented portion of the tomb chapel of Djehutihotep and the erased eldest son (second from the right)

3.2 *Documenting the Decoration Anew*

In order to address these shortcomings, we set out to document the tomb of Djehutihotep anew. In the first stage of this process a complete photographic record of the wall and ceiling decoration of its inner and outer chamber was produced. These photos were used to create orthorectified images of each wall surface in Agisoft Metashape (formerly PhotoScan), which were scaled using on-site measurements. Any areas which may have been blurred or distorted were manually corrected by aligning clearer photographs with the underlying orthophoto. In addition to the orthophotos, we use a large set of close-up images taken with a 60 mm macro lens on a DSLR, which at full size give a magnification of the decoration that is impossible to achieve with the naked eye. These close-ups allow us to really catch every detail that the ancient artist created with his brush.

The final orthophotos serve as base layers for on-site epigraphy following the digital Chicago House method developed by Krisztián Vértes.[35] In the same way that a practical and workable field methodology is essential for 3D scanning on site, this also goes for making facsimile drawings of the decoration. We use an iPad Pro (12.9 inch) embedded in a custom-made wooden drawing tablet, which provides the flexibility needed to make the base drawings in front of the tomb walls. Since the current iPad cannot handle the high-resolution large scale orthophotos directly—the largest complete wall surface in the tomb of Djehutihotep measures over 30 m^2—they are divided into smaller squares of 50 by 50 cm with an overlap of 2.5 cm on each side. Each square is then drawn separately in front of the original wall, to allow the epigrapher to change lighting conditions and his own perspective. The software used during this process is the iPad app Procreate, which allows using different layers to store information separately. In general, six layers are used:

1. Orthophoto
2. Painted decoration
3. Decoration in relief
4. Damage
5. Graffiti
6. Painted Coptic crosses

The resulting multilayer image forms the best possible documentation of these walls. This initial drawing (Figure 20.9a) is then collated by another Egyptologist who corrects errors and detects omissions, a process that is repeated until there is a consensus on the final base drawing. Once this is agreed upon, the squares are realigned, and a final 'inked' drawing is produced in Adobe Photoshop. At this stage the *trait de force* is added in case of relief decoration, a convention of varying line thicknesses to indicate raised or sunk relief, and the color variation in the decoration is rendered in different shades of grey (Figure 20.9b).

The digital facsimile that is obtained by this method is a reliable reproduction and permits the representation of the tomb of Djehutihotep throughout the different stages of its use: from an unobstructed representation of its Middle Kingdom decoration, through its Coptic reuse and into its present state.

35 Vértes 2017.

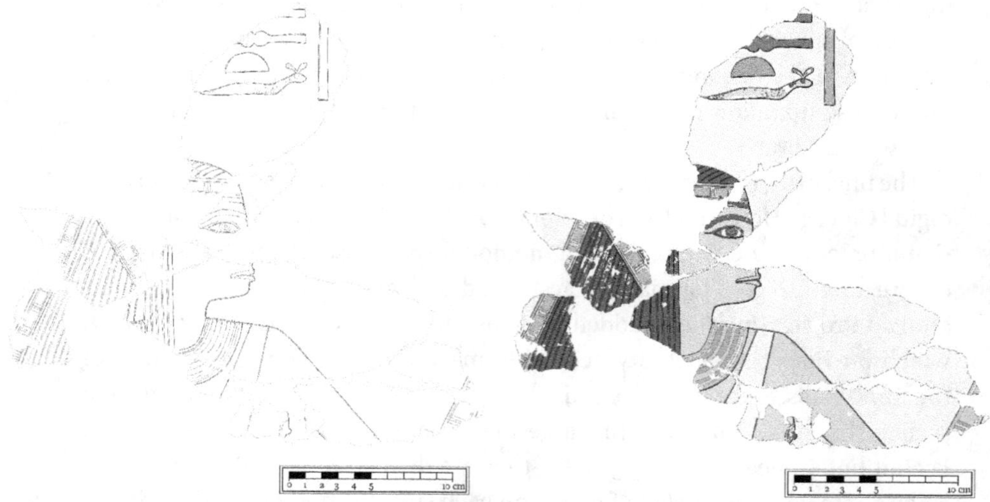

FIGURE 20.9A–B Initial line drawing (left, Figure 20.9a) and final inked drawing (right, Figure 20.9b) of a reconstructed part of the procession of ladies on the east wall in the inner chamber of Djehutihotep's funerary chapel (scale: 1:4)
Note: This reconstructed scene consists of fragments 1058/9–12, 14, 16–18 and 1290/144.

4 Work in Progress and Future Goals

Documenting a tomb, let alone an entire site, with the methods described above is a time-consuming effort and requires long term goals. While at present the 3D digitization of the standing architecture is to a large extent completed, many individual decorated wall fragments still await scanning. Of the digital epigraphic copy of Djehutihotep's tomb a first draft is complete at the time of writing, which still largely needs to be collated and corrected before being digitally inked. As a final step we plan to integrate the 2D digital drawings with the virtual 3D model of the tomb into one user interface, which will be made accessible online, allowing scholars to explore and study the tomb of Djehutihotep in the most detailed way possible.

The interactive virtual platform shows promising results in terms of its capability to handle multiple high-resolution models. Along the same lines, it has successfully served as a support tool for experts to digitally reassemble fractured fragments. However, its degree of automation is at the moment of writing still limited in the sense that the engine itself does not provide the user with automatic matching clues. Therefore, computational modules that suggest matches based on geometry are being developed, and the future goals aim towards matching and alignment algorithms to aid manual reassembly. In this regard several options are being explored, such as extraction and description in 3D of points of interest to estimate sets of correspondence points, extraction of relief-decoration to join continuous decoration lines or curves, and shape-based matching based on the decorated surface's contour. Additionally, machine learning techniques are also considered, so that the engine can progressively learn to align fragments based on the expert's operations.

Bibliography

Bassier, Maarten, Stan Vincke, Roberto de Lima, and Maarten Vergauwen. 2018. "An Overview of Innovative Heritage Deliverables Based on Remote Sensing Techniques." *Remote Sensing* 10 (10): 1607. https://doi.org/10.3390/rs10101607. Accessed 1-25-2022.

Berman, Lawrence M., Denise M. Doxey, and Harco Willems. 2009. "Discovering Deir el-Bersha." In *The Secrets of Tomb 10A: Egypt 2000 BC*, edited by Rita E. Freed, Lawrence M. Berman, Denise M. Doxey, and Nicholas S. Picardo, 91–103. Boston: Museum of Fine Arts.

Brovarski, Edward. 1981. "Ahanakht of Bersheh and the Hare Nome in the First Intermediate Period and Middle Kingdom." In *Studies in Ancient Egypt, the Aegean, and the Sudan. Essays in Honor of Dows Dunham on the Occasion of his 90th Birthday, June 1, 1980*, edited by William K. Simpson and Whitney M. Davis, 14–30. Boston: Department of Egyptian and Ancient Near Eastern Art, Museum of Fine Arts.

Chum, Ondřej, Jiří Matas, and Josef Kittler. 2003. "Locally optimized RANSAC." In *Pattern Recognition. DAGM 2003*, edited by Bernd Michaelis and Gerald Krell, 236–243 (Lecture Notes in Computer Science 2781). Berlin & Heidelberg: Springer. https://doi.org/10.1007/978-3-540-45243-0_31. Accessed 1-25-2022.

Clédat, Jean. 1902. "Notes archéologiques et philologiques." *Bulletin de l'Institut français d'archéologie orientale* 2: 41–70.

Cortebeeck, Kylie, and Harco Willems. 2015. "Newberry en zijn team / Newberry et son équipe." In *Djehoetihotep, 100 jaar opgravingen in Egypte / Djehoutyhotep, 100 ans de fouilles en Égypte*, edited by Marleen De Meyer and Kylie Cortebeeck, 61–85. Leuven: Peeters.

Davies, William V. 1999. "Djehutyhotep's Colossus Inscription and Major Brown's Photograph." In *Studies in Egyptian Antiquities. A Tribute to T.G.H. James*, edited by William V. Davies, 29–35. British Museum Occasional Paper 123. London: British Museum.

De Lima, Roberto, Toon Sykora, Marleen De Meyer, Maarten Vergauwen, and Harco Willems. 2018. "On Combining Epigraphy, TLS, Photogrammetry, and Interactive Media for Heritage Documentation: The Case Study of Djehutihotep's Tomb in Dayr al-Barsha." In *GCH 2018—Eurographics Workshop on Graphics and Cultural Heritage*, edited by Robert Sablatnig and Michael Wimmer, 225–229. Vienna: The Eurographics Association. https://doi.org/10.2312/gch.20181367. Accessed 1-25-2022.

De Lima, Roberto, and Maarten Vergauwen. 2018. "From TLS Recoding to VR Environment for Documentation of the Governor's Tombs in Dayr al-Barsha, Egypt." In *2018 IEEE International Symposium on Mixed and Augmented Reality Adjunct* (*ISMAR-Adjunct*), 293–298. Munich. DOI: 10.1109/ISMAR-Adjunct.2018.00089, https://ieeexplore.ieee.org/document/8699291. Accessed 1-25-2022.

De Lima, Roberto, Stan Vincke, Maarten Bassier, Lukas Mattheuwsen, Jens Derdeale, and Maarten Vergauwen. 2019. "Puzzling Engine: A Digital Platform to Aid the Reassembling of Fractured Fragments." *International Archives of the Photogrammetry, Remote Sensing & Spatial Information Sciences* XLII-2/W15: 563–570. https://doi.org/10.5194/isprs-archives-XLII-2-W15-563-2019. Accessed 1-25-2022.

De Meyer, Marleen. 2015. "De Campagne van George Andrew Reisner / La campagne de George Andrew Reisner." In *Djehoetihotep, 100 jaar opgravingen in Egypte / Djehoutyhotep, 100 ans de fouilles en Égypte*, edited by Marleen De Meyer and Kylie Cortebeeck, 101–128. Leuven: Peeters.

De Meyer, Marleen, and Kylie Cortebeeck, eds. 2015. *Djehoetihotep, 100 jaar opgravingen in Egypte / Djehoutyhotep, 100 ans de fouilles en Égypte*. Leuven: Peeters.

De Meyer, Marleen, and Harco Willems. 2016–2017. "The Regional Supply Chain of Djehutihotep's Ka-Chapel in Tjerty." In *Sésostris III et la fin du Moyen Empire: Actes du colloque des 12–13 décembre 2014 Louvre-Lens et Palais des Beaux-Arts de Lille*, edited by Guillemette Andreu-Lanoë and Fleur Morfoisse, 33–56. Cahier de Recherches de l'Institut de Papyrologie et d'Égyptologie de Lille 31. Lille: Université de Lille.

Griffith, Francis L., and Percy E. Newberry. 1895. *El Bersheh II*. Archaeological Survey of Egypt 4. London: Egypt Exploration Fund.

Irby, Charles L., and James Mangles. 1823. *Travels in Egypt, Nubia, Syria and Asia Minor during the years 1817–1818*. London: T. White and co.

James, Thomas Garnet Henry. 2001. *Howard Carter: The Path to Tutankhamun*. Revised 2nd ed. London: Taurus Parke.

Lepsius, Karl Richard. 1849. *Denkmäler aus Aegypten und Aethiopien, Text Volume II: Mittelaegypten mit dem Faijum*. Berlin: Nicolai.

L'Hôte, Nestor. 1840. *Lettres écrites d'Égypte, en 1838 et 1839: contenant des observations sur divers monuments égyptiens nouvellement explorés et dessinés*. Paris: Firmin Didot Frères.

Luft, Ulrich. 2010. "Die Stele des *Sn-nfr* in Deir el-Bersha und ihr Verhältnis zur Chronologie des Neuen Reiches." In *Millions of Jubilees, Studies in Honor of David P. Silverman*, Vol. I, edited by Zahi Hawass and Jennifer Houser Wegner, 333–374. Cahiers supplémentaires des Annales du Service des Antiquités de l'Égypte 39. Cairo: Supreme Council of Antiquities.

Newberry, Percy E. 1894. *El Bersheh I: The Tomb of Tehuti-hetep*. Archaeological Survey of Egypt 3. London: Egypt Exploration Fund.

Nicoll, Benjamin, and Brendan Keogh. 2019. *The Unity Game Engine and the Circuits of Cultural Software*. s.l.: Palgrave Pivot.

Pommerening, Tanja, and Harco Willems. 2021. “Unravelling Daressy’s Excavations of the Five Shafts in Front of the Tomb of Djehutihotep at Dayr al-Barshā” In *Remove that Pyramid! Studies on the Archaeology and History of Predynastic and Pharaonic Egypt in Honour of Stan Hendrickx*, edited by Wouter Claes, Marleen De Meyer, Merel Eyckerman, and Dirk Huyge, 871–898. Orientalia Lovaniensia Analecta 305. Leuven: Peeters.

Reeves, Carl Nicholas and John H. Taylor. 1992. *Howard Carter Before Tutankhamun*. London: British Museum Press.

Rusinkiewicz, Szymon and Marc Levoy. 2001. “Efficient variants of the ICP algorithm.” In *Proceedings Third International Conference on 3-D Digital Imaging and Modeling*, 145–152. Quebec: IEEE. doi: 10.1109/IM.2001.924423.

Sicard, Claude. 1717. *Nouveaux mémoires des missions de la compagnie de Jésus dans le Levant. Tome II*. Paris: Nicolas Le Clerc.

Smith, William Stevenson. 1951. “Paintings of the Egyptian Middle Kingdom at Bersheh.” *American Journal of Archaeology* 55 (4): 321–332.

Sykora, Toon. 2015. “Djehoetihotep, de laatste nomarch van de Hazengouw / Djehoutihotep, le dernier nomarque du Nome du Lièvre.” In *Djehoetihotep, 100 jaar opgravingen in Egypte / Djehoutyhotep, 100 ans de fouilles en Égypte*, edited by Marleen De Meyer and Kylie Cortebeeck, 22–38. Leuven: Peeters.

Van Loon, Gertrud J.M. 2016. “Établissements chrétiens au Spéos Artémidos.” In *Études Coptes XIV. Seizième journée d’études (Genève 19–21 juin 2013)*, edited by Anne Boud’hors and Catherine Louis, 13–30. Cahiers de la Bibliothèque Copte 21. Paris: de Boccard.

Vértes, Krisztián. 2017. *Digital Epigraphy*. Second Edition. Chicago: The Epigraphic Survey of the Oriental Institute, University of Chicago. Accessed 1-25-2022.

Wansleben, Johann Michael. 1678. *The Present State of Egypt: Or, a New Relation of a Late Voyage into that Kingdom: Performed in the Years 1672 and 1673, Wherein you have an exact and true Account of many Rare and Wonderful Particulars of that Ancient Kingdom, English ed. by M.D.D.B.* London: John Starkey.

Willems, Harco. 1983–1984. “The Nomarchs of the Hare Nome and Early Middle Kingdom History.” *Jaarbericht van het Vooraziatisch-egyptisch Genootschap Ex Oriente Lux* 28: 80–102.

Willems, Harco. 1988. *Chests of Life. A Study of the Typology and Conceptual Development of Middle Kingdom Standard Class Coffins*. Mededelingen en Verhandelingen Ex Oriente Lux 25. Leiden: Ex Oriente Lux.

Willems, Harco. 2021. “Djoser’s Complex as a Source of Inspiration for the Decoration of Private Coffins in the Middle Kingdom.” In *Middle Kingdom Palace Culture and Its Echoes in the Provinces: Regional Perspectives and Realities*, edited by Alejandro

Jiménez-Serrano and Antonio Morales, 462–481. Centre de Recherches d'Histoire et de Philologie de la IVe Section de l'École Pratique des Hautes Études, II. Hautes Études Orientales 12. Leiden: Brill.

Willems, Harco, Marleen De Meyer, David Depraetere, Christoph Peeters, Stan Hendrickx, Tomasz Herbich, Dietrich Klemm, Rosemarie Klemm, Lies Op De Beeck, and Mark Depauw. 2004. "Preliminary Report of the 2002 Campaign of the Belgian Mission to Deir al-Barsha." *Mitteilungen des Deutschen Archäologischen Instituts, Abteilung Kairo* 60: 237–283.

Willems, Harco, Lies Op de Beeck, Troy Leiland Sagrillo, René Van Walsem, and Stefanie Vereecken. 2007. *Dayr al-Barshā 1: The Rock Tombs of Djehutinakht (No. 17K74/1), Khnumnakht (No. 17K74/2), and Iha (No. 17K74/3), with an Essay on the History and Nature of Nomarchal Rule in the Early Middle Kingdom.* Orientalia Lovaniensia Analecta 155. Leuven: Peeters.

Online Resources

Recording Djehutihotep. Digital epigraphy in a Middle Kingdom governor's tomb at Dayr al-Barsha. Accessed on 5-5-2020.

Part 1: De Meyer, Marleen: http://www.digital-epigraphy.com/projects/recording-djehutihotep-digital-epigraphy-in-a-middle-kingdom-governors-tomb-at-dayr-al-barsha-part-1

Part 2: Sykora, Toon: http://www.digital-epigraphy.com/projects/recording-djehutihotep-digital-epigraphy-in-a-middle-kingdom-governors-tomb-at-dayr-al-barsha-part-2

Griffith Institute, Oxford. "The Middle Kingdom Tombs at Deir el-Bersha. Reconstruction of tomb wall-scenes using watercolours from the Griffith Institute Archive." Accessed on 5-5-2020. http://www.griffith.ox.ac.uk/archive/GI-watercolours/Deir-el-Bersha/GI_wd_Deir_el_Bersha.html

CHAPTER 21

Project *Croato-Aegyptica* (2002–2020)

Mladen Tomorad and Goran Zlodi

Abstract

The Croato-Aegyptica Database (CADb), which covers institutional and private collections in Croatia, is presented. This database includes not only ancient Egyptian artifacts, but also various other objects related to travelers to Egypt, Egyptian Revival and Egyptomania in our collections.

The CADb is based on the guidelines set by ICOM-CIDOC (International Committee for Documentation) and ICOM-CIPEG (International Committee for Egyptology). Each artifact is analyzed, described, and implemented in the database with their full bibliography and several images. The project is currently working on a thesaurus, maps, 3D images of several collections, and the implementation of the new collections from the Archaeological museums in Dubrovnik, Split and Zagreb and a number of minor collections that we discovered recently. After full implementation, the CADb will cover almost 5000 artifacts.

Keywords

Croatia – database – Egyptomania – collection history – history of Egyptology

1 Introduction

The basic aim of the *Croato-Aegyptica* project is to produce a database of relevant material related to the cultural influences of the Egyptian civilization in the historical territory of Croatia. Egyptian artifacts are today kept in numerous institutional and private collections which can be found in all regions of Croatia.

The CADb database (http://croato-aegyptica.starapovijest.eu/en) (Figure 21.1), accessible on the website *Stara povijest* (English: *Ancient History*), provides domestic and foreign researchers and cultural institutions insight into the richness of Croatian Egyptological material. At the same time, it offers a

 | DOI:10.1163/9789004501294_022

FIGURE 21.1 Homepage of the bilingual online catalog of the *Croato-Aegyptica database* (CADB)
HTTP://CROATO-AEGYPTICA.STARAPOVIJEST.EU/EN

quick and easy tool which enables searches through its catalog, with the data available in both Croatian and English.

The idea to launch a project in which all the collections of Ancient Egyptian material in Croatia would be researched, and through which the artifacts in those collections would then be systematically scientifically and professionally elaborated, was conceived during the second half of 2001. Beginning from October 2003, when financial support for the project commenced, and continuing until Spring 2020, all museum collections in Croatia were preliminarily processed. During this period, the majority of the material was thoroughly analyzed and processed within the project. There remains, however, considerable material which that has only been preliminarily processed and analyzed for the purpose of this project.

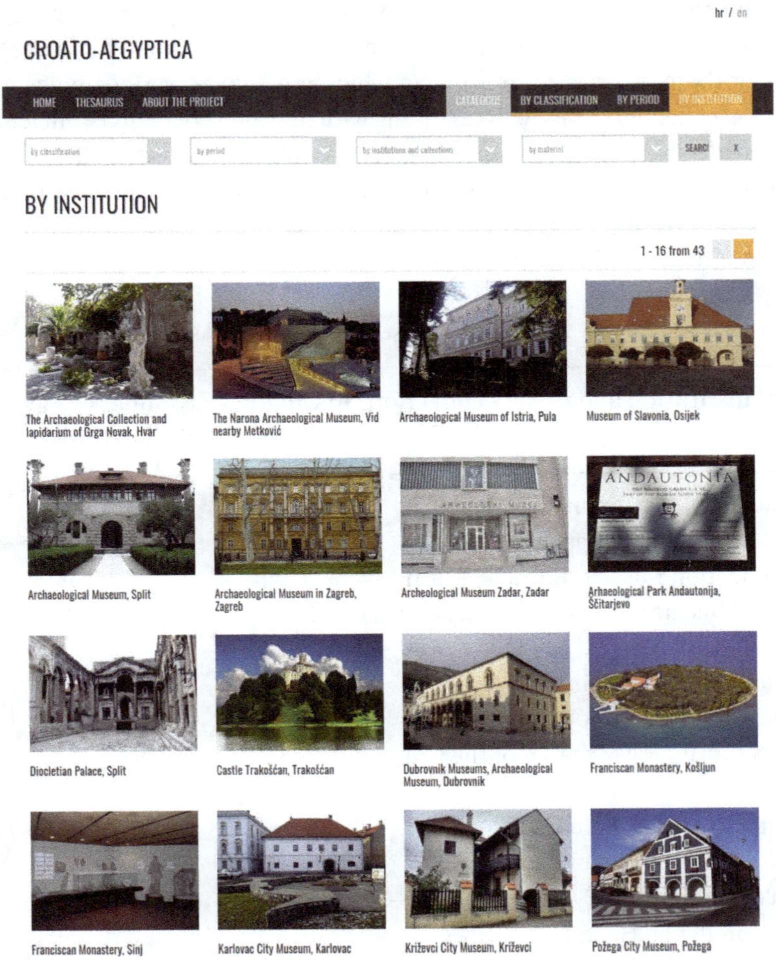

FIGURE 21.2 Access to information on institutions, collections and their objects HTTP://CROATO-AEGYPTICA.STARAPOVIJEST.EU/EN/CATALOGUE/BY%20INSTITUTION

2 Background: Egyptian Collections in Croatia

Modern Egyptian collections in Croatia were formed from numerous private collections, mostly during the 19th and 20th centuries. Today they are stored in more than 35 museums and other institutional collections in Croatia. Objects from those collections can be perused easily on the *Croato-Aegyptica* website (Figure 21.2).

According to the research conducted by M. Tomorad from 1999 to early 2019, at least 4,854 artifacts related with Egypt can be found in Croatia[1] (Table 20.1), and among them at least 4,336 artifacts which are kept in various types of institutional collections[2] (Table 20.2). Along with museum collections, ancient Egyptian artifacts can also be found in a great number of private collections scattered around the entire territory of Croatia. Their number is still uncertain, but we can assume that between 500 and 1,000 artifacts can be found in private collections[3] (Table 20.3).

TABLE 20.1 Egyptian artifacts in institutional, museum and private collections in Croatia

Museum, institution and private collections in Croatia	Number of Egyptian artifacts
Archaeological Museum in Zagreb	3279
Mimara Museum in Zagreb	505
Private collection Kovačić-Bošnjak, Kutina/Ivanić Grad	500?
Dubrovnik Museums—Archaeological Museum in Dubrovnik	228
Archaeological Museum in Split	132 (140)
Archaeological Museum of Istria in Pula	37
Maritime and History Museum of the Croatian Littoral in Rijeka	20 (30?)
Archaeological Museum in Osijek	18
Archaeological Museum in Zadar	16
Museum of Contemporary Art in Zagreb—Archaeological collection Benko Horvat	11
Monastery of St. Euphemia in Kampor at the island of Rab—Archaeological collection	10 (21?)
Franciscan Monastery at Košljun—Archaeological collection	10
City Museum Varaždin—Archaeological collection	9
Museum of Arts and Crafts in Zagreb	10
Diocletian Palace in Split	7
Franciscan Monastery in Sinj—Franciscan Monastery	6
Museum of Croatian History	6

1 Tomorad 2019b, 172–173.

2 Tomorad 2019b, 174.

3 Tomorad 2019b, 182.

TABLE 20.1 Egyptian artifacts in various collections in Croatia (*cont.*)

Museum, institution and private collections in Croatia	Number of Egyptian artifacts
Private collection Marović, Split	6
Cemetery Mirogoj, Zagreb	4
Trakošćan Castle, Trakošćan	4
City Museum Šibenik	3
Croatian Academy of Sciences and Arts—Department of Croatian Literature, Theatre and Music in Zagreb	3
Regional Museum Varaždinske Toplice—*Aquae Iasae* Archaeological park	3
City Museum Split	2
Archaeological Park *Andautonia*, Ščitarjevo	2
ZOO, Zagreb	2
Archaeological Museum *Narona*, Vid near Metković	1
Archaeological collection and Lapidarium "Grga Novak", Center for Preservation of Cultural Heritage, Hvar	1
City Museum "Dr. Zlatko Dragutin Tudjina," Pregrada	1
City Museum in Senj—Lapidarium	1
City Museum Karlovac	1
City Museum Koprivnica	1
City Museum Križevci	1
City Museum Požega	1
City Museum Zagreb	1
Croatian State Archive, Zagreb	1
Family House Glimb, Nin	1
Lapidarium Novigrad (Istria)	1
Lapidarium Veliki Brijun	1
Museo del territorio Parentino, Poreč	1
Museum of Croatian Archaeological Monuments, Split	1
Park Maksimir, Zagreb	1
Perish Dorm, Kruševo	1
Private collection Domančić, Split	1
Private collection Kaštelančić	1
Private collection Lik, Kazale	1
Regional Museum Imotski—Archaeological collection	1
Total number of artifacts	4854–4873?

TABLE 20.2 Egyptian artifacts in various types of institutional collections in Croatia

Institutional collection	**Number of Egyptian artifacts**
Archaeological Museum in Zagreb	3279
Mimara Museum in Zagreb	505
Dubrovnik Museums—Archaeological Museum in Dubrovnik	228
Archaeological Museum in Split	132 (140)
Archaeological Museum of Istria in Pula	37
Maritime and History Museum of the Croatian Littoral in Rijeka	20 (30?)
Archaeological Museum in Osijek	18
Archaeological Museum in Zadar	16
Museum of Contemporary Art in Zagreb—Archaeological collection Benko Horvat	11
Monastery of St. Euphemia in Kampor at the island of Rab—Archaeological collection	10 (21?)
Franciscan Monastery at Košljun—Archaeological collection	10
Museum of Arts and Crafts in Zagreb	10
City Museum Varaždin—Archaeological collection	9
Diocletian Palace in Split	7
Franciscan Monastery in Sinj—Franciscan Monastery	6
Museum of Croatian History	6
Trakošćan Castle, Trakošćan	4
City Museum Šibenik	3
Croatian Academy of Sciences and Arts—Department for Croatian Literature, Theatre and Music in Zagreb	3
Regional Museum Varaždinske Toplice—*Aquae Iasae* Archaeological park	3
City Museum Split	2
Andautonia Archaeological Park, Ščitarjevo	2
Narona Archaeological Museum, Vid near Metković	1
Archaeological collection and Lapidarium "Grga Novak," Center for Preservation of Cultural Heritage, Hvar	1
City Museum "Dr. Zlatko Dragutin Tudjina," Pregrada	1
City Museum in Senj—Lapidarium	1
City Museum Karlovac	1

TABLE 20.2 Egyptian artifacts in various types of institutional collections in Croatia (*cont.*)

Institutional collection	Number of Egyptian artifacts
City Museum Koprivnica	1
City Museum Križevci	1
City Museum Požega	1
City Museum Zagreb	1
Croatian State Archive, Zagreb	1
Lapidarium Novigrad (Istria)	1
Lapidarium Veliki Brijun	1
Museo del territorio Parentino, Poreč	1
Museum of Croatian Archaeological Monuments, Split	1
Perish Dorm, Kruševo	1
Regional Museum Imotski—Archaeological Collection	1
Total number of artifacts	4,337 (4,366)

TABLE 20.3 Registered and scientifically analyzed private collections in Croatia

Private collection	Number of artifacts
Private collection Kovačić-Bošnjak, Kutina/Ivanić Grad	500?
Private collection Marović, Split	6
Family House Glimb, Nin	1
Private collection Domančić, Split	1
Private collection Kaštelančić	1
Private collection Lik, Kazale	1
Total number of artifacts	510?

Chronologically, Egyptian artifacts in Croatia can be dated from the oldest, a Predynastic cup from the Badarian culture,[4] to modern 19th and 20th century products of Egyptomania and Egyptianizing public monuments (Tables 20.1

4 Varaždin: *The Varaždin City Museum*, inv. no. AO-6521.

and 20.4). The ancient Egyptian artifacts can be dated from the Predynastic period of the previously mentioned Badari culture to the Coptic artifacts from the 5th to the 7th centuries CE.[5] In Croatia, various Egyptian revival artifacts and monuments can also be found.[6] The majority of them are in Zagreb and the surrounding area. In the Mirogoj cemetery, four modern obelisks can be found. At the beginning of the 20th century, they stood in front of the National Theater, but just before the First World War they were transferred to Mirogoj.[7] Sadly, all four of them were badly damaged during devastating earthquakes in March 2020. In the Zagreb Zoological Park, two modern sculptures of ancient Egyptian pharaohs stood. They were most likely made by famous Croatian artist Jozo Turkalj in the 1920s.[8] In the Maksimir Park (Zagreb), another obelisk was erected in 1843.[9]

TABLE 20.4 *In situ* Egyptomania in Zagreb

***In situ* Egyptomania**	**Number of artifacts**
Cemetery Mirogoj, Zagreb	4
ZOO, Zagreb	2
Maksimir Park, Zagreb	1
Total number of artifacts	7

Most of the pharaonic Egyptian artifacts currently stored in Croatian museum and private collections are small objects. The collections are dominated by scarabs, amulets, *shabtis*, tablets of various materials, jewelry, fertility symbols, statuettes of various deities, but also footwear, diverse types of vessels, canopic jars, animal depictions, and male and female statues. Among the numerous artifacts, it is also possible to identify a larger collection of statuettes of various gods and goddesses, wooden and stone stelas with inscriptions and painted depictions of mostly funerary character, inscriptions on papyrus or linen wrappings, and the Book of the Dead.[10] Sarcophagi for the burial of

5 Zagreb: *The Mimara Museum*, inv. nos. 246, 340, 349, 351, 355–359, 361, 1719 and 3582.
6 Tomorad 2019d.
7 Tomorad 2019d, 246; Bagarić 2019, 250–251.
8 Tomorad 2019d, 246.
9 Tomorad 2019d, 245–246.
10 Tomorad 2019b, 171.

human and animal remains, as well as mummies and mummy wrappings, can be found in the Archaeological Museum in Zagreb.[11]

The thirteen sphinxes or their fragments, once part of decoration of Diocletian's Palace in Split, is another group of objects. Remains of Diocletian's Split palace also include several stone pillars made from the various types of Egyptian stone and one pharaonic statue from the Ptolemaic period. The collection of glass from the Mimara Museum in Zagreb stores many artifacts of ancient Egyptian production crafted during the Pharaonic, Hellenistic, and Roman periods. This is the only collection of glass artifacts from Egypt in our collections.

The only collection of Coptic objects is in the Mimara Museum in Zagreb. Croatian collections also include numerous coins of Ptolemaic kings and the Roman emperors minted in Alexandria, along with Roman coins with depictions of ancient Egyptian deities.[12] Their largest number (1,014 coins) are kept in the Numismatic Department of the Archaeological Museum in Zagreb.[13] The other numismatic collections with these types of coins are mostly scattered in various archaeological and city museums in Croatia (e.g. Pula, Rijeka, Split, Varaždin, Vinkovci, etc.).[14] These coins are mostly still unpublished, and so their number is still uncertain.

In museum institutions in Croatia and *in situ*, objects related to the modern Egyptomania can be found. Most such objects are mainly various vessels, candle sticks, clocks, furniture and the modern imitations of sphinx and grave tombs which were created during the 19th and 20th centuries.[15] Their greatest number is kept in the Museum of Arts and Crafts in Zagreb,[16] but similar objects can be found in Varaždin City Museum and Trakoščan castle. Mostly likely more examples of Egyptomania are to be found in other institutional collections in Croatia.

The last group of artifacts is related to archival documents, paintings, photographs, and diaries related to Croatian travelers to Egypt mostly during the 19th and early 20th centuries, which are scattered in various institutions all over Croatia.[17]

11 Uranić 2009; Tomorad 2019b, 176–177.
12 Tomorad 2019b, 171.
13 Mirnik 2016.
14 Tomorad 2019b, 171.
15 Tomorad 2019b, 171, 174.
16 Brdar Mustapić 2019; Lovrić Plantić 2019.
17 Tomorad 2019b, 174.

The ancient Egyptian artifacts in Croatian collections can be divided into two main groups, according to their origin. 1) Artifacts of ancient Egyptian origin, obtained in Egypt from various antiquarians, and 19th and early 20th century travelers. Artifacts of this type were mostly purchased for the private collections (e.g. the Koller collection in 1868) and later bought by or donated to the museums.[18] 2) Artifacts discovered in the Croatian territory during archaeological excavations of the pre-Roman or the Roman settlements in Croatia mainly, during the 19th or 20th centuries.[19] This type of material is mostly related to the diffusion of ancient Egyptian or the other syncretistic cults developed during the Greco-Roman period. The very popular cults of Isis, Serapis, Osiris and the other Egyptian divinities arrived in Croatian territory mainly through intermediaries (sailors, merchants, soldiers), spreading new religious beliefs into regions of ancient Istria, and the roman provinces of Dalmatia and Pannonia (today modern Croatia).[20] Such artifacts have been acquired by Croatian museums or private collectors mostly through archaeological excavations, or in some cases by purchase.[21] The material from this second category has posed problems, especially in the second half of the 19th and the early 20th centuries when they were often considered to be fakes. Fortunately, it has been recently established that such artifacts originated from the Roman antiquity and in present time these antiquities are considered genuine.[22]

The ancient Egyptian collections have only recently been fully systemized and mostly analyzed. During the last fifteen years scholars (M. Tomorad, I. Uranić, I. Vilogorac Brčić, P. Selem, V. Jurkić-Girardi, I. Mirnik) who have been working on *Croato-Aegyptica* project under our guidelines and supervision managed to finally publish most of the collection in the online database *Croato-Aegyptica* (http://croato-aegyptica.starapovijest.eu/en). Before 1999, when M. Tomorad started to study Egyptian collections in Croatia, most of the collections were not systemized, scientifically analyzed and were completely unknown to general public and scholars. Each artifact was catalogued according ICOM—CIDOC Guidelines for Museum Object Information, CIDOC Core Data Standard for Archaeological Objects and various standard categorizations of Egyptological artifacts (e.g. Petrie Museum of Egyptian Archaeology, Egyptian Treasures in Europe project etc.). The final categorization of the artifacts from Croatian collections were based on the M. Tomorad Ph.D. thesis (Univer-

18 Tomorad 2019a.
19 Tomorad 2019a, 166–169.
20 Tomorad 2018.
21 Tomorad 2018; Tomorad 2019a, 166–169.
22 Tomorad 2015; Tomorad 2018; Tomorad 2019c, 102.

sity of Zagreb, 2006)[23] which divided them into eight main categories (sculptures, religious and funerary equipment, artifacts of daily life, inscriptions and inscribed artifacts, molds and casts, coinage, artifacts from glass and glass paste, and modern Egyptomania).

Before the *Croato-Aegyptica* project, the only Croatian institutions which had published their Egyptian material in exhibition catalogs and scientific periodicals were the Archaeological Museum in Zagreb and, partly, the Archaeological Museum in Split.[24] Since the end of 1980s the collections from the Mimara Museum in Zagreb were published only partly in various museum catalogs and articles. The collections in the Archaeological Museums in Dubrovnik and Pula have only recently been partly analyzed and published. The minor collections in museums and other institutions have been mostly examined during the last twenty years. The exceptions are the collections of the Franciscan monastery in Sinj, the Archeological Museum in Zadar, the Archaeological Museum in Osijek and the Museum of the city of Varaždin which were previously known and partly published. During the last twenty years they have been completely analyzed and published through the work on the *Croato-Aegyptica* project.

The private collections are still mostly unknown and rarely systemized and published. During the exhibition "Egypt in Croatia" (Zagreb, September–October 2018) several unnamed private collectors contacted the authors of this chapter and asked about the value of various, mostly smaller, ancient Egyptian antiquities. Sadly, at this moment it is very hard to determine how many objects can be found in such collections, their origin, authenticity, and cultural and historical value, mostly because their owners have not registered their collections.

3 The History and Development of the Project

The first phase of the project's development started in late 2001 when the project was presented in the paper "Egyptian Antiquities in Croatia: Computerized Processing of Sources and Literature" at the international scientific conference "Historical Research, Study of History and Computerization" (December 10–12, 2001) at the Faculty of Humanities and Social Sciences, University of Zagreb, Zagreb, Croatia.[25] During the next two years (2002–2003) the first project team was formed, consisting of M. Tomorad, I. Uranić, G. Zlodi, H. Gra-

23 Tomorad 2006.

24 The complete list of publications related various collections in Croatia in: Tomorad 2020b, 169–266.

25 Tomorad and Zlodi 2015, 161; Tomorad 2020a, 141.

čanin, I. Malus Tomorad, and M. Gračanin.[26] During the same period, with the help of the late professor P. Selem (1936–2015), we gained support from all the museums and other institutions to start studying their collections, and for future publishing of all material online as a digital catalog.[27] After the successful application (2003) to the Ministry of Science, Education, Technology, and Sport of the Republic of Croatia, we obtained sufficient funds to start the project.[28]

Our team then started the second phase of project development. During the first half of 2004, the first project website, titled *Croato-Aegyptica Electronica*, was created. M. Gračanin designed and programmed the whole site.[29] A few hundred pages of texts written by M. Tomorad, I. Uranić, H. Gračanin, and I. Malus Tomorad were added to the site.[30]

In May 2004, the Croatian version of the web site was launched (http://infoz.ffzg.hr/cae (inactive after 2014)).[31] The English version, which was slightly different from the Croatian version, was launched at the end of June 2004.[32] The first database, entitled CAE, has been accessible online since July 2004.[33] In September 2004, the site obtained its own domain (www.croato-aegyptica.hr (inactive after 2014)) and the first version of the database was incorporated in it with a selection of 120 artifacts from the Egyptian Department of the Archaeological Museum.[34] At the same time, the CAE website became the first Croatian educational and information web portal about the history of any civilization in the world. It included information about the project, the history of Egyptian collections in Croatia, and numerous texts related to the Ancient Egyptian history, culture, and Egyptology. It also had a special news database about the most relevant current research in Egypt, exhibitions, lectures, and conferences all over the world. This database was made by the free software programs PHP and My SQL, powerful open-source programs which we used to develop our daily updated news.[35] By the end of 2004, the *Croato-Aegyptiaca* website had become very popular, and it was nom-

26 Tomorad and Zlodi 2015, 161; Tomorad 2020a, 141.
27 Tomorad and Zlodi 2015, 161.
28 Tomorad and Zlodi 2015, 161; Tomorad 2020a, 142.
29 Tomorad and Zlodi 2015, 161.
30 *Croato-Aegyptica Electronica* (www.croato-aegyptica.hr) (2004–2011); Tomorad 2020a, 142.
31 Tomorad and Zlodi 2015, 161; Tomorad 2019e, 274–275; Tomorad 2020a, 142.
32 Tomorad and Zlodi 2015, 161; Tomorad 2020a, 143.
33 Tomorad and Zlodi 2015, 161.
34 Tomorad and Zlodi 2015, 161.
35 Tomorad and Zlodi 2015, 161.

inated for the Croatian scientific award for the popularization of science for the year 2004.[36]

In 2006, the project unfortunately lost its required financial support. Soon after the project news part of the web site was hacked.[37] The introduction of the new regulations at CARNet (*Croatian Academic and Research Network*) forced us to lose our own domain (www.croato-aegyptica.hr (inactive after 2014)).[38] At the same time in early 2010 the server of the Faculty of Humanities and Social Sciences, where it was implemented, crashed; for almost a year, the web site and database were unavailable.[39] In April 2011, M. Tomorad moved to the new position of the professor of ancient history at the Department of History at the Croatian Studies in Zagreb (today the Faculty of Croatian Studies). The new position afforded him the opportunity to move the web site to the new address at the website of the Croatian Studies (http://www.hrstud.unizg.hr/sites/cae/ (inactive after 2014)).[40] This portal, however, was finally shut down during Spring 2014.[41]

In the same time, we had been working on the new web site entitled *Stara povijest*. With the financial support by the University of Zagreb, *Stara povijest* web site [www.starapovijest.eu] was designed and launched in Spring 2014.[42] The new website was also based on a new concept: it covered the whole of ancient history and archaeology. One of its parts was dedicated to the project, which in the meantime adopted the new name *Croato-Aegyptica*.[43] During 2015/2016, the completely new online digital catalog CADB was developed and prepared by G. Zlodi and M. Tomorad.[44] In that time the online catalog was mainly written in Croatian, with a selection of artifacts which were translated into English.

In early 2016, the M++ database used for CADB metadata was upgraded to enable the implementation of the Modulor++ platform. During 2016 and the first half of 2017, the first phase of data collection and digital representations from several databases and their joint presentation within *Stara povijest* web-

36 Tomorad and Zlodi 2015, 161.
37 Tomorad and Zlodi 2015, 162.
38 Tomorad and Zlodi 2015, 162.
39 Tomorad and Zlodi 2015, 162.
40 Tomorad and Zlodi 2015, 162.
41 Tomorad 2020a, 143.
42 *Stara povijest—Moderan magazin za staru povijest.* [www.starapovijest.eu]; Tomorad 2020a, 143.
43 *Stara povijest—Croato-Aegyptica.* [https://www.starapovijest.eu/category/croato-aegyptica/]
44 *Croato-Aegyptica Database.* [http://croato-aegyptica.starapovijest.eu/]; Tomorad 2020a, 143.

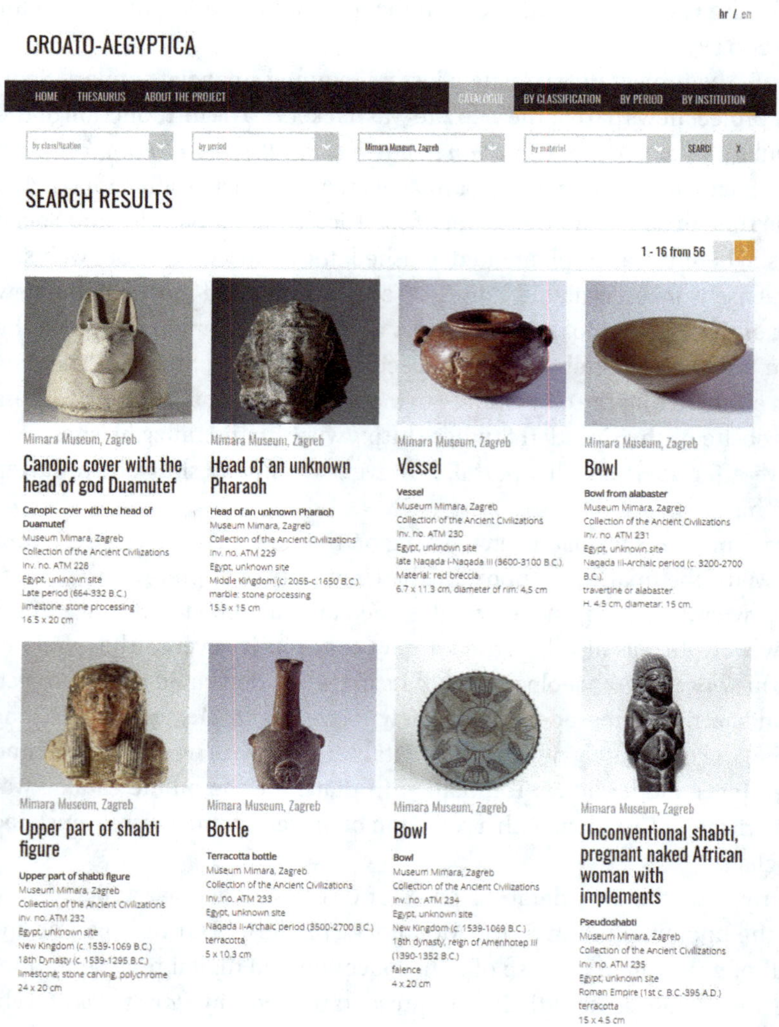

FIGURE 21.3 Display of search results in the online catalog
HTTP://CROATO-AEGYPTICA.STARAPOVIJEST.EU/EN/PRETRAGA

site was completed.[45] Between the end of 2017 and the September of 2018, most of the artifacts from Croatian collections related to ancient Egypt or Egyptomania were documented in the CADB, with almost 4000 artifacts.[46] Currently, 869 items from 43 collections are searchable through online catalog (Figure 21.3).

45 Tomorad and Zlodi 2016; Tomorad 2019.
46 Tomorad 2019e, 275.

4 Aggregation, Cataloguing and Publishing in the Modulor++ Collection Management and Publishing System

The Modulor++ platform, implemented as a web application, provides an integrated approach to key museum information system layers: it enables museum objects management and cataloguing (via access to the M++ database), museum documentation and archive management (physical and digital assets catalogued in the S++ database) and integration of bibliographic information (in K++ library information system).

For the Croato-Aegyptica project, by upgrading to the Modulor++ platform, a more advanced workflow is provided in which museum objects descriptions, depending on the agreement with the institution, can be aggregated from existing institutional databases or entered directly via the Modulor++ platform. This has created a cooperative environment that allows data to be sent and received, and if necessary, additional cataloging and linking with bibliographic information. The LIDO harvesting scheme was used to exchange metadata. LIDO is a flexible event-based metadata schema developed to support the exchange of rich and highly structured museum objects information, along with data from controlled vocabularies. LIDO is harmonized with the SPECTRUM collection management standard which ensure interoperability with various collection management systems. LIDO is also aligned with the CIDOC-CRM (Conceptual Reference Module) formal ontology for cultural heritage information, which offers the Croato-Aegyptica project the perspective of further deriving information and knowledge from existing datasets. LIDO is used in many practical projects, often for metadata exchange, between institutions' collections database and various online services, such as portals or catalogs of resources aggregated from different organizations.

The Modulor++ application database uses hierarchical thesauri for all key metadata categories. Thesauri application module enables establishing links to concepts from the Getty Thesaurus for Art and Architecture (AAT) and the Multilingual Egyptological Thesaurus, which originated in the mid-1990s. In 2006, M. Tomorad created the basic Croatian Egyptological thesaurus on which the entire catalog is based, the result of many years of work and the preparation of a database based on international experience. The terminology for the needs of the database is completely adapted to the needs of Croatian curators and Egyptologists, and the thesaurus is being constantly updated.

An important step forward occurred when the database was upgraded to a new Modulor++ system that supports the use of multilingual thesauri. Currently, all concepts in thesauri are implemented bilingually, in Croatian and English, and some concepts are already enriched with terms in French, German and Italian (the process is carried out using semantic technologies that allow access to the AAT thesaurus via SPARQL queries). The Modulor++ platform uses the Linked Open Data (LOD) approach as a key opportunity to link local micro-thesauri to standardized controlled vocabularies such as The Art & Architecture Thesaurus or to link to structured content available in Wikidata or DBpedia projects, in order to establish multilingualism and metadata interoperability at the global level. Furthermore, possibilities of linking to the important current developments of the THOT project (Thesauri and OnTology for documenting ancient Egyptian textual resources)[47] are currently being considered.

Reuse and presentation of digital content is enabled through a special module of the Modulor++ platform that allows publication of selected metadata and selected digital representations of museum objects in the online public access catalog. Since one set of museum objects in the database is not fully cataloged and not yet reviewed, only objects that meet the agreed level of cataloging, and that are accompanied with metadata in English, are selected for online public display in catalog.

In the public catalog, each item is presented with key metadata categories (Object type, Title, Origin of the item, Location, Dating, Material, Dimensions, Name of the institution and collection, Inventory number) and an extensive item description that includes data on acquisition, provenance, and context, and provides a link to similar objects in other institutions. Each object appears with a gallery of photographs, drawings, and 3D objects (Figure 21.4).

An extensive bibliography accompanies each museum item. Advanced search of the online catalog is enabled by Object type, Material, Period and Present location (museums and collections). Catalog provides browsing via interactive timeline based on a hierarchical thesaurus of periods (Figure 21.5).

Currently, 869 items from 43 collections from almost all parts of Croatia are publicly available through this online catalog. A large number of presented objects are currently not available otherwise (objects in depots, at conservation or restoration treatment), which makes this catalog a unique opportunity for researchers and broader audiences to view those objects.

47 THOT—Thesauri and OnTology for documenting ancient Egyptian textual resources. (https://thot.philo.ulg.ac.be/).

FIGURE 21.4 Display of an individual object metadata, description, image gallery and 3D model—Canopic jar of Ketjen in the form of god Imsety, GMV AO-5280, Varaždin City Museum HTTP://CROATO-AEGYPTICA.STARAPOVIJEST.EU/EN/ITEM/1461

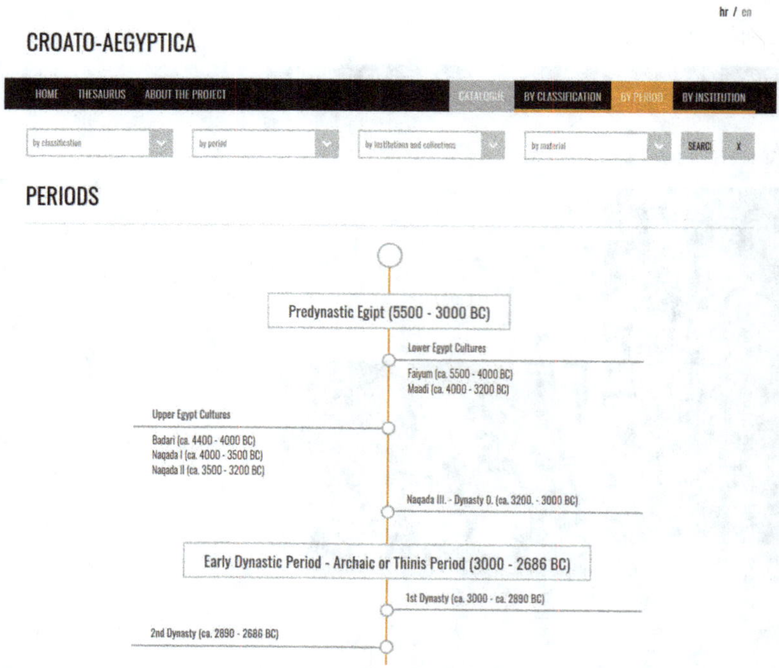

FIGURE 21.5 Interactive timeline based on a hierarchical thesaurus of periods
HTTP://CROATO-AEGYPTICA.STARAPOVIJEST.EU/EN/CATALOGUE/BY%20PERIOD

This project certainly increases the visibility of museums that take care of this valuable material—each museum object in the catalog appears with a link to the official website of the museum. To increase the visibility and accessibility of the material in the international context, it should be emphasized once again that his project pays special attention to multilingualism, which is implemented at the interface level, at the metadata level, and through translation of all available texts on website. The project intends to provide online public access to the entire catalog of Egyptian artifacts in Croatia, in as many languages as possible. The next important step of the project is to publish the data itself in structured, interlinked, and open form to enable their automated reuse.

Bibliography

Bagarić, Marina. 2019. "Egypt as Imaged by 19th and 20th-Century Zagreb: Buildings, Monuments and Street Furniture." In *Egypt in Croatia—Croatian Fascination with Ancient Egypt from Antiquity to Modern Times*, edited by Mladen Tomorad, Sanda Kočevar, Zorana Jurić Šabić, Sabina Kaštelančić, Marina Kovač, Marina Bagarić, Vanja Brdar Mustapić and Vesna Lovrić Plantić, 249–260. Oxford: Archaeopress.

Brdar Mustapić, Vanja. "Influences of Egyptian Style in Furniture Making: Examples from the Museum of Arts and Crafts, Zagreb." In *Egypt in Croatia—Croatian Fascination with Ancient Egypt from Antiquity to Modern Times*, edited by Mladen Tomorad, Sanda Kočevar, Zorana Jurić Šabić, Sabina Kaštelančić, Marina Kovač, Marina Bagarić, Vanja Brdar Mustapić and Vesna Lovrić Plantić, 269–271. Oxford: Archaeopress.

Lovrić Plantić, Vesna. "'Egyptomania' and Clock Production: Examples from the Museum of Arts and Crafts, Zagreb." In *Egypt in Croatia—Croatian Fascination with Ancient Egypt from Antiquity to Modern Times*, edited by Mladen Tomorad, Sanda Kočevar, Zorana Jurić Šabić, Sabina Kaštelančić, Marina Kovač, Marina Bagarić, Vanja Brdar Mustapić and Vesna Lovrić Plantić, 272–273. Oxford: Archaeopress.

Mirnik, Ivan. 2016. *Sylloge Nummorum Graecorum Croatia—Zagreb: The Archaeological Museum Numismatic Collection, Vol. VIII. Aegyptus: Ptolemaei—Roman Provincial Coinage*. Zagreb: Archeological Museum in Zagreb.

Tomorad, Mladen. 2004a. "Croato-Aegyptica Electronica. Database of the Egyptian antiquities in Croatian museum and private collections." *PalArch, series archaeology of Egypt/Egyptology* 3 (2): 7–11.

Tomorad, Mladen. 2006. *Model računalne obrade i prezentacije staroegipatskih predmeta u muzejskim zbirkama u Hrvatskoj*. PhD diss., University of Zagreb.

Tomorad, Mladen. 2007. "Croato-Aegyptica Electronica—model obrade i analize staroegipatskih predmeta u muzejskim i privatnim zbirkama u Hrvatskoj: dokumentacijski i komunikacijski pristup." *Muzeologija* 41–42 (2004/2005): 218–239.

Tomorad, Mladen. 2009. "Croato-Aegyptica Electronica-Database of the Egyptian antiquities in Croatian museum and private collections." In *Proceedings of the Third Central European Conference of Young Egyptologists. Egypt 2004: Perspectives of Research. Warsaw 12–14 May 2004*, edited by Joanna Popielska-Grzybowska, 183–190. Pułtusk: Institute of Anthropology and Archaeology.

Tomorad, Mladen. 2015. "The Early Penetration of Ancient Egyptian Artifacts and Dissemination of the Cults of Egyptian Divinities in Istria and Illyricum (1st Millennium B.C.-1st Century A.D.)." In *A History of Research into Ancient Egyptian Culture conducted in Southeast Europe*, edited by Mladen Tomorad, 165–200. Oxford: Archaeopress.

Tomorad, Mladen. 2017a. "The Ancient Egyptian Collections in Croatia and the Project

Croato-Aegyptica Electronica." In *Collections at Risk—New Challenges in a New Environment*, edited by Claire Derricks, 237–268. Atlanta: Lockwood Press.

Tomorad, Mladen. 2017b. *Staroegipatska civilizacija sv. II: Uvod u egiptološke studije*. Zagreb: Hrvatski studiji.

Tomorad, Mladen. 2018. "Various manifestations of Isis cult in the region of Istria, Illyricum and Pannonia and its diffusion to the Central European Region (2nd c. B.C.-4th c. A.D.)." *Shodoznanvstvo* 81: 58–99.

Tomorad, Mladen. 2019a. "Early Collectors and the Genesis of the Most Important Egyptian Collections in Croatia (1820–1950s)." In *Egypt in Croatia—Croatian Fascination with Ancient Egypt from Antiquity to Modern Times*, edited by Mladen Tomorad, Sanda Kočevar, Zorana Jurić Šabić, Sabina Kaštelančić, Marina Kovač, Marina Bagarić, Vanja Brdar Mustapić and Vesna Lovrić Plantić, 163–170. Oxford: Archaeopress.

Tomorad, Mladen. 2019b. "Egyptian Collections in Croatia." In *Egypt in Croatia—Croatian Fascination with Ancient Egypt from Antiquity to Modern Times*, edited by Mladen Tomorad, Sanda Kočevar, Zorana Jurić Šabić, Sabina Kaštelančić, Marina Kovač, Marina Bagarić, Vanja Brdar Mustapić and Vesna Lovrić Plantić, 171–187. Oxford: Archaeopress.

Tomorad, Mladen. 2019c. "Egyptian shabtis discovered in various regions in Europe." *Journal of Ancient Egyptian Interconnections* 23, no. 1 (September): 100–126.

Tomorad, Mladen. 2019d. "Egyptian Revival and Modern Egyptomania in Croatia." In *Egypt in Croatia—Croatian Fascination with Ancient Egypt from Antiquity to Modern Times*, edited by Mladen Tomorad, Sanda Kočevar, Zorana Jurić Šabić, Sabina Kaštelančić, Marina Kovač, Marina Bagarić, Vanja Brdar Mustapić and Vesna Lovrić Plantić, 245–248. Oxford: Archaeopress.

Tomorad, Mladen. 2019e. "Project Croato-Aegyptica." In In *Egypt in Croatia—Croatian Fascination with Ancient Egypt from Antiquity to Modern Times*, edited by Mladen Tomorad, Sanda Kočevar, Zorana Jurić Šabić, Sabina Kaštelančić, Marina Kovač, Marina Bagarić, Vanja Brdar Mustapić and Vesna Lovrić Plantić, 274–277. Oxford: Archaeopress.

Tomorad, Mladen. 2020a. "Digitalizacija, obrada i međunarodna dostupnost staroegipatskih spomenika u Hrvatskoj: baza Croato-Aegyptica—prošlost, sadašnjost, budućnost." *Muzeologija* 56: 138–159.

Tomorad, Mladen. 2020b. *Egipat u Hrvatskoj: hrvatska fascinacija staroegipatskom kulturom*. Zagreb: Fakultet hrvatskih studija.

Tomorad, Mladen and Goran Zlodi. 2005. "Croato-Aegyptica Electronica-Database of the Egyptian Antiquities in Croatian Museum and Private Collections: Documentation Vs. Communication Approach." In *Documentation & Users—Proceedings of the CIDOC Annual Conference*, edited by Maja Šojat-Bikić, 1–18. Zagreb: Muzejski dokumentacijski centar.

Tomorad, Mladen and Goran Zlodi. 2015. "Development of the 'Croato-Aegyptica' Database." In *A History of Research into Ancient Egyptian Culture conducted in Southeast Europe*, edited by Mladen Tomorad, 161–164. Oxford: Archaeopress.

Tomorad, Mladen and Goran Zlodi. 2016. "Projekt Croato-Aegyptica Electronica: iskustva okupljanja građe, interoperabilnosti metapodataka i razvoja egiptološkog nazivlja u višejezičnom okruženju." *Šesti festival hrvatskih digitalizacijskih projekata, Zagreb 21.–22. travanj 2016*, 31–32. Zagreb: Nacionalna sveučilišna knjižnica.

Uranić, Igor. 2009. *Aegyptiaca Zagrabiensis*. Musei Archeologici Zagrebiansis Catalogi et Monographiae, Vol. 4. Zagreb: Archaeological Museum in Zagreb.

Online Resources

Gračanin, H., I. Malus Tomorad, M. Tomorad, P. Selem & I. Uranić. 2004–2014. *Croato-Aegyptica Electronica*. Edited by M. Tomorad. No longer accessible.

Tomorad, M. ed. 2014–present. *Stara povijest*. http://www.starapovijest.eu

CHAPTER 22

Virtual Reality Storytelling: Pedagogy and Applications

Julia Troche and Eve Weston

Abstract

The Spirit of Egypt is a Virtual Reality project being developed by Exelauno. This project, which recreates an ancient Egyptian funeral procession and transition into the Hereafter, is being designed with instructional pedagogy in mind and, thus, can be an example through which to discuss real-world, best practices. If employed correctly, Virtual Reality is able not only to help students learn about ancient Egypt but also make it more memorable.

While there are great benefits to bringing technology into the classroom, real criticisms exist: the learning curve for "digital immigrants" can be steep; developing technology can diverge from Universal Design standards; and screens can negatively affect learning and retention for all students. Our assertion is that technology needs to be brought into the classroom in thoughtful, purposeful, and pedagogically sound ways. Uniquely, compared to other technologies that put something (i.e., a screen) between students and the information, VR in essence does the opposite—completely immersing students in the world they are studying.

Keywords

Virtual reality – Extended Reality – Augmented Reality – education – pedagogy – classroom technology – Hatshepsut – Deir el-Bahri – Universal Design standards

1 Introduction

Research has shown that Virtual/Augmented Reality simulators, immersive and digital learning environments, and technological applications in the classroom broadly defined can greatly benefit student learning and retention.[1] Just

1 e.g., Dede 1995; Byrne 1996; Troche and Jacobson 2010.

as all resources we bring into our classes need to be thoughtfully considered, new technologies similarly need to be assessed. Too often, we find ourselves, like flies to a bright light, running to the new shiny, tech thing. Indeed, technology, if not deployed intentionally and with pedagogy in mind, can easily become a barrier to learning, rather than an aid. This is perhaps particularly true when considering classroom applications (and other educational contexts, such as museums) of Extended Reality, or XR, which is an umbrella term encompassing all forms of immersive media.[2]

Quality needs to be considered alongside appropriateness of use. Johnson-Glenberg, following a *MediaX* presentation given by Bailenson in 2016, identifies four instances in which XR is most advantageous: when it enables something that is otherwise (1) impossible (e.g. flying above an ancient Egyptian funeral procession as the deceased's ethereal *ba*-aspect); (2) too expensive (e.g. flying an entire class to an ancient Egyptian mortuary temple); (3) too dangerous; or (4) counterproductive.[3] Beyond quality and appropriateness, the authors argue that the pedagogical utility of technology employed in learning environments must also be critically considered. There is certainly room for technology to work counterproductively to some instructional goals; but technology, especially if developed with pedagogy explicitly in mind, can have profound, positive effects on teaching and learning. Specifically, the authors argue that storytelling, humor, and experiential-embodied instructional design[4] are three particularly effective pedagogical tools that can be incorporated into XR design.

To test this theory, and to act as a proof-of-concept, authors Julia Troche and Eve Weston developed a Virtual Reality (VR) demonstration of the XR game *The Spirit of Egypt* with these three constructs in mind. A game was chosen, rather than some other modality its immersive and embodied learning capabilities.[5]

2 Extended Reality (XR) is a broad term used to refer to numerous immersive technologies that attempt, in different ways, to integrate physical and virtual experiences. The most common of these include Augmented Reality (AR) and Virtual Reality (VR). AR describes an experience in which the user walks and moves within the real, lived world, but a digital addition is overlaid to create a new experience. A popular example would be the video game *Pokémon Go*, in which digital creatures can be found roaming public streets. VR is a more complete immersive experience, in which the user participates in a wholly digital world.

3 Johnson-Glenberg 2018, 11.

4 As defined by Troche, in progress. In brief, it refers to the intersecting pedagogies of experiential learning and embodied learning.

5 This game was not built explicitly with *archaeogaming* in mind, but following Michael Shanks' definition of archaeology, as quoted by Reinhard in defining *archaeogaming*, this project fits the definition: "Archaeology is largely a set of experiences" (Reinhard 2018, 10).

This paper first considers the challenges and benefits of using technology in the classroom, though its discussion can be more broadly applied to technology used in a wide variety of educational settings (e.g. including the museum). *The Spirit of Egypt* is then introduced, followed by an analysis of these three key pedagogical tools, and a discussion of how these tools were considered in the design of *The Spirit of Egypt*.

2 Challenges and Benefits of Educational Technologies

Technology is most often found in the modern classroom in three areas: as teaching tools (e.g. PowerPoint), as student learning tools (e.g. digital flashcards), and as information and communication technologies (e.g., email, Banner by Ellucian, and course websites and learning management systems, such as Canvas).[6] Students are increasingly experiencing their world through technology and screens: accessing readings, taking notes, and submitting work on their computers, phones, and tablets (perhaps even more now as a result of the ongoing COVID-19 pandemic). Indeed, technology has infected our classrooms and it is not going away, nor should it. Technology has proven itself to be an incredibly useful, educational ally: it enables clear communication; it provides greater accessibility to more students with different learning styles, backgrounds, and abilities; it can be employed to encourage active learning and student participation in large classrooms where it would otherwise be difficult. Despite these easily identifiable benefits, technology can also present real challenges in the classroom, especially for so-called digital immigrants.

Marc Prensky coined the terms "digital native" and "digital immigrant" in a 2001 article for the education journal *On the Horizon*. Digital immigrants are in essence learning a new language, and language learned as an adult, Prensky suggests, is stored differently in our brains than language learned as a child.[7] Digital natives, on the other hand, are students who have been raised around digital technology and for whom using this technology is innate. While many of our students may be digital natives, our educational system, which was built by digital immigrants, has not entirely caught up to our students. Prensky begins his article with a bold statement—literally bold and italic, in the version published on his personal website: "***Our students have changed***

6 The "teaching machine" being perhaps among the first instructional tool of its kind; see Pressey 1926.

7 Prensky 2001, 3.

***radically. Today's students are no longer the people our educational system was designed to teach.*"[8]

However, assuming our students are digital natives because of their age is short-sighted. Problematically, nowhere in Prensky's highly-cited article does he address privilege, ability, wealth, or any number of myriad personal and social variables that affect one's capacity to interact with and access technologies. Especially in underprivileged communities and in rural academic settings, we cannot assume our students (no matter their age) have access to and experience with digital technology to the point that they are "natives." Furthermore, younger students may be "natives" when it comes to certain digital technologies, but social media expertise does not automatically translate to other technologies, such as Extended Reality. This has become apparent as universities rushed to remote instruction in Spring 2020 in response to the growing COVID-19 pandemic and huge numbers of students confronted remote-learning technologies with varying degrees of success. At the beginning of the Fall 2020 semester, Troche informally surveyed her introductory, mostly freshman, world history class at Missouri State University about the students' comfort with learning technologies, including Blackboard and Zoom, with about half expressing at least moderate concern over their ability to effectively use learning technologies, as well as over home access to stable internet and to necessary programs on home devices. While anecdotal, this does suggest that many students, at all ages, are not completely confidant in their ability to effectively use, access, and learn from educational technologies.[9] Indeed, for those who are digital immigrants, for whatever reason, new technologies can become a hindrance to teaching and learning, to the point even of being insurmountable.

Technology should never be employed for technology's sake. It is easy for digital materials built with the best intentions to veer away from Universal Design principles[10] in ways that the end-user cannot predict or fix. The internet provides students across the world with almost-incredible access to knowledge,

8 Prensky 2001, 1; https://www.marcprensky.com/writing/Prensky%20-%20Digital%20Natives,%20Digital%20Immigrants%20-%20Part1.pdf

9 It would be interesting to see if this has changed now that technology has increasingly been thrust upon students during the pandemic. In terms of access, while universities offer assistance in the way of computer labs and similar, during the pandemic (at least anecdotally according to Troche's experience) more and more students are living at home/away from campus and are taking online classes, without reliable access to these on-campus resources.

10 Universal Design refers to a paradigm in which things, spaces, and services are designed to be accessible and usable by all people, regardless of ability. See Preiser and Smith 2011.

both popular and academic. However, the seductive ease of "copy-paste" also makes cheating and plagiarism astonishingly tempting, especially when our university students are rarely receiving training in best practices of using digital resources or technologies in their secondary education. There are also pedagogical concerns that should be considered when implementing any technology, but in particular digital technology, in the classroom. People tend to read less actively when they read from a screen as compared to a physical, printed text, leading to diminished information retention.[11] Specifically, Mueller and Oppenheimer have shown that writing notes out in longhand may help students retain knowledge more effectively, due to typing's "shallower processing."[12]

Despite all of these potential challenges, the authors still advocate for the robust use of digital technology in the classroom, because there are real benefits as well. We simply urge caution and encourage critical examinations of new technologies before they are incorporated into classroom (or other) education curricula. We should always ask what our student learning objectives are, and ask if, or how, digital technology will help students achieve these goals. The use of digital technology is also justified when it makes our lives as educators easier, or enables us to be more equitable, or communicate more effectively to more people.

The bottom line is that we must consider the pedagogical purpose of every technology (digital or analog) that we bring into any given learning environment and weigh this against the potential hurdles that it may create for our students, especially for those who are most vulnerable (e.g., digital immigrants, students with access to fewer digital technologies either due to associated cost or lack of supporting infrastructure—notably, those in more rural communities, etc.). However, not all technology is created equal, and we should be careful not to throw the baby out with the bath water. There are certain technologies and technologically assisted practices which, studies have shown, possess significant, general pedagogical benefits.

Mayer, Heiser, and Lonn have argued that visual images in concert with speech are better for student learning than images accompanied with text.[13] This means that the use of images in a PowerPoint presentation while lecturing is better than having students look at those same images and reading about them in a book, because in the latter scenario there is cognitive competition

---

11 Mueller and Oppenheimer 2014.

12 Mueller and Oppenheimer 2014, 1159; Glass and Kang 2019 further show that screens can divide attention and reduce class performance.

13 Mayer, Heiser, and Lonn 2001.

between the text and images.[14] That being said, one can easily imagine a case in which an overabundance of images, and animations in particular, can distract and cause adverse results for students.

Immersive environments, either real or digital, have been found to greatly benefit student learning and retention.[15] For example, Krokos, Plaisant, and Varshney have shown that virtual worlds projected within immersive environments (i.e., those that surround the viewer, such as a projected Virtual Reality world or Augmented Reality) aid in student retention over virtual worlds explored in 2D "flat" environments, including those that are digital.[16] This means that if you have a Virtual Reality world, it is better for students to physically walk through a projection of it than to passively "walk" through as an avatar on a computer screen (or, for example, to look at a plan of the space on the printed page). In immersive Virtual and Augmented Reality, the screen—which in other media can act as a barrier—acts as a door. A student can actually get closer to their objects of study than physically possible in the real world. A student can manipulate a digitally projected molecule, or observe a supernova explosion, or walk along an ancient Egyptian funeral procession.

3 *The Spirit of Egypt*

We have built a proof-of-concept demonstrator model and developed plans for a fully realized Virtual Reality experience that uses Extended Reality to provide a unique first-person experience. A participant in *The Spirit of Egypt* will experience ancient Egyptian funerary rites and the imagined afterlife as the spirit of the recently deceased King Hatshepsut, who was the first female Pharaoh of Egypt, ruling during the New Kingdom's Eighteenth Dynasty, c. 1500 BCE.[17] Extended reality provides a unique experience opportunity, as it allows for an embodied encounter that archaeology and other media cannot provide. Furthermore, XR promotes a sense of presence and mindfulness that heightens the participant's experience as they inhabit bodies, objects, and realms that were historically never available to living humans.[18]

14 Mayer, Heiser, and Lonn 2001.

15 e.g., Dede 1995; Byrne 1996; McLellan 1996; Osberg 1997; Winn 1997; Youngblut 1998; Troche and Jacobson 2010.

16 Krokos, Plaisant and Varshney 2018; also, Jacobson 2008. Notably, though, it is not necessarily about the user's ability to move with six degrees of freedom, but is about the relative positioning of the virtual world vis-à-vis the user.

17 Roehrig et al. 2005; Cooney 2014.

18 Navarro-Haro et al. 2017.

The Spirit of Egypt will take place in four main gameplay zones, identified as the Theban Funeral Procession, Hatshepsut's Tomb, her Mortuary Temple at Deir el-Bahri, and the imagined Hereafter. Of these four zones, Hatshepsut's Tomb and part of the Hereafter were developed for our demonstrator model, with the other two zones currently existing only as part of our plans for future development. All four of these zones, and other aspects of *The Spirit of Egypt*, are not meant to be historical models that replicate with complete accuracy the places and events described. This is different than many other archaeological models. Fisher and Unwin explain that many in archaeology expect VR models to "act as surrogates, digital replacements for original artefacts, structures and landscapes" and that the "faithfulness with which these surrogates approximate their original referents is regarded as a direct correlate of the quality and volume of data that had gone into their creation."[19] They warn, though, that this "uncritical orthodoxy" has resulted in creators being "too often encouraged to sit back and admire the resultant models than to do anything useful with them."[20]

It is worth noting that these goals of historical accuracy and useful application need not be mutually exclusive, and that elsewhere in this volume there are clear examples of virtual models achieving both. For *The Spirit of Egypt*, education-driven application goals were the driving forces behind the decisions the authors made about its design. To this end, the authors intend to build the game as an educational exemplar that is broadly accessible and easily implemented across K-12 classrooms and in museums for general public consumption. It is a manufactured reality, based on historical evidence, but modified and simplified for accessibility, ease of educational applications, and for practical gameplay. There are obvious criticisms of this approach—how can something be educational if it is not completely accurate? To answer this, a brief anecdote: when Troche was in first grade, her teacher, using the once ubiquitous overhead projector, showed that if you have two straws, and you take away three, you have taken away all of the straws, leaving zero behind. Troche later learned more about integers and negative numbers, but this complexity was not necessary for the goal of her first-grade teacher, who was only trying to teach the fundamentals of subtraction. *The Spirit of Egypt* is not intended to be a teaching tool in graduate classrooms nor is it meant to be used by scholars for research. It is meant to introduce a general audience to some basic ancient Egyptian concepts (e.g., kingship, conceptions of the ancient Egyptian dead,

19 Fisher and Unwin 2002, 18.
20 Fisher and Unwin 2002, 18.

the separation of royal tombs and mortuary temples in the New Kingdom, etc.). Indeed, while the game is self-contained, for the most robust educational outcomes, we recommend that the game, when it is finished, be used as curricular enrichment in concert with exhibitions, lectures, or readings that contextualize the experience and provide further historical background. Eventually, these additional readings and learning guides will be provided alongside the game at purchase, though the complete game and associated materials are still hypothetical at this point.

The Spirit of Egypt was co-developed by authors Julia Troche and Eve Weston. The game was built using Unity. The authors contributed some to the game's build, took advantage of open-source Sketchfab models, and largely relied on freelancers and independent contractors, hired through Weston's company Exelauno, for the build engineering.

The first level of the game, which has not yet been developed, will begin with the participant experiencing the funeral procession as Hatshepsut's disembodied *ba*-aspect,[21] which will be able to fly freely and eavesdrop on conversations occurring among attendants—e.g., the royal family, priests, musicians, wailers, bodyguards, and average Egyptians who came to witness the royal burial. These conversations would provide an opportunity for the participant to learn about a variety of ancient Egyptian history topics: everyday life, funerary rites, social status, and relationships, and how the dead in ancient Egypt prepared for the journey into the Hereafter.

The second level of gameplay occurs at the zone called Hatshepsut's Tomb, which has been partially engineered for our demonstration. Here, the participant—still playing Hatshepsut's *ba*—will witness their own Opening of the Mouth ritual.[22] Only after this ritual is complete can the user engage in the rest of the game world. In order to be successful in the rest of the game, the participant must fill their "power tanks" of strength, equipment, and knowledge. These "power tanks" are based on emic Egyptian ideas of what makes the ancient Egyptian dead effective. The dead were often described as able by the ancient Egyptian term *iqer* (*jqr*), which is represented by the "strength" tank.

21 The *ba*-aspect was one of multiple spiritual aspects of the ancient Egyptian supernatural self. The term *ba* (*bꜣ*) is difficult to translate perfectly but it refers to the manifestation of the impression someone left on the world. The *ba*-aspect was imaged as a bird with a human head. It could thus fly and travel in ways that humans could not. The most extensive study remains Žabkar 1968.

22 A recent discussion of this ritual, as known by the well-documented Late Period example of Padiamenope, can be found in Engelmann-von Carnap 2018; a more focused discussion, but one that is deals with material contemporary to Hatshepsut, can be found in Bohleke and Strudwick 2017.

The dead were also described as being equipped, that is *aper* (*ʿpr*), with both knowledge and all the trappings of a proper burial, i.e., *ḫt nbt nfrt* "all good things," which is represented by the "equipment" and "knowledge" tanks respectively.[23]

Sources of all three types of power will be found in both the Hatshepsut's Tomb and the Mortuary Temple zones once they are fully developed. In Hatshepsut's Tomb, for example, the participant will be able to collect food, beer, clothing, jewelry, weapons, and amulets and they will glean knowledge from the *Am Duat* scenes which are painted upon the walls of the tomb.[24] In *The Spirit of Egypt* experience, "strength" will give the participant the energy needed to travel between zones; it is derived from collected items, and from "resting," which occurs when the *ba*-aspect rejoins its mummy, resulting in an otherworldly, meditative rest. The "equipment" tank is refilled by collecting items such as weapons and amulets, while "knowledge" is obtained by reading scrolls, deciphering what is depicted on the tomb walls, and listening to the living who visit the mortuary temple.

While knowledge fills the participant's "power tank," much of it must also be remembered and implemented to succeed in the experience. For example, in the proof-of-concept model developed for the 2019 University of Indiana-Bloomington *Ancient Egypt and New Technology* conference, the participant begins in Hatshepsut's tomb facing an *Am Duat* scene in which the names of a number of gods are recorded in hieroglyphic text. Specifically, part of the middle register of Hour 1 of the *Am Duat* is depicted.[25] These scenes are drawn as exemplars; that is, the scenes are not necessarily exact replicas of those found in Hatshepsut's tomb (KV20) nor are the scenes represented in their entirety with 100 % accuracy. The intent is to create an accessible approximation of the ancient Egyptian *Am Duat* text that works for the gameplay (such as being easily readable).

After users study this image—which includes Osiris' name written in hieroglyphs above his head in the middle of the scene—they move on to the first part of the Hereafter zone, where they will pass through a series of gates, each with a different activity. To pass through the first gate, for example, participants must place the correct hieroglyphic name tag on a figure of Osiris (see Figure 22.2). While there is no ancient text that identifies Osiris as the gatekeeper in Hour 1

23 Troche 2013.

24 Scenes are drawn as exemplars (meaning they are not replicated completely accurately nor are they precisely those known from Hatshepsut's tomb, KV20) based on those in Hornung and Abt 2007.

25 Based on Hornung and Abt 2007, 16–17.

of the *Am Duat*, the gameplay is based broadly on the role of Osiris as protector and guide in the Hereafter as evinced in the First Hour of the *Am Duat*: in the lower, middle register the sun-god in the form of Khepri (*ḫprj*) is positioned on a bark, flanked by two figures identified as Osiris (*Wsjr*), and later the sun-god calls upon these deities to "open wide your gates."[26] The knowledge gained in the tomb is used practically by the participant in the Hereafter in order to pass a mini-game within the experience.

At the third zone, the Mortuary Temple, which will be modelled on Hatshepsut's historical temple at Deir el-Bahri, the participant may interact with the living.[27] The Mortuary Temple zone, which has not yet been developed, is the only place where this can happen after the funeral procession is over. The participant will hear mourners and supplicants, offering gifts and asking for favors or blessings in return. Some visitors may practice the ritual of incubation, in which they sleep at a religious site in the hopes that a god (or in this case the dead divinized king) may come to them.[28] Thus, at the mortuary temple, the participant can gain knowledge by entering the visitors' dreams, or from listening to the visitors recount their adventures. For example, inspired by the journey's depiction upon the temple wall, a visitor might recount their trip accompanying Hatshepsut to Punt.[29]

The fourth main location of *The Spirit of Egypt* is the imagined Hereafter, which was partially developed for the demonstration. In the Hereafter zone, the participant will encounter three chronological sections representing the twelve symbolic hours of the Egyptian *Duat*. In each—Descending (hours 1–4), Deep (hours 5–8), and Ascending (hours 9–12)—participants will need to accomplish a task, or tasks, in order to pass through and continue on their journey to

26 *snšn=j ʿrryt=ṯn*, Hornung and Abt 2007, 26; 34–35.

27 This part of the game has not yet been developed. There are currently models that exist of Hatshepsut's mortuary temple (e.g. one built by the Polish Academy of Sciences in Warsaw, which the authors would hope to use if allowed). Otherwise, a simple model will be built based on videos, photographs, and publications of the site.

28 There is no clear evidence that incubation specifically occurred at Hatshepsut's mortuary temple. The practice, though, of sleeping at a tomb is known from at least the First Intermediate Period. The practice of incubation, in this case the act of sleeping near a tomb so that the dead can speak to the petitioner in a dream, is made explicit upon two First Intermediate Period Letters to the Dead, one now in the Michael C. Carlos Museum (2014.033.001) and the other known as the Payprus to Meru (Simpson 1966). Since the royal, valley tombs were inaccessible, presumably this practice would instead occur at a mortuary temple. Incubation is poorly studied for most of Pharaonic Egypt, with most studies instead focusing on the Greco-Roman periods during which time the practice becomes more ubiquitous, e.g. Renberg 2016.

29 See Ann Macy Roth in Roehrig 2005.

defeat Apophis and save themselves and Egypt, manifest in the sun's successful rising. These tasks may require strength, equipment, and/or knowledge, both stored and practical. Once the sun is pushed up over the horizon and is in the morning sky in the form of Khepri, participants have succeeded in completing their tasks and have effectively "won" the game. While there are other rules and details,[30] this is the basic outline of *The Spirit of Egypt* and enough context for us to talk about this experience as a case study for effective immersive pedagogical techniques.

4 Pedagogy: Ideas and Application

The authors have identified three high-impact teaching strategies with proven effectiveness in traditional, seated classrooms, and that, it stands to reason, will be similarly effective in XR: storytelling-as-pedagogy, humor, and experiential-embodied instructional design.

Storytelling is not a novel pedagogical tool. Despite it being one of the oldest forms of teaching, it remains pedagogically effective, especially when combined with narrative inquiry.[31] In short, learning that stems from a developed and well-performed story is remembered more accurately, and notably longer, than learning from facts and figures alone.[32] This is not without criticism, though. Narratives and storytelling-as-pedagogy can also incorporate implicit bias and give consumers a sense that what is before them is the "whole story," since the complexities and unknowns of the historical moment are left out. Thus storytelling, as with any teaching tool, must be employed in thoughtful, intentional ways to avoid these pitfalls.[33] Indeed, Egan has argued that storytelling "is not just some casual entertainment; it reflects a basic and powerful form in which we make sense of the world and experience."[34] Deniston-Trochta argues in favor of storytelling-as-pedagogy in part because "our students' lived experience is the foundation for their learning."[35] Storytelling, if aptly constructed, is therefore one way in which an instructor can help make the material

30 Specifically, *The Spirit of Egypt* is designed to include the four defining traits of games, as defined by McGonigal 2011: a goal, rules, a feedback system, and voluntary participation.

31 e.g. Coulter, Michael, and Poynor 2007.

32 Neuhauser 1993.

33 Thanks are due to an anonymous reviewer who encouraged us to develop this important point beyond the remarks in this chapter's original draft.

34 Egan 1988, 2.

35 Deniston-Trochta 2003, 104.

they are presenting more relatable. The historic narrative can become part of the students' own lived experience.

In *The Spirit of Egypt*, the students' lived experience comes to include the posthumous journey of Hatshepsut's *ba*-aspect, tasked nightly with defeating Apophis and rising the morning sun. This in itself is a story; still, within it, there are other opportunities to employ storytelling and enjoy its pedagogical benefits. For example, at the funeral procession, student participants will be able to overhear conversations between various groups of people. They may hear the bodyguards discussing their career anxieties; now that Hatshepsut is gone, will the next Pharaoh keep them on at the palace? May they, like Sinuhe, fear retribution or responsibility for her death? Participants might eavesdrop on Hatshepsut's daughter Neferure,[36] who has a completely different reaction; Hatshepsut's passing has inspired in her both a deep determination to continue the Ahmosid family line and also a sense of paranoia about her Thutmosid step-family who may eventually wish her dead. In a VR experience like *The Spirit of Egypt*, history ceases to be far removed: it is happening all around the participant.

Humor as a pedagogical tool is more controversial, especially in elite, academic circles. Humor can be construed as "not serious" and therefore inappropriate in certain academic environments. Humor is also highly subjective. However, we all know that the best lectures are the ones that make us laugh a bit. Not only do we enjoy these learning experiences on an emotional level, but they linger with us longer. This is not just observational but is backed up by science: Goel and Dolan's neuroscience research reveals that humor systematically activates the brain's dopamine reward system, and cognitive studies show that dopamine is important for both goal-oriented motivation and long-term memory.[37] Furthermore, Banas et al. indicate that correctly-used humor can improve retention in students from kindergarten through college.[38] So, humor is a not only a legitimate pedagogical tool, but a particularly effective one that helps with information retention, student motivation, and, anecdotally we have found, attentiveness.

Though humor can be potentially controversial or even offensive, perhaps a more notable academic critique is that humor is subjective, which makes it difficult to build into a lesson plan. There are, however, certain conditions

36 Historical sources are unclear on whether Neferure was alive at the time of her mother's death or if she had died before Hatshepsut. See Pawlicki 2007, 124–125.

37 Goel and Dolan 2001.

38 Banas et al. 2010.

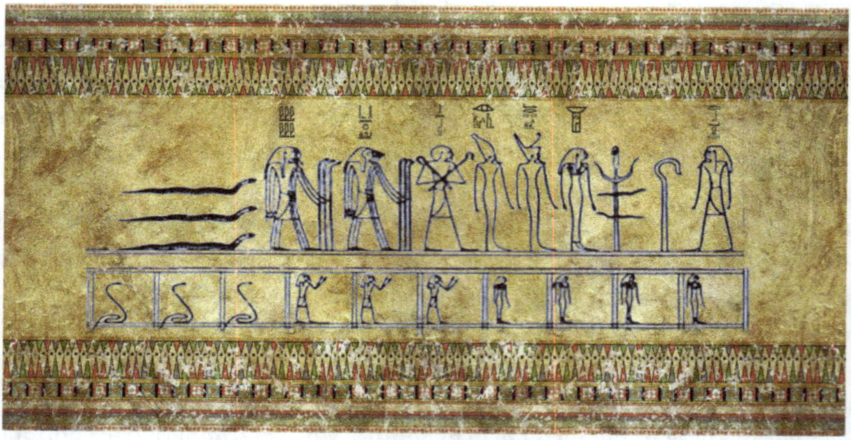

FIGURE 22.1 Image of the *Am Duat* scene included in *The Spirit of Egypt*. Scene is based on Hour 1 of the *Am Duat* as published in Hornung and Abt 2007, 16–17 and is digitally rendered by Eve Weston.

that typically make something funny, which McGraw, in a number of publications, has identified in what he calls the benign violation theory.[39] This theory hypothesizes that "people experience humor when they simultaneously appraise a violation as being benign."[40] McGraw and Warner have also acknowledged that there is a certain *je ne sais pas* quality when it comes to humor, or rather being a good humorist—meaning, some people are just naturally better humorists.[41] This pedagogical device, then, might not be for everyone, but humor-as-pedagogy also need not be shunned or diminished, and should rather be recognized as a legitimate academic, educational tool.

There are unlimited opportunities for humor in *The Spirit of Egypt*. One simple example was realized in our model and is located in the Hereafter zone at a gate that the deceased must pass in their journey through the *Duat*. Participants are required to demonstrate knowledge of the name of the deity standing before the gate (in this case, Osiris) in order to pass through the gate.[42] While in the tomb, participants will have had the opportunity to read the god's

39 e.g. Warren and McGraw 2016.
40 Warren and McGraw 2016, 3, fig. 1.
41 McGraw and Warner 2014, 210.
42 To avoid any perceived commentary on ancient Egyptian ethnicity and/or race, Troche and Weston intentionally elected to have the Osiris avatar mirror the *Am Duat* "stick figure" style (as seen in Hornung and Abt 2007), and to be a slightly shimmery, transparent outline, rather than a more human avatar.

FIGURE 22.2
Image of Hello My Name is Osiris used in *The Spirit of Egypt* gameplay.

FIGURE 22.3
Image of Hello My Name Is Queen Bee used in *The Spirit of Egypt* gameplay.

name, which is written in hieroglyphs as part of an *Am Duat* scene that depicts numerous gods from the middle register of Hour 1 (see Figure 22.1).[43] If they do not remember this information, they can go back into Hatshepsut's Tomb and study the scene, before returning to the Hereafter zone. This game utilizes memory-matching, so no philological knowledge of hieroglyphs is required. Participants choose from several names (all in hieroglyphs) on a sheet of "Hello My Name Is" stickers (Figures 22.2 and 22.3). The unexpected use of this contemporary convention in an ancient, spiritual place produces an incongruity that is humorous without undermining the information being conveyed.[44] References to popular culture or recent history add to this humous experience. A modern hieroglyph, the circular logo of the Obama campaign, is on one sticker. When players select this sticker, Osiris humorously dispels the absurd myth that Obama was Middle Eastern and not American. Another sticker depicts the hieroglyph for a goddess with the hieroglyph of a bee (Figure 22.3). When this is placed incorrectly on Osiris' chest, he proclaims "I am not Beyoncé, Queen B. Though I'd really like to be." This otherworldly deity, now idolizing a con-

43 Hornung and Abt 2007, 16–17.

44 An anonymous reviewer also noted here that this could "serve a double purpose, first trigger a positive emotional response, and second, provide an experience of disruption … which highlights that the experience is mediated." Indeed, this incongruity is not only humorous but helps to avoid the pitfalls mentioned above in regard to storytelling that can give a false sense of a complete narrative. This reminds students that this is a constructed and imagined world, while the use of hieroglyphs and truthful renderings of the *Am Duat* simultaneously give credibility to the experience and knowledge.

temporary popstar, is suddenly relatable and at the same time absurd. These comedic choices make this moment, and its lessons, that much more memorable.

In addition to utilizing storytelling and humor, *The Spirit of Egypt* was built with the benefits of both experiential and embodied learning in mind—what Troche has termed "experiential-embodied instructional design."[45] Though similar in many regards, separate scholarship has emerged on experiential learning and embodied learning. The theory of experiential learning was developed by David Kolb in 1984 and has since been expanded upon by himself and others.[46] Kolb defines experiential learning as "the process whereby knowledge is created through the transformation of experience. Knowledge results from the combination of grasping and transforming experience."[47] Experiential learning is, in essence, "learning by doing." It emphasizes stages of experience, reflection, synthesis, and application—ideally to real world situations. In doing so, it puts weight on the learning process and invites students to participate actively in the process (versus passively, as is typical when listening to a class lecture). The "transformation of experience" calls upon all of the bodily senses; of course, classroom learning often involves sight and hearing, but experiential learning also calls upon the senses of touch, smell, and more rarely taste. In this way, the body is stimulated, but the experience need not necessarily include physical movement. In embodied learning, derived from the field of "embodied cognition," the student physically moves as they learn.[48] This can include full body movement (e.g. walking on a treadmill while reading) or partial body movement (e.g. gesturing while learning a multiplication table). Troche's concept of "experiential-embodied instructional design" integrates both of these approaches into a singular pedagogical method. In short, instruction includes movement, sensory activation, and stages of reflection and synthesis to create a robust memory and learning experience.

The efficacy of experiential and embodied learning has been noted in multiple publications and scientific studies. Paulo Freire's 1970 seminal work *Pedagogy of the Oppressed* was, among other things, an early proponent of experiential learning. Other scholars of education and pedagogy had similarly long

45 Troche is still in the process of finalizing these ideas for publication; the main ideas as relevant to *The Spirit of Egypt* are summarized here.

46 Kolb 1984; Kolb, Boyatzis, and Mainemelis 2001.

47 Kolb 1984, 41.

48 e.g., Lindgren and Johnson-Glenberg 2013; Johnson-Glenberg 2018; Skulmowski and Reg 2018.

called for more active student engagement in the classroom.[49] More recently, research on embodied approaches to cognition have shown a positive correlation between kinesthetic movement and studying/retention.[50] In fact, compelling conclusions from a 2018 study by Johnson-Glenberg further confirm these claims. Johnson-Glenberg set out to test the hypothesis that "active and embodied learning in mediated educational environments results in significantly higher learning gains."[51] She developed a mixed reality lesson on electric fields in which embodied learning was scaled.[52] She concludes that "when learners perform actions with agency and can manipulate content during learning, they are able to learn even very abstract content better than those who learn in a more passive and low embodied manner."[53] She thus concluded that embodied learning was most effective.[54]

Experiential, active learning also helps us learn deeper and remember what we learned longer. According to Dale, after two weeks, we tend to remember 10 percent of what we read, 20 percent of what we hear, and 90 percent of what we do.[55] While this data may be slightly outdated, the broader conclusion continues to, at least anecdotally, ring true. Though artificial or imagined in certain ways, the experiences of XR are very much tangible and physical. As Bailenson writes in his seminal book on VR, "a VR experience is often better understood *not as a media experience, but as an actual experience*, with the attendant results for our behavior."[56] There can be no doubt, then, that immersive Virtual and Augmented Reality constitute the next stage in experiential-embodied learning.

Previously, experiential learning was limited by the physical limitations of our world, including those of time and space. This is no longer the case with the rise of XR. We can now recreate scenarios that would previously have been

---

49 e.g. Dewey 1938.

50 e.g. Madan and Singhal 2012.

51 Johnson-Glenberg 2018, 9.

52 Her four scaled groups were: a control group; a low embodied group; a high embodied group; a high embodied group with narrative. Her inclusion of two high embodied groups—one with narrative and one without, is notable and potentially a point for further consideration in XR gaming. Cf. Koenig 2008 who found narrative games produced more *enjoyment* from their users but not statistically significant better learning.

53 Johnson-Glenberg 2018, 11.

54 Similarly, research at the Johnson and Johnson Institute using VR for surgical training confirms the utility of this approach, highlighting that 83% of VR trained surgeons were able to perform their surgery with minimal guidance (compared to 0% of the traditionally trained surgeons); see Logishetty, Rudran, Gofton and Beaulé 2020.

55 Dale 1969.

56 Bailenson 2018, 45.

impossible. Students will not just witness these scenarios but participate in them. For example, to fill up their "power tanks" the participant of *The Spirit of Egypt* will have to physically walk around and search for relevant artifacts and items, such as bread to fill up their "strength tank" or a scroll to fill up their "knowledge tank." As the Mortuary Temple zone is not yet built, this remains hypothetical, but our plan includes requiring those who want to travel between the Tomb and Mortuary Temple zones to flap their arms like a *ba*-bird. These engaging kinesthetics aid in learning and VR's ability to add embodied movement presents a pedagogical advantage.

5 Conclusions

We suggest that a real benefit to instruction and student learning can be found at the intersection of "traditional" classroom pedagogy and immersive Extended Reality. *The Spirit of Egypt* consciously employs storytelling, humor, and experiential-embodied instructional design in an effort to show how they might now be effectively employed in an immersive XR experience that retains academic integrity and historicity. The authors are continuing to develop this game and hope that our demonstration model, and description of it, will stand as a proof-of-concept. Further, beyond this project specifically, we intend for this paper, and the model it is based upon, to act as a case study for ideal pedagogical approaches to implementing XR and other developing technologies in the classroom and in diverse learning environments.[57]

Bibliography

Bailenson, Jeremy. 2016. "The Trials and Tribulations of Narrative in VR." *MediaX*. Retrieved from: https://vhil.stanford.edu/news/2016/the-trials-and-tribulations-of-narrative-in-vr-mediax.

Bailenson, Jeremy. 2018. *Experience on Demand: What Virtual Reality Is, How It Works, And What It Can Do*. New York: W.W. Norton & Company.

Banas, John, Norah Dubar, Dariela Rodriguez, and Shr-Jie Liu. 2010. "A Review of Humor in Educational Settings: Four Decades of Research." *Communication Education* 60: 115–144.

57 This project, including participation in the Indiana-Bloomington symposium, was supported in part by Missouri State University, CHPA travel funding and the Department of History. Many thanks to the symposium organizers and participants, volume editors, and reviewers for their feedback. Many thanks as well to the Unity engineers hired by Exelauno to work on this project's build.

Barslaou, Lawrence. 1992. *Cognitive Psychology*. Hillsdale: Lawrence Erlbaum Associates.

Bohleke, Briant and Nigel Strudwick. 2017. "A label for Opening of the Mouth implements from the burial of Senneferi (TT99) and remarks on the ritual." *Études et Travaux* 30: 105–123.

Byrne, Christine. 1996. *Water on Tap: The Use of Virtual Reality as an Educational Tool*. PhD diss., University of Washington.

Cooney, Kara. 2014. *The Woman Who Would Be King*. New York: Penguin Random House.

Coulter, Cathy, Charles Michael, Leslie Poynor. 2007. "Storytelling as Pedagogy: An Unexpected Outcome of Narrative Inquiry." *Curriculum Inquiry* 37 (2): 103–122.

Dale, Edgar. 1969. *Audiovisual Methods in Teaching*. New York: Dryden Press.

Dede, Chris. 1995. "The Evolution of Constructivist Learning Environments: Immersion in Distributed, Virtual Worlds." *Educational Technology* 35 (5): 46–52.

Deniston-Trochta, Grace. 2003. "The Meaning of Storytelling as Pedagogy." *Visual Arts Research* 29 (57): 103–108.

Dewey, John. 1938. *Experience and Education*. New York: Macmillan.

Dimock, Michael. 2019. "Defining Generations: Where Millennials end and Generation Z begins." *Pew Research Center*. PewResearch.org. 17 January 2019. https://www.pewresearch.org/fact-tank/2019/01/17/where-millennials-end-and-generation-z-begins/.

Egan, Kieran. 1988. *Teaching as storytelling: An alternative approach to teaching and the curriculum*. London: Routledge.

Engelmann-von Carnap, Barbara. 2018. "Zum Mund öffnungsritual im Grab des Padiamenope (TT 33)." *Bulletin of the French Institute of Oriental Archeology* 118: 127–141.

Fisher, Peter and David Unwin. 2002. *Virtual Reality in Geography*. New York: CRC Press.

Freire, Paulo. 2005. *Pedagogy of the Oppressed*. Translated by M.B. Ramos. New York: Continuum.

Glass, Arnold. and Mengxue Kang. 2019. "Dividing attention in the classroom reduces exam performance." *Educational Psychology* 39 (3): 394–408.

Goel, Vinod and Raymond Dolan. 2001. "The functional anatomy of humor: segregating cognitive and affective components" *Nature Neuroscience* 4: 237–238.

Hornung, Erik and Theodor Abt. 2007. *The Egyptian Amduat: The Book of the Hidden Chamber*. Translated by David Warburton. Revised and Edited. Zurich: Living Human Heritage Publications.

Jacobson, Jeffrey. 2008. *Ancient Architecture in Virtual Reality; Does Visual Immersion Really Aid Learning*? PhD diss., University of Pittsburgh.

Johnson-Glenberg, Mina. 2018. "Immersive VR and Education: Embodied Design Principles That Include Gesture and Hand Controls." *Frontiers in Robotics and AI* 5 (81). Accessed 29 September 2021. Doi: https://doi.org/10.3389/frobt.2018.00081.

Koenig, Alan. 2008. "Exploring effective educational video game design: The interplay between narrative and game-schema construction." PhD diss., Arizona State University.

Krokos, Eric, Catherine Plaisant and Amitabh Varshney. 2018. "Virtual memory palaces: immersion aids recall." *Virtual Reality* 23 (1): 1–15.

Kolb, David. 1984. *Experiential Learning: Experience as the source of learning and development*. New York: Prentice-Hall.

Kolb, David, Richard E. Boyatzis and Charalampos Mainemelis. 2001. "Experiential Learning Theory: Previous Research and New Directions." In *Perspectives on Cognition, Learning, and Thinking Styles*, edited by J. Sternberg and L.F. Zhang, 227–248. Mahwah, NJ: Lawrence Erlbaum Associates Inc. Publishers.

Lindgren, Robb and Mina Johnson-Glenberg. 2013. "Emboldened by embodiment: Six precepts for research on embodied learning and mixed reality." *Educational Researcher* 42: 445–452.

Logishetty, Kartik, Branavan Rudran, Wade Gofton and Paul Beaulé. 2020. "Fully Immersive Virtual Reality for Total Hip Arthroplasty: Objective Measurement of Skills and Transfer of Visuospatial Performance After a Competency-Based Simulation Curriculum." *Journal of Bone and Joint Surgery* 102 (6). Accessed 29 September 2021. doi: 10.2106/JBJS.19.00629.

Madan, Christopher and Anthony Singhal. 2012. "Using actions to enhance memory: effects of enactment, gestures, and exercise on human memory." *Frontiers in Psychology* 3, Article 507. doi: 10.3389/fpsyg.2012.00507.

Mayer, Richard, Julie Heiser and Steve Lonn. 2001. "Cognitive Constraints on Multimedia Learning: When Presenting More Material Results in Less Understanding." *Journal of Educational Psychology* 93 (1): 187–198.

McGonigal, Jane. 2011. *Reality is Broken: Why Games Make Us Better and How They Can Change the World*. New York: Penguin Random House.

McGraw, Peter and Joel Warner. 2014. *The Humor Code: A Global Search for What Makes Things Funny*. New York: Simon & Shuster.

McLellan, Hilary. 1996. "Virtual Realities." In *Handbook of Research for Educational Communications and Technology*, edited by David Jonassen, 461–497. New York: Simon & Shuster Macmillan.

Mueller, Pam and Daniel Oppenheimer. 2014. "The Pen is Mightier than the Keyboard: Advantages of Longhand over Laptop Note Taking." *Psychological Science* 25 (6): 1159–1168.

Navarro-Haro, María, Yolanda López-del-Hoyo, Daniel Campos, Marsha Linehan, et al. 2017. "Meditation experts try Virtual Reality Mindfulness: A pilot study evaluation of the feasibility and acceptability of Virtual Reality to facilitate mindfulness practice in people attending a Mindfulness conference." *PLoS ONE* 12 (11). doi: 10.1371/journal.pone.0187777.

Neilsen Global Media. 2020. “Millennials: Breaking the Myths.” *Nielsen.com*. Accessed 10 September 2020. https://www.nielsen.com/us/en/insights/report/2014/millennials-breaking-the-myths/. Accessed 3-23-22.

Neuhauser, Peg. 1993. *Corporate legends & lore: The power of storytelling as a management tool.* New York: McGraw Hill.

Osberg, Kimberley Marie. 1997. “Constructivism in Practice: The Case for Meaning-making in the Virtual World.” PhD diss., University of Washington.

Pawlicki, Franciszek. 2007. “Princess Neferure in the temple of Queen Hatshepsut at Deir el-Bahari: failed heiress to the pharaoh’s throne?” *Études et Travaux* 21: 109–127.

Preiser, Wolfgang F.E. and Korydon H. Smith, eds. 2011. *Universal Design Handbook*. Second Edition. New York: McGraw-Hill.

Prensky, Marc. 2001. “Digital Natives, Digital Immigrants Part 1.” *On the Horizon* 9 (5): 1–6.

Pressey, Sidney. 1926. “A simple apparatus which gives tests and scores—and teaches.” *School and Society* 23 (586): 373–376.

Reinhard, Andrew. 2018. *Archaeogaming: An Introduction to Archaeology in and of Video Games*. New York: Berghahn Books.

Renberg, Gil. 2016. *Where Dreams May Come: Incubation Sanctuaries in the Greco-Roman World.* 2 volumes. Religions in the Graeco-Roman World 184. Leiden: Brill.

Roehrig, Catherine et al., eds. 2005. *Hatshepsut: From Queen to Pharaoh*. The Metropolitan Museum of Art. New Haven: Yale University Press.

Simpson, William Kelly. 1966. “The letter to the dead from the tomb of Meru (N 3737) at Nag’ ed-Deir.” *Journal of Egyptian Archaeology* 52: 39–52.

Skulmowski, Alexander and Günter Daniel Rey. 2018. “Embodied learning: introducing a taxonomy based on bodily engagement and task integration.” *Cognitive Research: Principles and Implications* 3 (1): 6. doi: 10.1186/s41235-018-0092-9.

Troche, Julia. 2013. “Describing the Dead: Socio-religious significance of the terms *akh iqr* and *akh apr*.” Paper presented at the American Research Center in Egypt Annual Meeting, Cincinnati, Ohio, April.

Troche, Julia and Jeffery Jacobson. 2010. “An Exemplar of a Ptolemaic Egyptian Temple.” Paper presented at the 38th Annual Meeting of Computer Applications in Archaeology, Granada. http://publicvr.org/publications/TrocheCAA2010.pdf.

Warren, Caleb and Peter McGraw. 2016. “Differentiating What is Humorous from What is Not.” *Journal of Personality and Social Psychology* 10: 407–430.

Winn, William. 1997. “The Impact of Three-Dimensional Immersive Virtual Environments on Modern Pedagogy.” *HIT Lab Technical Report* R-97–15. Seattle: Human Interface Technology Laboratory.

Youngblut, Christine. 1998. “Educational uses of virtual reality technology.” *IDA Document* D-2128. Alexandria, VA: Institute for Defense Analyses.

Žabkar, Louis V. 1968. *A Study of the* Ba *Concept in Ancient Egyptian Texts*. Chicago: University Chicago Press.

CHAPTER 23

Cleo: The Artificial Intelligence Egyptology Platform

Heleen Wilbrink and Joshua Aaron Roberson

Abstract

Many museum collections and other resources on ancient Egypt have been digitized over the last few decades, but they are not yet used to their full potential. Collections are not searchable in their entirety and do not follow the same standards. They are written in different languages, are often not user friendly, and do not offer the opportunity for image-based searching. In order to address these issues and more, Cleo was created. Cleo is an online platform connecting at present four major international museum collections, including over 45,000 objects, which leverages the power of Artificial Intelligence (AI) for "smart" searching.[1] The available objects from all collections can be searched currently in English or Dutch, on any device (e.g., mobile, table, or laptop). Objects can be explored by text, image, and location. AI searches can be used to present the user with objects similar to photos uploaded by the user or with images selected from the platform. Cleo was created by Aincient, founded by Heleen Wilbrink, with the help of many others.[2] Following Wilbrink's presentation of the beta version of Cleo at the Bloomington conference on "Ancient Egypt and New Technology," Joshua Roberson (assistant professor at the University of Memphis) proposed a collaboration with his university's Institute of Egyptian Art & Archaeology and the Institute for Intelligent Systems, to research long-term hosting solutions and to incorporate other institutions and collections into the Cleo platform over time.

Keywords

Artificial intelligence – AI – digital collections – Cleo

1 https://cleo.aincient.org/pages/en/ Accessed 2-9-2022.
2 https://aincient.org/ Accessed 2-9-2022.

1 Barriers to Overcome

The material culture of ancient Egypt constitutes one of the best preserved and most robust archaeological and linguistic corpora to survive from antiquity, as well as one of the most popular avenues for public engagement with the humanities. Unfortunately, the global scattering of Egyptian artifacts among disparate museums and other institutions limits public access to this important facet of our shared, global heritage. At the same time, the lack of easy access to far-flung collections also hinders academic study and research into the linguistic and archaeological remains of this vitally important ancient culture. These problems have been mitigated partially through the digitization of museum collections over the last few decades. However, none of these efforts have exploited the full potential for interconnectivity that the Internet has to offer. Persistent issues may be summarized as follows:

- **Lack of connectivity.** Virtually all digitized collections of Egyptian material (currently about 50) exist as isolated websites with no ability to search between them or simultaneously. Google and similar web-crawler search engines are capable of locating only a small proportion of the data that exists within the museums' databases. Furthermore, most currently digitized museum collections remain unconnected to relevant external resources, such as the UCLA Encyclopedia of Egyptology,[3] Giza Archives,[4] et al.
- **Insufficient metadata.** Descriptions of material are frequently too brief or incorrect, leading to difficulties in locating specific and relevant objects within a given collection.
- **Lack of standardization.** Egyptological terminology standards, such as those provided by Thot thesauri (discussed below), have not yet been implemented widely.
- **Data unavailable in common language(s).** While search functions and metadata for some digital collections are available in multiple languages, many other museums describe their objects primarily or exclusively in their local language. Given the global distribution of Egyptian material, users might be forced to contend with metadata or even basic search functionality

3 https://escholarship.org/uc/nelc_uee Accessed 2-9-2022.

4 http://www.gizapyramids.org/ Accessed 2-9-2022.

in French,[5] German,[6] Italian,[7] Danish,[8] Dutch,[9] or other languages. Translated versions into a common language, e.g., English or Arabic, are often not provided or not employed throughout the site. Furthermore, the accuracy of widely available automated translators, such as Google Translate, remains insufficient for reliably rendering technical terminology, object descriptions, and metadata.

Several cultural aggregator sites have been created to address some of these issues, including Europeana,[10] Artstor,[11] and Google Arts and Culture.[12] However, those services still pose difficulties for culturally specific disciplines such as Egyptology, including difficulty finding relevant objects in large datasets spanning multiple cultures; inability to execute full-text searches of the metadata in common languages; aggregator focus on collection highlights rather than complete collections (thus, e.g., Google); and a lack of global focus for some sites (thus, e.g., Europeana). The Global Egyptian Museum attempted to overcome some of these barriers.[13] However, that site includes only a relatively small selection of objects (14,975) and has not been updated in over a decade. Furthermore, technological progress—particularly with regard to Artificial Intelligence and "smart," image-based searching—offers new opportunities for more efficient research.

2 Possibilities of AI

AI developed for computer vision and image search has been proposed as a partial solution to the problems outlined above, specifically those surrounding metadata. However, existing, large-scale AI solutions for image search, such

5 e.g., the Louvre: http://cartelfr.louvre.fr/cartelfr/visite?srv=crt_frm_rs&langue=fr&initCritere=true, including French results exclusively. Accessed 2-9-2022.

6 e.g., Staatliche Museen zu Berlin: https://www.smb.museum/en/home.html, including mixed language results in English version of the site. Accessed 2-9-2022.

7 e.g. Archaeological Museum of Bologna: http://www.museibologna.it/archeologicoen/, including Italian language results in English version of the site. Accessed 2-9-2022.

8 e.g., National Museum of Denmark: https://samlinger.natmus.dk/AS/asset/10331, including Danish object categories with English search results. Accessed 2-9-2022.

9 e.g., National Museum of Antiquities, the Netherlands: https://www.rmo.nl/en/collection/search-collection/, including Dutch object categories and descriptors with English search results. Accessed 2-9-2022.

10 https://www.europeana.eu/portal/en Accessed 2-9-2022.

11 https://www.artstor.org/ Accessed 2-9-2022.

12 https://artsandculture.google.com/ Accessed 2-9-2022.

13 http://www.globalegyptianmuseum.org/ Accessed 2-9-2022.

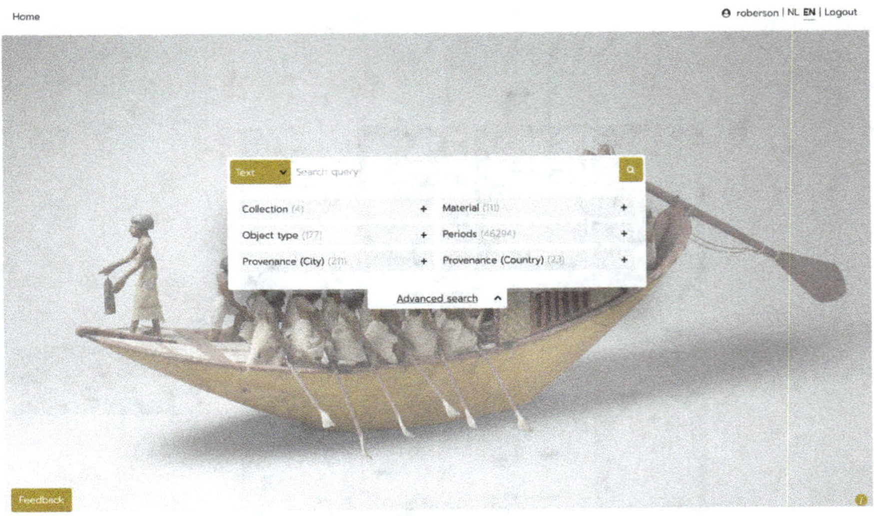

FIGURE 23.1 Cleo search interface with "Advanced search" tab expanded

as Google, Microsoft Bing, or the TinEye reverse image search,[14] do not work well for specialized disciplines such as Egyptology, because their algorithms for image recognition have not been trained specifically on ancient Egyptian corpora. Similarly, other promising experiments within the digital humanities, such as Replica[15] and the those of the Bibliothèque nationale de France,[16] have not yet created definitive solutions focused on the specific problems and priorities of ancient Egyptian material.

3 The Cleo AI Egyptology Platform

To address this deficit, in 2018, Wilbrink launched the beta version of Cleo, the Artificial Intelligence Egyptology platform.[17] The platform was developed together with software company Goldmund, Wyldebeast & Wunderliebe. Cleo stands as the first and, to date only, "smart" museum collections aggregator, developed specifically to address the needs of scholars and the public seeking to engage with and explore the vast material culture of ancient Egypt. Cleo features intelligent image search and integrated metadata translation capabilities, with standardized terminology.

14 https://tineye.com/ Accessed 2-9-2022.
15 https://www.epfl.ch/labs/dhlab/projects/replica/ Accessed 2-9-2022.
16 https://data.bnf.fr/ Accessed 2-9-2022.
17 https://www.cleo.aincient.org/pages/en/ Accessed 2-9-2022.

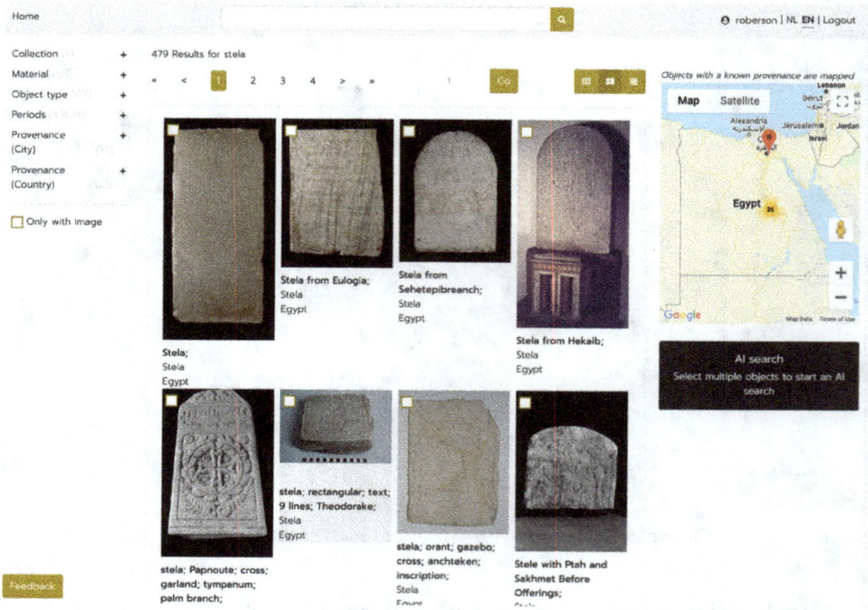

FIGURE 23.2 Sample search results for search term "stela" illustrating World Map and AI search functionality

Cleo was developed originally under the open source Apache License, Version 2.0, as a platform to connect four major international collections of Egyptian artifacts with over 45,000 objects from the online collections of the National Museum of Antiquities (The Netherlands),[18] the Brooklyn Museum,[19] The Metropolitan Museum of Art,[20] and the Walters Art Museum (Baltimore).[21] After registering an account, the user can search available objects from all collections, on any device (mobile, tablet, laptop, or desktop), via a multilingual interface, available currently in either English or Dutch (Figure 23.1).[22]

The researcher can start with a text search or upload an image of a specific object and find similar objects, initiating the AI search capability, discussed below. Results can be filtered, e.g. by "Material" or "Periods," can be studied in detail (always with a link to the original collection website), or analyzed on a world map, where objects are plotted based on their original provenance (Figure 23.2). Every search can be expanded by selecting several objects of interest

18 https://www.rmo.nl/en/ Accessed 2-9-2022.
19 https://www.brooklynmuseum.org/ Accessed 2-9-2022.
20 https://www.metmuseum.org/ Accessed 2-9-2022.
21 https://thewalters.org/ Accessed 2-9-2022.
22 Credits for all object photos illustrated at figs. 1–3 may be found on the Cleo website.

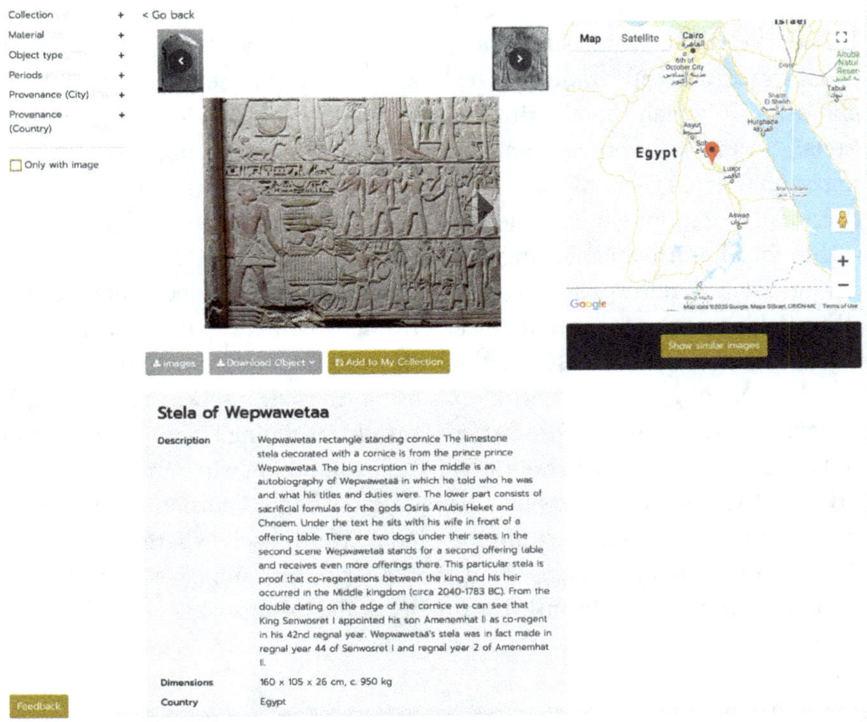

FIGURE 23.3 Sample object metadata (excerpted here for length) and additional user options for image download and personal collections

and performing an (additional) image search. In addition, a personal collection of objects can be created for later reference or descriptions and images can be downloaded immediately (Figure 23.3).

The Cleo team standardized the Egyptological terminology using Thot thesauri, a "wide range of multilingual thesauri related to documentary and textual metadata" pertaining to Egyptology,[23] in order to improve text search functionality. Most importantly, Cleo incorporates an innovative AI algorithm, created using Keras and TensorFlow, which has been trained specifically on Egyptian corpora, and from which the platform was able to suggest objects relevant to the initial text or image search query, which the user might otherwise have overlooked.[24]

23 http://thot.philo.ulg.ac.be/project.html Accessed 2-9-2022.

24 For detailed discussion of the programming workflow, see https://cleo.aincient.org/pages/en/ai/ Accessed 2-9-2022.

The Cleo AI algorithm offers several innovative features that permit unparalleled levels of access to linked collections. The AI image search can be used in two ways: 1.) Alongside traditional text-based searches, a user might also upload photos, from which the algorithm suggests a typological identification (e.g., "stela," "scarab"), in conjunction with a selection of similar objects. This classification process is based solely on the AI analysis of the uploaded image; 2.) Users can search for similar objects by selecting two or more objects indexed within the platform and clicking "AI search."

The results that the platform returns are based on both object metadata and all available images. At present, the algorithm recognizes 23 types or groups of objects, consisting of at least 400 objects per group. In addition to the AI image search, users might search entire texts using translations, available presently in English and Dutch. This process consists of translating Egyptological words and concepts by means of Thot thesauri, in conjunction with automatic translation of the remaining texts by means of the Google Translation API. These translations are contained within the Cleo platform itself, where they may be checked for accuracy and sense by native speakers and Subject Matter Experts (SMES), improving significantly the quality of the results.

4 History

Wilbrink presented her initial ideas for an integrated AI Egyptology search platform at the International Conference of Egyptologists (ICE) in 2015, in Florence, Italy. She created a first prototype in collaboration with AI company SynerScope, Google Cloud, and the National Museum of Antiquities in the Netherlands. The platform's prototype was unveiled the next year at the sixty-seventh annual meeting of the American Research Center in Egypt, held in Atlanta, Georgia, in 2016, followed by a press conference at the Netherland's National Museum of Antiquities, in 2017.[25] Additional presentations of the Cleo prototype were held at the Digital Humanities Benelux conference in Utrecht, Netherlands (2017); and Dutch-Flemish Egyptology Day in Nijmegen, Netherlands (2017). A design sprint was executed with a team of international Egyptologists and AI experts in Leiden (2017), facilitated by the University of Leiden, the National Museum of antiquities and Google. Wilbrink founded Aincient in order to create the Cleo platform, which was supported by grants and

25 https://www.telegraaf.nl/nieuws/1331584/razendsnelle-database-voor-museum-oudheden Accessed 2-9-2022.

investments from the SIDN fund (Netherlands), Google, and the National Museum of the Netherlands.

The official launch of Cleo's pilot phase was presented at the meeting of the International Committee for Egyptology and International Council of Museums (CIPEG-ICOM), in Swansea, UK (2018); followed by a poster presentation at the annual meeting of the American Schools of Oriental Research (ASOR), in Denver, CO (2018); a presentation at the conference of the Knowledge Institute for Culture and Digitization (DEN), in Rotterdam (2019); a presentation at the XIIth International Congress of Egyptologists (ICE) and at the CIPEG-ICOM meeting at the British Council, both in Cairo (2019).

Since the launch of Cleo's beta version in September 2018, more than 4,800 individuals have visited the site, with more than 1000 registered users. At least three universities in the US, Canada, and the Netherlands have so far incorporated Cleo into their curricula for Egyptology. The source code for Cleo's first phase has been made freely available under the Apache License, Version 2.0, via GitHub.[26]

In March of 2019, Wilbrink presented the Cleo platform at the conference on "Ancient Egypt and New Technology," held at Indiana University, Bloomington. During the Q&A session, representatives from numerous national and international museums, housing Egyptian collections of all sizes, expressed interest in making their data available through Cleo. The overwhelming consensus that emerged in discussions was that Wilbrink's platform, with its intelligent and multivariate search capabilities, represented the best solution to integrate disparate museum collections. In these discussions, Wilbrink emphasized the urgent need for institutional partnership, as a means of sustaining the Cleo platform and increasing its footprint, through the integration of additional museum collections.

Also present at this discussion was Joshua Roberson (University of Memphis), who proposed a collaboration with the Department of Art, Institute of Egyptian Art & Archaeology, and https://www.memphis.edu/iis/ Institute for Intelligent Systems of the University of Memphis. In Spring 2019, Wilbrink and Roberson began investigating opportunities for external grant funding, with assistance from Dr. Kathryn Piquette (University College London) and programming and AI specialists from the Institute for Intelligent Systems of the University of Memphis. The primary goals of this partnership are to: 1.) Expand the searchable database with additional museum collections and secondary data from relevant third-party partners; 2.) Refine and improve Cleo's AI algo-

26 https://github.com/Aincient/cleo Accessed 2-9-2022.

rithms and image recognition capabilities and to improve the end-user search experience for both scholars and the public; 3.) Improve Cleo's metadata translation abilities and expand available languages to include Modern Standard Arabic; and 4.) Incorporate Cleo into relevant curricula at the K–12 and university levels, as well as professional training curricula in Egyptology at the graduate level.

5 Expansion of Cleo

Expansion of Cleo's existing dataset will be accomplished through the addition of new collections of Egyptian artifacts, their images, and metadata. These will be added using the methods developed and tested already on the 45,000+ objects incorporated into Cleo's beta phase. This process may be summarized briefly as: downloading the images and metadata using either a museum-developed API or a data dump; standardizing the data using Thot thesauri; and automatically translating the metadata into a common language.

The present (beta) incarnation of Cleo stands already as the largest and most sophisticated aggregator of Egyptian artifacts currently online. For Cleo's expansion, nine international museum collections and five online Egyptological platforms have expressed their intent to share data on Cleo, which would add at least 100,000 objects to the platform. The selection of new museum partners has been based mainly upon the collections data provided by the Egyptological museum search tool.[27] The expansion, which will include many thousands of new exemplars for existing object classes as well as entirely new classes of objects, will permit the software engineers to train the AI image recognition algorithm to recognize objects more effectively. Furthermore, the introduction of new and more varied metadata will permit our Egyptologists and language specialists to refine the platform's translation capabilities on a variety of collections, both large and small.

Alongside the collections of artifacts, the next phase of Cleo will also expand the sorts of information that Cleo might return for a given object query, to include more robust linguistic information in the case of inscribed artifacts, as well as secondary literature, where available, and other data as aggregated by relevant third-party platforms with an Egyptian focus, such as the UCLA Encyclopedia of Egyptology, Trismegistos,[28] et al.

---

27 http://static.egyptology.ru/varia/mus.php?help=help Accessed 2-9-2022.
28 https://www.trismegistos.org/ Accessed 2-9-2022.

Improvement of the AI image search will be achieved on two separate fronts. Firstly, the existing algorithm will be extended by searching the input space of the model for better features, as well as searching the topology space of the neural network model itself to arrive at a more performant system. Those results can be quantified using standard measures (precision, recall, and F1 scores). The second improvement will involve extracting salient details from images and artifacts using machine vision techniques and training a separate, clustering-based model. The result will be a system capable of finding both direct and subtle connections between data points.

One benefit to this improvement in the AI will be significantly increased accuracy of search results from user-generated object photos, comparable to results from scanned or professional images. Of course, any changes to the system's back end must carefully consider usability on the front end, ensuring that both common and novel processes readily suggest their presence and operational parameters. User experience (UX) testing will accomplish this through a series of use case time trials, efficiency comparisons, and affective analyses within a range of participants from likely user populations (researchers, educators, students, public).

Cleo's translation abilities currently offer options for translation of metadata into English, the most widely used language for academic research and public humanities programming, and Dutch, the mother tongue of Cleo's creator. The goal for Cleo's translation capabilities is to include Modern Standard Arabic (MSA). This is a vital addition to the platform's functionality, as Arabic is now the fourth most popular global language.[29] In addition, and perhaps most importantly, the inclusion of Arabic will facilitate access to the artifacts by the Egyptian public, for whom these collections—scattered globally, far beyond their original home—represent a vital cultural inheritance. Warfare, colonialism, and unregulated archaeological excavation facilitated the industrial-scale export of ancient Egyptian material culture prior to the 1970 UNESCO Convention. As a result, many modern Egyptians have no means or opportunity to engage with artifacts held in foreign collections. By expanding Cleo's interface to accommodate MSA, the platform will help to fulfill an ethical obligation that museums and other cultural institutions around the globe owe to the people whose past has been mined for foreign benefit.

---

29 https://medium.muz.li/web-design-for-right-to-left-languages-the-basics-287329d508cf Accessed 2-9-2022.

6 Long-Term Benefits to Research, Education, and Public Programming

At the recent conference in Bloomington, numerous attendees—representing Egyptologists and digital humanists from leading international institutions and all levels of instruction—expressed their enthusiastic support for the Cleo platform as a tool for research, education, and teaching. This enthusiasm reflects several major, long-term benefits to academic research, education, and public programming in the humanities, which the large-scale implementation of Cleo promises to fulfill. Above all, there is currently no other solution for searching object classes and comparanda across multiple museum collections, with the purpose-built capability to recognize Egyptian artifacts. Thus, anyone seeking information concerning specific objects or more general object types would benefit directly from this innovative platform.

User testing during the beta phase has included Egyptologists and other academics, graduate students, museum professionals, and interested members of the public. User feedback during this pilot period has been overwhelmingly positive, with particular enthusiasm expressed for the use of Cleo in the classroom, common language searching, and location of objects via AI search. Our intention is that, as Cleo becomes more widely recognized during its implementation phase, users will expand to include teachers and students and at every level of education, given the prominent place that Egypt holds in curricula nationally and internationally, at the primary, secondary, and college/university levels. To better reach this wider audience, the Cleo team will solicit users from educators at each of these levels, to inform them of Cleo's features and possibilities, so that they might incorporate it directly into their curricula and teaching. In fact, any program, from elementary education through graduate training, that focuses on Egypt, the Near East, or the ancient world in general stands to benefit from Cleo's integrated search features and intelligent image recognition algorithms.

In addition, there is a clear long-term benefit to museum professionals of all sorts, working with collections that might include an Egyptian component. Cleo will allow researchers who might not have direct access to the narrow expertise of an Egyptological specialist to bridge the gap between their local institution and the vast collections and metadata of the largest museums in the world, alongside the no-less important small collections, whose pieces might be overlooked in a conventional search of so-called "masterpieces." Thus, the expansion of the Cleo dataset to include the numerous Egyptian collections and diverse secondary literature aggregators, etc., will establish Cleo as the premier digital reference for image- and text-based research on Egyptian artifacts, comparable in scale and importance to such services as the UCLA Encyclopedia of Egyptology[30] and the Online Egyptological Bibliography (Oxford).[31]

As a final note regarding the long-term benefits of Cleo, it is critical to look beyond the Egyptian material to the broader fields of language and literature, (art) history, religion, and similar areas of humanities research and public programming, which might incorporate image- and text-based object searches. In fact, any institution utilizing image—and object-based data for research or public programming stands to benefit tremendously from the intelligent search capabilities that Cleo offers. Because the platform's code has been and will continue to be made available as open source via GitHub, other developers can build upon Cleo's foundations to address their own collection-specific needs. This broader humanities significance is evident already in requests from numerous users to adapt the Cleo code for other, presently unlinked museum websites, as well as requests to extend the platform to include other ancient cultures (e.g., Greek, Roman, Indonesian).[32] In fact, approximately 80 percent of Cleo's code could be adapted with relatively little modification to accommodate the written and material culture of virtually any ancient civilization.

7 Closing Thoughts: Sustainability

Financial sustainability forms a major challenge for any online platform, in digital humanities or otherwise. At the March 2019 conference on Ancient Egypt and New Technology, several scholars raised the issue of financial sustainability for platforms like Cleo, highlighting lessons learned from other online initiatives, such as Trismegistos, Giza Archives, and the UCLA Encyclopedia of Egyptology. The consensus was that those projects weathered their initial years almost entirely through grants and other external subventions but struggled subsequently with the issue of self-sustainability. By contrast, one of the very few online projects within Egyptology to become self-sustaining is the Online Egyptological Bibliography (OEB), the survival of which depends entirely upon the institutional and private annual subscriptions.

Learning from these examples, the Cleo team has opted to employ a "freemium" model, whereby some features or functionalities are available at no cost,

30 https://escholarship.org/uc/nelc_uee Accessed 2-9-2022.

31 http://oeb.griffith.ox.ac.uk/ Accessed 2-9-2022.

32 For a similar initiatives being developed for Classical artifacts, compare Athena's Repository, (https://www.athenasrepository.org/, viewed Feb. 9, 2022), created by Danielle Bennett (San Diego State University).

with others available at a subscription rate. The challenges, as well as the clear benefits of such a model are now well known. In brief, any premium feature must be sufficiently enticing—and clearly defined—to encourage subscription. At the same time, the Creative Commons licenses that govern the use of museums' metadata typically require that all shared materials remain freely available to all users. In order to navigate between these two extremes, Cleo's approach resembles and will resemble that of the successful freemium model provided by Dropbox, whereby free accounts retain access to all features, subject to monthly data limits, while premium accounts are given access to more data per month.

Along similar lines, the Cleo solution will be to permit anyone to register a free account, granting full access to Cleo's search functionality but with a limited number of searches per month. "Power users," professional researchers, etc., will have the option to upgrade their free account to provide extra numbers of searches with an individual subscription. In addition, a higher, institutional subscription rate will permit unlimited searches for an unlimited number of users logging in from their institutional email. The income from this model will ultimately help to defer or eliminate the already modest cost of long-term hosting and maintenance while the ability to retain a free account will fulfill the project's philosophical commitment to open access for the public.

General Index

www.ingramcontent.com/pod-product-compliance
Lightning Source LLC
LaVergne TN
LVHW010637300926
849161LV00005B/966
9789004501287